NanoBioMedicine

Shailendra K. Saxena • S. M. Paul Khurana
Editors

NanoBioMedicine

Springer

Editors
Shailendra K. Saxena
Centre for Advanced Research
King George's Medical University
Lucknow, India

S. M. Paul Khurana
Amity University of Biotechnology
Amity University Haryana
Gurgaon, India

ISBN 978-981-32-9897-2 ISBN 978-981-32-9898-9 (eBook)
https://doi.org/10.1007/978-981-32-9898-9

This Springer imprint is published by the registered company Springer Nature Singapore Pte Ltd.
The registered company address is: 152 Beach Road, #21-01/04 Gateway East, Singapore 189721, Singapore

Dedicated
to the ***Parents*** *&* ***Family***
who believed in academics
as the way forward for an intelligent mind
and
to the ***Teachers***
who introduced us to the subject
and nurtured over interest in it.

Foreword

सत्यमेव जयते

प्रोफेसर (डा.) बलराम भार्गव, पदम श्री

एमडी, डीएम, एफआरसीपी (जी.), एफआरसीपी (ई.), एफएसीसी,
एफएएचए, एफएएमएस, एफएनएएस, एफएएससी, एफ.एन.ए., डी.एस.सी.

सचिव, भारत सरकार
स्वास्थ्य अनुसंधान विभाग
स्वास्थ्य एवं परिवार कल्याण मंत्रालय एवं
महानिदेशक, आई सी एम आर

Prof. (Dr.) Balram Bhargava, Padma Shri
MD, DM, FRCP (Glasg.), FRCP (Edin.),
FACC, FAHA, FAMS, FNASc, FASc, FNA, DSc

Secretary to the Government of India
Department of Health Research
Ministry of Health & Family Welfare &
Director-General, ICMR

icmr
INDIAN COUNCIL OF MEDICAL RESEARCH
Serving the nation since 1911

भारतीय आयुर्विज्ञान अनुसंधान परिषद
स्वास्थ्य अनुसंधान विभाग
स्वास्थ्य एवं परिवार कल्याण मंत्रालय
भारत सरकार
वी. रामलिंगस्वामी भवन, अंसारी नगर
नई दिल्ली - 110 029

Indian Council of Medical Research
Department of Health Research
Ministry of Health & Family Welfare
Government of India
V. Ramalingaswami Bhawan, Ansari Nagar
New Delhi - 110 029

Foreword

With the recent advancement in the field of medicine, nanotechnology has been widely accepted and utilized for the management of human health. The application of nanotechnology in disease diagnosis and treatment is defined as the nanomedicne which is a promising multidisciplinary area of research involving physics, chemistry, biology and medical sciences. This evolving multidisciplinary area of science has the remarkable change in medical science. Utilization of molecular machine systems in the field of nanomedicine will be used to address complex medical challenges, and will use knowledge of molecular medicine to maintain and improve human health at the targeted molecular level. The nanomedicine has extensive and surprising implications for the medical professionals in diagnosis and treatment of diseases.

Nanotechnology has made the medical sciences possible to analyze biological systems at cellular and sub-cellular levels. The advantage of nanotechnology has improved the diagnosis and therapy of various complex human diseases. The development made in the area of nanotechnology has provided several advantages including bioavailability, efficacy, dose-response, personalized medicine, targeting abilities and safety factors. The designing of multifunctional nanoparticle complexes is now most exciting and beneficial which can deliver the diagnostics as well as therapeutic agents to the target sites. These characteristics are astonishing which may be further developed towards the management of patients with best possibilities. In this context, the development of targeted, exceedingly specific nanoparticles is of crucial significance. Although nanotechnology has diverse application in medicine, the development of nanoparticle-based therapeutics has a key impact on translational medicine. Investigation of nanomaterials for designing nanoformulations has several physical advantages such as solubility, reduced systemic toxicity, reduction in degradation or physiologic clearance rates, and enhanced clinical efficacy.

Authors have given an interesting insight on the various perspectives of nanotechnology in medicine by which complex diseases such as cancer, infectious diseases neurodegerative diseases could be diagnosed and effectively treated. They have discussed the application of nanotechnology in various aspect of medicine including its associated effectiveness and toxicity. The information in this book will attract the attention of global scientific community to accept nanomedicine and also to explore their therapeutic and diagnostic potential in the treatment of complex diseases. The subject area has general interest to wide variety of clinicians, researcher's representatives, pharmaceutical industries, microbiologists, scientists, policy makers etc. and needs to be highlighted to eradicate the complex diseases from the face of this earth. I am confident that this book might increase the interest in this field of medical research and that the readers will find it helpful for their clinical usage, investigations and management of complex diseases.

(Balram Bhargava)

Preface

Although we are in the twenty-first century having most of the advanced technologies in hand, yet communicable and noncommunicable diseases are majorly responsible for mortality due to scarcity of potential therapeutics and effective differential diagnostics. This book, *NanoBioMedicine*, provides a comprehensive overview of recent trends in various nanotechnology-based therapeutic and associated challenges. It focuses on various aspects and properties of communicable and noncommunicable diseases including nanotechnology-based approaches for development of effective therapeutics and differential diagnostics, which is imperative for safeguarding the human race from more loss of resources and economies due to disease burden. To overcome these issues and fill the gap, we hope *NanoBioMedicine* will provide a readily available resource in this area.

The quick development of nanoscience in the last few years has given an abundance of information into the physicochemical properties and biological performance of nanoscale materials that can be considered while utilizing nanotechnology for medicinal applications. The designing and application of nanoscale materials to alleviate complex diseased condition are of great interest. Nanoscale materials exhibit novel properties that can be exploited in various ways to design diagnostics and effective therapeutics. Nanomedicine involves multidisciplinary areas of sciences such as nanotechnology, nanoengineering and nanoscience interacting with the field of life sciences.

The elemental proposition upon which nanomedicine is pursued is nanoparticles, which are introduced into the body as foreign bodies and are subjected to the full armory of the body's defense system labeled with antibody molecules targeting specific cells. Advances in the field of nanomedicine have come a long way since the time it was envisioned to be studied. The objectives and goal to establish global roadmaps in nanomedicine are guided by the need to take care of life-threatening clinical issues. Recent advancement in the field of nanomedicine has developed various tools for early diagnostics, treatment of complex diseases with high efficacy and specificity, and personalized therapy to improve the quality of life. This book is focusing on the various fields of medicine such as personalized medicine, cancer biology, stem cell therapy, regenerative medicine, infectious diseases, molecular diagnostics, biosensors, and nanotoxicology.

The editors of this book hope that this work might increase the interest in this field of research and that the readers will find it useful for their investigations, management, and clinical usage. The editor and contributors report no conflict of interest.

Lucknow, India Shailendra K. Saxena
Gurugram, India S. M. Paul Khurana

Acknowledgements

This book is conceptualized to present an updated account of the several state-of-the-art examples of the application of nanotechnology in the field of biomedical research. It also discusses the future prospects of targeting cancer stem cells through nanotherapeutic approaches for cancer treatment. It provides an updated account of the development of the virus-based nanoparticle system as vaccine therapy for the prevention and treatment of various infectious diseases. The book provides an unparalleled reservoir of information about nanotoxicology.

All these aspects are imperative for safeguarding human race from more loss due to various diseases. To overcome these issues and fill the gap, we hope *NanoBioMedicine* shall provide a readily available resource in this area. The aim of this book is to provide a comprehensive overview of the recent trends in various nanotechnology-based therapeutics and challenges associated with its development.

We are overwhelmed in all humbleness and gratefulness to acknowledge our debt to all the contributors who trusted and supported us for this work. We hope they are as proud of this book as we are. We would also like to thank Springer Nature Publisher to consider this book for publication. All the reports cited in this book are taken with proper citation. However, any missed information is just unintentional and explicable.

Our research fellows and students are central to all our research and academic work. They are the motivating force behind anything constructive we do. They are truly brilliant and have a bright future. We would like to specially thank and express our gratitude to our mentors, teachers, and students who gave us strength to accomplish this. Also, we would like to thank the colleagues, family, and friends who gave a lot of encouragement and support during the work on this book.

A happy environment at home is essential for any kind of growth, and we thank our families especially talented wives and children.

Lucknow, India — Shailendra K. Saxena
Gurugram, India — S. M. Paul Khurana

About this Book

This book provides a comprehensive overview of the recent trends in various nanotechnology-based therapeutics and challenges associated with its development. Nanobiotechnology is an interdisciplinary research that has wide applications in the various fields of biomedical research. The book discusses the various facets of the application of nanotechnology in drug delivery, clinical diagnostics, nanomedicine, and treatment of infectious and chronic diseases. The book also highlights the recent advancements on important devices and applications that are based on nanotechnology in medicine and brief the regulatory and ethical issues related to nanomedical devices. It also reviews the toxicological profile of various nanomaterials and emphasizes the need for safe nanomaterials for clinical use. Finally, the book discusses the recent developments of potential commercial applications of nanotechnology.

Contents

About the Editors

Professor (Dr.) Shailendra K. Saxena is Vice Dean and Professor at King George's Medical University, Lucknow. His primary research interest is to understand molecular mechanisms of host defense during human viral infections and to develop new strategies for predictive, preventive, and therapeutic strategies for them. Prof. Saxena's work has been published in reputed international journals with high impact factor. His work has been highly cited by numerous investigators globally and honored by several prestigious national and international awards, fellowships, and scholarships in India and abroad, including various Young Scientist Awards, BBSRC India Partnering Award, and named as the Global Leader in Science by The Scientist magazine (USA) and International Opinion Leader/Expert involved in the vaccination for JE by IPIC (UK). Prof Saxena has been elected Fellow of The Royal Society of Biology UK (FRSB); The Royal Society of Chemistry UK (FRSC); The Academy of Environmental Biology, India (FAEB); Indian Virological Society (FIVS); The Biotech Research Society, India (FBRS); and the (European) Academy of Translational Medicine Professionals, Austria (FacadTM). He has been awarded Dr. JC Bose National Award of Department of Biotechnology (DBT, Min. of Science & Technology, Govt. of India) in biotechnology and has active collaboration with US universities.

Professor (Dr.) S. M. Paul Khurana completed his PhD in 1969 on papaya viruses and 2 years of Post-Doctoral Research (1970–72) in Adv Plant Virology at Kyushu University, Fukuoka, Japan, with Prof. Zyun Hidaka as JSPS fellow. He also availed GOI DBT overseas fellowship for one year at University of Minnesota, St Paul, USA, with Prof EE Banttari and specialized in Immunodiagnostics (March 1987 to April 1988). Prof. Khurana worked at Central Potato Research Institute (CPRI), Shimla, since 1973 as Scientist/Senior Scientist; Principal Scientist; Head, Virus/Seed Pathology (1976–82/1988); Principal Scientist and Head, Plant Pathology Division (1988–93); Project Coordinator AICRP-Potato (1994–2004); and Director (2002–2004) CPRI, Shimla, and as Vice-Chancellor of Rani Durgavati University, Jabalpur (2004–2009). He also served as visiting Consultant for CIP/FAO (1992, 1996, and 1997) and is Fellow of many professional societies, has bagged 40+ awards. has been Chair of three RACs and QRTs of three ICAR Research Instts.

Prof. Khurana served as the Director of Amity Institute of Biotechnology, AUUP, Noida (2009–10) and moved to Amity University Haryana at Gurgaon in August 2010 for establishing Institute of Biotechnology (2010–2015) and served as Dean of Science, Engineering and Technology (2013–2016 and continues as prof of biotechnology) at Amity University Haryana, Gurgaon.

Prof. Khurana an internationally recognized virologist/pathologist and biotechnologist having 53 years of experience, written 230 research papers, garnered 100+ reviews, guided 16 PhDs, and edited 18 books.

Part I

NanoBioMedicine: Revolutionary Interdiscipline

Current Advances in Nanotechnology and Medicine

1

Shailendra K. Saxena, Rajni Nyodu, Swatantra Kumar, and Vimal K. Maurya

Abstract

One of the most imperative questions that arise is why we are indulging ourselves in nanotechnology and medicine. The rationale is its size in nanometer range which makes it easy to enter into cells of human body making it beneficial for special types of cell target therapy such as efficient delivery of drugs to the target cell and competent detection of diseases. Studies have come up with one more beneficial factor, that is, the nanoparticles can also protect drug from degradation because of the shield-like properties. The elemental proposition upon which nanomedicine is pursued is nanoparticles, which are introduced into the body as foreign bodies and are subjected to the full armory of the body's defense system labeled with antibody molecules targeting specific cells. Diverse types and forms of nanoparticles are used in medicine. Advances in the area of nanomedicine have come a long way since the time it was envisioned to be studied. The objectives and goal to establish global roadmaps in nanomedicine are guided by the need to take care of life-threatening clinical issues. Nanomedicine has a potential to combat several human diseases including cancer as well as infectious, neurological, musculoskeletal, cardiovascular diseases. In this chapter we will be discussing the ongoing progress of nanotechnology and its application in various fields of medicine.

Keywords

Nanomedicine · Nanoparticles · Liposomes · Magnetic Nanoparticles · Molecular Imaging · NDDS · Quantum dots · Diagnostics · Therapeutics · Regenerative Medicine · Cancer

S. K. Saxena (✉) · R. Nyodu · S. Kumar · V. K. Maurya
Centre for Advanced Research, Faculty of Medicine,
King George's Medical University (KGMU), Lucknow, India
e-mail: shailen@kgmcindia.edu

S. K. Saxena, S. M. P. Khurana (eds.), *NanoBioMedicine*,
https://doi.org/10.1007/978-981-32-9898-9_1

1.1 Introduction

We have entered into an era of the notion, "The smaller the better". Nano refers to things that are very small, smaller than the smallest. Hence, in the bare-bone version of definition, "Nanotechnology is the investigation and utilization of structures between 1 nanometer (nm) and 100 nanometers in size" (Logothetidis 2006). One-billionth of a meter is defined as a nanometer which is too microscopic to be seen by any usual optical microscope. Nanotechnology has the potential to transform the way the medical and healthcare solutions are developed and delivered through its application toward the diagnosis, treatments, or prevention of diseases at the cellular level, and hence, the application is termed as *nanomedicine* (Bayford et al. 2017). The two fundamental areas in nanomedicine are nanomedicine-based diagnostics and nanotechnology-based therapies. The quick development of nanoscience in the last few years has given an abundance of information into the biological performance and physicochemical properties of nanoscale materials that can be considered while utilizing nanotechnology for medicinal applications, for example, in atomic imaging applications (diagnostics), thermal initiated annihilation of tumors (therapy), and, furthermore, disintegration of ineffectively soluble medications generally utilized inside the pharmaceutical businesses (Fig. 1.1) (Rizzo et al. 2013). Nanomedicine is defined as the comprehensive examination, control, development, repair, resistance, and enhancement of all human natural frameworks, working from the atomic dimension, utilizing designed nano-devices and nanostructures. Nanomedicine is also known as the science and innovation of diagnosis,

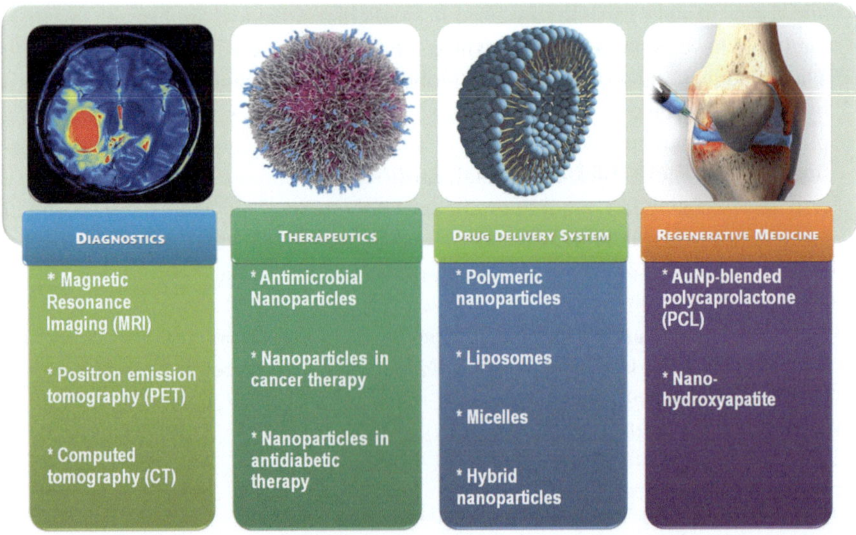

Fig. 1.1 Applications of nanotechnology in medicine. This figure depicts emerging field of nanomedicine which involves various multidisciplinary areas of medicine such as diagnostics, therapeutics, drug delivery system and regenerative medicine etc.

therapeutics, and anticipating infection and awful damage, of assuaging torment, and of safeguarding and enhancing human well-being, utilizing subatomic instruments and subatomic learning of the human body (Saxena et al. 2012).

1.2 Nanotechnology in Disease Diagnosis

With the advancement in the era of clinical diagnostics, a definitive objective is to empower physicians to detect a disease as early of schedule as could be expected under the circumstances. Nanotechnology is required to make detection/diagnosis conceivable at the cellular and sub-cellular level. With the assistance of nanomedicine, early detection as well as prevention, precise diagnosis, effective treatment and follow-up of infections are conceivable (Archakov 2010). Nanotechnology guarantees efficient and incredibly precise apparatuses for in vitro and in vivo diagnostics. The enhanced diagnostics leads to better treatments and improved health outcomes. Nanotechnology has been utilized broadly in the field of restorative diagnostics (Zhang et al. 2018a). Cancer is one field of ailment in which nanotechnology is probably going to have the greatest effect as enhanced diagnostics and treatment. EGFR which is an epidermal development factor receptor present outwardly of their layers where nonmalignant growth cells have significantly less EGFR. Upon binding of gold nanoparticles (AuNps) to an antibody for EGFR (anti-EGFR), the capacity to tie the nanoparticles to the malignant growth cells has been achieved. When bound, the malignant growth cells show distinctive light dissipating and assimilation spectra than favorable cells. Pathologists may utilize these findings to distinguish harmful cells in biopsy tests (Zitka et al. 2012). Certain nanoscale particles are utilized as labels or tags. Point of care and imaging technologies are two specific fields that have the advantage of nanoparticle applications. From the last few decades, imaging has turned into a basic apparatus in the diagnosis of disease.

1.2.1 Nanoparticles for In Vivo Diagnostics

Molecular imaging is outlined as the in vivo visual images, activity, and characterization of biological phenomenon at the molecular and cellular level. Such imaging helps in unfolding the possibility of early detection of underlying diseases and to probe for the stage of disease and, therefore, helps in effective image-guided treatment. This imaging technique frequently includes the employment of imaging probes that turn out signals by means of nuclear reaction together with different molecules to image via sound waves (ultrasound), magnetism (MRI), or light (optical techniques of luminescence and fluorescence) (Miller 2016). Advancement in molecular imaging needs the application of refined probes that square measurable to discover biological processes on the cellular and molecular level. That is when the need of nanoparticles arises considering the point that probes for molecular imaging need to possess certain features for the efficacy. Such probes need to possess the characterization properties to be able to get accumulated at the site of interest and furthermore and eventually be able to get imaged (Baptista 2014). Different nanosystems, with unique physical and chemical properties, are continuously being proposed with distinct functions and targets. And such nanoparticles-based molecular imaging probes and contrast agents have been studied to be significant over all the other single molecule-based contrast probes. Nanoparticulate-based imaging contrast agents involve the use of appropriate contrast-generating materials such as fluorescent, radioactive, paramagnetic, and super paramagnetic or electron dense (Emeto et al. 2017). These significantly efficient probes have led to the development of plethora of noble metal nanoparticles and quantum dots and also a profusion of nanoparticulate-based contrast agents as micelles, liposomes, polymerosomes, dendrimers, carbon nanoparticles, and magnetic nanoparticles (iron oxides, metal alloys) (Naseri et al. 2018).

1.2.1.1 Magnetic Resonance Imaging (MRI)

MRI is a non-invasive method of anatomic images with spatial resolution which is based on NMR principle and is processed with the help of several NMR active nuclei. However, sometimes these active nuclei (also, termed as inherent contrast) are inadequate for the proper characterization of tissue, and, hence, the use of contrast agents is required (Tognarelli et al. 2015). MRI contrast agents are referred to as T1 and T2, with respect to the relaxation time that they predominantly affect. T1 agents have a propensity of paramagnetic compounds (as lanthanides), and T2 agents are commonly super-paramagnetic agents, as iron oxides. Nanosystem-based contrast agents referred as T1 agents are micelles, liposomes, polymerosomes, dendrimers, and carbon nanoparticles, whereas T2-agents are magnetic nanoparticles (iron oxides, metal alloys) (Sands and Levitin 2004). As MRI contrast agents, several of the nanoparticles have been tested which are FDA approved and are of great relevance such as super paramagnetic iron oxide nanoparticles (SPION). Research and studies are conducted to develop nanoparticles as T1 contrast agents in order to overcome the drawbacks of T2 contrast agents (Kim et al. 2018).

1.2.1.2 Positron Emission Tomography (PET)

PET is a quantitative imaging procedure which uses biomolecule release electrons that interacted with radionuclides emitting positrons. PET imaging is used mostly in early detection of tumorous cells. One of the most widely used imaging probes is the metal oxide nanoparticles which help in the construction of PET imaging (Jones and Townsend 2017). Liposomes have also been used in combination with positron emitting radionuclides for contrast imaging. Furthermore, polymeric micelles and hydrogels are also used for tumor detection and delivery (Silindir et al. 2012).

1.2.1.3 Computed Tomography (CT)

Extensive research is being carried out in the area of molecular imaging which focuses on the development of CT contrast agents. The concentration of contrast agents is required in millimolar for CT imaging of desired organs. Nevertheless, nanoparticle contrast agents can amplify the contrast which reduces the comparatively high exposure radiation in CT (Goldman 2007). AuNPs have been shown to enhance the visibility xenografted human breast tumors of mm-sized where an anti-Her2 antibody was shown to be 1.6-fold efficiently targeted. There were 22-fold higher uptakes of gold nanoparticles in the tumor fringe recorded. Glucose AuNPs have also been found to act as a CT agent that allows differentiation between inflammatory process and cancerous condition (Cole et al. 2015).

1.2.2 Recent Advancement of Nano-Based Molecular Diagnostics

A number of nanoparticles are being used as diagnostic tools. The foremost technologies are quantum dots (QDs), magnetic nanoparticles (MNPs), and gold nanoparticles (GNPs or AuNps).

1.2.2.1 Gold Nanoparticles (GNPs or AuNPs)

Gold nanoparticles have got much attention because of its physiochemical properties. Gold nanoparticles play a crucial role in the identification of genetic diseases based on biomarkers, SNP genotyping, and detection of nucleic acids in infectious conditions (Singh et al. 2017). Noble metal nanoparticles have been used extensively as tags for nucleic acid probes, and such AuNps can bind to small pieces of DNA of size not larger than 13 nm in diameter. Several FDA-approved products use AuNps as probes for diagnostic purposes (Mieszawska et al. 2013).

1.2.2.2 Quantum Dots (QDs)

Quantum dots (QDs) are inorganic semiconductor nanocrystals with a typical diameter of 2–10 nm. QDs are capable of emitting at very well-defined wavelength upon excitation and have been successfully used for imaging tumor target cells in animal models. Studies have unfolded great relevance of quantum dots in the field of early diagnosis and, also, in the field of locating the cancer tumors in patients and for carrying diagnostic tests in samples (Matea et al. 2017). The recent studies in quantum dots are on its composition of cadmium which is considered to be ounce of toxic, the reason behind the limitations on its use in vivo, and hence, study has been proposed to manufacture quantum dots of silicon as it's considered to be less toxic. Quantum dot is combined with microscopy to look for cells of live animals. The carcinoma marker Her2 is immunofluorescently labeled with the particular cancer antibodies covalently coupled with polyacrylate cap and carbohydrate detectable luminescence is useful for cancer imaging purposes (Zhang et al. 2008a). Another application of QDs is for infectious agent identification. Rapid and sensitive identification of respiratory syncytial virus (RSV) is imperative for the management and development of antiviral. Nanoparticles conjugated with antibody rapidly have been developed which specifically identify RSV (Bawage et al. 2013). When viral particles or infected cells come into the contact with QDs they attach to their surfaces.

1.2.2.3 Magnetic Nanoparticles

The magnetic properties and ease of derivatization of magnetic nanoparticles (MNPs) have been crucial for separation of analytes or to potentiate immobilization prior to recognition step. Detection of trace levels of prostate-specific antigen (PSA) and amyloid-beta derived diffusible ligands (ADDLS) in clinical samples by bio-barcode assay are the prominent applications of MNPs (Cardoso et al. 2018). Within the blood stream, the MNPs attach to form microvesicles that help to identify the target cells. The early diagnosis can be then achieved by using NMR which discovers the microvesicles/magnetic nanoparticle clusters (Zhang et al. 2018b).

1.3 Nanotechnology-Based Therapeutics

1.3.1 Antimicrobial Nanoparticles

One of the encapsulated peptide LL37 has been made to make the nanoparticles with enhanced antimicrobial activity. The combination of LL37 and A1 has been used to develop the first solid lipid nanoparticle (SLN) formulation which can deliver it in precise ratios resulting in enhancing antibacterial activity against *E. coli* and *S. aureus* (Fumakia and Ho 2016). One of the earliest uses of nano-surfaces is nanocrystals which can be utilized as the antimicrobials. Silver ions are released up to 7 days from the surface of nanocrystal and thereby reducing a broad range of microorganisms, including drug-resistant bacteria such as vancomycin-resistant *Enterococcus* and methicillin-resistant *Staphylococcus aureus*. Similarly, zinc oxide nanoparticles, single-wall carbon nanotubes, and antibiotics-coated nanoparticles are also under the major listings for research and study worldwide (Biswaro et al. 2018).

1.3.2 Nanotechnology in Cancer Therapy

The enhancement in carcinogenic treatment depends on the upgraded enhancement and retention (EPR) impact of the vasculature encompassing tumors. For the effective treatment of cancer there is a need to develop targeted drug delivery methods. Tumor focusing with nanoparticles can be acknowledged through active and passive way. The particles can be adjusted with different kinds of materials including biomolecule. The active way depends on binding of ligand-coordinated nanoparticles to receptors expressed by tumor cells (Misra et al. 2010). These ligands include peptides, antibodies, aptamers, nucleic acids, sugars, and tiny molecules. The important factors for the anticancer nanoparticles are the size of the nanoparticles, surface properties (for example hydrophobicity), and focusing on ligands. Nanoparticles intended for tumor focused on treatments comprise of different segments, a nanocarriers and a functioning operator. Nanoparticles carrying drugs are considered as submicroscopic colloidal structure that act as the drug vehicles, either as nanocapsules (repositories in which the drug is kept in hydrophobic or hydrophilic center encompassed by a single polymeric layer) or nanospheres (framework in which the drug is scattered) (Zhao et al. 2018a). Nanoparticle carriers are typically made out of gold, iron oxides, dendrimers biodegradable polymers, lipid based transporters, for example, liposomes and micelles, viral nanoparticles and organometallic mixes (Zhang et al. 2008b).

1.3.3 Nanoparticles in Antidiabetic Therapy

Diabetes is a kind of chronic disorder which is characterized by high circulating glucose. Several of the ions have been reported to be effective in the maintenance of blood sugar levels such as vanadium, chromium, magnesium, and zinc (DiSanto and Subramanian 2015). Zinc oxide nanoparticles have been shown to be effective via oral administration which shows higher serum insulin (70%), improved glucose tolerance, reduced blood glucose (29%), reduced triglycerides (48%), and reduced nonesterified fatty acids (40%) (Woldu and Lenjisa 2015). Nanoparticles are systemically absorbed resulting in elevated zinc levels in the adipose tissue, liver and pancreas. Amplified secretion of insulin and activity of superoxide dismutase have been observed in rat insulinoma (RIN-5F) cells (Xie and Xie 2018).

1.3.4 Nanoparticles in Regenerative Medicine

Regeneration of damaged tissue or organ has taken a turn with the development of nanotechnology and its application in medicine. The damaged tissue can be repaired and reproduced with the use of nanomedicine. These artificially reproduced cells are then used in tissue engineering revolutionizing the artificial implants and transplantation of organs. In vivo regeneration or, alternatively, in vitro development of a scaffold-based complex functional organ is the aim of regenerative medicine. Various nano-structured materials have been used to regenerate cartilage, bone, muscle, vascular, nervous system, bladder, skin and other tissues in the past few years (Van Rijt and Habibovic 2017). Orthovita's Vitoss is used as bone void filler; a primary example of nano-hydroxyapatite (component of bone) can be used in osseointegration. Mesenchymal stem cells (MSCs) have been shown to be significantly proliferated and differentiated into cardiac cells by AuNp-blended polycaprolactone (PCL) scaffolds which can be used in myocardial infarction repairmen (Jain 2008). Magnetic nanoparticles are used for the isolation and grouping of stem cells, and in addition, carbon nanotubes, fluorescent CNTs, and fluorescent MNPs along with quantum dots have been used for tracing of stem cells, molecular imaging, and gene or drug delivery into stem cells. Stem cells proliferation and differentiation can be regulated by the combination of nanocarriers with biological molecules. Applications of nanotechnology in stem cell biology research are opening the newer path in the area of regenerative medicine (Pan et al. 2017).

1.4 Nanoparticles as Drug Delivery System

Targeted drug delivery can be achieved via nanoparticles mostly due to its high surface area and volume ratio. The specific drug dose is employed, and side effects are considerably reduced since the drug is exclusively deposited within the morbid region. By carrying drugs within, the local drug concentration can be modulated by using nanoparticles and controlling its release upon binding to targets (Patra et al. 2018). Nanoparticles interact with the target cells in variety of ways primarily depending on whether the source materials are non-biological such as gold or cadmium or biological components such as phospholipids. Since the drug needs to be transported and released, the bio-degradable nanoparticle formulations are suitable. Thus various forms of nanoparticles such as nano porous materials and dendrimers have been designed (Castro and Kumar 2013). Micelles which are derived from co-polymers are used for encapsulation of drug. Minute drug molecules can be transported to the specified target sites. The drug consumption as well as the treatment expenses can be reduced to a significant level in the site specific drug delivery, creating the treatment of patients cost effective. With the recent development in nanotechnology, various forms of inorganic nanoparticles have been generated (gold nanoparticles, iron oxide nanoparticles and fullerenes) as an efficient drug carrier or vehicles and few organic nanoparticles (liposomes, micelles) as drug reservoirs (Kumar et al. 2018). Among these liposomes and organic nanoparticles have entered the market, and micelles are yet to be and are still in the clinical trials. Furthermore, polymeric nanoparticles have also gained attention as drug delivery systems (Fig. 1.2).

1.4.1 Impact of Nanoparticles in Drug Delivery System

The present time shows many drugs failing the clinical trial phase because of the poor delivery factor precisely at the targeted site without having an interacting with the nonspecific target sites and organs (Wang et al. 2017). To overcome this problem, NDDS has been majorly studied and, hence, unfolded various remedial options for the efficacy in specific targeting, which is cost-effective and also lowers drug toxicity. Worldwide novel methods are being developed for effective drug delivery methods. NDDS is now expected to be globally accepted which is now making nearly 80% of the drug delivery market. Improvement in human health by tackling the lack of treatment efficacy can be overcome through the utilization of nanoparticles in drug delivery system up to a higher extent comparatively to where it is now (Suri et al. 2007).

1.4.2 Journey of Nanoparticles-Based Delivery

The most consistent route to deliver NDDS (nanoparticle drug delivery system) is usually through intravenous injection although there are various other routes such as pulmonary or oral. Post-injection, the drugs are transported from the injected site

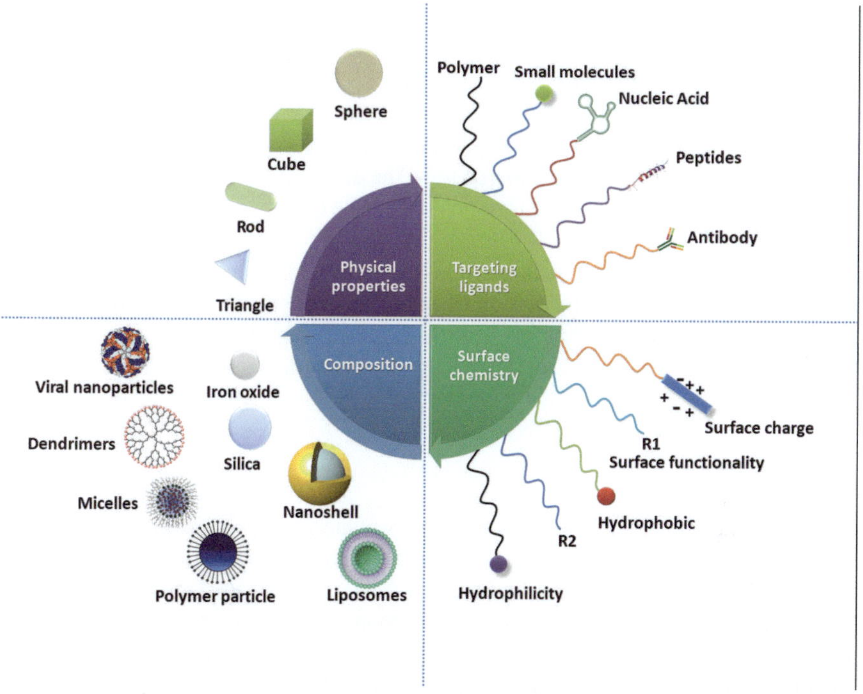

Fig. 1.2 **Nanoparticle drug delivery system (NDDS).** This figure demonstrates the various forms of nanotechnology-based drug delivery methods. Ligands are of various types such as peptides, antibodies, aptamers, nucleic acids, sugars, and tiny molecules. Targeted drug delivery also can be achieved by considering the size of the nanoparticles, surface properties (e.g., hydrophobicity), and focusing on ligands. Physical properties of nanoparticles are also crucial to determine its binding and absorption to the target site. Nanoparticle carriers are typically made out of gold, iron oxides, dendrimers biodegradable polymers, lipid-based transporters, for example, liposomes and micelles and viral nanoparticles

via the venous network to the heart (Serda et al. 2011). NDDS interacts with proteins present in plasma, and the interaction is monitored by the physiochemical properties of NDDS, and the interactions affect the time of circulation as well as the tissues deposition. The sheathing of NDDS with PEG diminishes the deposition of non-specific protein and complement activation of NDDS.

1.4.3 Nanoparticles-Based Drug Release

Factors to be considered while developing drug delivery system based on nanoparticles are both drug release and polymer degradation. The rate of drug release depends upon various factors such as drug diffusion through the nanoparticles matrix, solubility, desorption of the adsorbed drug, degradation of nanoparticle matrix and the combination of diffusion and degradation process (Rizvi and Saleh

2018). The particle release regulates solubility, diffusion and biodegradation of the drug. For nanoparticles, the release of drugs is displayed by erosion or matrix spread. In case of polymer-coated nanoparticles, the release from the polymeric membrane is controlled by the diffusion of the drug, and the covering of the membrane is used to prevent drug release. Hence, the diffusion and solubility of drug in or across the membrane are considered to be the determining factors for drug release (Okada and Toguchi 1995). The foremost objective of delivery systems is to release their payload at the target site. To accomplish this objective, the key approaches are active and passive targeting. Nanoparticles-based drug delivery is a passive approach as the drugs encapsulate in nanoparticles and drugs linked with macromolecules can target tumors passively due to improved permeability with retention effect. Liposomes are further useful for delivering pharmaceuticals agents (Yu et al. 2016).

1.4.4 Liposomes

Liposomes demonstrated possibility for conveyance of a wide scope of therapeutics, since their payload can be encapsulated in their interior fluid compartment or inserted inside the phospholipid bilayer (Jain 2008; Akbarzadeh et al. 2013). Clinical utilizations of liposomes in the conveyance of anticancer specialists for the treatment of various malignant growth signs are entrenched. Stealth liposomes can inactively collect in strong tumors due to their inalienably broken vasculature and faulty lymphatic seepage. Doxil, Caelyx, and Myocet are nano-meter-sized liposome frameworks (exemplifying doxorubicin in their fluid center) which have been utilized for Kaposi's sarcoma, ovarian disease, and various myelomas (Cattel et al. 2003). DepoCyt (cytarabine-containing multivesicular liposomes) with a sustained-discharge profile has likewise been endorsed for malignancy treatment. Reported treatment is done by the use of Cremophor EL in case of head and neck cancer that allows the paclitaxel to be delivered intravenously. Carbon nanoparticles have been replaced at the place of toxic Cremophor, use less paclitaxel, and show reduced side effects along with improved drug targeting. Similarly, an albumin-bound paclitaxel, namely, Abraxane has been studied in case of the breast cancer and non-small cell lung cancer (Olusanya et al. 2018). In addition, doxorubicin has been delivered by using a bound nanoparticle chain in case of breast cancer cells. Within the tumor post-penetration of the nano chains was performed with the help of magnetic nanoparticles (Zhao et al. 2018b). Tumor growth has been shown to be halted significantly by nanotechnology than the conventional treatment and is less harmful to healthy cells.

1.4.5 Hybrid Nanoparticles in Drug Delivery System

To control a targeted delivery, some of the nanoparticles are manipulated by the external magnetic field gradient, and such nanoparticles are termed as hybrid nanoparticles that are reported to be highly efficient in drug delivery system. Such magnetite-polymer hybrid nanoparticles can be attached to fluorescence groups; these hybrid nanoparticles are proficient tracers of drug transporters due to the enrichment of fluorescence molecules. This allows the magnetic nanoparticles to be more widely applied in site specific drug delivery system (Madni et al. 2017). Multicomponent hybrid nanoparticles may possess multiple functionalities for various applications that are complicated to attain with nanoparticles of single component. Noble metal/iron oxide hybrid nanoparticles not only show unique optical properties but also exhibit magnetic resonance. Hybrid nanoparticles can be of great help in drug delivery system especially in cases of cancer and, precisely, in the treatment of multidrug-resistant cancer (Date et al. 2018).

1.5 Conclusions

Considering the various biological, pharmaceutical molecules and structures that operate in the living cells, the size scale is of 100 nanometers and less, and hence, the application of nanotechnology in healthcare presents some stimulating potential outcomes. Prescription nanotechnology may change the way we detect and treat human body and illness damage in the future, and several procedures that have just been planned a few years earlier are becoming a marvelous ground for substances. Nanotechnology is a useful asset for making these "smart" materials. This methodology is challenging and is still a long way from being accomplished. Nanoplatforms have been seen to be taking lead in the development of targeted nanomedicine for anticancer therapy in the upcoming time. Detection, prevention and treatment of various forms of cancers are possible using such nanoplatform techniques based targeted anticancer therapy without toxicity and with enhanced biocompatibility of nanoplatform. Diabetics can become totally free of dietary regulations and the strict systemic regime if nanotechnology is at the forefront. Some devices are so adjustable that people with diabetes are no longer dependent on insulin injections, and at this moment their blood glucose concentrations are adjusted to their glucose levels. This would allow them to live a normal life, particularly young people who are always active. It helps the patient to feel more confident and mentally safe as well as cost-effective in other respects, because fewer resources are required to achieve a significantly more effective result.

1.6 Future Perspectives

One of the most crucial factors which have led to the designing and advancement of various nano-based drugs is the small size that enables the successful delivery of the drugs in the specific site. A significant advancement in the area of nanotechnology-based drug delivery has been achieved with the remaining question of associated toxicity on human body. The impact of nanoparticles on biological systems depends on the size, chemical nature and composition, solubility, shape, surface structure, and aggregation. These factors can modulate the cellular uptake, translocation from entry to the target sites, target affinity and possibility of causing injury to tissues. Pharmacodynamics of nanoparticles depends upon the route of exposure such as skins, GI tract, systemic administration for diagnosis and therapeutic applications. Personalized medicine can be developed based on nanotechnology by several means. Nanotechnology-based diagnostics may improve the present restrictions of detection limits as well as enhances the molecular imaging. Nanotechnology can be integrated in detection of biomarkers, biochips, point of care diagnosis and biosensors which are crucial for developing personalized nanomedicine. Ethical consideration is crucial for all nanotechnology-based applications in medicine. Nanotechnology has not raised new ethical issues so far; it is advisable to keep the associated ethical considerations while developing and applying nano based therapeutics. Moreover, there is a need to develop unique ethical questionnaire distinct from conventional form of medicines.

Acknowledgments The authors are grateful to the Vice Chancellor, King George's Medical University (KGMU), Lucknow, for the encouragement and support for this work. SK Saxena is also supported by CCRH, Government of India, and US NIH grants: R37DA025576 and R01MH085259. The authors have no other relevant affiliations or financial involvement with any organization or entity with a financial interest in or financial conflict with the subject matter or materials discussed in the manuscript apart from those disclosed.

References

Akbarzadeh A, Rezaei-Sadabady R, Davaran S, Joo SW, Zarghami N, Hanifehpour Y, Samiei M, Kouhi M, Nejati-Koshki K (2013) Liposome: classification, preparation, and applications. Nanoscale Res Lett 8(1):102

Archakov AI (2010) Nanobiotechnologies in medicine: nanodiagnostics and nanodrugs. Biomed Khim 56(1):7–25

Baptista PV (2014) Nanodiagnostics: leaving the research lab to enter the clinics? Diagnosis (Berl) 4:305–309

Bawage SS, Tiwari PM, Pillai S, Dennis V, Singh SR (2013) Recent advances in diagnosis, prevention, and treatment of human respiratory syncytial virus. Adv Virol 2013:595768

Bayford R, Rademacher T, Roitt I, Wang SX (2017) Emerging applications of nanotechnology for diagnosis and therapy of disease: a review. Physiol Meas 38(8):R183–R203

Biswaro LS, da Costa Sousa MG, Rezende TMB, Dias SC, Franco OL (2018) Antimicrobial peptides and nanotechnology, recent advances and challenges. Front Microbiol 9:855

Cardoso VF, Francesko A, Ribeiro C, Bañobre-López M, Martins P, Lanceros-Mendez S (2018) Advances in magnetic nanoparticles for biomedical applications. Adv Healthc Mater 7(5): 1–35, 1700845

Castro E, Kumar A (2013) Nanoparticles in drug delivery systems. In: Nanomedicine in drug delivery. pp 1–22

Cattel L, Ceruti M, Dosio F (2003) From conventional to stealth liposomes: a new frontier in cancer chemotherapy. Tumori 89(3):237–249

Cole LE, Ross RD, Tilley JM, Vargo-Gogola T, Roeder RK (2015) Gold nanoparticles as contrast agents in x-ray imaging and computed tomography. Nanomedicine (Lond) 10(2):321–341

Date T, Nimbalkar V, Kamat J, Mittal A, Mahato RI, Chitkara D (2018) Lipid-polymer hybrid nanocarriers for delivering cancer therapeutics. J Control Release 271:60–73

DiSanto RM, Subramanian V, Gu Z (2015). Recent advances in nanotechnology for diabetes treatment. Wiley Interdiscip Rev Nanomed Nanobiotechnol. 7(4): 548–564

Emeto TI, Alele FO, Smith AM, Smith FM, Dougan T, Golledge J (2017) Use of nanoparticles as contrast agents for the functional and molecular imaging of Abdominal Aortic Aneurysm. Front Cardiovasc Med 4:16

Fumakia M, Ho EA (2016) Nanoparticles encapsulated with LL37 and Serpin A1 promotes wound healing and synergistically enhances antibacterial activity. Mol Pharm 13(7):2318–2331

Goldman LW (2007) Principles of CT and CT technology. J Nucl Med Technol 35(3):115–128

Jain KK (2008) Nanomedicine: application of nanobiotechnology in medical practice. Med Princ Pract 17(2):89–101

Jones T, Townsend D (2017) History and future technical innovation in positron emission tomography. J Med Imaging (Bellingham) 4(1):011013

Kim D, Kim J, Park YI, Lee N, Hyeon T (2018) Recent development of inorganic nanoparticles for biomedical imaging. ACS Cent Sci 4(3):324–336

Kumar S, Maurya VK, Dandu HR, Bhatt ML, Saxena SK (2018) Global perspective of novel therapeutic strategies for the management of NeuroAIDS. Biomol Concepts 9(1):33–42

Logothetidis S (2006) Nanotechnology in medicine: the medicine of tomorrow and nanomedicine. Hippokratia 10(1):7–21

Madni A, Tahir N, Rehman M, Raza A, Mahmood MA, Khan MI, Kashif PM (2017) Hybrid Nano-carriers for potential drug delivery. Adv Technol Deliv Ther (Book): 53–87. https://doi.org/10.5772/66466. IntechOpen

Matea CT, Mocan T, Tabaran F, Pop T, Mosteanu O, Puia C, Iancu C, Mocan L (2017) Quantum dots in imaging, drug delivery and sensor applications. Int J Nanomed 12:5421–5431

Mieszawska AJ, Mulder WJ, Fayad ZA, Cormode DP (2013) Multifunctional gold nanoparticles for diagnosis and therapy of disease. Mol Pharm 10(3):831–847

Miller AD (2016) Nanomedicine therapeutics and diagnostics are the goal. Ther Deliv 7(7):431–456

Misra R, Acharya S, Sahoo SK (2010) Cancer nanotechnology: application of nanotechnology in cancer therapy. Drug Discov Today 15(19–20):842–850

Naseri N, Ajorlou E, Asghari F, Pilehvar-Soltanahmadi Y (2018) An update on nanoparticle-based contrast agents in medical imaging. Artif Cells Nanomed Biotechnol 46(6):1111–1121

Okada H, Toguchi H (1995) Critical reviews in therapeutic drug carrier systems. Crit Rev Ther Drug Carr Syst 12(1):1–99

Olusanya T, Haj Ahmad R, Ibegbu D, Smith J, Elkordy A (2018) Liposomal drug delivery systems and anticancer drugs. Molecules 23(4):907

Pan S, Yu H, Yang X, Yang X, Wang Y, Liu Q, Jin L, Yang Y (2017) Application of nanomaterials in stem cell regenerative medicine of orthopedic surgery. J Nanomater 2017:1–12, 1985942

Patra JK, Das G, Fraceto LF, Campos EVR, Rodriguez-Torres MDP, Acosta-Torres LS, Diaz-Torres LA, Grillo R, Swamy MK, Sharma S, Habtemariam S, Shin HS (2018) Nanobased drug delivery systems: recent developments and future prospects. J Nanobiotechnol 16(1):71

Rizvi SA, Saleh AM (2018) Applications of nanoparticle systems in drug delivery technology. Saudi Pharm J 26(1):64–70

Rizzo LY, Theek B, Storm G, Kiessling F, Lammers T (2013) Recent progress in nanomedicine: therapeutic, diagnostic and theranostic applications. Curr Opin Biotechnol 24(6):1159–1166

Sands MJ, Levitin A (2004) Basics of magnetic resonance imaging. Semin Vasc Surg 17(2):66–82

Saxena SK, Tiwari S, Nair MP (2012) Nanotherapeutics: emerging competent technology in neuroAIDS and CNS drug delivery. Nanomedicine (Lond) 7(7):941–944

Serda RE, Godin B, Blanco E, Chiappini C, Ferrari M (2011) Multi-stage delivery nano-particle systems for therapeutic applications. Biochim Biophys Acta 1810(3):317–329

Silindir M, Özer AY, Erdoğan S (2012) The use and importance of liposomes in positron emission tomography. Drug Deliv 19(1):68–80

Singh L, Kruger HG, Maguire GEM, Govender T, Parboosing R (2017) The role of nanotechnology in the treatment of viral infections. Ther Adv Infect Dis 4(4):105–131

Suri SS, Fenniri H, Singh B (2007) Nanotechnology-based drug delivery systems. J Occup Med Toxicol 2(1):16

Tognarelli JM, Dawood M, Shariff MI, Grover VP, Crossey MM, Cox IJ, Taylor-Robinson SD, McPhail MJ (2015) Magnetic resonance spectroscopy: principles and techniques: lessons for clinicians. J Clin Exp Hepatol 5(4):320–328

Van Rijt S, Habibovic P (2017) Enhancing regenerative approaches with nanoparticles. J R Soc Interface 14(129). pii: 20170093

Wang YF, Liu L, Xue X, Liang XJ (2017) Nanoparticle-based drug delivery systems: what can they really do in vivo? F1000Res 6:681

Woldu MA, Lenjisa JL (2015) Nanoparticles and the new era in diabetes management. Int J Basic Clin Pharmacol 2014(3):277–284

Xie F, Xie L (2018) Serum zinc level is associated with liver dysfunction caused by white smoke inhalation. Gastroenterol Rep 6(4):304–307

Yu X, Trase I, Ren M, Duval K, Guo X, Chen Z (2016) Design of Nanoparticle-Based Carriers for targeted drug delivery. J Nanomater. pii: 1087250

Zhang H, Yee D, Wang C (2008a) Quantum dots for cancer diagnosis and therapy: biological and clinical perspectives. Nanomedicine (Lond) 3(1):83–91

Zhang L, Gu FX, Chan JM, Wang AZ, Langer RS, Farokhzad OC (2008b) Nanoparticles in medicine: therapeutic applications and developments. Clin Pharmacol Ther 83(5):761–769

Zhang W, Wang W, Yu DX, Xiao Z, He Z (2018a) Application of nanodiagnostics and nanotherapy to CNS diseases. Nanomedicine (Lond) 13(18):2341–2371

Zhang H, Liu XL, Zhang YF, Gao F, Li GL, He Y, Peng ML, Fan HM (2018b) Magnetic nanoparticles based cancer therapy: current status and applications. Sci China Life Sci 61(4):400–414

Zhao CY, Cheng R, Yang Z, Tian ZM (2018a) Nanotechnology for Cancer Therapy Based on Chemotherapy. Molecules 23(4). pii: E826

Zhao N, Woodle MC, Mixson AJ (2018b) Advances in delivery systems for doxorubicin. J Nanomed Nanotechnol 9(5). pii: 519

Zitka O, Ryvolova M, Hubalek J, Eckschlager T, Adam V, Kizek R (2012) From amino acids to proteins as targets for metal-based drugs. Curr Drug Metab 13(3):306–320

Nanobiotechnology: Paving the Way to Personalized Medicine

2

Qaneeta Haseeb, Syed Damin Abbas Hamdani, Aasma Akram, Duaa Ahmad Khan, Tausif Ahmed Rajput, and Mustafeez Mujtaba Babar

Abstract

Current medical treatments have been addressing the medical needs of patients for centuries. These drugs target various cellular surface receptors and intracellular moieties to exhibit their effects. The integration of the nanobiotechnology tools in pharmaceutical sciences has proved that the therapeutic effectiveness of the drug molecules and other bioactive molecules can be considerably increased. Moreover, the drug molecules can be targeted to the particular site of action causing a significant decrease in the appearance of the adverse drug reactions. The development in omics sciences has further supported the development of biotechnology as an efficient means to detect, diagnose, and treat various diseased conditions. Conversely, nanotechnology along with the developments in the fields of materials sciences, bioengineering, and systems biology has made it possible to view, model, fabricate, manipulate, and modify the anatomical, biochemical, and physiological patterns within a living cell and in turn tissue, organ, and, ultimately, an organism. The merger of the two sciences, nanotechnology and biotechnology, has given rise to nanobiotechnology which can serve as a means to attain the ultimate goal of the healthcare system by being predictive, preventive, personalized, and participatory. The current chapter reviews the interdisciplinary nature of the field of nanobiotechnology and the employment of its principles in the developing safe, stable, efficient, and cost-effective personalized treatment options.

Q. Haseeb · S. D. A. Hamdani · D. A. Khan · T. A. Rajput · M. M. Babar (✉)
Shifa College of Pharmaceutical Sciences, Shifa Tameer-e-Millat University, Islamabad, Pakistan

A. Akram
Shifa College of Pharmaceutical Sciences, Shifa Tameer-e-Millat University, Islamabad, Pakistan

Government College University, Faisalabad, Pakistan

S. K. Saxena, S. M. P. Khurana (eds.), *NanoBioMedicine*,
https://doi.org/10.1007/978-981-32-9898-9_2

Keywords
Nanomedicine · Personalized medicine · Omics approaches · P4 medicine · Nanobiotechnology

Abbreviations

ADRs	adverse drug reactions
BBB	blood-brain barrier
CNS	central nervous system
DNA	deoxyribonucleic acid
EPR	enhanced permeation retention
FDA	food and drug authority
LMW	low molecular weight
NCs	nano carriers
nm	nanometer
PTMs	post-translational modifications
RNA	ribonucleic acid
TAA	tumor-associated antigens

2.1 Introduction

Nanobiotechnology, an interlink of nanotechnology and biotechnology, involves the utilization of nanotechnology tools to learn, comprehend, and intervene biological processes at the molecular level. The development and application of nanoparticles, nano-devices, nanosensors, and other nanoscale entities for analytical, diagnostic, prophylactic, therapeutic, and interventional procedures falls under the scope of nanobiotechnology. Structures obtained from biological sources or those synthetically prepared to mimic them are used for performing specific biofunctions in nanobiotechnology (Steele et al. 2017). Materials that effectively interact with biological systems at the molecular level, measuring up to a few nanometers in one or multiple dimensions, are known as nanomaterials (Tietjen et al. 2018). Researchers use these materials to facilitate the diagnosis and treatment of different pathological conditions in humans, animals, and plants (Rocha et al. 2017). The role of nanoparticles is much like a link between complex materials and molecular structures (Wu et al. 2017). In pharmaceutical context, nanotechnology is associated with the synthesis of nanocarriers for drug delivery purpose varying in size from 10 to 1000 nm (Sonali et al. 2018). Nano-sized biomolecules have been an area of interest for almost half a century now; however, nanobiology or nanobiotechnology emerged as a discipline only after researchers realized that the scope of this area includes solutions for many complicated biological problems. Some of the subclasses of this discipline include, but are not limited to, nanobiological structures and systems, nano-interfacial biology, nanoscale biology, biomimetics, and nanomedicine. Nanobiological structure and system is a discipline which employs nanotechnology for detection, measurement, or investigation of biological systems (Das et al. 2013). Nano-interfacial biology is study of nanotechnology in relation to material sciences and biochemistry (Thangavelu et al. 2016). Nanoscale biology revolves around biological research performed on the nanoscale, or carried out with the help of nanoscale technologies. Biomimetics is the investigation of principles of design of highly complex natural materials and engineering models at different scales to mimic the natural biological materials (Naik and Singamaneni 2017). In the field of medicine, practical application of biotechnology is nanobiotechnology. Figure 2.1 provides a review of the role of nanobiotehnology in various fields of biological and medical sciences. The current chapter starts with a review of the employment of nanobiotechnology in diagnostic and therapeutic aspects of sciences. It then reviews the major challenges that are currently being faced by the medicinal and healthcare systems and how nanobiotechnology can address these aspects. Toward the end of the chapter, the challenges in integrating nanobiotechnology in the mainstream medicine are discussed.

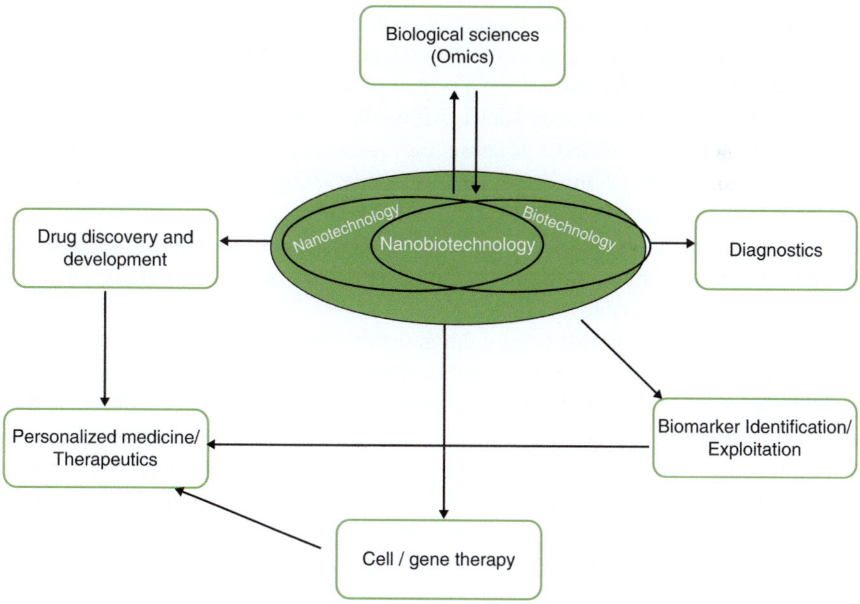

Fig. 2.1 The multidisciplinary nature of nanobiotechnology and its role in enhancing the personalized medicine

2.2 Nanobiotechnology in Nanotheranostics and Nanomedicine

Nanotheranostics has emerged as a new discipline over the past few decades, where nanotechnology is employed to diagnose and treat various diseases. Numerous nanomaterials for example polymer based nanoparticles are being produced as a result of rapid development in the field of nanotechnology. Consequently, these innovative nanoparticles have opened new aspects for nanotheranostics applications. Accurate diagnosis of diseases and detection at early stages and cost-effectiveness only add up to the growing popularity of nanotheranostics (Selvan and Narayanan 2016). For instance, in case of cancer where each type of cancer is variable on the basis of its biomolecular makeup and the possible treatment, nanobiotechnology tools are used for the identification of biomarkers that can not only serve as a means of diagnosing the conditions but can also potentially serve as a therapeutic target (Zarei and Aalaie 2019). Biomarkers are identified with the help of nanoparticles that tend to get targeted to particular diseased tissues. Nanotheranostics have found their application in the discovery and targeted imaging of biomarkers and treating disease on the basis of the biomarker distribution (Kim et al. 2013). Moreover, nanobiotechnology has many promising biological and engineering applications. Its tools can be used as molecular sensors for diagnosis, as a drug molecule and a vehicle for delivering different drugs. It can also be employed for imaging cells and living animals (Selvan and Narayanan 2016). With nanotheranostics it has become possible to incorporate different drug components which can include, but are not limited to, controlled-release drug delivery systems, targeted drug delivery system and diagnosing via nano-scale system. The advent of "intelligent" bioresponsive nanotheranostics is stimulated either by the biological environment or pathology of disease because they are only activated through specific triggers or biomarkers (Wang et al. 2017). The intersection of biology and engineering helps in detecting, monitoring and localized targeting of the cancer cells. Enzymes, nucleic acids, ions, and small biomolecules are the internal triggers for activating nanotheranostics while providing real time assessment of the therapy without a need for invasive testing.

Nanomedicine, in contrast to nanotheranostics, employs the application of nanotechnology primarily for the therapeutic purposes. Nanomedicine, with a multidisciplinary basis, has been the center of attention of scientists from different fields. These nanomeds have improved control of drug delivery in comparison to simple drug molecules resulting in an Enhanced Permeation and Retention (EPR) effect (Srinivasarao et al. 2019; Zhang et al. 2019). The absorbance and emissions of nanoparticles are substantially greater than that of small molecular probes. Consequently, better local contrast in biological imaging and sensing can be achieved by the use of nanoparticles. Much similar to other drug delivery systems, the nanosystems must be able to carry out diagnosis and treatment with a very good safety profile (Zhang 2017). Furthermore, for clinical effectiveness, nanoparticles must fall within a particular size range implying thereby that they should not be very small that they are rapidly cleared out through the kidneys (>10 nm) or very large that they are cleared up directly by the reticuloendothelial system (<200 nm) (Pietersz et al. 2017). Appropriate consideration of size and development method must, hence, be given while developing an appropriate nanobiotechnology-based theranostic or medical approach.

2.3 Nanobiotechnology in P4 Medicine

The traditional focus of medicine has been set on the treatment of diseases and symptoms after they begin to appear in a patient. The current aim of healthcare systems and organizations revolves around providing treatment through medicine and other healthcare products to patients after they fall sick and/or are suffering from a disease. However, the integration of an interdisciplinary approach has lead the scientists to conceive a more holistic approach, an in-depth model in medicine known as P4 medicine representing the predictive, preventive, personalized and participatory medicine (Noell et al. 2018). P4 medicine integrates different data types from information of various stages of biological system such as tissues, cells, metabolites, proteins, RNA, DNA, and epigenetic changes (Davis et al. 2019). This provides an opportunity to be specific and, hence, reduces the challenges associated with large data sets or, in simple terms, biological data. This is done by eliminating all sorts of technical errors and irrelevant information from different aspects of biology that are unwanted in case of a specific area. It, hence, enables the scientists in creating effective, predictive, and applicable models to guide through the treatment of patients (Hood and Friend 2011).

Systems biology can be defined as the study of biological systems in terms of collections of networks at multiple levels, ranging from the molecular level, through cells, tissues, and organisms, to the population level (Flores et al. 2013). With respect to the systems approach to disease, need for developing new modern technologies for exploring new dimensions of data space arises. Genomics, organomics, interactomics, cellomics, proteomics, metabolomics, high-throughput phenotypic measurements, and imaging both in vitro and in vivo help in deciphering the biological systems from molecular to organismal level (Leonavicius et al. 2019). Significant work is currently being carried out in the fields of microfluidics and nanotechnology so that complicated chemical processes can be miniaturized, parallelized, and automatized. The need of comprehending the various biological processes such as disease pathways must be the driving force for these new technologies. This helps to amplify the ability of generating large quantities of personal data and interpreting it in digital form. However, this massive information still needs to be translated to be understood and used (Cirillo and Valencia 2019). Over the last decade, genome-wide sequencing of organisms has led to a "data explosion" that provides an opportunity to explore new dimensions of therapeutics (Hood and Flores 2012). Development of quantified systems for evaluating the biological networks is a new concept and carries superiority over the traditional reductive science, particularly when it comes to understanding of a disease through a single gene or protein. Collaboration of biologists with engineers, computational scientists, mathematicians, earth scientists, and physicists has made it possible to control and predict function of biological systems (Flores et al. 2013). Advances in various sciences have yet to actualize in case of P4 medicine. In this regard, a lot of efforts are still required such as developing techniques to determine individualized genomes, molecular imaging, individual cell analysis and microfluidic techniques (Van Den Berg et al. 2019). These techniques would help to better understand the distresses caused by treatments in biological processes by employing various new computational and mathematical methods such as dynamic networks (Sobradillo et al. 2011). Additionally, there is a need to create awareness of P4 medicine and its benefits among healthcare workforce as well as patients. They would have to be trained to understand complex perspective derived from new methods of biomedicine system and its ethical and legal consequences (Wang et al. 2019). In the near future, all domains of healthcare industry including healthcare providers, medical diagnostic laboratories, pharmaceutical companies, and insurance companies have to adopt the modern approaches in order to support global collaboration between administrations, industry, and academics for the facilitation and adaptation of the P4 medicine.

2.4 Personalizing Nanomedicine

The healthcare approach in which specific treatments for each patient/ patient group are developed keeping in consideration environmental, genetic, and phenotypic factors is known as personalized medicine. These factors can have a substantial effect on the efficacy and safety of the treatment (Fornaguera and García-Celma 2017). Nanomedicine is growing popular in the field of personalized medication as it helps in providing targeted treatment for each patient or group of patients having similar characteristics with respect to their genome (Fornaguera and García-Celma 2017). Contrary to different issues faced with the use of conventional medicine, the fusion of nanotechnology and personalized medicine has adopted a more intelligent approach, i.e., it offers novel smart drug-nanocarriers [NCs] compatible with the biomolecular makeup and genome of individual patients and effective combination of multiple drugs with targeted action. This helps in improving the therapeutic and safety profile of personalized nanomedicine (Zhou et al. 2019). Therefore, through nanomedicine, scientists have been able to develop better treatment therapies for the management of various diseases (Kaushik et al. 2018). Personalized medicine is an approach where an individual patient is diagnosed, treated, and monitored for diseases for improving their health-care services and quality of life. For cancer, there is a considerable genotypic and phenotypic difference in tumors; thus, treatment plans should be devised on the basis of the tumor morphology and preferences of the patient (Xie et al. 2019). It would then require materials and methods that help to study tumor and its etiology comprehensively, and the conventional imaging does not comply with such needs. Only personalized nanomedicine can help in improving the overall outcomes in such a condition. Figure 2.2 represents the current challenges of personalized medicine and their solution through nanobiotechnology.

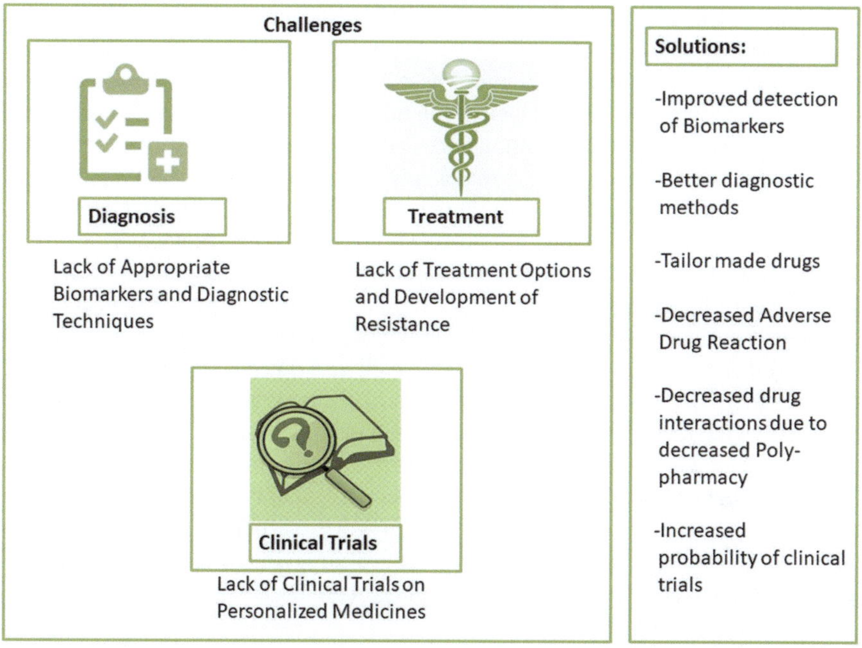

Fig. 2.2 Challenges of personalized medicine and their solutions through nanobiotechnology

The current therapeutics have been based on the idea that only mode of action, dosage regimen, and route used for administrating the medication are the modifiable parameters of a therapeutic agent. However, with nanomedicine, changes can be made on even smaller levels based on the need of a particular patient. Nanobiotechnology allows us to carry out monitoring of bio-distribution, accumulation at the target site, localization of the drug, and drug release, without any invasive procedure with the help of imaging moieties. Nanobiotechnology when used with theranostic nanomedicines can help to actualize the idea of personalized nanomedicine. With the help of tools, which help to oversee delivery and release of drug and drug efficacy, patients can be prescreened for identifying tumors which can be treated with nanomedicine. Consequently, with image-guided and individualized treatments, the personalized nanomedicine appears to be an excellent solution for the treatment of not only different types of cancers but many other diseases as well (Theek et al. 2014). Currently, the focus of the healthcare sector is set on the treatment of disease instead of effectively managing health. Thereby, implying that majority of treatment options are based on symptomatic relief rather than identifying and completely eliminating the pathological factors.

Development of the personalized nanomedicine is complex but not impossible. The most efficient way to attain this is by reverse engineering of the underlying pathology. Reverse engineering is a tried and tested approach used for understanding many complex systems (Alonso et al. 2019). The reverse engineering process includes breaking down of the system into its components, and then the function of each component is analyzed and understood; these components are then reintegrated into the system. Under different conditions, the effects of these components on different processes are simulated. These factors include specialized cells and organs, signaling pathways, and their regulation under normal and stress conditions. With the help of these models, simulations can be conducted for every patient using their unique variables which is a major advantage of this practice. Personalized medicine is one of the most significant applications of nanomedicine (Galetti et al. 2019). Personalization is based on molecular conception and molecular precision. Targeted therapy is used as a basis for research into the molecular precision (Reineke 2018). A two-library approach may be used for a nanomedicine to work in conjunction with a surface modifier for providing the proper retention ligand, preferential distribution, and release kinetics for a positive clinical outcome for various stratified patient subgroups using the systematic characterization and proper knowledge of the particokinetics of a nanomedicine when used alongside critical biomarkers (Siegrist et al. 2019). This type of personalized medicine does not come without additional regulatory concerns, but the general regulatory concerns related with nanomedicine have already been addressed. In order to acquire a successful personalized outcome there is a need to apply stratification to the nanomedicine platform in personalized nanomedicine along with the patient groups (Reineke 2018).

2.5 Omics and Nanobiotechnology

Genomics, proteomics, transcriptomics, and metabolomics constitute the basic components of omics technologies. These sciences have undergone rapid development over the past several years and have made the personalization of medicine possible. The advances in the medical sciences through these technologies have been employed in clinical practice, and they are growing popular at a fairly fast pace (Liu et al. 2019b). The disadvantage associated with these technologies is that any one of them cannot individually understand the complex pathology of many diseases in humans. Thus, several technologies working along with one another have to be used for getting a complete picture of the biology and pathology of the disease under consideration (Karczewski and Snyder 2018).

Omics-based nanobiotechnology works with the objective of detecting diseases at early stage and identifying the underlying pathophysiology for targeted therapy. Nanogenomics is a relatively new field, focusing on diagnosis and treatment of diseases that were difficult to tackle through conventional medicine (Hirani et al. 2016). Proteomics and metabolomics, with their innovative applications in personal genomics, have helped in addressing the issues related to the human disease. Due to these developments newer post-genomics innovations are attaining constituting a crossroad of omics and nanotechnology (Kobeissy et al. 2014). Following the first use of mass spectroscopy for characterization of multiple proteins and their post-translational modifications (PTMs), there has been an increased interest in the field of proteomics to study the protein/peptide biomarkers and their discovery (Liu et al.

2019a). The application of proteomics is widespread in the field of medical science; it is used to decipher the pathogenesis of diseases on a molecular level, identification of potential diagnostic and prognostic biomarkers, and the characterization of novel drug targets. This technology enables us to identify and quantify proteins associated with a particular disease, through analysis of changes in their levels of expression and PTMs in patients suffering from that disease (López et al. 2012).

The clinical applications of proteomics are revolutionizing the modern medicine by providing the information related to the molecular players of the diseases. Great progress has been made by means of identification of exclusive patterns of protein expression, or biomarkers linked with a disease. The outcomes of these tools help in the timely and reliable diagnosis of the disease and understanding the progression of disease. The sensitivity and specificity of the treatment increases many folds when multiple biomarkers are measured and analyzed rather than any single biomarker. It is thought that the content with highest level of information is present in the low molecular weight (LMW), low abundance fraction of biological fluids (Patil et al. 2019). With nanotechnology, harvesting of low abundant panels of biomarkers becomes an achievable task. With the integration of biotechnological techniques in the nanotechnology, omics data has become more accessible, and newer methods to develop and test the simulated systems for personalizing the treatment options have been attained (López et al. 2012). Because of the ever-increasing mortality and morbidity owing to various communicable and non-communicable diseases, the therapeutic cost for nanoparticles has been increased two to threefolds over 10 years. The development of drug tolerance and resistance over the past several decades has further led to a rise in the dose and cost of the treatment. These factors further invigorate the improvement of customized and precise atomic techniques to battle against different resilient medical and therapeutic issues (Devasahayam 2019). Owing to the recent advancements in biomedical and bioengineering sciences, humanity has entered the period of “nano” and “omics” sciences. In collaboration with surface sciences, natural sciences and atomic sciences, myriad nanoparticles have been developed to deliver the drug payload to the diseased tissue. The outer layers of the nano-particles comprise polymers, metals, micelles, and quantum speck. Reports have shown that these nanoparticles have the capability to address the issue of multidrug resistance and have higher specificity, more strength and lower toxicity. Moreover, nano-chips and other nano-gadgets have been developed that can deliver medication, with significant efficiency (Elim and Chiang 2019). The “omics” science evolved from the development of genomics and later stretched out to four general territories: genomics, transcriptomics, proteomics, and metabolomics. The clinical aspect of omics considers each patient to be composed of multiple data points and, hence, digitalized them into “big data” (Lakkireddy and Bazile 2019). The medication selection process is ought to considerably change if it is based on genomic medicine principles. The nanobiotechnology can, hence, contribute to addressing the medical issues of individual cells by providing safe and effective diagnostic and prognostic options in an efficient manner.

2.6 Clinically Viable Nanomedicine

Nanomedicine offers a unique opportunity to treat human diseases. However, the physiological and pathological barriers within the diseased tissue frequently obstruct its viability. Nanoparticles comprising surface markers are particularly subject to these restraints. Moreover, cancer immunotherapies have helped to unveil the paths of nanoimmune communications and to create therapeutically effective medicine that can exploit the body's innate capability to target a diseased tissue. In this manner, a thorough comprehension of the imperative local procedures that manage the destiny of nanomedicine is vital for the development of individualized regimens for different patients (von Roemeling et al. 2017).

One of the fundamental advantages of nanomedicine is that nanoparticles can be altered to dodge the current barriers in therapeutics (Zhao et al. 2019). They can help in the transport and deposition of different anti-cancer agents inside tumor tissue, thereby, preventing the Adverse Drug Reactions (ADRs). Moreover, by incorporating in the nanoparticles, the circulatory half-life of the drugs can be increased by prevention of degradation and metabolism before it reaches target site. This can, hence, help in addressing one of the mechanisms of development of drug resistance. Moreover, the nanoparticles do not only deliver conventional drug molecules but also biomolecules like nucleic acids, peptides, and proteins (Busatto et al. 2019). Additionally, delivery of nucleic acids through nanoparticles enables us to control immune responses, which can prove to be very useful in cancer immunotherapy. Food and Drug Authority (FDA) has already approved several nanomedicines, which are being used for treatment of different types of cancer. Table 2.1 provides a few of the registered nano-drugs. Though these nanoparticles have excellent delivery and therapeutic profile, the disadvantages of nanomedicine, like drug distribution and clearance, must be resolved to promote the use of this type of medicine for treatment of complex diseases such as cancer (von Roemeling et al. 2017).

Table 2.1 FDA-approved nanoparticulate drugs

Formulation name	Diseased condition	Further reading
Doxil (doxorubicin by Janssen)	Neoplasia	Barenholz (2012)
DepoCyt (liposomal cytarabine)	Meningitis	McClune et al. (2005)
Adynovate (antihemophilic factor by Shire)	Hemophilia	Turecek et al. (2016)
Krystexxa (pegloticase by Horizon)	Chronic gout	Alconcel et al. (2011)
Plegridy (IFN by Biogen)	Multiple sclerosis	Chaplin and Gnanapavan (2015)
Avinza (morphine by Pfizer)	Psychostimulant	Rauck et al. (2006)
Megace ES (megestrol by Par)	Antianorexic	Munshi et al. (2017)
Dexferrum (iron dextran by American Regent)	Iron deficiency in CKD	Szybowicz et al. (2015)
Rapamune (sirolimus by Wyeth)	Immunosuppressant	Musalem et al. (2018)
Ontak (denileukin diftitox by Eisai)	T-cell lymphoma	Fuentes et al. (2015)

While developing nanoparticles that are clinically viable, a number of factors have to be considered. For the delivery of nanoparticles across the blood-brain barrier, many research studies have demonstrated that the diameter should be between 20 and 70 nm for being ideal for transport (Chaudhuri and Straubinger 2019). Similarly, the surface charges should be ideally distributed for transport across the physiological barriers. Similarly, the nanoparticle morphology, geometric shape, and porosity also contributed to its candidature for being developed as an effective therapeutic agent. The metabolic fate, tissue distribution and deposition also had to be considered while developing nanomedicine. For instance, in case of nanoparticles targeted to the CNS, they are not promptly cleared by physiological procedures, which could cause the accumulation of drugs and, hence, toxicity. It is, therefore, imperative that a thorough investigation of pharmacological and biopharmaceutical properties of a candidate nanoparticle has to be studied before forwarding them to the next step of drug discovery pipeline (Pelaz et al. 2017).

2.7 Prospects and Challenges

Over the past two decades, there has been a stable yet constant growth and development in the field of nanotechnology and nanomedicine. Public and private sector, throughout the world, has been financing research in nanotechnology. This has, ultimately, led to concentrated efforts by the pharmaceutical scientists, scientific experts, material scientists, engineers, physicists and clinicians for the incorporation of nanotechnology in their respective fields. New tools and approaches have, therefore, been introduced to ensure the effective drug delivery process. Successful product launches have further contributed to enhancing the investment in the research and development of nanoparticles and nanomedicine (Rana and Sharma 2019). However, the regulatory procedures associated with the approval of drug molecules and drug delivery systems is a long and costly process. Over the past 20 years, by and large, only 30 new medications were endorsed by the US FDA every year, and just a couple of nanomedicine-containing medications are included in the FDA-approved list (von Roemeling et al. 2017).

Nanotechnology has been successfully employed to address the issues of various diseases and syndromes. A number of nanotech-based approaches have been developed to address the more resilient disorders. For instance, by using the techniques of nanomedicine there is an increasing potential to further improve immune cell activation or sensitization of immune-resistant tumors through the simulation of the tumor mass by employing combined immunomodulatory treatment. The salient features of "nanovaccine" that set it apart from conventional cancer vaccines include optimization of in vivo targeted delivery specially to dendritic cells, their protective potential for the antigens against proteolytic degradation, and the combinatorial packaging of multiple antigens, cytokines, oligonucleotides, and immunostimulatory antibodies all of which contribute to decreasing the tumor-induced immunosuppression (Theek et al. 2014). Nanomedicine, as opposed to the conventional drug forms, can target the tumors both actively and passively. Their enhanced permeation and retention (EPR) effect on the tumor tissues enables the nanoparticles to reach the tumor sites. The ligands attached to the surface of the nanoparticles have the ability to attach to molecular structures or Tumor Associated Antigens (TAA), thereby improving their overall therapeutic efficacy (Pietersz et al. 2017). In general, the nanoparticles can be engineered to act as diagnostic, therapeutic, and theranostic agents. In case of ex vivo diagnostics, for instance, especially in case of cancer or neurological diseases, diagnosis can be made at an early stage with the help of nanomaterial-based bio-marker assay. Surface functionalization of nanomaterials with DNA segments or antibodies is employed for detecting the expression of a gene or protein biomarker of a disease in a specific cell type or biological fluid (Dyawanapelly et al. 2019). Similarly, recent studies have shown that the superparamagnetic nanomaterials can be used for in vivo imaging of cancer cell development. As discussed earlier, the nanobiotechnological products can be used for carrying out noninvasive detection and monitoring of in situ changes in the expression of biomarkers in tumor sites (Pillai 2019). Moreover, a single nanovector can be used for the delivery of the imaging and therapeutic components which can provide the real time data regarding the therapeutic efficacy of a nano-cargo. On the other hand, from the clinical perspective, the development of personalized medicine components is of prime importance. In this case, the drug and dosage customization is done considering different host factors including their genetic, phenotypic and environmental profiles, which can have a crucial impact on the outcome of a treatment. By inculcating the bionanotechnology principles, patient-specific medication can be designed by manipulating particle size, shape, surface, and other properties of the drug molecule which ensure that the safety and effectiveness of nano-medicine is maintained. The knowledge enables the researchers and the clinicians to develop an understanding of the safety profile of a drug molecule (Halappanavar et al. 2018). The nanoparticles are quite potent and can have harmful effects on the cellular and molecular level as well. Their physicochemical makeup affects their in vivo bioactivity. By employing the toxicity assays, it has been observed that the nanoparticles approved for clinical use are biocompatible and biodegradable. However, newer toxicity assays have to be developed that can determine the exact safety profile of the nanoparticles.

Though the minute size of the nanoparticles is of significance in improving their therapeutic efficacy, their shape is to be considered as well while devising the nanomedicines. The size of a nanoparticle varies between 1 and 100 nm; their shape can be anything from a sheet to a cube, to a sphere to a fiber or a tubular structure; and there is a high level of variation in their surface as well depending on their crystalline structure, impurities, present charge, and presence of coatings. Furthermore, the difference in optical, physical, electrical, mechanical, and chemical properties of different materials makes it rather difficult to test these materials. These variations, therefore, make it difficult to conduct toxicity testing of every nanomaterial variant. The fact that nanomaterials can be involved in many biological functions and pathways which are not fully understood yet is also a challenge. Additionally, time and money required for these tests is another issue that needs attention along with many ethical considerations (Maestri et al. 2019; Paradise 2019; Sunshine and

Paller 2019). Despite, the extensive synthesis and use of nanomaterials, there is still a need for an accepted strategy for safety testing of nanomaterials/nano-enabled products and human health risk assessment. There is, therefore, a need for newer tools that effectively address the issues associated with the nanomaterial research, and for systems that evaluate the toxicity of a nanoparticle based on its physicochemical properties (Halappanavar et al. 2018). Despite the fact that nanomedicine has achieved significant progress in disease management, it is important to be cautious and ensure the protection of the health and safety of patients, personnel, the public and the environment. Some aspects that require urgent attention include the study of the toxic effects of free nanoparticles, genetic modification because of the mode of action at the chromosomal level and issues of safety of nanoparticles at the workplace and in environment. Concentrated efforts, hence, need to be made by the researchers and clinicians to develop safe and effective options for targeting various infectious and metabolic diseases.

2.8 Conclusions

There has been a significant advancement in the field of nanotechnology over the past decades. Combined with tools of biotechnology, it has contributed to the drug discovery and development process as well as clinical medicine. A few drugs have been able to get approval by the FDA and other regulatory bodies. Dozens are at various phases of clinical trials throughout the world. There is, however, a need to develop appropriate platforms, tools, and techniques for evaluating the pharmacological and biopharmaceutical profiles of the nanobiotechnological products. Omics tools help in identification of the specific targets that can be used by the nanomedicine as diagnostic, prognostic, and theranostic agents, thereby improving their capability to be used in contributing to the personalization of medicine.

References

Alconcel SN, Baas AS, Maynard HD (2011) FDA-approved poly (ethylene glycol)–protein conjugate drugs. Polym Chem 2(7):1442–1448

Alonso SG, de la Torre Díez I, Zapiraín BG (2019) Predictive, personalized, preventive and participatory (4P) medicine applied to telemedicine and eHealth in the literature. J Med Syst 43(5):140. (1–10)

Barenholz YC (2012) Doxil®—the first FDA-approved nano-drug: lessons learned. J Control Release 160(2):117–134

Busatto S, Pham A, Suh A, Shapiro S, Wolfram J (2019) Organotropic drug delivery: synthetic nanoparticles and extracellular vesicles. Biomed Microdevices 21(2):46. (1–17)

Chaplin S, Gnanapavan S (2015) Plegridy for the treatment of RRMS in adults. Prescriber 26(9):29–31

Chaudhuri TR, Straubinger RM (2019) Nanoparticles for brain tumor delivery. In: Nervous system drug delivery. Elsevier, Amsterdam, pp 229–250

Cirillo D, Valencia A (2019) Big data analytics for personalized medicine. Curr Opin Biotechnol 58:161–167

Das S, Mitra S, Khurana SP, Debnath N (2013) Nanomaterials for biomedical applications. Front Life Sci 7(3–4):90–98

Davis JD, Kumbale CM, Zhang Q, Voit EO (2019) Dynamical systems approaches to personalized medicine. Curr Opin Biotechnol 58:168–174

Devasahayam S (2019) Nanotechnology and nanomedicine in market: a global perspective on regulatory issues. In: Mohapatra SS, Ranjan S, Dasgupta N, Mishra RK, Thomas S (eds) Characterization and biology of nanomaterials for drug delivery. Elsevier, Cambridge, MA, pp 477–522

Dyawanapelly S, Mehrotra P, Ghosh G, Jagtap DD, Dandekar P, Jain R (2019) How the surface functionalized nanoparticles affect conformation and activity of proteins: exploring through protein-nanoparticle interactions. Bioorg Chem 82:17–25

Elim HI, Chiang LY (2019) Nanochip medicine: physical chemistry engineering. Sci Nat 2(1):086–089

Flores M, Glusman G, Brogaard K, Price ND, Hood L (2013) P4 medicine: how systems medicine will transform the healthcare sector and society. Pers Med 10(6):565–576

Fornaguera C, García-Celma MJ (2017) Personalized Nanomedicine: a revolution at the nanoscale. J Pers Med 7(4):12. (1–20)

Fuentes AC, Szwed E, Spears CD, Thaper S, Dang LH, Dang NH (2015) Denileukin diftitox (Ontak) as maintenance therapy for peripheral T-cell lymphomas: three cases with sustained remission. Case Rep Oncol Med 2015:1–5

Galetti M, Rossi S, Caffarra C, Gerboles AG, Miragoli M (2019) Innovation in nanomedicine and engineered nanomaterials for therapeutic purposes. In: Exposure to engineered nanomaterials in the environment. Elsevier, Amsterdam, pp 235–262

Halappanavar S, Vogel U, Wallin H, Yauk CL (2018) Promise and peril in nanomedicine: the challenges and needs for integrated systems biology approaches to define health risk. Wiley Interdiscip Rev Nanomed Nanobiotechnol 10(1):1–7

Hirani A, Lee AGYW, Pathak Y, Sutariya V (2016) Nanotechnology for Omics-based ocular drug delivery. In: Khosrow-Pour M (ed) Oncology: breakthroughs in research and practice: breakthroughs in research and practice. IGI Global, Hershey, pp 283–298

Hood L, Flores M (2012) A personal view on systems medicine and the emergence of proactive P4 medicine: predictive, preventive, personalized and participatory. New Biotechnol 29(6):613–624

Hood L, Friend SH (2011) Predictive, personalized, preventive, participatory (P4) cancer medicine. Nat Rev Clin Oncol 8(3):184–187

Karczewski KJ, Snyder MP (2018) Integrative omics for health and disease. Nat Rev Genet 19(5):299–310

Kaushik A, Jayant RD, Nair M (2018) Nanomedicine for neuroHIV/AIDS management. Future Med 2018:669–673

Kim TH, Lee S, Chen X (2013) Nanotheranostics for personalized medicine. Expert Rev Mol Diagn 13(3):257–269

Kobeissy FH, Gulbakan B, Alawieh A, Karam P, Zhang Z, Guingab-Cagmat JD, Mondello S, Tan W, Anagli J, Wang K (2014) Post-genomics nanotechnology is gaining momentum: nanoproteomics and applications in life sciences. Omics J Integr Biol 18(2):111–131

Lakkireddy HR, Bazile DV (2019) Nano-carriers for drug routeing–towards a new era. J Drug Target 27(5–6):525–541

Leonavicius K, Nainys J, Kuciauskas D, Mazutis L (2019) Multi-omics at single-cell resolution: comparison of experimental and data fusion approaches. Curr Opin Biotechnol 55:159–166

Liu C, Knudsen GM, Pedley AM, He J, Johnson JL, Yaron TM, Cantley LC, Benkovic SJ (2019a) Mapping post-translational modifications of de novo purine biosynthetic enzymes: implications for pathway regulation. J Proteome Res 18(5):2078–2087

Liu X, Luo X, Jiang C, Zhao H (2019b) Difficulties and challenges in the development of precision medicine. Clin Genet 95(5):569–574

López E, Madero L, López-Pascual J, Latterich M (2012) Clinical proteomics and OMICS clues useful in translational medicine research. Proteome Sci 10(1):35. (1–11)

Maestri E, Marmiroli N, Song J, White JC (2019) Ethical issues of engineered nanomaterials applications and regulatory solutions. In: Marmiroli N, White JC, Song J (eds) Exposure to engineered nanomaterials in the environment. Elsevier, Amsterdam, pp 315–330

McClune B, Buadi F, Aslam N, Przepiorka D (2005) Intrathecal liposomal Cytarabine (Depocyt) is safe and effective for prevention of meningeal disease in patients with acute lymphoblastic leukemia and high-grade lymphoma treated with the HyperCVAD regimen. Blood 106(11):4594. (1–2)

Munshi L, Sun S, Brito M (2017) Megace (Megestrol Acetate) induced hypogonadism in a male patient-a case report. Endocr Pract 23:192. (1)

Musalem HM, Alshaikh AA, Tuleimat LM, Alajlan S (2018) Outcome with topical sirolimus for port wine stain malformations after unsatisfactory results with pulse dye laser treatment alone. Ann Saudi Med 38(5):376–380

Naik RR, Singamaneni S (2017) Introduction: bioinspired and biomimetic materials. ACS Chem Rev 117:12581–12583

Noell G, Faner R, Agustí A (2018) From systems biology to P4 medicine: applications in respiratory medicine. Eur Respir Rev 27(147):170110. (1–15)

Paradise J (2019) Regulating Nanomedicine at the Food and Drug Administration. AMA J Ethics 21(4):347–355

Patil A, Mishra V, Thakur S, Riyaz B, Kaur A, Khursheed R, Patil K, Sathe B (2019) Nanotechnology derived nanotools in biomedical perspectives: an update. Curr Nanosci 15(2):137–146

Pelaz B, Alexiou C, Alvarez-Puebla RA, Alves F, Andrews AM, Ashraf S, Balogh LP, Ballerini L, Bestetti A, Brendel C (2017) Diverse applications of nanomedicine. ACS Nano 11(3):2313–2381

Pietersz GA, Wang X, Yap ML, Lim B, Peter K (2017) Therapeutic targeting in nanomedicine: the future lies in recombinant antibodies. Nanomedicine 12(15):1873–1889

Pillai G (2019) Nanotechnology toward treating Cancer: a comprehensive review. In: Mohapatra SS, Ranjan S, Dasgupta N, Mishra RK, Thomas S (eds) Applications of targeted Nano drugs and delivery systems. Elsevier, Amsterdam, pp 221–256

Rana V, Sharma R (2019) Recent advances in development of Nano drug delivery. In: Mohapatra SS, Ranjan S, Dasgupta N, Mishra RK, Thomas S (eds) Applications of targeted Nano drugs and delivery systems. Elsevier, Amsterdam, pp 93–131

Rauck RL, Bookbinder SA, Bunker TR, Alftine CD, Ghalie R, Negro-Vilar A, Gershon S (2006) The ACTION study: a randomized, open-label, multicenter trial comparing once-a-day extended-release morphine sulfate capsules (AVINZA) to twice-a-day controlled-release oxycodone hydrochloride tablets (OxyContin) for the treatment of chronic, moderate to severe low back pain. J Opioid Manag 2(3):155–166

Reineke J (2018) Terminology matters: there is no targeting, but retention. J Control Release 273:180–183

Rocha M, Chaves N, Báo S (2017) Nanobiotechnology for breast cancer treatment. In: Van Pham P (ed) Breast cancer-from biology to medicine. InTech Open, Rijeka, pp 411–432

Selvan ST, Narayanan K (2016) Introduction to Nanotheranostics. In: Selvan ST, Narayanan K (eds) Introduction to Nanotheranostics. Springer, Singapore, pp 1–6

Siegrist S, Cörek E, Detampel P, Sandström J, Wick P, Huwyler J (2019) Preclinical hazard evaluation strategy for nanomedicines. Nanotoxicology 13(1):73–99

Sobradillo P, Pozo F, Agustí Á (2011) P4 medicine: the future around the corner. Archivos de Bronconeumología (English Edition) 47(1):35–40

Sonali MKV, Singh RP, Agrawal P, Mehata AK, Datta Maroti Pawde N, Sonkar R, Muthu MS (2018) Nanotheranostics: emerging strategies for early diagnosis and therapy of brain Cancer. Nano 2(1):70–86

Srinivasarao DA, Lohiya G, Katti DS (2019) Fundamentals, challenges, and nanomedicine-based solutions for ocular diseases. Wiley Interdiscip Rev Nanomed Nanobiotechnol 11(4):e1548. (1–26)

Steele JF, Peyret H, Saunders K, Castells-Graells R, Marsian J, Meshcheriakova Y, Lomonossoff GP (2017) Synthetic plant virology for nanobiotechnology and nanomedicine. Wiley Interdiscip Rev Nanomed Nanobiotechnol 9(4):c1447. (1–18)

Sunshine JC, Paller AS (2019) Which Nanobasics should be taught in medical schools? AMA J Ethics 21(4):337–346

Szybowicz M, Koralewski M, Karon J, Melnikova L (2015) Micro-Raman spectroscopy of natural and synthetic ferritins and their mimetics. Acta Phys Pol A 127(2):534–536

Thangavelu RM, Gunasekaran D, Jesse MI, SU MR, Sundarajan D, Krishnan K (2016) Nanobiotechnology approach using plant rooting hormone synthesized silver nanoparticle as "nanobullets" for the dynamic applications in horticulture–an in vitro and ex vitro study. Arab J Chem 11(1):48–61

Theek B, Rizzo LY, Ehling J, Kiessling F, Lammers T (2014) The theranostic path to personalized nanomedicine. Clin Transl Imaging 2(1):67–76

Tietjen GT, Bracaglia LG, Saltzman WM, Pober JS (2018) Focus on fundamentals: achieving effective nanoparticle targeting. Trends Mol Med 24(7):598–606

Turecek P, Romeder-Finger S, Apostol C, Bauer A, Crocker-Buqué A, Burger D, Schall R, Gritsch H (2016) A world-wide survey and field study in clinical haemostasis laboratories to evaluate FVIII: C activity assay variability of Adynovate and Obizur in comparison with ADVATE. Haemophilia 22(6):957–965

Van Den Berg A, Mummery CL, Passier R, Van der Meer AD (2019) Personalised organs-on-chips: functional testing for precision medicine. Lab Chip 19(2):198–205

von Roemeling C, Jiang W, Chan CK, Weissman IL, Kim BY (2017) Breaking down the barriers to precision cancer nanomedicine. Trends Biotechnol 35(2):159–171

Wang J, Tao W, Chen X, Farokhzad OC, Liu G (2017) Emerging advances in Nanotheranostics with intelligent bioresponsive systems. Theranostics 7(16):3915

Wang Q, Peng W-X, Wang L, Ye L (2019) Toward multiomics-based next-generation diagnostics for precision medicine. Pers Med 16(2):157–170

Wu J-c, Meng Q-c, Ren H-m, Wang H-t, Wu J, Wang Q (2017) Recent advances in peptide nucleic acid for cancer bionanotechnology. Acta Pharmacol Sin 38(6):798

Xie X, Zhang Y, Li F, Lv T, Li Z, Chen H, Jia L, Gao Y (2019) Challenges and opportunities from basic cancer biology for nanomedicine for targeted drug delivery. Curr Cancer Drug Targets 19(4):257–276

Zarei M, Aalaie J (2019) Profiling of nanoparticle–protein interactions by electrophoresis techniques. Anal Bioanal Chem 411(1):79–96

Zhang X (2017) Nanomedicine, bioimaging and biological microenvironment monitoring. Nanomedicine 1(1):3

Zhang YR, Lin R, Li HJ, He W l, Du JZ, Wang J (2019) Strategies to improve tumor penetration of nanomedicines through nanoparticle design. Wiley Interdiscip Rev Nanomed Nanobiotechnol 11(1):e1519. (1–12)

Zhao Y, Sultan D, Liu Y (2019) Biodistribution, excretion, and toxicity of nanoparticles. In: Cui W, Zhao X (eds) Theranostic bionanomaterials. Elsevier, Amsterdam, pp 27–53

Zhou J, Kroll AV, Holay M, Fang RH, Zhang L (2019) Biomimetic nanotechnology toward personalized vaccines. Adv Mater 1901255:1–22

Part II

NanoBioMedicine in Cancer Diagnosis and Therapy

Antibody-Targeted Nanoparticles for Cancer Treatment

3

Dhruv Kumar, Neelam Sharma-Walia, Sonia Kapoor, and Simran Tandon

Abstract

One of the major scientific breakthroughs of the twenty-first century would be the field of nanotechnology and more specifically the use of engineered nanoparticles in the area of healthcare. These tiny structures of less than 100 nm have the ability to be conjugated to drugs, antibodies, or other chemical compounds, leading to their targeted delivery to the cells of interest. With the improvement in healthcare, the average lifespan has increased; however, advancing age has also led to the higher incidence of various diseases such as cancer. Therapeutics which can effectively halt the progress of this disease is the need of the hour. In this chapter we have given an overview of the recent advances in nanoparticles conjugated to antibodies for cancer treatment.

Keywords

Cancer · Cancer therapy · Angiogenesis · Tumor microenvironment · Nanoparticles · Antibodies

D. Kumar · S. Kapoor · S. Tandon (✉)
Amity Institute of Molecular Medicine & Stem Cell Research (AIMMSCR), Amity University Uttar Pradesh, Noida, Uttar Pradesh, India
e-mail: standon1@amity.edu

N. Sharma-Walia
Department of Microbiology & Immunology, Chicago Medical School, Rosalind Franklin University of Medicine and Science, North Chicago, IL, USA

S. K. Saxena, S. M. P. Khurana (eds.), *NanoBioMedicine*, https://doi.org/10.1007/978-981-32-9898-9_3

3.1 Introduction

Cancer has the ominous reputation of being the major cause of death followed by heart-related diseases. The biology of cancer progression has been elucidated, and therefore it is now important to recognize and implement specific therapeutic approaches for cancer patients. Antibody-targeted cancer therapeutic approaches have garnered significant attention over conventional therapy in the last few years. Several strategies for utilizing the full potential of monoclonal antibodies (mAbs) for targeting specific antigens present on cancer cells (Jarboe et al. 2014) or as tools for delivering cytotoxic agents as antibody-drug conjugates (ADCs) (Leal et al. 2014) or in the recruitment of cytotoxic T cells so as to enhance the immune system (Honeychurch et al. 2015) have been successfully tested in the clinic.

Recent developments in nanotechnology have led to a treasure house, referred to as nanoparticles (NPs), which have unique biological, chemical, and physical properties. The concept of using NPs for drug delivery was first proposed more than 40 years ago by Widder, Senyi, and colleagues (Widder et al. 1978). Exciting alternatives for the utility of Abs to target nanosized drug delivery vehicles, which can be then harnessed for the development of specific cancer medicines, have come up. Several NPs have been translated successfully in the clinical trial setting using targeting ligands. The high specificity of antibodies (Abs) coupled with their role in modern cancer therapeutics has led to them becoming widely studied for targeting ligands and specific targets (Widder et al. 1983; Bertrand et al. 2014).

This chapter aims to describe the different aspects of antibody-nanoparticle conjugates which are currently used in clinical, preclinical, and translational cancer medicine. Recent advancement in the field of antibody-nanoparticle conjugate development and its role in cancer therapy will also be discussed.

3.2 Cancer

Cancer is a disease which arises because of the genomic alteration(s). These genomic changes lead to genetic disorders and oncogene generation, which include point mutation, deletion, amplification, chromosomal translocation, and insertion activation (King et al. 1985; Joensuu and Dimitrijevic 2001; Heinrich et al. 2002; Thomas et al. 2007). Genomic alteration in genes (e.g., p53) leads to the formation of proteins that have a protuberant role leading to modulation of molecular process of cellular signaling. Abnormalities in events acting at the cellular, molecular, and biological levels lead to generation of cancerous cells. Under normal conditions, p53 acts as a tumor suppressor gene and plays a pivotal role in cell survival, division, death, senescence, differentiation, angiogenesis, and DNA metabolism. Mutations in p53 alter the cellular signaling pathways, which transform the ability of p53 to modulate other signaling molecules, such as CDK1-P2 and CDC2 which keep cancer cells in G1 and G2 phases of cell cycle (Chae et al. 2011).

3.2.1 Solid Tumor

Solid tumors are abnormal growth of the tissues that usually arise because of genomic alterations and can be classified as either benign or malignant. Tumors which are of a benign nature are noncancerous, whereas malignant tumors are cancerous. Solid tumors are named based on the type of cells and tissues that form them (e.g., sarcomas, carcinomas, and lymphomas).

3.2.1.1 Characteristics of Solid Tumors

The physiology of solid tumors is vastly different from its normal counterpart. The major difference lies in the vascularity and tissue microenvironment. In comparison to the normal tissue where regular, ordered vasculatures are seen, the tumors display highly abnormal blood vessels. It has also been reported that the tissue microenvironment is totally different in tumors compared to the normal tissues. Tumor growth also requires continuous angiogenesis and constant supply of several factors from surrounding tissues. These physiological differences impact cancer treatment, and it has been seen that lowered oxygen present in the solid tumor niche leads to both radio- and chemoresistance. However, such significant features can be also be utilized successfully for selective cancer treatment (Brown and Giaccia 1998).

Signaling pathways in cancer system are highly complicated. Mainly two types of signaling pathways are involved in cancer, oncogenic and tumor suppressors. Oncogenic signaling pathways play a pivotal role in the development and progression of solid tumors. The mechanism by which oncogenic and tumor suppressor signaling exerts its mode of action and how the key mutations in these signaling pathways lead to cancer development is well studied. Oncogenic signaling promotes cancer progression by modulating key aspects of cellular behavior such as proliferation, growth, and apoptosis in solid tumors, whereas signaling by tumor suppressors leads to inhibition of cancer progression by controlling cell proliferation, cell cycle, and apoptosis. Oncogenic signaling mainly exerts its mode of action via phosphorylation and activation of various transcription factors, and thereby controls expression of target genes, which are involved in tumorigenesis. Depletion or inhibition of key kinases in these oncogenic signaling pathways has been shown to suppress cancer cell growth and is associated with induction of apoptosis.

Regulation of oncogenic and tumor suppressor signaling is highly dependent on the surrounding environment, which is called the tumor microenvironment. Tumor microenvironment has been reported to play a critical role in the cancer cell biology by controlling growth, proliferation, DNA synthesis, and survival. It also leads to epithelial mesenchymal transition (EMT) and also helps in metastasis (Steelman et al. 2011). Amplification of oncogenic proteins and suppression of tumor suppressor proteins exert an important role in tumorigenesis. When the amplification of oncogenes occurs, it plays a leading role in cancer cell function by suppressing the tumor suppressor genes and thereby promoting cancer cell growth. Most of the oncogenic proteins serve as transcription factors which bind to the promoter regions of growth retarding genes such as the tumor suppressors and the various coactivators and corepressors, to suppress the function of tumor suppressor gene.

In solid tumors, the activation, amplification, and mutations of several genes occur at different stages which control important cellular signaling pathways including Ras-MAPK, PI3K/Akt, PDGF, NOTCH, Wnt, EGFR, JAK/STAT, mTOR, MYC, Hedgehog (Hh), c-Met, FGFR, p53, etc.

3.2.2 Nonsolid Cancer

The general concept of a tumor is that it is a distinct and solid lump or mass. However, cancer cells can reside anywhere in the body, swimming in our bloodstream, hiding out in our lymphatic system, and leading to the nonsolid blood cancer types. The origin of blood cancers, unlike solid organ tumors such as breast, colon, pancreas, prostate, etc., is the bone marrow or the lymphatic system.

3.2.2.1 Characteristics of Nonsolid Cancer

Blood cancers affect thousands of lives annually. Most of these cancers initiate in the bone marrow where blood is formed. In most of the blood cancers, the normal blood cell development process is disturbed by uncontrolled growth of an abnormal type of blood cell. These abnormal blood cells, or cancerous cells, prevent several functions of the blood cells like fighting off infections or preventing serious bleeding.

Blood cancers can be categorized into three main types: (1) Leukemia is cancer of the blood and bone marrow, which is caused by the rapid production of abnormal white blood cells. These abnormal white blood cells are unable to fight infection and impair the ability of the bone marrow to produce red blood cells and platelets. (2) Lymphoma affects the lymphatic system, which removes excess fluids from the body and produces immune cells. Abnormal lymphocytes become lymphoma cells, which reproduce and collect in lymph nodes and other tissues. (3) Myeloma is a cancer of the plasma cells. Plasma cells are white blood cells that produce disease- and infection-fighting antibodies in the body system. Myeloma cells prevent the normal production of antibodies, leaving the body's immune system damaged and susceptible to infection.

3.2.3 Tumor Microenvironment

Tumors are not just cancer cells but complex organs, in which many other cells are recruited and can be associated with the transformed cells. Interactions between cancer and non-transformed cells create the tumor microenvironment (TME). The noncancerous cells of the TME have a dynamic and often tumor-promoting function at all stages of carcinogenesis (Hanahan and Weinberg 2011; Hanahan and Coussens 2012). Cellular communication is driven by a complex and vibrant network of chemokines, cytokines, growth factors, and inflammatory and matrix remodeling enzymes against a background of major agitations to the physical and chemical properties of the tissue. The evolution, structure, and actions of the cells in the TME have many counterparts with the processes of wound healing and inflammation, but cells such as macrophages are also found in cancers that have no known association with chronic inflammatory conditions (Mantovani et al. 2008; Grivennikov et al. 2010). One reason for this is that inflammatory and wound-healing processes are activated downstream of oncogenic mutations in the malignant cells (Mantovani et al. 2008). The cells of the immune system, the lymphatics, and the tumor vasculature, pericytes, fibroblasts, and adipocytes are the major non-malignant cell types which are found in the TME. The common features of many TMEs suggest that targeting the nonmalignant cells, or mediators of their communication, has applications across different tumor types and could also complement other treatment options.

3.2.4 Tumor Vasculature

The tumor vasculature is vital for keeping the tumor alive and enabling its growth. Tumor cells must be within a certain distance of a perfused blood vessel to receive sufficient oxygen and nutrients for their survival. It is for this reason that solid tumors must become angiogenic and recruit their own vasculature to grow beyond 1–2 mm in diameter (Gimbrone Jr et al. 1972). This is a complex process that requires interaction between different cell types, the extracellular matrix, and several cytokines and growth factors. Abnormal angiogenesis is associated with excessive growth-promoting signals and a lack of adequate cues to spatially and temporally coordinate vessel growth, maturation, stabilization, and remodeling. Cancer research is realizing the importance of angiogenesis and developing new perspectives regarding disease processes and targeting angiogenesis as cancer therapeutics. The capillary plexus gets remodeled by sprouting, microvascular growth, and fusion into a mature and functional vascular bed during angiogenesis (Yancopoulos et al. 2000; Ferrara et al. 2003; Jain 2003). Angiogenesis maintains physiological homeostasis and tissue integrity during inflammation, wound healing, and endometrial growth in the adult (Ferrara et al. 2003).

Hypoxia is one of the key triggers of blood vessel formation or angiogenesis. Significant progress has been made in this decade to elucidate the molecular mechanisms that mediate hypoxic responses in angiogenesis. Cells respond to low O2 supply by stimulating several hypoxia-inducible factors (HIFs) and other molecules including endoplasmic reticulum-associated kinases, mTOR, and soluble guanylate cyclase that mediate O2 homeostasis which are responsible for the development of the vasculature (Simon and Keith 2008). Expression of HIF1α is induced in endothelial cells, resulting in vascular endothelial growth factor A (VEGFA) and vascular endothelial growth factor receptor 2 (VEGFR2) expression under hypoxic conditions (Tang et al. 2004).

3.3 Nanoparticles

Nanoparticles (NPs) are defined as any naturally occurring or synthetic particulate material of dimensions between 1 and 100 nm in size, present by itself or as an aggregate or agglomerate (Kreyling et al. 2010). NPs are often defined by their size and functional properties which are unique and not shared with other particles with the same chemical composition (Auffan et al. 2009). The extensive diversity of functional properties is facilitated by the variety of materials available to synthesize NPs. The resulting adaptability offers attractive translational potential for a number of biomedical applications such as targeted delivery of treatment, innovative imaging techniques, and novel therapeutics for cancer. In the last few decades, nanoparticles have received growing attention for their tremendous potential in both the diagnosis and treatment of cancer. A list of various nanomaterials (Sanna et al. 2014; Ventola 2017) which have been tested along with drugs against specific cancers is given in Table 3.1.

3.3.1 Nanoparticle Variability

3.3.1.1 Gold Nanoparticles

Gold nanoparticles (AuNPs) have been extensively used in biological sciences including bio-nanotechnology based on their exceptional properties and multiple

Table 3.1 List of nanomaterials for cancer therapeutics

S. no	Nanomaterial	Bioactive compound	Cancer type
1.	Abraxane	Paclitaxel	Breast cancer
2.	CRLX101	Camptothecin	Non-small cell lung cancer
3.	DaunoXome	Daunorubicin	Kaposi's sarcoma
4.	DepoCyt	Cytarabine	Leukemia
5.	Doxil/Caelyx	Doxorubicin	Breast cancer, ovarian cancer, multiple myeloma, Kaposi's sarcoma
6.	Genexol-PM	Paclitaxel	Breast cancer, lung cancer, ovarian cancer
7.	Myocet	Doxorubicin	Breast cancer
8.	NC-4016	Oxaliplatin	Various solid tumors
9.	NC-6004	Cisplatin	Pancreatic cancer
10.	NK105	Paclitaxel	Gastric cancer
11.	NK911	Doxorubicin	Various solid tumors
12.	NL CPT	Irinotecan	Glioma
13.	Oncaspar	Asparagine-specific enzyme	Acute lymphoblastic leukemia
14.	Onco TCS	Vincristine	Non-Hodgkin's lymphoma
15.	Opaxio	Paclitaxel	Lung cancer, ovarian cancer
16.	Paclical	Paclitaxel	Ovarian cancer
17.	ProLindac	DACH-Pt	Ovarian cancer
18.	SPI-77	Cisplatin	Ovarian cancer
19.	ThermoDox	Doxorubicin	Liver cancer, breast cancer

surface functionalities. The ease of AuNP functionalization provides a useful platform for nanobiological assemblies with protein, peptides, oligonucleotides, and antibodies. Bioconjugates of AuNPs have also turned out to be promising candidates in the design of novel biomaterials for the investigation of biological systems (Yeh et al. 2012; Samanta and Medintz 2016). The therapeutic utility of AuNPs has been documented in diagnostic imaging and therapeutic applications. AuNP-based diagnostic strategies have been established, exploiting gold interaction with near-infrared radiations, based on surface plasmon scattering (SPS), or surface plasmon resonance (SPR), which can be used to detect NPs at concentrations as low as 10–16 M (El-Sayed et al. 2005). For therapeutic purposes, AuNPs have been used as photosensitizers, emitting heat to bring about cell death when excited by minimal near-infrared radiation (Huang et al. 2006). AuNPs also act as a practical platform for therapeutic mediators, with their high surface area allowing a dense appearance of multifunctional moieties (Alexander et al. 2011; Sun et al. 2014).

3.3.1.2 Quantum Dots

Quantum dots (QDs) are a class of semiconductor crystal nanomaterials with good optical properties (Zhao and Zeng 2015). QDs have exceptional optical and electrical properties due to its size effect and quantum effect. Nanosized particles can cause quantum confinement effect, dielectric confinement effect, size effect, surface effect, and macroscopic quantum effect. The fluorescent properties of QDs which can be modified as a result of changing the diameter or material of the particle make them a perfect tool for biological applications (Dubertret et al. 2002; Smith et al. 2008; Aillon et al. 2009; Gonda et al. 2010).

3.3.1.3 Liposomes

Liposomes are spherical vesicles comprising of a lipid bilayer that condenses around an aqueous phase, in which other biomolecules and chemical molecules can be stored. The size of the liposome varies from 400 nm to 2.5 mm. The development of liposomal vesicular carriers has been explored and tested for drug delivery (Malam et al. 2009; Çağdaş et al. 2014). Liposomal drug formulations can ensure longer drug half-lives as well as personalized drug release profiles, reducing high peak plasma concentrations (Ramos-Cabrer and Campos 2013).

3.3.1.4 Dendrimers

Dendrimers are well-organized, highly branched, three-dimensional, nanoscopic macromolecules (typically 5000–500,000 g/mol), which possess a low polydispersity index. Dendrimers play an essential role in the emerging field of drug delivery, drug packaging, and nanomedicine. The name dendrimer has been derived from the Greek word "dendron" meaning "tree," which typifies its unique treelike branching structure. Dendrimers are characterized by sheets between each cascade point popularly known as "Generations." The complete structure of dendromer can be classified into the inner core moiety followed by radially attached generations that possess chemical functional groups at the exterior terminal surface (Florence and Hussain 2001; Mattheolabakis et al. 2012; Pandita et al. 2014).

3.3.1.5 Polymeric Micelles

Polymeric micelles are core-shell structures made up of aggregates of amphiphilic polymers constituted into hydrophobic interiors, surrounded by a halo of hydrophilic polymeric chains exposed to the aqueous environment. These micelles are generated when the concentration of the polymer in solution exceeds a certain threshold which is known as the critical micellar concentration and above a certain threshold temperature known as the critical micellar temperature. Encapsulation of several anticancer drugs through polymeric micelles has been developed through conjugation to the hydrophobic core. Polymeric micelles also exhibit other favorable characteristics such as lessening toxicities and increasing the extent of delivery to the compromised leaky tumor vasculature sites, affecting the therapeutic efficacy of chemotherapeutic drugs (Rapoport 2007; Kedar et al. 2010).

3.3.1.6 Polymeric Nanoparticles

Polymeric nanoparticles (PNPs) have gained significant interest over the last few years due to their exclusive properties and behaviors resulting from their small size. These PNP structures incorporate both nanospheres and nanocapsules. Nanospheres are among the simplest NP formulations, consisting simply of a solid matrix of polymer, while the nanocapsule contains an aqueous core (Crucho and Barros 2017). The formulation used is totally dependent on the solubility of the drug molecules; less water-soluble drugs are more readily encapsulated within the nonaqueous environment in nanospheres, while water-soluble and labile drug substances, such as proteins/peptides and DNA/RNA, are more easily encapsulated within nanocapsules (Sokolova and Epple 2008; Kwon et al. 2014).

3.3.2 Targeting Nanoparticles for Cancer Treatment

The recent advancement of nanotechnology in medicine is providing substantial opportunities and new perceptions for novel and effective treatments in many diseases including cancer (Sanna et al. 2014). In the last few decades, a significant number of nano-tools have been developed based on various components, ranging from metals to proteins and DNA/RNA, including lipids, polymers, dendrimers, carbon, metal oxides, silica oxides, nanocrystals, and quantum dots (Fig. 3.1). Nanoparticles acquire exceptional characteristics, including structural properties, large surface area, and long circulation time in bloodstream compared with small molecules. Nanoparticles have emerged as an attractive therapeutic candidate for optimized therapy through stratified and personalized medicine. Cancer is becoming one of the major causes of death followed by heart disease. Despite therapeutic advancement, the survival rate of cancer is still low largely due to the heterogeneous and idiosyncratic nature of individual cancers (Hanahan and Weinberg 2011) and the inability to target therapeutics to neoplastic areas without harming normal tissues (Haber et al. 2011). Generally the antitumor agents administered by authorized therapeutic protocols are systemically distributed throughout the body without giving preference to the cancer tissues. This widespread biodistribution of

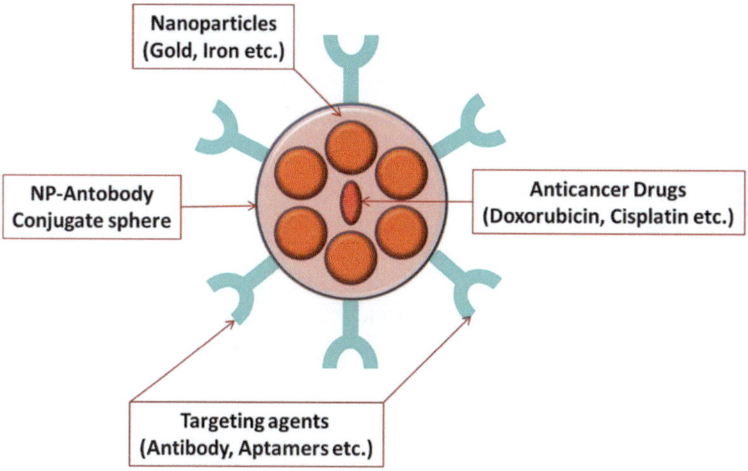

Fig. 3.1 Cancer cell targeting NP-antibody conjugate

chemotherapeutic agent results in both anticancer effects and off-target adverse effects. There are several nanotherapeutic materials which are specific to the tumor tissues and are only approved by the US Food and Drug Administration (FDA) for clinical use, although majority of the materials are currently under preclinical and clinical development (Table 3.2) (Jain and Stylianopoulos 2010; Kim et al. 2010; Wicki et al. 2015; Pelaz et al. 2017; Shi et al. 2017).

3.4 Antibodies in Cancer Therapy

Owing to the specificity and affinity to biological targets, antibodies have gained a pride of place in the area of healthcare. By conjugating them to drugs, their application in the clinical setting has ventured out into a whole new realm, and now with the advent of nanoparticles, the possibilities of delivering a pharmacological agent, selectively and in therapeutic concentrations, with lowered systemic toxicity to a specific tissue or cell have opened up the field of cancer therapeutics. Table 3.2 shows a list of FDA-approved antibodies against cancer.

3.5 Antibody-Nanoparticle Bioconjugation

Several methods have been used for conjugating the antibody to nanoparticle surface (Sapsford et al. 2013; Manjappa et al. 2011). The major concern during the bioconjugation process is the requirement that the antibody-nanoparticle conjugate retains its optimal activity (Fig. 3.1). The functionality of the antibody is especially critical and must be retained for active targeting. In addition, the bioconjugation

Table 3.2 List of FDA-approved therapeutic antibody for cancer

S. no	Name	Trade name	Target	Cancer
1.	Abciximab	ReoPro	CD41 (integrin alpha-IIb)	Platelet aggregation inhibitor
2.	Alemtuzumab	Lemtrada, Campath	CD52	Multiple sclerosis
3.	Atezolizumab	Tecentriq	PD-L1	Cancer
4.	Avelumab	Bavencio	PD-L1	Cancer
5.	Belimumab	Benlysta	BAFF	Non-Hodgkin's lymphoma, etc.
6.	Bemarituzumab		FGFR2	Gastric cancer or gastroesophageal junction adenocarcinoma
7.	Brentuximab vedotin	Adcetris	CD30 (TNFRSF8)	Hodgkin's lymphoma
8.	Capromab pendetide	ProstaScint	Prostatic carcinoma cells	Prostate cancer (detection)
9.	Catumaxomab	Removab	EpCAM, CD3	Ovarian cancer, malignant ascites, gastric cancer
10.	Cemiplimab	Libtayo	PCDC1	Cutaneous squamous cell carcinoma
11.	Cetuximab	Erbitux	EGFR	Metastatic colorectal cancer and head and neck cancer
12.	Cixutumumab		IGF-1 receptor (CD221)	Solid tumors

process must be chosen in a manner that it does not adversely affect physiochemical properties of the nanoparticles as these determine their biological behaviors after systemic administration (Walkey et al. 2012; Barua et al. 2013). The methods that are utilized for conjugation can mainly be classified into two classes: (i) physical methods, wherein the antibodies are randomly adsorbed on the nanoparticles, and (ii) chemical conjugation methods, in which the antibodies are covalently linked to the external surface of nanoparticles.

3.5.1 Physical Methods

The physical adsorption of the antibodies on the nanoparticle surfaces may occur via hydrophobic interactions, electrostatic adsorption, van der Waals forces, and hydrogen bonding. In other cases, a combination of two or several forces may be involved. This is a fast and relatively inexpensive process which bypasses the need of adding any chemical. Moreover, the electrostatic interaction between the antibody and nanoparticle can easily be modulated by varying the environmental pH of the antibody (Walkey et al. 2012). The electrostatic charges on the protein will vary with the surrounding pH. Physical immobilization of antibodies is especially preferred for metallic nanoparticles including gold nanoparticles. For instance, gold nanoparticles have been reported to be conjugated with antibodies against epidermal growth factor receptor (EGFR), prostate-specific antigen, and others (Sokolov et al. 2003; Tanaka et al. 2006, Chen et al. 2007; Wang et al. 2012; Barua et al. 2013). Despite several advantages, physical immobilization of antibody on nanoparticles has certain limitations. Adsorption is a random process whereby an antibody may adsorb in a manner that some of the antigen-binding sites are occluded, thus decreasing the overall affinity of the bioconjugated nanoparticles toward the target site. Also, there is a possibility of displacement of the adsorbed antibody from the nanoparticle surface by serum proteins during circulation or extracellular tissue proteins at the target site.

3.5.2 Chemical Methods

In order to overcome the limitations due to physical adsorption of antibodies on nanoparticles, the chemical conjugation methods may be utilized. Depending on the nature of the nanoparticle and the antibody (whole or fragment), various chemistries are exploited in the synthesis of the conjugated materials. Chemical bioconjugation methods may further be categorized into two main types, namely, (a) random conjugation and (b) site-specific conjugation.

3.5.2.1 Random Conjugation Methods

These include the methods in which the conjugation of antibodies is rather random and no particular site is attached to the nanoparticles, although it is desirable that the conjugation occurs at a site which is distinct and away from the antigen-binding domain or the domains which are involved in any other biological effector functions. One of the examples of random conjugation methods includes thiolation (thiol conjugation) in which the disulfide linkages present in the antibody are reduced to sulfhydryl groups and any of the thiol groups can then be attached to the surface groups of nanoparticles directly or through thiol-reactive linkers.

3.5.2.2 Site-Specific Conjugation Methods

Site-specific conjugation is the controlled coupling of antibody to nanoparticle surface in a predetermined orientation and most often utilizes coupling through the constant or the "Fc" region of the antibody. This allows the site of the antibody which is involved in ligand binding to remain free for interaction with the targeted ligand. This also has an added advantage that the "Fc" region of the antibodies, which is required for eliciting immune responses in host, does not remain free to interact with the host immune components upon administration of the conjugate. Moreover, the "Fc" region of the antibodies is highly glycosylated as compared to the rest of the regions. Therefore, if required, glycosylation chemistry can be exploited to conjugate the antibody to nanoparticle surface (Kumar et al. 2008; Li and Ng 2012; Makaraviciute and Ramanaviciene 2013).

The process of bioconjugation may involve direct coupling between the antibody and nanoparticle surface, or it may involve the use of linker arms/adaptors between the antibody and the nanoparticles. At times, the direct coupling of antibody to nanoparticle surface may limit its accessibility especially when antibody fragments are conjugated. The use of linker arms may assist in increasing the flexibility of the conjugated antibody. However, the use of linkers may require additional derivatization of the antibodies.

3.5.3 Direct Coupling of Antibody and Nanoparticles

Chemically, antibodies represent the polymers of amino acids. As a result several functional groups of amino acids are available including the primary amine groups (such as in glutamine, asparagine), thiol groups (in cysteine), carboxylic groups (such as in glutamic acid, aspartic acid), aldehyde groups (which may be obtained by the oxidation of hydroxyl groups in serine), and functional groups of carbohydrate moieties as antibodies are glycosylated proteins. These functional groups may easily be utilized for conjugating antibody to the nanoparticle surface. The most commonly utilized chemistry is the formation of a covalent amide bond between the primary amine and the carboxylic acid groups, one of which may be present on the nanoparticle, while the other is contributed by the antibody. For instance, the conjugation of nanoparticle with antibody can be designed through 1-ethyl-3-(3-dimethylaminopropyl)carbodiimide-N-hydroxysuccinimide (EDC-NHS) coupling. Quantum dots (QDs), gold nanoparticles, and graphene-derived nanoparticles can be covalently conjugated with antibody by slight modification of the process (Liu et al. 2010). The limitations of this method include self-cross-linking between different antibody molecules and high probability of random orientation of antibody conjugation due to the availability of several amine coupling groups. Moreover, there is a high dependence of amide linkage formation on pH of the medium which makes the reaction sensitive to pH conditions. Therefore, alternatively, the sulfhydryl groups (freely available or engineered into the antibody by reducing disulfide linkages using agents such as mercaptoethanol, 2-mercaptoethylamine, and DTT (dithiothreitol)) can be conjugated to nanoparticles after they are modified by thiol reacting groups including iodoacetyl and maleimide. For instance, trastuzumab, the HER2 receptor targeting antibody, is conjugated to HSA nanoparticles by using thiol modification approach (Steinhauser et al. 2006). By controlling the degree of reduction of the disulfide groups prior to coupling, reasonably controlled site specificity can be tailored into the protein-nanoparticle coupling reaction. Conjugation of a particular site of antibody to the nanoparticle can also be achieved by selectively modifying the "Fc" region of antibodies. This approach does not affect the region which is involved in binding of antigen with protein. The polysaccharides present in the Fc region of the antibody can be selectively oxidized using sodium periodate or boronic acid derivatives (Lin et al. 2009; Tao et al. 2012). The antibodies bioconjugated to nanoparticles through polysaccharide moieties have been shown to better retain the activity of the bioconjugate as compared to the bioconjugate formed using amide coupling chemistry (Puertas et al. 2010). Metallic nanoparticles, liposomes, and polymeric (PLGA-based, chitosan-based) nanoparticles have been coupled with mAb 2C5, antinucleosome mAb 2C5, and rituximab for targeting breast, colon, and Lewis lung carcinoma (Elbayoumi and Torchilin 2008; Koren et al. 2012; Voltan et al. 2013; Abdelghany et al. 2013).

3.5.4 Conjugation of the Antibody to Nanoparticles Through Adaptor Molecules

Certain adaptor biomolecules, which bind to both nanoparticles and the antibody simultaneously, can be exploited to facilitate antibody-nanoparticle bioconjugation. These adaptor molecules may include biotin, “Fc” binding proteins, and nucleic acids. The antibody and nanoparticle may need to be derivatized in order to use the adaptor-based conjugation. For instance, the antibody can typically be modified with biotin, while nanoparticle will need to be functionalized with avidin. The resulting bioconjugate formed by non-covalent interactions between biotin and streptavidin is highly stable. Moreover, a high density of antibody molecules per nanoparticle can be achieved because of the presence of up to four biotin-binding sites in each streptavidin molecule (Wang et al. 2010). Biotinylated anti-CD3 antibodies conjugated to NeutrAvidin-functionalized gelatin nanoparticles have been shown to exhibit high anticancer activity against T-lymphocytic leukemic cells (Dinauer et al. 2005). A similar approach utilizes specific adaptor proteins which bind to the Fc region of antibodies such as protein A and protein G, and Fc receptors (FcγRs) can be attached to the nanoparticle surface which subsequently binds to the immunoglobulin of interest. An advantage of this approach is that the Fc region of antibody is involved in conjugation and the antigen-binding region remains freely available for interaction with the target site. The third type of adaptor molecules includes the nucleic acids. Since single-stranded DNA can bind to another single-stranded DNA which has complementary base pair, one of the single strands is attached to the nanoparticle, while the complementary strand is attached to the antibody. Conjugation between nanoparticle and antibody can occur through complementary base pairing.

3.6 Clinical Applications of Targeting Antibody-Nanoparticle Bioconjugate for Cancer Therapy

As already discussed, nanoparticles (NPs) are diverse with physical properties that can be engaged in cancer medicine. Targeting NPs using antibodies and antibody fragments could bypass many limitations of current targeted cancer therapies. It may provide benefits such as small size, vigor of platform, and synthetic production enabling scale-up production (Carter et al. 2016).

3.6.1 Targeting Solid Tumors

3.6.1.1 Breast Cancer

The most commonly used monoclonal antibodies in clinical practice are bevacizumab, cetuximab, panitumumab, trastuzumab, and tocilizumab which are covalently linked to a near-infrared antibody-IRDye800 bioconjugate. These antibodies when injected intravenously in the nude mice with human breast cancer flank tumors demonstrated ligand-specific binding. Imaging at various time points could confirm the presence or absence of tumor, which could aid in the differentiation of subclinical segments of disease using fluorescence histology, thus demonstrating their potential in clinical translation (Korb et al. 2014).

Another approach had been targeting cell surface receptors, other than the ones classically associated with breast cancer and the use of small molecules apart from folate and carbonic anhydrase IX ligands (Jiang et al. 2018), commonly used for active targeting. In one such scenario, a unique small molecule is conjugated with a highly cytotoxic warhead, DM4 (a maytansinoid). This DM4 has the ability to specifically bind the overexpressed TrkC receptor seen in metastatic breast cancer. TrkC+ and TrkC− human breast cells were used in the study, which revealed that the conjugate had far better therapeutic efficacy in vivo as compared to DM4 alone, and furthermore it completely ablated orthotopic 4T1 breast tumor (Jiang et al. 2018).

Controlled release polymer nanoparticles using a poly(D,L-lactic acid)-poly(ethylene glycol) copolymer with a terminal maleimide functional group (PLA-PEG-MAL), conjugated to a polypeptide sequence with binding characteristics for the HER-2 receptor, have also been tested in breast and pancreatic cancers. These nanoparticles minimized their nonspecific clearance by the host immune system and showed less aggregation. Specific binding of the NPs followed by internalization by the HER2-expressing cells resulted in enhanced differential cytotoxicity as compared to cells not expressing HER2 (Sun et al. 2007). A recent study has shown advancement in the development of quantum dot (QD) nanotechnology, which aided the construction of a biomedical imaging platform for the study of cancer cell behavior (Sun et al. 2018). Water-soluble CuInS2/ZnS QDs conjugated to an anti-Ki-67 monoclonal antibody (QD-Ki-67 probes), when tested in breast cancer scenario, did not exhibit distinct toxic side effects, provided high quantum yield, and emerged as a potential candidate for bioimaging (Sun et al. 2018). Ki-67 is a nuclear protein associated with cell cycle and cell proliferation and overexpressed in breast cancer cells.

Polydopamine (pD)-based surface modification method to formulate novel nanoformulation of NP-aptamer bioconjugates (Apt-pD-DTX/NPs) for in vivo tumor targeting has been tested in breast cancer. Apt-pD-DTX/NPs effectively enhanced local active drug concentration on tumor sites and drug biodistribution and minimized side effects (Tao et al. 2016).

3.6.1.2 Cervical Cancer

Cervical cancer is the third most common malignancy among women. Cisplatin, carboplatin, paclitaxel, ifosfamide, and topotecan are the routinely used chemotherapeutics to treat advanced and metastatic cervical cancer (Ordikhani et al. 2016). Nanotechnology has contributed majorly in the enhanced drug delivery to cervical cancer cells with higher specificity, lowered systemic drug toxicity, and improved absorption rates. Derivatives of poly(lactide-co-glycolide) (PLGA) have shown sustained and controlled delivery along with greater efficacy of docetaxel in cervical cancer both in vitro and in vivo settings (Ordikhani et al. 2016). Folate receptor antibody-conjugated NPs also have immense therapeutic potential as human cervical cancer cells overexpress folate receptor (Dixit et al. 2015; Zhang et al. 2016).

Various modifications of NPs have been tested to enhance their efficacy and specificity, and few of these include conjugation with folic acid to L-tyrosine polyphosphate, gelatin, or chitosan (Ordikhani et al. 2016). These NPs, when loaded with chemotherapeutic drugs such as silver carbene complex, cisplatin, or carboplatin, led to a tenfold greater specificity compared to control NPs in cervical cancer cells (Ordikhani et al. 2016). Folate-targeted doxorubicin-loaded NPs exhibited improved targeting, in vivo antitumor activity by inhibiting tumor cells in cervical carcinoma and metastatic hepatocellular carcinoma (Rosenblum et al. 2018).

3.6.1.3 Colorectal Cancer

Colorectal cancer (CRC) is the fourth widely diagnosed cancer worldwide that leads to significant morbidity and mortality (Cisterna et al. 2016). Though systemic drug delivery strategies to specific molecular targets including various receptors expressed in various cancer cell types such as FGFR, EGFR, CD44, EpCAM, CA IX, PPAR, and COX-2 have proven to be efficacious, research to develop better nanoparticles has also been emphasized in the clinical field to tackle CRC and other cancers (Lin et al. 2015). Targeted nanoparticles with diverse molecules as surface ligands which associate with tumor cells have been used as drug delivery systems. These nanoparticles have been found as the most promising strategies to deliver cytotoxic drugs for CRC therapy (Cisterna et al. 2016).

3.6.1.4 Liver Cancer

Nanocarriers conjugated with SP94 peptides, identified by phage display, revealed superior binding affinity to human hepatocellular carcinoma (HCC) in comparison to hepatocytes as well as other noncancerous cell types. Using near-infrared fluorescence imaging, a differential tumor-targeting picture was seen between the nontargeted PEGylated liposomal doxorubicin (LD) and SP94-conjugated PEGylated liposomal doxorubicin (SP94-LD), in terms of better tissue distribution, along with the antitumorigenic activity in xenograft-bearing mice, which strikingly extended the survival of the mice (Wu et al. 2018a, b). AS1411, a 26-nucleotide guanine-rich DNA aptamer, which forms a guanine quadruplex structure, has shown promising results as a treatment for cancers in phase I and phase II clinical trials. Doxorubicin (Dox) conjugated to form a synthetic drug-DNA adduct (DDA), called AS1411-Dox, has shown efficacy in the treatment of HCC in vitro and in the murine xenograft model of hepatocellular carcinoma (Trinh et al. 2015).

3.6.1.5 Lung Cancer

In a very comprehensive study, the potential of gold nanoparticle contrast agents cetuximab (C225)-AuNPs, llama heavy-chain variable region Ab fragments (VHH)-AuNPs, and nontargeted control poly(ethylene glycol) or PEG-AuNPs for the imaging of EGFR-expressing lung tumors was compared. C225-AuNPs showed higher binding affinity in culture and much higher accumulation in tumors in the animal models but had the lowest half-life of the three groups. The AuNPs and PEG-AuNPs also showed similar accumulation in the tumor (Ashton et al. 2018). Anti-EGFR antibody-conjugated poly(lactide-co-glycolide) (PGLA) or cetuximab conjugated to docetaxel (DTX)-loaded PLGA NPs demonstrated tremendous in vivo efficacy and higher anti-proliferative activity and presents a promising active targeting carrier for tumor selective therapeutic treatment (Patel et al. 2018; Nadda et al. 2018).

3.6.1.6 Oral Cancer

Cancer of the oral cavity and oropharynx is a common and aggressive cancer type with high metastasis and mortality rate and is immune to therapeutic drugs (Calixto et al. 2014). Curcumin NPs (Cur-NPs) have been shown to induce apoptotic cell death through MDR1 signaling, production of reactive oxygen species (ROS), and activation of caspase-3 and caspase-9 in human oral cancer cells. Cur-NPs were less cytotoxic to normal human gingival fibroblasts and normal human oral keratinocytes, showing promise for a novel treatment against cisplatin-resistant human oral cancer (Chang et al. 2013). Gold nanoparticles (AuNPs) or anti-epidermal growth factor receptor (EGFR) antibody-loaded AuNPs have also shown potential as a potent technique for oral cancer diagnostics (El-Sayed et al. 2005; Kah et al. 2007).

3.6.1.7 Ovarian Cancer

Limited studies are cited for NP-conjugated Abs against ovarian cancers. In one report, folate receptor α (FRα), HER2, and TRAIL (TNF-related apoptosis-inducing ligand) receptor targeting NPs have shown efficacy in in vitro and in vivo ovarian cancer models (Lutz 2015; Langdon and Sims 2016). In another study, inorganic nanoparticles which were conjugated to human antibody fragment were used in epithelial ovarian cancer against the overexpressed folate receptors commonly seen in cancer cells. The results of the study revealed that the conjugated nanoparticles led to greater accumulation along with better retention at tumor site for a longer period of time (Quarta et al. 2015).

3.6.1.8 Pancreatic Cancer

Several studies have shown selective targeting of the highly aggressive pancreatic ductal carcinoma (PDA) which has a 5-year survival below 10%. In another study using monoclonal antibody against EGFR-1, gold nanoparticles and cadmium selenide and indium gallium phosphide quantum dots were conjugated to cetuximab (C225), which revealed selective cytotoxicity in the pancreatic cancer cell line Panc-1 cells as compared to breast cancer cells Cama- 1 (Glazer and Curley 2010).

An in vitro study in pancreatic cancer cell lines using a triple conjugate of cetuximab, gemcitabine, and magnetic Fe3O4 nanoparticles (NPs) showed selective and enhanced killing as revealed by viability and flow cytometry data (Wang et al. 2015). A very recent study using TAB004 antibody, targeting the hypoglycosylated form of mucin 1, overexpressed in pancreatic cancers, was conjugated to poly(lactic-co-glycolic acid) nanoparticles (PLGA NPs). Enhanced internalizing and cytotoxicity in PDA cells compared to the non-conjugated counterpart was seen (Wu et al. 2018a, b).

3.6.1.9 Prostate Cancer

The high incidence of prostate cancer coupled with resistance to conventional treatment has prompted research into the use of NPs, which may offer hope to the many inflicted with this form of cancer. Although tremendous breakthroughs have been made in the field of nanoparticles conjugated to antibodies and chemotherapeutic drugs against various types of cancers, limited studies are cited pertaining to prostate cancer. The use of superparamagnetic iron platinum nanoparticles (SIPPs) conjugated to prostate-specific membrane antigen (PSMA) antibody along with paclitaxel was carried out for human prostate cancer xenografts in the mouse model. The results revealed retardation in tumor growth which could be attributed to the selective uptake into the xenografts in the nude mouse model (Taylor and Sillerud 2012).

3.6.1.10 Skin Cancer

Melanoma, the malignant transformation of melanocytes, is the most aggressive type of skin cancer and is associated with high mortality. The higher incidence of multidrug resistance (MDR) commonly seen in melanoma is also associated with relapse and death. Nanofibers loaded with various drugs such as cobimetinib, ipilimumab, nivolumab, trametinib, etc. have been tested as the choice of drug delivery for melanoma skin cancer therapy. It has been seen that the ability of NPs to penetrate the skin barrier is governed by the size. NPs having a size of more than 10 nm are excluded even via localized massage, but NPs can enter the openings of the hair follicles and create a reservoir of NPs which offers an alternative route for therapeutic absorption (Naves et al. 2017).

3.6.1.11 Brain Tumors and Glioblastoma

The blood-brain barrier (BBB) is the most important challenge facing the effective delivery of chemotherapeutic drugs to the cancers of the brain. High doses of drugs are usually administered in order to penetrate the BBB, which results in significant systemic toxicity to the entire body. Moreover many of the drugs such as cisplatin are irreversibly bound to plasma proteins, leaving only a very small percentage which can then exert antitumor effects (Urien and Lokiec 2004). With the median survival of only 15 months, glioblastomas have the unenviable distinction of being one of the most common and aggressive primary brain tumors resulting in high rates of morbidity and mortality (Ostrom et al. 2014).

In order to specifically target the blood-brain barrier (BBB) and glioblastoma multiforme (GBM), etoposide-entrapped solid lipid nanoparticles were conjugated to a melanotransferrin antibody and the chemotherapeutic drug tamoxifen, resulting in greater killing against U87MG cells (Kuo and Wang 2016).

3.6.2 Receptor-Based Targeting

Making effective use of cell surface receptors for targeted delivery of drugs has led to NPs being conjugated with antibodies to selectively deliver the drug to the cell of choice. Cancer cells usually overexpress the growth factor ligands or the receptor which can be effectively selected using either mono- or polyclonal antibodies for strategic cell killing.

3.6.2.1 EGFR

ErbB family member EGFR is comprised of various domains which include the extracellular N- terminal ligand-binding domain, hydrophobic transmembrane region, and an intracellular C- terminal tyrosine kinase (TK) domain (Master and Sen Gupta 2012). This TK domain activates various signaling pathways, which are responsible for cell proliferation, angiogenesis, survival, and metastatic potential in tumors, which are instrumental in cancer progression and poor prognosis. EGFR-binding ligands have been used on the surface of nanovehicles to aid in tumor cell-specific delivery and also internalization (Wong 2005; Bhuvaneswari et al. 2009). EGFR- targeted nanoparticles involve receptor-blocking monoclonal antibody cetuximab. Drugs such as gemcitabine, paclitaxel, and doxorubicin have also been studied in vitro and in preclinical animal models (Kuo and Liang 2011; Liao et al. 2011). Cationic solid lipid nanoparticles have been conjugated with anti-EGFR antibodies for delivery of carmustine and gemcitabine in human glioblastoma (Kuo and Liang 2011) and non-small cell lung carcinoma animal model (Durr et al. 2007). Nanoparticles have been tested for molecular imaging of live cells (Davis 2009). Gold nanoparticles conjugated with anti-EGFR antibodies and surface coated with cyclodextrin, have been tested for β-lapachone delivery by glutathione-mediated release to cancer cells (Davis 2009). Among various gold NPs such as C225-AuNPs, VHH-AuNPs, and nontargeted control PEG-AuNPs, C225-AuNPs show higher binding affinity and accumulation in EGFR+ tumors, despite having the lowest half-life (Haugsten et al. 2010).

A native ligand of EGFR, 6-kDa protein EGF, has been tested in aerosol administrations of gelatin nanoparticles and showed specific accumulation in orthotopic lung adenocarcinomas in severe combined immunodeficiency (SCID) mice (Tseng et al. 2007, 2008). EGF conjugated to polymeric lipid-based nanoparticles for the delivery of paclitaxel demonstrated significant growth inhibition in vivo (Shimada et al. 2009). Conjugation with high-density lipoprotein (HDL)-mimicking NPs, poly(ethylene glycol) (PEG)-poly(e-caprolactone) (PCL) micelles, and iron oxide NPs was promising in vitro for both drug delivery and photothermal ablation (Shimada et al. 2009; Fonge et al. 2010; Zhang et al. 2010). Murine EGF-conjugated lipid NPs loaded with gemcitabine significantly reduced tumor volume in vivo (Sandoval et al. 2011).

3.6.2.2 FGFR

Tumors of the breast, papillary thyroid, prostate, bladder, and stomach overexpress various members of the fibroblast growth factor receptor (FGFR) family. These receptors are associated with tumor progression and poor patient prognosis (Haugsten et al. 2010). An interesting finding which offers a potential therapeutic target is that FGFR expression is found on the surface of noncancer cells also, which are present in the tissue around the tumor vicinity currently (Heinzle et al. 2011). Small molecule tyrosine kinase inhibitors and receptor-specific antibodies carrying nanoparticles have been developed and tested to target FGFR in many cancers (Wesche et al. 2011).

In infrared-induced thermal ablation which used FGFR-targeted gold nanoconjugates, high specificity and high rate of internalization were seen in FGFR-expressing cells. Internalization of these AuNPs reduced (down to 40%) tumor cell viability after irradiation with near-infrared light, whereas the proliferation of cells lacking FGFRs was not affected (Szlachcic et al. 2012).

3.6.2.3 HER2

Human epidermal growth factor receptor 2 (HER2), a type I transmembrane glycoprotein, plays a critical role in signaling cascades regulating cell proliferation, survival, and apoptosis in breast cancer. HER2 is gene-amplified in 20–25% of breast cancer patients and is linked to aggressive phenotype and poorer prognosis (Chen et al. 2018). The use of targeted therapies in human epidermal growth factor receptor 2 (HER2)-positive breast cancer has transformed the clinical approach to target this type of cancer. Trastuzumab, which is an anti-HER2 antibody, has been recognized as the gold standard for treatment of HER2+ breast cancer patients (Kanazaki et al. 2015; Sano 2017). Iron oxide nanoparticles (IONPs) impart remarkable contrast and high sensitivity and resolution in photoacoustic (PA) imaging. IONPs conjugated with various anti-HER2 moieties, which include the intact IgG, single-chain variable fragment, or peptide for HER2-targeted PA tumor, showed increased cellular uptake, in vivo biodistribution, high affinity, and specific binding to HER2-expressing cells in HER2+ tumors (Kanazaki et al. 2015).

3.6.2.4 VEGFR

In order for the tumor to grow, the requirement of nutrients is pivotal and impacts cancer progression. Therefore, targeting the cognate receptor for the potent angiogenic effector vascular endothelial growth factor (VEGF), namely VEGFR, which is overexpressed in a number of cancers, is an effective therapy. There are three subtypes of receptors: Flt-1 (VEGFR-1) and KDR/Flk-1 (VEGFR-2) which are involved in the process of angiogenesis and FLT-4 (VEGFR-3) which has been associated with the development of the lymphatic system (Karkkainen et al. 2002).

Monoclonal antibodies such as bevacizumab and IMC-1121B have been shown to be safe and effective against various cancers, and therefore conjugating them to nanoparticles shows a greater efficacy and is a promising area. In ovarian cancer cells, mesoporous silica nanoparticles (MSN) were conjugated to bevacizumab, and the results revealed that this conjugate induced apoptosis in a time- and concentration-dependent manner (Zhang et al. 2015).

In another promising study, a triple conjugate of bevacizumab (Bev), SiO2@ LDH nanoparticles, and DOX showed selective targeting and curative efficiency against neuroblastoma highlighting the potential of VEGF-targeting nanocarrier for therapy of VEGF-positive cancers (Zhu et al. 2017).

3.6.2.5 TGF-β

Transforming growth factor-β (TGF-β) is a major molecular player, implicated in various cellular processes such as cell cycle arrest, and apoptosis, preserves genomic stability, and thus acts as a potent anticancer agent regulating the proliferation of various cell types, including epithelial, endothelial, and hematopoietic. Oncogenesis induces major aberrations in the TGF-β pathway and converts TGF-β from a suppressor to a tumor promoter by upregulating growth, invasion, and tumor progression (Tian and Schiemann 2009). TGF-β1 receptor expression is upregulated concomitantly with many cancers, and NP-mediated delivery of a TGF-β1 antisense-expressing construct has been demonstrated as a potential strategy to target TGF-β1/TGF-β1 receptor-mediated tumorigenic pathways (Xu et al. 2014; Sun et al. 2012).

3.6.3 Cancer Stem Cell Targeting

The root of all evil as it may be is the cancer stem cell (CSC), tucked away deep in the tumor tissue, evading the chemotherapeutic drugs and waiting for the right time to overrun the organ. At one point of time, the notion of cancer stem cells was far-fetched, but with the recent advancements, the existence of these stem cells has become a chilling reality. These CSCs are distinct from the bulk of the tumor and possess chemo- and radioresistance enabling them to have a highly aggressive phenotype which contributes to malignant progression, metastasis, and cancer recurrence and confers poor survival outcomes (Singh and Settleman 2010). The ability of CSC to evade the conventional chemotherapeutic regimens stems from their slow-cycling phenotype, the upregulated expression of efflux pumps (ABC), anti-apoptotic proteins, efficient DNA response, and repair machinery (Louka et al. 2015). Therapeutics utilizing antibodies to stall aberrant signaling pathways has shown promise. Antibodies specifically targeting cancer stem cells take advantage of cell surface proteins such as CD44 which is associated with highly aggressive phenotype in various cancers including breast, pancreatic, urothelial, prostate, and gastric cancer as well as acute myeloid leukemia (AML) (Ghosh et al. 2012).

Research in the area of head and neck cancers conducted in mice using patient-derived cells which used monoclonal antibodies RO5429083/RG7356, which target the CSC marker, CD44, showed potential in inhibiting tumor progression by activating cytolytic natural killer (NK) cells (Perez et al. 2012). In fact, there are a large number of success stories of antibodies targeting CSC which have moved on to the clinical trial stage. However, many of the antibodies have shown limited efficacy in patients leading to tumor relapse driven by drug-resistant cancer cells. Therefore, multifunctional receptor-targeting antibodies are conjugated with NPs; nanoparticles for effective targeting of aberrant pathways are the way ahead for next-generation cancer therapy. An in-depth review on nanomedicine-mediated dual drug delivery to target CSCs and bulk cancer cells is discussed elsewhere (Singh et al. 2017).

Nanotechnology-based approaches have shown immense potential as they offer a multipronged approach of selectivity, along with greater bioavailability, especially if a triple conjugate of drug, antibody, and nanoparticle is used. Examples abound in literature and we have highlighted a few. Preclinical studies in mice showed that trastuzumab-conjugated gold nanoparticles (T-AuNPs) exerted a strong cytotoxic effect on both Tmab-sensitive and Tmab-resistant gastric cancer cells (Kubota et al. 2018).

The nonselective antitumor effect of magnetic nanoparticles (MNP) in breast and gastric cancer cells was overcome when anti-CD44 antibody was coated on the MNPs (Aires et al. 2016). Other surface markers which have been effectively targeted by nanoparticle-conjugated antibodies include CD133 which is a common cancer stem cell marker. The nature of the nanoparticles used by various groups includes single-walled carbon nanotubes against glioblastomas (Wang et al. 2011).

3.7 Potential of Targeted Antibody-Nanoparticle Bioconjugate for Cancer Therapy

Nanotechnology is progressively emerging and is revolutionizing cancer diagnosis and therapy specifically targeting angiogenesis, uncontrolled cell proliferation, and metastasis. This field is continuously supported by advancement in protein engineering and novel materials science leading to development of clinically approved therapeutic nanocarriers. Translating basic research to the cancer clinic is challenging (Byrne et al. 2008). Development of different kinds of nanoparticles requires multidisciplinary (engineering, biology, physics, and chemistry) expertise, "multiplex" detection devices, novel mathematical models, imaging dyes, and targeted therapeutic drugs (Sokolov et al. 2009; Ferrari 2005) (Fig. 3.2). Cancer nanotherapeutics are progressing, and this field has practiced an exponential growth since early 2000. There has been a continuous challenge in nanotechnology to be creative in order to accommodate the need of personalized medicine; to improve the efficacy, retention, and tolerability of new chemotherapeutic drugs; and to advance clinical outcome (Bertrand et al. 2014).

A novel therapeutic approach targeting B lymphoma cells with doxorubicin-conjugated liposomal nanoparticles displaying high-affinity glycan ligands (sialic acid-binding Ig-like lectin, siglec) of CD22 did not only show effective internalization but also could significantly extend life in a xenograft model of human B cell lymphoma (Boons 2010; Chen et al. 2010). These nanoparticles could interact and kill malignant B cells from peripheral blood samples obtained from patients with marginal zone lymphoma, hairy cell leukemia, and chronic lymphocytic leukemia. These encouraging studies highlight the prospective of using a carbohydrate recognition-based approach for competently targeting B cells in vivo and offer enhanced treatment options for patients with B cell cancers (Chen et al. 2010).

In one study, small poly(d-l-lactic acid) nanoparticles (PLA NPs) of about 170 nm in size were coated with anti-HER2 and anti-CD20 monoclonal antibodies to prepare mAb-NPs for specific tumor targeting (Cirstoiu-Hapca et al. 2007). The interaction between mAb-NPs and cancer cells was determined by confocal microscopy using Daudi lymphoma cells (overexpressing CD20) and SKOV-3 human

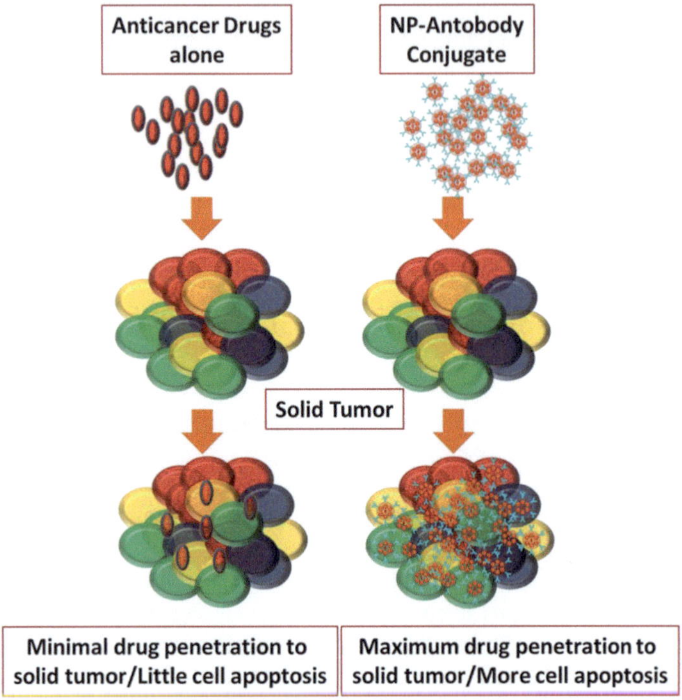

Fig. 3.2 Model represents the NP-antibody conjugate which increases the cell penetration power in solid tumor and increases cell death in solid tumor

ovarian cancer cells (overexpressing HER2). Highly selective targeting of mAb-NPs to tumor cells overexpressing the specific antigen was observed. Anti-CD20-labeled NPs were bound to, but remained at the cellular surface, whereas anti-HER2-labeled NPs were internalized efficiently, further endorsing the promise of NPs in cancer therapy. Many nanoparticle systems have been approved by either the FDA or EMA, and significant efforts are made to implicate their use in clinical care (Anselmo and Mitragotri 2016). A number of clinical trials investigating the role of nanoparticle systems are ongoing to get approval for active targeting in cancer therapies. Overall, nanoparticle drug delivery systems seem promising in cancer therapy (Ruoslahti et al. 2010).

3.8 Challenges of Targeted Antibody-Nanoparticle Bioconjugate for Cancer Therapy

The past decade has seen advancement in research for the development of antibody-nanoparticle bioconjugates as targeted chemotherapeutic agents; however the horizon for achieving effective cancer treatment still seems quite distant. Nanoparticles (NPs) can be engineered by conjugating them to various antibodies either monoclonal or polyclonal, making them extremely selective in binding to the target, thereby leading to increased efficacy along with lowered toxicity. However, the challenges in achieving the required success rate are dependent on various factors which need to be addressed and which are basically an interplay between the nature and size of the NP and ligand. The size of the NP is a key factor which dictates the NP trafficking within the body. Although small NPs have the advantage in that they can passively target the tumor, however, their small size is also problematic as it can be very quickly cleared by the kidneys. The disadvantage with the larger NPs is that their size limits their availability (Silva et al. 2016).

Owing to the high surface area and free surface energy of NPs, colloidal stability represents a major challenge. The use of surfactants, polymers, and proteins to enhance colloidal stability has been studied (Mout et al. 2012). Freeze-drying of nanoparticle-conjugated antibodies has shown promise as it was able to show colloidal stability and retained biological activity (Hamalya et al. 2018). However, the greatest challenge is how to deliver the antibody-conjugated NPs by the oral route bypassing the intravenous route. Once this is achieved, the number of patients who will be benefited will be enormous (Morishita and Peppas 2006).

Another challenge is how to specifically target the slow-cycling cancer stem cells which are one of the major reasons for relapse. Yet another challenge is therapeutics targeting brain-related cancers which are restricted by the blood-brain barrier, and, therefore, ensuring NP-conjugated antibodies target such tumors is a challenge.

The adverse biological effects of NPs at cellular, tissue, organ, and organism levels can also result in nanotoxicity. In this regard, the biophysical features such as size and surface properties affect their distribution in vivo, and this may in turn impact signaling pathways and affect biological functions. Various studies have shown the negative impact of different NPs on the liver, kidney, and skin by upregulating the inflammatory pathway (Zhang et al. 2015; Desai 2012).

Studies using NPs in mice model have shown an increase in epigenetic changes including histone posttranslational modifications and DNA methylation (Ha et al. 2015). Yet another study has shown that silver NPs led to the posttranslational modifications to certain enzymes which play a role in chromatin remodeling (Dubey et al. 2015).

In-depth studies are however required to fully understand the mechanisms behind the possible toxicity of nanoparticles in the biological system without which the applicability of NP-conjugated derivatives would be limited.

3.9 Conclusions

In order to find an effective treatment against cancer, a combinatorial tumor-targeted therapy is needed. In this regard the promise of nanotechnology, specifically nanoparticle-conjugated antibodies, offers hope to patients afflicted with cancer. With the advancement in the field of nanomedicine, the day is not far away when not only conjugated NPs be used for therapy but also aid in the diagnosis and also offer hope in terms of theragnostics, wherein nanoparticles will be so engineered that not only will they be therapeutically targeting the tumor mass but will also serve to image the disease status, for the purpose of diagnosis as well as assessment of disease progression and prognosis.

References

Abdelghany SM, Schmid D, Deacon J et al (2013) Enhanced antitumor activity of the photosensitizer meso-tetra(N-methyl- 4-pyridyl) porphine tetratosylate through encapsulation in antibody-targeted chitosan/alginate nanoparticles. Biomacromolecules 14(2):302–310

Aillon KL, Xie Y, El-Gendy N, Berkland CJ, Forrest ML (2009) Effects of nanomaterial physicochemical properties on in vivo toxicity. Adv Drug Deliv Rev. https://doi.org/10.1016/j.addr.2009.03.010

Aires A, Ocampo SM, Simões BM, Rodríguez MJ, Cadenas JF, Couleaud P, Spence K, Latorre A, Miranda R, Somoza Á (2016) Multifunctionalized iron oxide nanoparticles for selective drug delivery to CD44-positive cancer cells. Nanotechnology 27:065103

Alexander CM, Maye MM, Dabrowiak JC (2011) DNA-capped nanoparticles designed for doxorubicin drug delivery. Chem Commun. https://doi.org/10.1039/c0cc04916f

Anselmo AC, Mitragotri S (2016) Nanoparticles in the clinic. Bioeng Transl Med 1(1):10–29. Epub 2016/06/03

Ashton JR, Gottlin EB, Patz EF Jr, West JL, Badea CT (2018) A comparative analysis of EGFR-targeting antibodies for gold nanoparticle CT imaging of lung cancer. PLoS One 13(11):e0206950

Auffan M, Rose J, Bottero JY, Lowry GV, Jolivet JP, Wiesner MR (2009) Towards a definition of inorganic nanoparticles from an environmental, health and safety perspective. Nat Nanotechnol. https://doi.org/10.1038/nnano.2009.242

Barua S, Yoo JW, Kolhar P et al (2013) Particle shape enhances specificity of antibody-displaying nanoparticles. Proc Natl Acad Sci U S A 110(9):3270–3275

Bertrand N, Wu J, Xu X, Kamaly N, Farokhzad OC (2014) Cancer nanotechnology: the impact of passive and active targeting in the era of modern cancer biology. Adv Drug Deliv Rev 66:2–25. Epub 2013/11/26. https://doi.org/10.1016/j.addr.2013.11.009

Bhuvaneswari R, Gan YY, Soo KC, Olivo M (2009) Targeting EGFR with photodynamic therapy in combination with Erbitux enhances in vivo bladder tumor response. Mol Cancer 8:94. Epub 2009/11/03

Boons GJ (2010) Liposomes modified by carbohydrate ligands can target B cells for the treatment of B-cell lymphomas. Expert Rev Vaccines 9(11):1251–1256. Epub 2010/11/20

Brown JM, Giaccia AJ (1998) The unique physiology of solid tumors: opportunities (and problems) for cancer therapy. Cancer Res 58(7):1408–1416

Byrne JD, Betancourt T, Brannon-Peppas L (2008) Active targeting schemes for nanoparticle systems in cancer therapeutics. Adv Drug Deliv Rev 60(15):1615–1626. Epub 2008/10/09

Çağdaş M, Sezer AD, Bucak S (2014) Liposomes as potential drug carrier systems for drug delivery. In: Application of nanotechnology in drug delivery. https://doi.org/10.5772/58459

Calixto G, Bernegossi J, Fonseca-Santos B, Chorilli M (2014) Nanotechnology-based drug delivery systems for treatment of oral cancer: a review. Int J Nanomedicine 2014:3719–3735

Carter T, Mulholland P, Chester K (2016) Antibody-targeted nanoparticles for cancer treatment. Immunotherapy 8(8):941–958. Epub 2016/07/07

Chae SW, Sohn JH, Kim DH et al (2011) Overexpressions of cyclin B1, cdc2, p16 and p53 in human breast cancer: the clinicopathologic correlations and prognostic implications. Yonsei Med J. https://doi.org/10.3349/ymj.2011.52.3.445

Chang PY, Peng SF, Lee CY, Lu CC, Tsai SC, Shieh TM et al (2013) Curcumin-loaded nanoparticles induce apoptotic cell death through regulation of the function of MDR1 and reactive oxygen species in cisplatin-resistant CAR human oral cancer cells. Int J Oncol 43:1141–1150

Chen ZP, Peng ZF, Zhang P, Jin XF, Jiang JH, Zhang XB, Shen GL, Yu RQ (2007) A sensitive immunosensor using colloidal gold as electrochemical label. Talanta 72(5):1800–1804

Chen WC, Completo GC, Sigal DS, Crocker PR, Saven A, Paulson JC (2010) In vivo targeting of B-cell lymphoma with glycan ligands of CD22. Blood 115(23):4778–4786. Epub 2010/02/26

Chen F, Ma K, Madajewski B, Zhuang L, Zhang L, Rickert K et al (2018) Ultrasmall targeted nanoparticles with engineered antibody fragments for imaging detection of HER2- overexpressing breast cancer. Nat Commun 9(1):4141. Epub 2018/10/10

Cirstoiu-Hapca A, Bossy-Nobs L, Buchegger F, Gurny R, Delie F (2007) Differential tumor cell targeting of anti-HER2 (Herceptin) and anti-CD20 (Mabthera) coupled nanoparticles. Int J Pharm 331(2):190–196. Epub 2007/01/02

Cisterna BA, Kamaly N, Choi WI, Tavakkoli A, Farokhzad OC, Vilos C (2016) Targeted nanoparticles for colorectal cancer. Nanomedicine 11(18):2443–2456. Epub 2016/08/17

Crucho CIC, Barros MT (2017) Polymeric nanoparticles: a study on the preparation variables and characterization methods. Mater Sci Eng C. https://doi.org/10.1016/j.msec.2017.06.004

Davis ME (2009) The first targeted delivery of siRNA in humans via a self-assembling, cyclodextrin polymer-based nanoparticle: from concept to clinic. Mol Pharm 6(3):659–668. Epub 2009/03/10

Desai N (2012) Challenges in development of nanoparticle-based therapeutics. AAPS J 14:282–295

Dinauer N, Balthasar S, Weber C, Kreuter J, Langer K, von Briesen H (2005) Selective targeting of antibody-conjugated nanoparticles to leukemic cells and primary T-lymphocytes. Biomaterials 26(29):5898–5906

Dixit N, Vaibhav K, Pandey RS, Jain UK, Katare OP, Katyal A et al (2015) Improved cisplatin delivery in cervical cancer cells by utilizing folate-grafted non-aggregated gelatin nanoparticles. Biomed Pharmacother 69:1–10. Epub 2015/02/11

Dubertret B, Skourides P, Norris DJ, Noireaux V, Brivanlou AH, Libchaber A (2002) In vivo imaging of quantum dots encapsulated in phospholipid micelles. Science (80-). https://doi.org/10.1126/science.1077194

Dubey P, Matai I, Kumar SU, Sachdev A, Bhushan B, Gopinath P (2015) Perturbation of cellular mechanistic system by silver nanoparticle toxicity: cytotoxic, genotoxic and epigenetic potentials. Adv Colloid Interface Sci 221:4–21

Durr NJ, Larson T, Smith DK, Korgel BA, Sokolov K, Ben-Yakar A (2007) Two-photon luminescence imaging of cancer cells using molecularly targeted gold nanorods. Nano Lett 7(4):941–945. Epub 2007/03/06

Elbayoumi TA, Torchilin VP (2008) Tumor-specific antibody mediated targeted delivery of Doxil reduces the manifestation of auricular erythema side effect in mice. Int J Pharm 357(1–2):272–279

El-Sayed IH, Huang X, El-Sayed MA (2005) Surface plasmon resonance scattering and absorption of anti-EGFR antibody conjugated gold nanoparticles in cancer diagnostics: applications in oral cancer. Nano Lett 5(5):829–834. https://doi.org/10.1021/nl050074e

FDA approved therapeutic antibody for cancer. https://www.fda.gov/Drugs/default.htm

Ferrara N, Gerber H-P, LeCouter J, Kerbel RS (2003) Angiogenesis as a therapeutic target. Nature. https://doi.org/10.1038/nature04483

Ferrari M (2005) Cancer nanotechnology: opportunities and challenges. Nat Rev Cancer 5(3):161–171. Epub 2005/03/02

Florence AT, Hussain N (2001) Transcytosis of nanoparticle and dendrimer delivery systems: evolving vistas. Adv Drug Deliv Rev. https://doi.org/10.1016/S0169-409X(01)00184-3

Fonge H, Lee H, Reilly RM, Allen C (2010) Multifunctional block copolymer micelles for the delivery of 111In to EGFR-positive breast cancer cells for targeted Auger electron radiotherapy. Mol Pharm 7(1):177–186. Epub 2009/11/21

Ghosh SC, Neslihan Alpay S, Klostergaard J (2012) CD44: a validated target for improved delivery of cancer therapeutics. Expert Opin Ther Targets 16:635–650

Gimbrone MA Jr, Leapman SB, Cotran RS, Folkman J (1972) Tumor dormancy in vivo by prevention of neovascularization. J Exp Med. https://doi.org/10.1084/jem.136.2.261

Glazer ES, Curley SA (2010) Radiofrequency field-induced thermal cytotoxicity in cancer cells treated with fluorescent nanoparticles. Cancer 116(13):3285–3293. https://doi.org/10.1002/cncr.25135

Gonda K, Watanabe TM, Ohuchi N, Higuchi H (2010) In vivo nano-imaging of membrane dynamics in metastatic tumor cells using quantum dots. J Biol Chem. https://doi.org/10.1074/jbc.M109.075374

Grivennikov SI, Greten FR, Immunity KM (2010) Inflammation and Cancer. Cell. https://doi.org/10.1016/j.cell.2010.01.025

Ha SW, Jang HL, Nam KT, Beck GR Jr (2015) Nano-hydroxyapatite modulates osteoblast lineage commitment by stimulation of DNA methylation and regulation of gene expression. Biomaterials 65:32–42

Haber DA, Gray NS, Baselga J (2011) The evolving war on cancer. Cell. https://doi.org/10.1016/j.cell.2011.03.026

Hamalya MA, Abulateefeha SR, Al-Qaoudb KM, Alkilanya AM (2018) Freeze-drying of monoclonal antibody-conjugated gold nanorods: colloidal stability and biological activity. Int J Pharm 550:269–277

Hanahan D, Coussens LM (2012) Accessories to the crime: functions of cells recruited to the tumor microenvironment. Cancer Cell. https://doi.org/10.1016/j.ccr.2012.02.022

Hanahan D, Weinberg RA (2011) Review hallmarks of cancer: the next generation. Cell 144(5):646–674. https://doi.org/10.1016/j.cell.2011.02.013

Haugsten EM, Wiedlocha A, Olsnes S, Wesche J (2010) Roles of fibroblast growth factor receptors in carcinogenesis. Mol Cancer Res 8(11):1439–1452. Epub 2010/11/05

Heinrich MC, Blanke CD, Druker BJ, Corless CL (2002) Inhibition of KIT tyrosine kinase activity: a novel molecular approach to the treatment of KIT-positive malignancies. J Clin Oncol. https://doi.org/10.1200/JCO.20.6.1692

Heinzle C, Sutterluty H, Grusch M, Grasl-Kraupp B, Berger W, Marian B (2011) Targeting fibroblast-growth-factor-receptor-dependent signaling for cancer therapy. Expert Opin Ther Targets 15(7):829–846. Epub 2011/03/08

Honeychurch J, Cheadle EJ, Dovedi SJ, Illidge TM (2015) Immuno-regulatory antibodies for the treatment of cancer. Expert Opin Biol Ther. https://doi.org/10.1517/14712598.2015.1036737

Huang X, Jain PK, El-Sayed IH, El-Sayed MA (2006) Determination of the minimum temperature required for selective photothermal destruction of cancer cells with the use of Immunotargeted gold nanoparticles. Photochem Photobiol. https://doi.org/10.1562/2005-12-14-RA-754

Jain RK (2003) Molecular regulation of vessel maturation. Nat Med. https://doi.org/10.1038/nm0603-685

Jain RK, Stylianopoulos T (2010) Delivering nanomedicine to solid tumors. Nat Rev Clin Oncol. https://doi.org/10.1038/nrclinonc.2010.139

Jarboe J, Gupta A, Saif W (2014) Therapeutic human monoclonal antibodies against cancer. Methods Mol Biol. https://doi.org/10.1007/978-1-62703-586-6_4

Jiang Z, Yang Z, Li F, Li Z, Fishkin N, Burgess K (2018) Targeted maytansinoid conjugate improves therapeutic index for metastatic breast cancer cells. Bioconjug Chem 29(9):2920–2926. Epub 2018/08/14

Joensuu H, Dimitrijevic S (2001) Tyrosine kinase inhibitor imatinib (STI571) as an anticancer agent for solid tumours. Ann Med. https://doi.org/10.3109/07853890109002093

Kah JC, Kho KW, Lee CG et al (2007) Early diagnosis of oral cancer based on the surface plasmon resonance of gold nanoparticles. Int J Nanomedicine 2(4):785–798

Kanazaki K, Sano K, Makino A, Shimizu Y, Yamauchi F, Ogawa S et al (2015) Development of anti-HER2 fragment antibody conjugated to iron oxide nanoparticles for in vivo HER2- targeted photoacoustic tumor imaging. Nanomedicine 11(8):2051–2060. Epub 2015/08/05

Karkkainen MJ, Makinen T, Alitalo K (2002) Lymphatic endothelium: a new frontier in metastasis research. Nat Cell Biol 4:E2–E5. https://doi.org/10.1038/ncb0102-e2

Kedar U, Phutane P, Shidhaye S, Kadam V (2010) Advances in polymeric micelles for drug delivery and tumor targeting. Nanomed Nanotechnol Biol Med. https://doi.org/10.1016/j.nano.2010.05.005

Kim BYS, Rutka JT, Chan WCW (2010) Ph D. nanomedicine. N Engl J Med. https://doi.org/10.1056/NEJMra0912273

King CR, Kraus MH, Aaronson SA (1985) Amplification of a novel v-erbB-related gene in a human mammary carcinoma. Science (80-). https://doi.org/10.1126/science.2992089

Korb ML, Hartman YE, Kovar J, Zinn KR, Bland KI, Rosenthal EL (2014) Use of monoclonal antibody-IRDye800CW bioconjugates in the resection of breast cancer. J Surg Res 188(1):119–128. Epub 2013/12/24

Koren E, Apte A, Jani A, Torchilin VP (2012) Multifunctional PEGylated 2C5-immunoliposomes containing pHsensitive bonds and TAT peptide for enhanced tumor cell internalization and cytotoxicity. J Control Release 160(2):264–273

Kreyling WG, Semmler-Behnke M, Chaudhry Q (2010) A complementary definition of nanomaterial. Nano Today. https://doi.org/10.1016/j.nantod.2010.03.004

Kubota T et al (2018) HER2-targeted gold nanoparticles potentially overcome resistance to trastuzumab in gastric cancer. Nanomedicine: NBM 14:1919–1929. https://doi.org/10.1016/j.nano.2018.05.019

Kumar S, Aaron J, Sokolov K (2008) Directional conjugation of antibodies to nanoparticles for synthesis of multiplexed optical contrast agents with both delivery and targeting moieties. Nat Protoc 3(2):314–320

Kuo YC, Liang CT (2011) Inhibition of human brain malignant glioblastoma cells using carmustine-loaded catanionic solid lipid nanoparticles with surface anti-epithelial growth factor receptor. Biomaterials 32(12):3340–3350. Epub 2011/02/08

Kuo YC, Wang IH (2016) Enhanced delivery of etoposide across the blood-brain barrier to restrain brain tumor growth using melanotransferrin antibody- and tamoxifen-conjugated solid lipid nanoparticles. J Drug Target 24(7):645–654. https://doi.org/10.3109/1061186X.2015.1132223. Epub 2016 Jan 15

Kwon I, Kato M, Xiang S et al (2014) Phosphorylation-regulated binding of RNA polymerase ii to fibrous polymers of low-complexity domains. Cell. https://doi.org/10.1016/j.cell.2014.01.002

Langdon SP, Sims AH (2016) HER2-targeted antibody treatment for ovarian cancer future opportunities. J Mol Pharm Org Process Res 4:1

Leal M, Sapra P, Hurvitz SA et al (2014) Antibody-drug conjugates: an emerging modality for the treatment of cancer. Ann N Y Acad Sci. https://doi.org/10.1111/nyas.12499

Li J, Ng CK (2012) Methods for nanoparticle conjugation to monoclonal antibodies. In: Pathak Y, Benita S (eds) Antibody-mediated drug delivery systems. John Wiley & Sons, Inc, Hoboken, pp 191–207

Liao C, Sun Q, Liang B, Shen J, Shuai X (2011) Targeting EGFR-overexpressing tumor cells using Cetuximab-immunomicelles loaded with doxorubicin and superparamagnetic iron oxide. Eur J Radiol 80(3):699–705. Epub 2010/09/03

Lin PC, Chen SH, Wang KY, Chen ML, Adak AK, Hwu JR, Chen YJ, Lin CC (2009) Fabrication of oriented antibody-conjugated magnetic nanoprobes and their immunoaffinity application. Anal Chem 81(21):8774–8782

Lin C, Ng HL, Pan W, Chen H, Zhang G, Bian Z et al (2015) Exploring different strategies for efficient delivery of colorectal cancer therapy. Int J Mol Sci 16(11):26936–26952. Epub 2015/11/17

Liu Y, Li K, Liu B, Feng SS (2010) A strategy for precision engineering of nanoparticles of biodegradable copolymers for quantitative control of targeted drug delivery. Biomaterials 31(35):9145–9155

Louka M, Boutou E, Bakou V, Pappa V, Georgoulis A, Stürzbecher H-W, Vorgias CE, Vlachodimitropoulos D (2015) DNA damage response/repair in cancer stem cells—potential vs. controversies, advances in DNA repair. IntechOpen, Arizona. https://doi.org/10.5772/61355

Lutz RJ (2015) Targeting the folate receptor for the treatment of ovarian cancer. Transl Cancer Res 4(1):118–126

Makaraviciute A, Ramanaviciene A (2013) Site-directed antibody immobilization techniques for immunosensors. Biosens Bioelectron 50:460–471

Malam Y, Loizidou M, Seifalian AM (2009) Liposomes and nanoparticles: nanosized vehicles for drug delivery in cancer. Trends Pharmacol Sci. https://doi.org/10.1016/j.tips.2009.08.004

Manjappa AS, Chaudhari KR, Venkataraju MP et al (2011) Antibody derivatization and conjugation strategies: application in preparation of stealth immunoliposome to target chemotherapeutics to tumor. J Control Release 150(1):2–22

Mantovani A, Allavena P, Sica A, Balkwill F (2008) Cancer-related inflammation. Nature. https://doi.org/10.1038/nature07205

Master AM, Sen Gupta A (2012) EGF receptor-targeted nanocarriers for enhanced cancer treatment. Nanomedicine 7(12):1895–1906. Epub 2012/12/20

Mattheolabakis G, Rigas B, Constantinides PP (2012) Nanodelivery strategies in cancer chemotherapy: biological rationale and pharmaceutical perspectives. Nanomedicine. https://doi.org/10.2217/nnm.12.128

Morishita M, Peppas NA (2006) Is the oral route possible for peptide and protein drug delivery? Drug Discov Today 11:905–910

Mout R, Moyano DF, Rana S, Rotello VM (2012) Surface functionalization of nanoparticles for nanomedicine. Chem Soc Rev 41:2539–2544

Nadda M, Tullayakorn P, Kesara NB (2018) Application of active targeting nanoparticle delivery system for chemotherapeutic drugs and traditional/herbal medicines in cancer therapy: a systematic review. Int J Nanomedicine 13:3921–3935

Naves LB, Dhand C, Venugopal JR, Rajamani L, Ramakrishna S, Almeida L (2017) Nanotechnology for the treatment of melanoma skin cancer. Prog Biomater 6(1–2):13–26. Epub 2017/03/18

Ordikhani F, Erdem Arslan M, Marcelo R, Sahin I, Grigsby P, Schwarz JK et al (2016) Drug delivery approaches for the treatment of cervical cancer. Pharmaceutics 8(3):23. Epub 2016/07/23

Ostrom QT, Gittleman H, Liao P, Rouse C, Chen Y, Dowling J et al (2014) CBTRUS statistical report: primary brain and central nervous system tumors diagnosed in the United States in 2007–2011. Neuro-Oncology 16(Suppl 4):iv1–i63

Pandita D, Poonia N, Kumar S, Lather V, Madaan K (2014) Dendrimers in drug delivery and targeting: drug-dendrimer interactions and toxicity issues. J Pharm Bioallied Sci. https://doi.org/10.4103/0975-7406.130965

Patel J, Amrutiya J, Bhatt P, Javia A, Jain M, Misra A (2018) Targeted delivery of monoclonal antibody conjugated docetaxel loaded PLGA nanoparticles into EGFR overexpressed lung tumour cells. J Microencapsul 35(2):204–217. https://doi.org/10.1080/02652048.2018.1453560

Pelaz B, Alexiou C, Alvarez-Puebla RA et al (2017) Diverse applications of nanomedicine. ACS Nano. https://doi.org/10.1021/acsnano.6b06040

Perez A, Neskey DM, Wen J, Goodwin JW, Slingerland J, Pereira L, Weigand S, Franzmann EJ (2012) Targeting CD44 in head and neck squamous cell carcinoma (HNSCC) with a new humanized antibody RO5429083. Cancer Res 72(8 Supplement). https://doi.org/10.1158/1538-7445.AM2012-2521

Puertas S, Moros M, Fernández-Pacheco R, Ibarra M, Grazú V, De La Fuente J (2010) Designing novel nano-immunoassays: antibody orientation versus sensitivity. J Phys D Appl Phys 43:474012

Quarta A, Bernareggi D, Benigni F, Luison E, Nano G, Nitti S, Cesta MC, Di Ciccio L, Canevari S, Pellegrino T, Figini M (2015) Targeting FR-expressing cells in ovarian cancer with Fab-functionalized nanoparticles: a full study to provide the proof of principle from in vitro to in vivo. Nanoscale 7(6):2336–2351. https://doi.org/10.1039/c4nr04426f

Ramos-Cabrer P, Campos F (2013) Liposomes and nanotechnology in drug development: focus on neurological targets. Int J Nanomedicine. https://doi.org/10.2147/IJN.S30721

Rapoport N (2007) Physical stimuli-responsive polymeric micelles for anti-cancer drug delivery. Prog Polym Sci. https://doi.org/10.1016/j.progpolymsci.2007.05.009

Rosenblum D, Joshi N, Tao W, Karp JM, Peer D (2018) Progress and challenges towards targeted delivery of cancer therapeutics. Nat Commun 9(1):1410. Epub 2018/04/14

Ruoslahti E, Bhatia SN, Sailor MJ (2010) Targeting of drugs and nanoparticles to tumors. J Cell Biol 188(6):759–768. Epub 2010/03/17

Samanta A, Medintz IL (2016) Nanoparticles and DNA-a powerful and growing functional combination in bionanotechnology. Nanoscale. https://doi.org/10.1039/c5nr08465b

Sandoval YH, Li Y, Anand-Srivastava MB (2011) Transactivation of epidermal growth factor receptor by enhanced levels of endogenous angiotensin II contributes to the overexpression of Gialpha proteins in vascular smooth muscle cells from SHR. Cell Signal 23(11):1716–1726. Epub 2011/06/30

Sanna V, Pala N, Sechi M (2014) Targeted therapy using nanotechnology: focus on cancer. Int J Nanomedicine. https://doi.org/10.2147/IJN.S36654

Sano K (2017) Development of molecular probes based on iron oxide nanoparticles for in vivo magnetic resonance/photoacoustic dual imaging of target molecules in tumors. Yakugaku Zasshi 137(1):55–60. Epub 2017/01/05

Sapsford KE, Algar WR, Berti L et al (2013) Functionalizing nanoparticles with biological molecules: developing chemistries that facilitate nanotechnology. Chem Rev 113(3):1904–2074

Shi J, Kantoff PW, Wooster R, Farokhzad OC (2017) Cancer nanomedicine: Progress, challenges and opportunities. Nat Rev Cancer. https://doi.org/10.1038/nrc.2016.108

Shimada T, Ueda M, Jinno H, Chiba N, Wada M, Watanabe J et al (2009) Development of targeted therapy with paclitaxel incorporated into EGF-conjugated nanoparticles. Anticancer Res 29(4):1009–1014. Epub 2009/05/06

Silva AL, Soema PC, Slutter B, Ossendorp F, Jiskoot W (2016) PLGA particulate delivery systems for subunit vaccines: linking particle properties to immunogenicity. Human Vac Immunother 12:1056–1069

Simon MC, Keith B (2008) The role of oxygen availability in embryonic development and stem cell function. Nat Rev Mol Cell Biol. https://doi.org/10.1038/nrm2354

Singh A, Settleman J (2010) EMT, cancer stem cells and drug resistance: an emerging axis of evil in the war on cancer. Oncogene 29:4741–4751

Singh D, Minz AP, Sahoo SK (2017) Nanomedicine-mediated drug targeting of cancer stem cells. Drug Disc Today 22(6):952–959

Smith AM, Duan H, Mohs AM, Nie S (2008) Bioconjugated quantum dots for in vivo molecular and cellular imaging. Adv Drug Deliv Rev. https://doi.org/10.1016/j.addr.2008.03.015

Sokolov K, Follen M, Aaron J, Pavlova I, Malpica A, Lotan R, Richards-Kortum R (2003) Real-time vital optical imaging of precancer using anti-epidermal growth factor receptor antibodies conjugated to gold nanoparticles. Cancer Res 63(9):1999–2004

Sokolov K, Tam J, Tam J, Travis K, Larson T, Aaron J et al (2009) Cancer imaging and therapy with metal nanoparticles. Conf Proc IEEE Eng Med Biol Soc 2009:2005–2007. Epub 2009/12/08

Sokolova V, Epple M (2008) Inorganic nanoparticles as carriers of nucleic acids into cells. Angew Chem – Int Ed. https://doi.org/10.1002/anie.200703039

Steelman LS, Chappell WH, Abrams SL et al (2011) Roles of the Raf/MEK/ERK and PI3K/PTEN/Akt/mtor pathways in controlling growth and sensitivity to therapy- implications for cancer and aging. Aging (Albany NY). https://doi.org/10.18632/aging.100296

Steinhauser I, Spänkuch B, Strebhardt K, Langer K (2006) Trastuzumab-modified nanoparticles: optimisation of preparation and uptake in cancer cells. Biomaterials 27(28):4975–4983

Sun J, Zhu MQ, Fu K, Lewinski N, Drezek RA (2007) Lead sulfide near-infrared quantum dot bioconjugates for targeted molecular imaging. Int J Nanomedicine 2(2):235–240. Epub 2007/08/29

Sun DX, Liu Z, Tan XD, Cui DX, Wang BS, Dai XW (2012) Nanoparticle-mediated local delivery of an antisense TGF-β1 construct inhibits intimal hyperplasia in autogenous vein grafts in rats. PLoS One 7(7):e41857

Sun TM, Wang YC, Wang F et al (2014) Cancer stem cell therapy using doxorubicin conjugated to gold nanoparticles via hydrazone bonds. Biomaterials. https://doi.org/10.1016/j.biomaterials.2013.10.011

Sun G, Xing W, Xing R, Cong L, Tong S, Yu S (2018) Targeting breast cancer cells with a CuInS2/ZnS quantum dot-labeled Ki-67 bioprobe. Oncol Lett 15(2):2471–2476. Epub 2018/02/13

Szlachcic A, Pala K, Zakrzewska M, Jakimowicz P, Wiedlocha A, Otlewski J (2012) FGF1-gold nanoparticle conjugates targeting FGFR efficiently decrease cell viability upon NIR irradiation. Int J Nanomedicine 7:5915–5927. Epub 2012/12/12

Tanaka R, Yuhi T, Nagatani N, Endo T, Kerman K, Takamura Y, Tamiya E (2006) A novel enhancement assay for immunochromatographic test strips using gold nanoparticles. Anal Bioanal Chem 385(8):1414–1420

Tang N, Wang L, Esko J et al (2004) Loss of HIF-1α in endothelial cells disrupts a hypoxia- driven VEGF autocrine loop necessary for tumorigenesis. Cancer Cell. https://doi.org/10.1016/j.ccr.2004.09.026

Tao L, Zhang K, Sun Y, Jin B, Zhang Z, Yang K (2012) Anti-epithelial cell adhesion molecule monoclonal antibody conjugated fluorescent nanoparticle biosensor for sensitive detection of colon cancer cells. Biosens Bioelectron 35:186

Tao W, Zeng X, Wu J, Zhu X, Yu X, Zhang X et al (2016) Polydopamine-based surface modification of novel nanoparticle-aptamer bioconjugates for in vivo breast cancer targeting and enhanced therapeutic effects. Theranostics 6(4):470–484. Epub 2016/03/05

Taylor RM, Sillerud LO (2012) Paclitaxel-loaded iron platinum stealth immunomicelles are potent MRI imaging agents that prevent prostate cancer growth in a PSMA-dependent manner. Int J Nanomedicine 7:4341–4352. https://doi.org/10.2147/IJN.S34381. Epub 2012 Aug 6

Thomas RK, Baker AC, DeBiasi RM et al (2007) High-throughput oncogene mutation profiling in human cancer. Nat Genet. https://doi.org/10.1038/ng1975

Tian M, Schiemann WP (2009) The TGF-β paradox in human cancer: an update. Future Oncol 5(2):259–271. PMC2710615

Trinh TL, Zhu G, Xiao X, Puszyk W, Sefah K, Wu Q et al (2015) A synthetic aptamer-drug adduct for targeted liver cancer therapy. PLoS One 10(11):e0136673. Epub 2015/11/03

Tseng CL, Wang TW, Dong GC, Yueh-Hsiu Wu S, Young TH, Shieh MJ et al (2007) Development of gelatin nanoparticles with biotinylated EGF conjugation for lung cancer targeting. Biomaterials 28(27):3996–4005. Epub 2007/06/16

Tseng CL, Wu SY, Wang WH, Peng CL, Lin FH, Lin CC et al (2008) Targeting efficiency and biodistribution of biotinylated-EGF-conjugated gelatin nanoparticles administered via aerosol delivery in nude mice with lung cancer. Biomaterials 29(20):3014–3022. Epub 2008/04/26

Urien S, Lokiec F (2004) Population pharmacokinetics of total and unbound plasma cisplatin in adult patients. Clin Pharmacol 57(6):756–763

Ventola CL (2017) Progress in Nanomedicine: approved and investigational Nanodrugs. P T. https://doi.org/10.1016/j.psychres.2007.07.030

Voltan R, Secchiero P, Ruozi B et al (2013) Nanoparticles engineered with rituximab and loaded with Nutlin-3 show promising therapeutic activity in B-leukemic xenografts. Clin Cancer Res 19(14):3871–3880

Walkey CD, Olsen JB, Guo H, Emili A, Chan WC (2012) Nanoparticle size and surface chemistry determine serum protein adsorption and macrophage uptake. J Am Chem Soc 134(4):2139–2147

Wang J, Tian S, Petros RA, Napier ME, Desimone JM (2010) The complex role of multivalency in nanoparticles targeting the transferrin receptor for cancer therapies. J Am Chem Soc 132(32):11306–11313

Wang CH, Chiou SH, Chou CP, Chen YC, Huang YJ, Peng CA (2011) Photothermolysis of glioblastoma stem-like cells targeted by carbon nanotubes conjugated with CD133 monoclonal antibody. Nanomedicine 7:69–79

Wang JL, Tang GP, Shen J et al (2012) A gene nanocomplex conjugated with monoclonal antibodies for targeted therapy of hepatocellular carcinoma. Biomaterials 33(18):4597–4607

Wang L, An Y, Yuan C, Zhang H, Liang C, Ding F, Gao Q, Zhang D (2015) GEM-loaded magnetic albumin nanospheres modified with cetuximab for simultaneous targeting, magnetic resonance imaging, and double-targeted thermochemotherapy of pancreatic cancer cells. Int J Nanomedicine 10:2507–2519. https://doi.org/10.2147/IJN.S77642

Wesche J, Haglund K, Haugsten EM (2011) Fibroblast growth factors and their receptors in cancer. Biochem J 437(2):199–213. Epub 2011/06/30

Wicki A, Witzigmann D, Balasubramanian V, Huwyler J (2015) Nanomedicine in cancer therapy: challenges, opportunities, and clinical applications. J Control Release. https://doi.org/10.1016/j.jconrel.2014.12.030

Widder KJ, Senyei AE, Scarpelli DG (1978) Magnetic microspheres: a model system for site specific drug delivery in vivo. Exp Biol Med 158(2):141–146

Widder KJ, Marino PA, Morris RM, Howard DP, Poore GA, Senyei AE (1983) Selective targeting of magnetic albumin microspheres to the Yoshida sarcoma: Ultrastructural evaluation of microsphere disposition. Eur J Can Clin Oncol 19(1):141–147

Wong SF (2005) Cetuximab: an epidermal growth factor receptor monoclonal antibody for the treatment of colorectal cancer. Clin Ther 27(6):684–694. Epub 2005/08/25

Wu CH, Lan CH, Wu KL, Wu YM, Jane WN, Hsiao M et al (2018a) Hepatocellular carcinoma-targeted nanoparticles for cancer therapy. Int J Oncol 52(2):389–401. Epub 2017/12/06

Wu ST, Fowler AJ, Garmon CB, Fessler AB, Ogle JD, Grover KR, Allen BC, Williams CD, Zhou R, Yazdanifar M, Ogle CA, Mukherjee P (2018b) Treatment of pancreatic ductal adenocarcinoma with tumor antigen specific-targeted delivery of paclitaxel loaded PLGA nanoparticles. BMC Cancer 18(1):457. https://doi.org/10.1186/s12885-018-4393-7

Xu Z, Wang Y, Lu Z, Leaf H, Nano ACS (2014) Nanoparticle-delivered transforming growth factor-β siRNA enhances vaccination against advanced melanoma by modifying tumor microenvironment. ACS Nano 8(4):3636–3645. PMC4004320

Yancopoulos GD, Davis S, Gale NW, Rudge JS, Wiegand SJ, Holash J (2000) Vascular-specific growth factors and blood vessel formation. Nature. https://doi.org/10.1038/35025215

Yeh YC, Creran B, Rotello VM (2012) Gold nanoparticles: preparation, properties, and applications in bionanotechnology. Nanoscale. https://doi.org/10.1039/c1nr11188d

Zhang Z, Chen J, Ding L, Jin H, Lovell JF, Corbin IR et al (2010) HDL-mimicking peptide-lipid nanoparticles with improved tumor targeting. Small 6(3):430–437. Epub 2009/12/04

Zhang Y, Guo, Zhang XL, Li DP, Zhang TT, Gao FF, Liu NF, Sheng XG (2015) Antibody fragment- armed mesoporous silica nanoparticles for the targeted delivery of bevacizumab in ovarian cancer cells. Int J Pharm 496(2):1026–1033. https://doi.org/10.1016/j.ijpharm.2015.10.080. Epub 2015 Nov 2

Zhang C, Zhang Z, Zhao L (2016) Folate-decorated poly(3-hydroxybutyrate-co-3- hydroxyoctanoate) nanoparticles for targeting delivery: optimization and in vivo antitumor activity. Drug Deliv 23(5):1830–1837. Epub 2015/12/15

Zhao MX, Zeng EZ (2015) Application of functional quantum dot nanoparticles as fluorescence probes in cell labeling and tumor diagnostic imaging. Nanoscale Res Lett. https://doi.org/10.1186/s11671-015-0873-8

Zhu R, Wang Z, Liang P, He X, Zhuang X, Huang R, Wang M, Wang Q, Qian Y, Wang S (2017) Efficient VEGF targeting delivery of DOX using Bevacizumab conjugated SiO2@LDH for anti-neuroblastoma therapy. Acta Biomater 63:163–180. https://doi.org/10.1016/j.actbio.2017.09.009. Epub 2017 Sep 18

Nanomedicine in Cancer Stem Cell Therapy

4

Piyush Kumar Gupta, Gunasekaran Dharanivasan,
Ranjita Misra, Santosh Gupta, and Rama Shanker Verma

Abstract

It is now well established that most of the tumors are heterogeneous in nature that comprise a population of cancer stem cells (CSCs) and differentiated cancer cells. Like normal stem cells, CSCs have also self-renewal, proliferation, and differentiation capacities that are responsible for the development of drug resistance and relapse. Therefore, targeting CSCs is essential for the elimination of tumor recurrence condition. Although several anti-CSC therapeutics have been used in clinics, they are found to have limited efficacy due to poor solubility, lesser stability, and short circulation time in the blood. Therefore, tools in nanomedicines are being used to tackle these limitations. Recently, nanodrug carriers have been used to target CSCs and somewhat eliminate drug resistance by targeting CSC metabolism, inhibiting drug transporters, disturbing CSC survival pathways, etc. Even with these progress, the challenges for targeting CSCs by nanomedicines still remain and open up plenty of space for further development and improvement in synthesizing drug carriers with higher efficacy. In this chapter, we summarize about CSCs and their biological characterization toward resistance, then discuss several anti-CSC therapeutic approaches based on nanomedicines in the current state of research and development, and finally overview their future directions.

Keywords

Nanotherapy · Cancer therapy · Biomarkers · Drug resistance · Stem cells

P. K. Gupta · G. Dharanivasan · R. Misra · S. Gupta · R. S. Verma (✉)
Stem cell and Molecular Biology Laboratory, Bhupat and Jyoti Mehta school of Biosciences, Department of Biotechnology, Indian Institute of Technology-Madras,
Chennai, Tamilnadu, India
e-mail: vermars@iitm.ac.in

S. K. Saxena, S. M. P. Khurana (eds.), *NanoBioMedicine*,
https://doi.org/10.1007/978-981-32-9898-9_4

4.1 Introduction

Many historical studies have established that malignant cancers contain a subpopulation of rare cells that display self-renewal, proliferation, and differentiation capacities in new cancer cells, which are called cancer stem cells (CSCs) or tumor-initiating cells. These cells are responsible for metastasis, drug resistance, and tumor recurrence condition (Gupta et al. 2009). Like multidrug-resistant cells, CSCs also display the same phenotypic feature including overexpression of ABC transporters, metabolism reprogramming, and activation of survival pathways (Dean et al. 2005). As the presence of CSCs has been seen in many malignant tissues like the breast, brain, lung, colon, pancreatic, etc., that was concluded with the xenotransplantation of primary tumor into mice. Also, the treatment of tumor with conventional methods like chemotherapy, radiotherapy, and targeted therapy results in an increase of CSC fraction by which tumor cells survive and lead to the metastasis at distant sites (Ma et al. 2008). During the treatment cycles of chemotherapy, tumor recurrence condition is observed due to the presence of resistant CSCs. If these CSCs are targeted with different therapeutic modalities, then tumor-relapsed condition can be eliminated. Generally, tumors exhibit plasticity that means two types of tumor cell population, i.e., CSCs and non-CSCs. If CSCs are eliminated without killing non-CSCs, then complete cure cannot be seen. Therefore, there is a need for more preclinical and clinical studies to understand the CSC response during therapy.

Currently, several effective therapeutic agents are available to target and kill CSCs. Most of them are chemo- or radiotherapy drugs, therapeutic nucleic acids, targeted monoclonal antibodies, or small molecular inhibitors. In clinic, the therapeutic efficacy of these agents has decreased due to several limitations like lesser stability, poor water solubility, nonspecific biodistribution, short circulation time, or off-target effects (Chen 2010). Therefore, nanotechnology-assisted drug delivery systems, i.e., nanomedicine, have gained the significant attention to overcome these limitations (Davis et al. 2008; Rink et al. 2013). Usually, nanomedicines can be loaded with high payload of single or multiple drugs by controlling their size and surface properties. As a result, the pharmacokinetic and pharmacodynamic properties of nanomedicines have improved by reducing their side effects on healthy cells. In current state, the clinically approved anticancerous nanomedicines are Doxil (doxorubicin-encapsulated liposomes), Oncaspar (PEG-L-asparaginase), and Abraxane (albumin-paclitaxel conjugate). In current settings, several multifunctional nanoparticles have designed their cancer theranostic applications under the special consideration of CSC targeting (Sun et al. 2014a). Further, several proofs of concept studies have been designed to tackle CSC's associated challenges. Some of them have displayed inspiring results previously. For example, codelivery of both doxorubicin (Dox) and all-trans retinoic acid (ATRA) was carried out using polymeric nanoparticles for eliminating human breast CSCs with drug-resistant cancer cells and exhibited improved anticancer therapy compared to free agents (Sun et al. 2015). Also, SignPath Pharmaceutical Company developed curcumin-loaded nanodrug carriers called NanoCurc™ which significantly inhibited the growth of glioblastoma by reduction of CD133+ CSCs (Lim et al. 2011). In this setting, several CSC-targeting nanomedicines have been developed, and their efficacy was evaluated in various preclinical studies. However, many clinical challenges have to be addressed before their use in clinics. In this chapter, we briefly described about CSCs and their biological processes in the background of drug resistance, followed by brief discussion on CSC-targeting nanomedicine approaches in the context of delivery of different types of therapeutic agents. We also emphasized the future directions of anti-CSC nanomedicines including the consideration of most innovative therapeutic strategies and the development of highly efficient nanodrug carriers.

4.2 CSCs and Drug Resistance

4.2.1 CSCs and How Does It Lead to Drug Resistance or Tumor Recurrence Condition??

CSCs are well known to have many distinct properties such as self-renewal capacity, proliferative capability, and resistance to apoptosis (Vinogradov and Wei 2012). Furthermore, several studies have already proved that CSCs are associated with high invasiveness and metastasized tumorigenic potential leading to drug-resistant condition for the current conventional therapies in clinic (Liu et al. 2010). Therefore, CSCs have become an important target for the success of potential therapeutic approach in translational cancer research.

Currently available treatment modalities are able to kill the cancer cells only but are unable to eliminate the critical CSCs that are present in tumor cell population which escape by pertaining some specific resistance mechanisms. This survival of CSCs leads to disease relapse by developing tumors which are more malignant, highly invasive, and resistant to chemo- and radiotherapy. According to the traditional view, cancer cells lead to the development of a small population of drug-resistant cancer cells with repeated chemotherapeutic treatment that could result to the inactivation of drugs, alterations in drug targets, and reduced drug accumulation inside the cancer cells (Niero et al. 2014).

However, the current CSC concept states that the relapse condition is mainly contributed by the intrinsic and acquired resistance mechanisms of the CSC population, present in cancer cell mass. Moreover, the CSCs are drug resistant due to several other factors such as the tumor microenvironment which normally contains different kinds of proteins including growth factors and cytokines that could help in the activation of CSC survival pathways and the possible role of chemo- or radiotherapy to enhance the stemness property by converting cancer cells into CSCs. Recently, it was reported that the irradiation of breast cancer cells leads to increased number of CSC population and also found that some noncancerous cells attained the CSC phenotype (Atena et al. 2014). Studies have shown that human gastric cancer cell lines with 5-fluorouracil (5-FU) treatment become resistant in longer time and acquire the stem cell features such as stemness, tumorigenicity, and self-renewal capacity (Xu et al. 2015a). In this background, the involvement of CSCs in metastasis, tumor progression, drug resistance, and relapse was shown in Fig. 4.1.

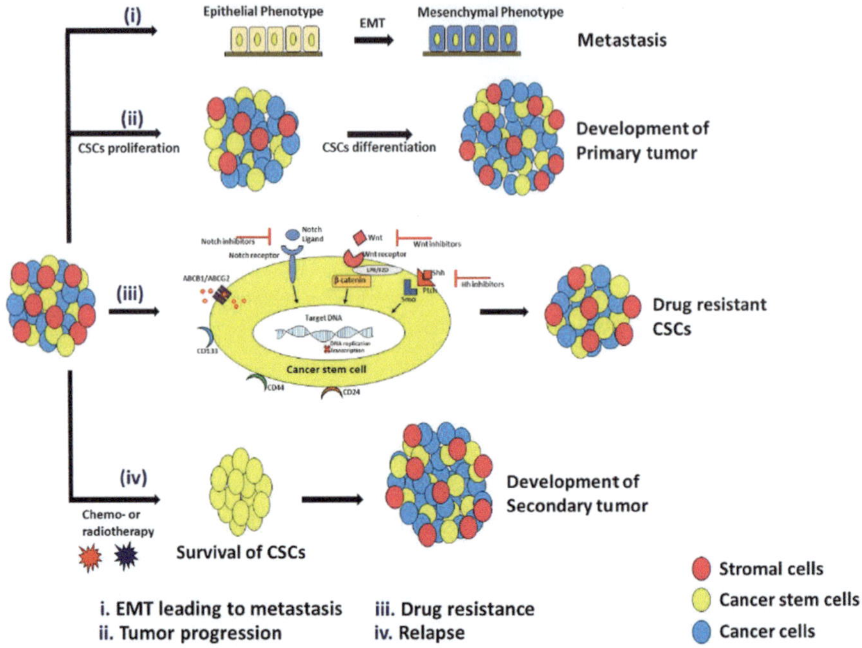

Fig. 4.1 Involvement of CSCs in metastasis, tumor progression, drug resistance, and relapse. **(i)** CSCs generally led to the local and distant metastasis via epithelial-mesenchymal transition (EMT) program; **(ii)** CSCs drive tumor progression using self-renewal, proliferation, and differentiation properties; **(iii)** CSCs develop multiple drug-resistant mechanisms to protect itself from conventional cancer treatment, leading to enrichment of CSCs within tumors; and **(iv)** CSCs have the capability to proliferate and differentiate after the success of initial treatment, leading to the development of relapsed condition

4.2.2 CSC Isolation and Characterization

It is well known that CSCs have different functions and phenotypic features from non-stem cancer cells (Abbaszadegan et al. 2017). These diverse features led to the development of several assays to isolate and characterize the CSCs. The CSCs can be identified by their fundamental properties such as self-renewal and lineage capacity. Moreover, these CSCs could be characterized by more specific phenotypic surface markers such as CD34+/CD38− in leukemia cells, CD44+/CD24− in solid tumors, or CD133+ in other tumors (Dragu et al. 2015). CD44 binds to a glycosaminoglycan (hyaluronic acid) which is present in extracellular matrix that helps in the CSC attachment contributing to proliferation and migration of the stem cells. It has been reported that the higher expression of CD44 is strongly associated with therapeutic drug resistance (Goodison et al. 1999). Similarly, CD133 is found to be highly expressed on CSCs of several cancers of different tissue origin that increased drug resistance. Thus, targeting of CD44 and CD133 molecules can be used to deliver chemotherapeutic drugs to eliminate CSCs.

CSCs can be isolated mainly by long-term cell culture using FACS (fluorescence-activated cell sorting) and MACS (magnetic-activated cell sorting) techniques. FACS can be used for CSC isolation and enrichment based on the expression of some specific cell surface markers such as CD24, CD34, CD44, and CD133. MACS is considered a standard method for CSC isolation based on specific stem cell markers as this technology aids in isolation of high-quality cells. At first, the cell surface markers are labeled with monoclonal antibody (mAb) or magnetic microbeads; then isolation is carried out. In the next step, marked cells are separated by positive selection removing the unmarked ones. Thus, isolation of target cells from a cell suspension can be efficiently done by one of the best and most direct methods called positive selection method. As mentioned above, CSCs express a set of cell surface markers that can be used for detection and separation by positive selection method.

One more functional method for CSC isolation and characterization includes colony-forming unit (CFU) assay. This quantitative and high-throughput method is believed to be analogous for in vivo transplantation. CFU assay is used to analyze the pattern of CSC proliferation and differentiation by their ability to form colonies in a semisolid medium. These colonies are formed by a particular number of input cells, which give the basic information about the proliferation and differentiation potential of CSCs. In brief, CSCs are non-adhesively cultured in serum-free medium supplemented with growth factors which can develop into tumorspheres. Usually cancer cells undergo anoikis (a suspension-induced apoptosis) in these conditions, whereas CSCs can survive and form tumorspheres on the colony basis. These tumorspheres exhibit higher CSC portions than the original tumor cells.

Similarly, CSCs can also be isolated based on the overexpression of drug efflux transporters specifically BCRP or ABCG2. For example, cell populations which efflux Hoechst 33342 dye maintain CSC properties in a variety of cancers, and presently this method is the most widely used for the isolation of CSCs.

Overall, currently a good set of tools is available to study the CSCs. A variety of in vitro assays must be used in combination, and due to their cost-effectiveness and high efficiency, these methods are very useful in the beginning of the study, and the results have to be carefully validated in vivo.

4.2.3 Dysregulated Pathways in Cancer Stem Cell's Survival

In quiescent stage, self-renewal and maintenance are key features of normal stem cells and CSCs. The cellular physiology of CSCs are regulated by different cellular signaling pathways in their tumor microenvironment. The regulation of such signaling pathways like Notch, Wnt, and Hedgehog plays an important role in regulating self-renewal, proliferation, and differentiation property of stem cells (Matsui 2016). The dysregulation of such pathways has been proved to attribute to the CSC growth, metastasis, and emergence of drug-resistant condition during cancer treatment. The dysregulation happens due to mutation or abnormal activation of such pathway's genes. For example, Notch pathway is one of the conservative signaling pathway of multicellular organisms that is linked to self-renewal and CSC survival. The dysregulation in Notch signaling has led to the development of various types of cancers such as breast cancer and glioblastoma. The Notch protein also contributes to drug resistance, because decrease in Notch expression leads to enhanced drug sensitivity of cancer cells, inhibiting tumor regrowth, and reduces migration and invasion of cancer cells (Wang et al. 2012a).

Normally, the activation of Notch signaling initiates with the binding of a transmembrane ligand to the Notch transmembrane receptor (NOTCH1/NOTCH2/NOTCH3/NOTCH4) on a neighboring cell. This leads to proteolytic cleavage of the Notch receptor, thereby releasing constitutively active intracellular domain of NOTCH (NICD) which further translocates into the nucleus, where NCID binds to transcription factors CSL (CBF1/RBPJ/suppressor of hairless/Lag-1) and coactivator to activate the transcription of Notch-responsive genes. The development of Notch pathway targeting therapeutics is a primary focus for inhibition of CSC growth. Monoclonal antibodies like demcizumab (OMP-21 M18), OMP-52 M51, and OMP-59R5 are designed to target Notch pathway and currently being used in clinical trials for inhibiting CSC growth (Smith et al. 2014). Furthermore, the direct link of Notch pathway was established with the metastasis potential of CSCs. Gemcitabine-resistant pancreatic cancer cells overexpress Notch-2 and its ligand Jagged-1 that helps in maintaining EMT and acquisition of the CSC phenotype. EMT is a crucial process in which CSCs move from tumor lesions into the blood, whereas the opposite process of transition of mesenchymal epithelium was believed as the main mechanism of invasion of CSCs into the healthy organs. It has been investigated that disruption in Notch signaling by using siRNA leads to reversal in the EMT phenotype partially. Therefore, Notch signaling activation and progression of EMT can be directly related to the resistance to gemcitabine in pancreatic cancer. Thus, the inhibition of Notch pathway could be a potent approach for overcoming drug resistance and metastasis in clinic.

Wnt signaling is also one of the important pathways that plays a vital role in embryogenesis and development of cancer. The Wnt proteins act as growth factors that help in the normal stem cell maintenance and proliferation. Mutations in Wnt/β-catenin pathway lead to the development of various types of cancers including leukemia, colon, epidermal, breast, and cutaneous carcinoma (Polakis 2012). Moreover, Wnt/β-catenin signaling pathway also plays a vital role in ABCB1/MDR-1 transcription factor-driven colorectal carcinogenesis (Correa et al. 2012). The dysregulation of such signaling pathway involves in the chemoresistance of pancreatic cancer (Cui et al. 2012). Studies demonstrated that silencing of Wnt activity using siRNA against β-catenin was able to efficiently inhibit the proliferation and drug resistance in lung cancer cells (Cai et al. 2017). Similarly, inhibition of Wnt activity leads to the reversal of 5-fluorouracil resistance in colon CSCs (Deng et al. 2010). Recently, studies have demonstrated that the Wnt pathway is associated in the maintenance of stemness such as the self-renewal capacity, and heterogeneity of breast CSCs is promoted by proliferating cell nuclear antigen-associated factor (PAF) (Wang et al. 2016). Further, the latent competent cancer cells have been isolated from lung and breast carcinoma cell lines and demonstrated that Sox2 or Sox9 expression induces DKK1 (a natural Wnt inhibitor) allowing the cells to enter a quiescent state. This results in lower expression of natural killer (NK) cell ligands and weak innate immunity, thereby conferring CSCs with the quiescent state for a long time (Malladi et al. 2016).

Similarly, activation of the Hedgehog (Hh) signaling pathway leads to the development of various types of cancer, such as basal-cell carcinoma, breast, brain, and pancreatic tumors. Hh signaling maintains the self-renewal property in glioblastoma, breast, and myeloma stem cells. Several studies have reported the role of the Hh pathway in the development of metastasis particularly through the EMT initiation via activation of some protein expressions such as MMP-9 (matrix metalloproteinase 9) and E-cadherin. In inactive state during the absence of Hh, smoothened (Smo) gets inhibited by Ptch1, a transmembrane receptor. During an active state, Hh is secreted by the adjacent cells; thereby, the Smo will be activated by the Ptch receptor. After Smo activation, Gli1/2 is to be translocated into the nucleus, which then leads to activation of h-associated genes. It has been reported that high expression of Hh signaling molecules such as Smo and Gli1 was attributed to the tamoxifen resistance in breast cancer (MCF-7 and T47D) cells (Villegas et al. 2016). Moreover, breast, colon, and pancreatic CSCs have shown sensitivity to the Hh pathway inhibitors. Numerous compounds targeting Hh pathway have demonstrated potential efficacy in preclinical studies and are now in phase I and II clinical trials (Takebe et al. 2015).

4.2.4 Molecular and Cellular Therapeutic Targets (Biomarkers) in Drug-Resistant CSCs

Current therapeutic strategies against cancer such as chemo- and radiotherapy have multiple limitations that frequently result into treatment failure and relapse in cancer patients. These therapies are not specific to target CSCs leading to toxicity in healthy tissues; thereby, the risk of disease relapse or recurrence increases in patients. Thus, CSC elimination is very crucial for preventing tumor relapse. Recently, multiple novel strategies have been investigated with the specific aim of killing CSCs and altering their niche. These specific targets include both slight differences in surface marker expression and altered signaling pathways. Several studies have concentrated on dysregulation of signaling pathways in CSCs to develop a new and advanced approach for cancer treatment. This way of finding might be promising because most of the cancers are associated with dysregulation of the same signaling cascades. In this context, CSCs can be diagnosed by expression of surface markers but also by the signals sent by them to the tumor microenvironment (Dragu et al. 2015). Investigators often use the surface markers as important targets for therapy. They choose the ligands or antibodies for surface markers and used them as an adjunct to chemotherapy, radiotherapy, and surgery. Most importantly, monoclonal antibody development is highlighted in targeting CSCs. At the moment, some therapeutic strategies are successfully used in clinic, while others are still under preclinical evaluation.

Some of the important CSC-based markers along with strategies for targeting them are as follows:

CD133 (Prominin-1) It is a cell surface glycoprotein, widely expressed on many types of CSCs in solid tumors including glioma, lung, and breast cancer. Patients with large CD133 subpopulation have shown poor clinical outcomes. For this reason, strategies for anti-CD133 therapy represent a promising approach for cancer treatment. Paclitaxel-loaded polymeric nanoparticles functionalized with CD133 antibody were investigated to efficiently reduce the number of cell and colony formation in colorectal adenocarcinoma Caco-2 cells. Moreover, these drug-loaded nanoparticles have shown better efficacy as compared to free paclitaxel in xenograft model (Swaminathan et al. 2013). Similarly, anti-CD133 scFv and pseudomona exotoxin 38 (PE-38)-based fusion construct (immunotoxin) exhibited tumor regression property after several intraperitoneal injections of anticancer drug for about 4–6 weeks in ovarian cancer xenograft model. This resulted in cancer-free survivors for a long period of time (Skubitz et al. 2013). These studies prove that anti-CD133 therapy is associated with drug delivery and drug antibody constructs that might increase the efficacy of CD133+ CSCs by abolishing them. Similarly, CD133+-based cell therapy showed antiproliferative activity resulting in reduced tumor-initiating ability in sarcoma CSCs (Stratford et al. 2013). Furthermore, this cell therapy also showed similar type of results in pancreatic and hepatic CSCs (Huang et al. 2013). Recently, anti-CD133-conjugated carbon nanotubes in combination with irradiation of near-infrared laser light could selectively kill the CD133+ glioblastoma cells (Wang et al. 2011).

CD44 It is a transmembrane protein and found to be overexpressed on different cancer cells such as breast, prostate, gastric, pancreas, ovary, colorectal, bladder, hepatocellular, head and neck, and leukemia CSCs. Thus, targeting CD44 by using monoclonal antibody proves as a promising strategy to kill CSCs. Further, the efficacy of combining antihuman CD44 monoclonal antibody with cyclophosphamide and doxorubicin is reported in preventing relapse of metastatic breast cancer (Marangoni et al. 2009). Anti-CD44 antibody has also shown much efficacy in killing leukemic stem cells in acute lymphoid leukemia (ALL) disease (Huang et al. 2017). Presently, numerous antibodies are approved by FDA and are being used for the treatment of different solid and hematological cancers in clinic such as anti-CD20 (rituximab), anti-EGFR (cetuximab), anti-ER2 (trastuzumab), anti-VEGF-A (bevacizumab), etc. (Kwiatkowska-Borowczyk et al. 2015).

Dysregulation of signaling pathways in CSCs is one of the mechanisms by which they are able to avoid or survive cancer therapeutics (Chen et al. 2013). Moreover, inactive proapoptotic and parallel active antiapoptotic pathways are sizzling points attracting researchers. Targeting Notch signaling by monoclonal antibodies has shown good results (Fischer et al. 2011). In addition, Notch1 inhibition also reduced the CD44+/CD24 CSC subpopulation, thereby inhibiting the condition of brain metastasis during breast cancer treatment (McGowan et al. 2011). Elevated levels of β-catenin have been shown to contribute CSC tumorigenicity property in colon cancer (Vermeulen et al. 2010). Numerous pharmaceuticals and monoclonal antibodies against Wnt signaling are under preclinical investigation for clinical trials and have shown promising efficacy in cancer treatment (Chen et al. 2013). Recently, it has been investigated that Hedgehog inhibitors have exhibited potential effect in inhibiting systemic metastases in pancreatic orthotopic xenograft mice models. Moreover, a large decrease in ALDH-positive cells was observed indicating reduction in tumor-initiating population in pancreatic cancer (Feldmann et al. 2008). Many researchers have proved the efficacy of an SMO signaling inhibitor (cyclopamine) of Hedgehog cascade in inhibiting the growth, invasion, and metastasis of prostate, breast, and brain tumors in both in vitro and in vivo conditions. Synergistic effect of both cyclopamine and gemcitabine was also reported to inhibit the growth of ALDH high pancreatic CSCs (Feldmann et al. 2007). Furthermore, the synergistic effect of cyclopamine and temozolomide (TMZ) has shown to reduce the cell number of glioma CSCs in in vivo condition (Clement et al. 2007).

4.2.5 Current Therapies and Challenges in Cancer Stem Cell Therapy

As mentioned earlier, curative therapies should target both CSCs and non-CSCs. In recent studies, multiple novel therapeutic strategies have been designed for killing CSCs. However, some of these drugs are in preclinical development, and some are in clinical trials that can specifically eliminate or suppress CSCs. The surface marker differences and alterations in signaling cascades are alluring therapeutic targets for CSC therapy. Moreover, researchers have also investigated some more potential CSC therapeutic targets including ABC transporter-binding protein, microenvironment niche protein, etc. As discussed earlier, CSCs can be eradicated by using treatment targeting the signaling pathways such as Notch, Wnt, and Hedgehog. However, most of the signaling pathways are common in both normal and cancer stem cells that become critical to target the cancer stem cells specifically without harming the normal stem cells. Luckily, CSCs seemed to have their own particular enhanced signaling pathways (Vinogradov and Wei 2012). Furthermore, as mentioned above, the surface markers exploited for isolation of CSCs are also vital targets for CSC treatment. Antibodies that are used in immunotherapy for targeting surface markers overexpressed on CSCs are often used in combination with conventional therapies for better efficacy. CSCs are also known to express ABC transporter proteins at high levels. These proteins help in protecting CSCs from therapeutic agents. Thus, downregulation of these proteins may prove as a promising approach for overcoming the drug resistance to current conventional cancer therapies. However, studies conducted in chemoresistant patients have not yet

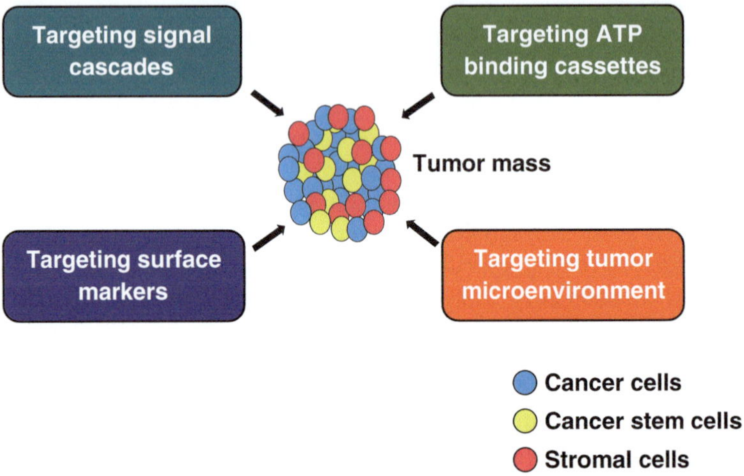

Fig. 4.2 Elimination of CSCs based on different targeting approaches such as tumor microenvironment, surface marker expression, dysregulated signal cascades, and ABC transporters to get rid of relapsed condition

shown the advantage of using these ABC transporter protein inhibitors for overcoming disseminated condition associated with CSCs. Therefore, there is a need to add some extra therapeutic agents in order to eradicate CSCs for better long-term therapy. Thus, inhibition of drug efflux activity in combination with conventional cancer therapies could be one of the promising strategies for CSC treatment in the near future. Last but not the least, CSC niche impairment might also be a potential approach for targeting CSCs. The tumor microenvironment helps to nurse and protect the CSCs from outside toxic agents. Several studies have investigated the role of stromal cells in bone marrow microenvironment and secondary lymphoid organs to favor the disease progression (Chen et al. 2013). Hence, the discussed major therapeutic approaches for CSC therapy are shown in Fig. 4.2.

In addition, tumor angiogenesis is also one of the important factors related to CSC survival and chemoresistance, which is initiated by vascular endothelial growth factor (VEGF). Earlier studies demonstrated that targeting VEGF with bevacizumab in mouse glioblastoma leads to normalization of tumor vasculature resulting to dramatic reduction in glioblastoma stem cell number (Burkhardt et al. 2012). Also, the combination of VEGFR2 antibody DC101 and cyclophosphamide has shown more efficiency against C6 glioma xenografts in vivo than the single therapeutic agent alone (Folkins et al. 2007).

The above discussed avenues represented some of the possible ways of therapeutically targeting CSCs. Over the past few years, many novel approaches have been designed for targeting CSCs, such as a nanoparticle functionalized with a targeting ligand specific to CSC containing an anticancer drug molecule to eliminate CSCs in combination with a chemosensitizer to overcome drug resistance (such as an ABC transporter inhibitor) and an imaging agent to facilitate tumor diagnostics. Such combinational approaches might exert the anti-tumor effect more effectively with lesser side-effects. Moreover, this approach would facilitate exact identification of the primary tumor localization and its metastases. Although all the alternative therapies are very effective, some approaches are not specific and might affect the normal tissue also. Moreover, CSCs have many ways to evade treatment as these cells reside in low oxygen area (Hypoxia) far from vascularized region, thereby preventing the efficient delivery of the therapeutic agents. Future challenges should involve the development of newer strategies for targeting CSCs specifically in an efficient manner by avoiding toxicity on normal healthy tissue stem cells.

Furthermore, these new strategies should also aid in easy delivery and retention of the drugs in the CSCs. As a result, these new therapies should increase the efficiency of the current drugs against cancer, thereby preventing tumor relapse and enhancing patient survival.

4.3 Nanomedicine-Based Cancer Stem Cell Therapy

4.3.1 Importance and Urgent Utility of Nanomedicine in Cancer Stem Cell Therapy

In clinic, cancer patients are generally treated through surgery, chemotherapy, and radiation therapy by individual drug or their combinations. In chemotherapy, most of drug-sensitive cancer cells are killed during treatment cycle, but unfortunately, tumor relapse condition was observed later due to the presence of cancer stem cells (CSCs) or tumor-initiating cells (TICs) (Shen et al. 2016). Targeting of CSCs has become a biggest challenge in modern medical science. It has reduced the therapeutic potential of current anticancer drugs. Therefore, the elimination of CSCs is an important aspect to prevent cancer drug-resistant condition by targeting drug efflux transporters (Singh and Settleman 2010), reprogramming of metabolic processes (Zhao et al. 2013a), and activation of antiapoptotic signaling pathways (Zhao et al. 2009).

In past few years, the identification of anticancer modalities which can kill CSCs has increased in significant numbers. Recently, the anticancer properties of antibiotics, bioactive compounds, therapeutic peptides, nucleic acids, and small molecule inhibitors have been broadly reviewed with their limitations. However, the clinical application of these anticancerous agents is limited due to their peculiar characteristic features like nonspecificity, poor water solubility, short circulation time or rapid clearance from blood, instability, and nonspecific biodistributions, which lead to low therapeutic values. Targeted killing of CSCs alone will not fulfill the complete eradication of cancer disease as cancer cells have plasticity and heterogeneity, which can reverse their phenotype into CSCs. Hence, there is urgent need to pay more attention toward the development of novel therapeutic approaches, which can kill both multiple drug-resistant CSCs and bulk malignant tumor cells simultaneously (Barenholz 2012).

Since the last two decades, nanotechnology has made several significant contributions in the field of biomedical science. Notably, nanoparticle-based drug delivery system (liposomes, dendrimers, metal oxide nanoparticles, polymeric nanomicelles, and carbon nanotubes) is one of them, which has gained substantial attention on translational point of view. Moreover, nanomedicines are smaller in size (~200 nm) which can easily cross through blood capillaries and reach the target site. The drug loading capacity, biocompatibility, and pharmacokinetic properties of these nanodrug carriers can be optimized by different modifications of nanoparticle's surface. Several other characteristic features like size control, tunable surface properties, and surface-to-volume ratios, attractive surface functional groups for bioconjugation, lesser nonspecific biodistribution, and minimal side effects have shown nanomedicine as a promisable tool to overcome aforementioned limitations. Recently, researchers have made attempts and formulated many nanodrug formulations for CSC therapy (Zhao et al. 2013a). Some of them were clinically approved for cancer treatment including doxorubicin-loaded liposomal formulation (Doxil), albumin-bound paclitaxel (Abraxane), and PEG-L-asparaginase (Oncaspar) (Barenholz 2012). Furthermore, novel and bioengineered multifunctional nanoparticles are being developed for CSC treatment, which are discussed further.

4.3.2 Examples of Nanomedicine for Cancer Stem Cell Therapy

Different types of nanomaterial such as polymeric nanoparticles, metal-based nanoparticles, carbon nanotube, magnetic nanoparticles, and liposome have been used to formulate targeted nanodrug carriers for CSC targeting using chemo-drugs, antibiotics, nucleic acids, peptides, and proteins. These therapeutic modalities target downstream cellular signaling pathways, CSC survival-associated genes, cell surface markers, and metabolic pathways, as shown in Fig. 4.3.

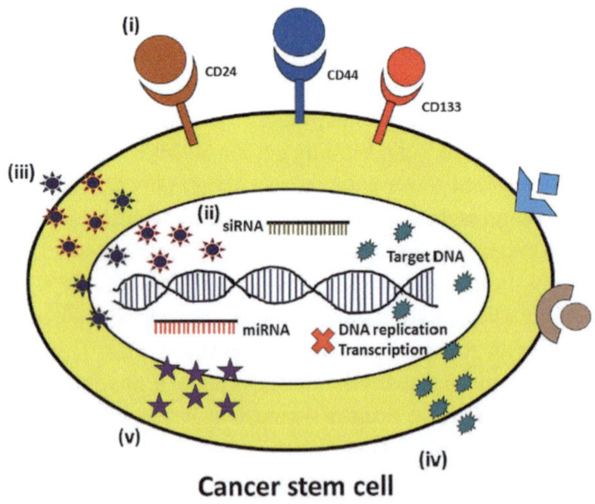

Fig. 4.3 CSC targeting by functionalized nanoparticles in combination with anticancer therapeutic agents: **(i)** targeting surface markers, **(ii)** targeting genes associated with CSCs, **(iii)** chemo- or radiotherapy drugs, **(iv)** small molecular inhibitors (Wnt, Notch, Hh), and **(v)** targeting metabolic targets (ROS)

4.3.2.1 Nucleic Acid-Loaded Nanomedicines against CSCs (miRNA, siRNA, Aptamer)

Nucleic acids have been used as therapeutic agents to treat several cancers of different origins with targeting CSCs also (Liu et al. 2011; Wu et al. 2011). Interestingly, wtp53 plasmid, miRNAs, siRNAs, and aptamers were used to modulate the gene expression by targeting cancer-specific oncogenic mRNAs for inhibiting cancer development, metastasis, and recurrence condition.

The mutations in tumor suppressor genes lead to the development of primary tumor. Restoration of such mutations in tumor suppressor genes can increase the survival rate of patients. For example, GBM is the most aggressive and lethal form of brain tumor in adults. GBM patients have poor prognosis and low survival rate due to the development of recurrence condition by CSCs and drug-resistant cancer cells. The frequent mutations in p53 gene are responsible for ~30% and ~65% risks of primary and secondary brain tumor (GBM), respectively. Temozolomide (TMZ) is an alkylating chemo-drug for the first line treatment of GBM with its methylation action at guanine bases to trigger unsuccessful mismatch repair leading to cell cycle arrest and apoptosis. O^6- methylguanine-DNA methyltransferase (MGMT) is a ubiquitous DNA repair enzyme that repairs DNA methylation and mismatch generated by TMZ. The research evidence suggests that wild-type p53 gene was found to be linked with MGMT expression. It negatively regulates the expression of MGMT in GBM-associated CSCs and drug-resistant cancer cells, which increases the therapeutic efficacy of TMZ. However, the effective delivery of wtp53 gene to target tumor across the blood-brain barrier is a notable challenge at present. Recently, researchers developed a novel nanocarrier based on cationic liposome (scL) [1,2-dioleoyl-3-trimethylammonium propane (DOTAP)/dioleoyl-phosphatidyl ethanolamine (DOPE)] for combinatorial delivery of wtp53 and TMZ specifically to GBM-associated CSCs and drug-resistant cancer cells using anti-transferrin receptor (TfR) single-chain antibody fragment (scFv). The scL-wtp53 does not only cause the cell cycle arrest and apoptosis in GBM but also increases sensitivity of GBM CSCs to TMZ in in vitro study. This approach was subjected in both preclinical and clinical study through systemic administration of scL nanodrug carriers (Kim et al. 2014).

Micro-RNA (miRNA) is a small noncoding RNA of 21–25 nucleotide length and is involved in the posttranscriptional regulation of genes. Earlier, several studies reported the alternations in miRNA expression leading to the development of cancer. Moreover, miRNA expression is usually found to be upregulated in cancer cells, called as oncogenic miRNA, used as biomarker for early diagnosis of cancer, while some miRNA expressions are found to be downregulated, called as tumor suppressor miRNA, used as novel therapeutic agents for removal of drug-resistant cancer and CSCs. Moreover, miRNA-based therapeutic approaches have several limitations in clinic such as higher instability, poor cellular uptake, lesser endosomal release, and risk of systemic toxicity. Therefore, the development of effective nanodrug carriers is essential for delivering therapeutic miRNAs to drug-resistant cancer and CSCs. For example, solid-lipid nanocarrier (DDAB) was used to deliver miRNA-34a in $CD44^+$-B16F10 CSCs. These vehicles protected miRNA-34a from nuclease degradation in serum and enhanced their bioavailability at target site. As a result, miSLNs-34a inhibited the migratory, invasive, and metastasis properties of B16F10 cells by attenuating CD44 expression and induced apoptosis at both in vitro and in vivo platform. Such nanoparticulate delivery system augmented the therapeutic effect of miRNA-34a for lung CSC therapy (Shi et al. 2013). In another approach, spherical nucleic acid (SNA) was used to determine the anticancerous activity of miRNA-182 in orthotopic GBM xenograft mice model. This nucleic acid-based nano-formulation was prepared through immobilization of miRNA-182 on gold nanoparticles that helped in crossing the blood-brain barrier (BBB) and tumor vascular network. Also, the systemic delivery of 182-SNA initiated the neutralization of Bcl2L12, c-Met, and HIF-2α oncogene expression in GBM, due to which the drug sensitivity of patient-derived glioma-initiating cells (GICs) increased to TMZ and receptor tyrosine kinase inhibitors (RTK-Is) by curbing stem cell-associated mRNA signatures. Further, antitumor property of 182-SNA is tested with glioma cells, GICs, and glioma-bearing mice. In result, 182-SNAs reduced GBM burden and increased animal life expectancy without any side effects (Kouri et al. 2015). Furthermore, a combinational drug delivery approach was developed to deliver chemo-drugs with miRNAs during drug-resistant cancer treatment. For example, solid-lipid nanoparticles (SLNs) were used for codelivery of both miRNA-200c and paclitaxel (PTX) to breast CSCs (BCSCs), where miRNA-200c restored the sensitivity to PTX by downregulating the expression of class III β-tubulin (TUBB3). As a result, the cellular cytotoxicity of PTX-loaded SLNs was increased against BCSCs. It revealed that combinational delivery is a novel therapeutic approach for CSC treatment (Liu et al. 2016).

Likewise, there is another class of therapeutic nucleic acid called small interfering RNA (siRNA) which is responsible for posttranscriptional gene silencing (PTGS) by RNA interference (RNAi) pathway. It is a double-stranded nucleic acid of 20–25 nucleotides long. It has been widely used in the treatment of many diseases by knocking down disease-specific gene. Generally, the delivery of siRNA to target tissue and cells is achieved through virus-like particles, metallic nanoparticles, lipidic and polymeric nanoparticles, and so on. In recent years, the application of siRNA has been broadly extended to drug-resistant cancers. Generally, cancer cells need more fuel (glucose) than healthy cells as they divide because cancer cells are more dependent on glycolysis for energy production rather than oxidative phosphorylation (in presence of oxygen) for building their biological macromolecules. Usually, cancer cells express more glucose transporters (GLUT) for glucose influx than normal cells. Among these, GLUT3 is highly expressed in GBM and particularly in GBM CSCs because it has fivefold higher affinities with glucose than GLUT1 which is predominantly found in normal tissues for key biological functions; therefore, knockdown of GLUT3 was targeted for GBM treatment using siRNA (siGLUT3) encapsulated with PEG-PLA cationic-lipid nanoparticles (as shown in Fig. 4.4), due to which the glucose uptake decreased in glioma cells leading to downregulation of stemness and proliferation of U87MG and U251 cells inhibited in a glucose-limited environment. Further, this nanoformulation was systemically administered through intravenous injection (i.v.) into U87MG xenograft

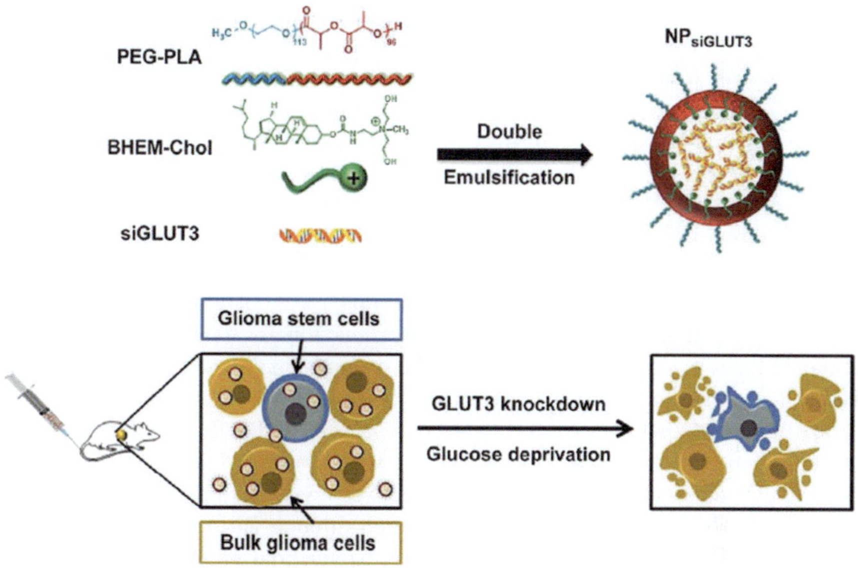

Fig. 4.4 Schematic illustration of the preparation of NPsiGLUT3 and the mechanism of NPsiGLUT3-mediated tumor growth inhibition in U87MG xenograft mice model. After intravenous injection, NPsiGLUT3-mediated GLUT3 knockdown caused glucose deprivation in glioma stem cells and bulk glioma cells, which inhibited the glucose metabolism and the growth of tumor. (Permission obtained from Elsevier press, Ref. no. 58)

murine model. As a result, NPsiGLUT3 significantly inhibited the growth of GBM in mice due to reduced expression of GLUT3 transporters and decreased the stemness of glioma cells. This study concluded that siRNA is a potential anticancer agent for cancer stem cell targeting (Xu et al. 2015b). Similarly, another study used siRNA-loaded nanocarriers to knockdown MDR1 gene-encoding drug efflux transporters. In this study, both MDR1 siRNA and paclitaxel (PTX) were encapsulated in cross-linked lipid nanocomplex composed of PEI-lipid in 1:16 ratio. Such nanocarrier system reduced the MDR1 expression in CD133+ human colon CSCs resulting in increased chemosensitivity to paclitaxel drug. Moreover, the synergistic action of siMDR1 and paclitaxel increased the therapeutic efficacy compared to drug alone (Liu et al. 2009). Recently, the siPlk1-encapsulated cationic lipid-polymeric nanoparticles were developed to target TGF-β signaling pathway for BCSC therapy. The nanoparticles carrying siPlk1 efficiently eliminated breast CSCs derived from MDA-MB-231 cells in vitro. However, LY364947 (an inhibitor of TGF-β type 1 receptor) increased the permeation of these polymeric nanoparticles in tumor tissue via vascular leakage leading to nanoparticle accumulation in drug-resistant cancer and CSCs. The synergistic action of LY364947 and siPlk1inhibited the tumor growth and reduced BCSC population in vivo *(*Zuo et al. 2016*)*. In another study, siPLK1/SSB-encapsulated HA-PEI/PEG nanocarriers were developed to target CD44 receptors in resistant A549 lung cancer cells. The nanodrug carriers showed dose-dependent therapeutic effect and targeted specific gene knockdown at both in vitro and in vivo platform. Further, these HA-based nanosystems can deliver any siRNA for systemic targeting of CD44 overexpressed on cancers (Ganesh et al. 2013). Overall, nanocarriers have gained considerable attention among the researchers for delivery of siRNAs, miRNAs, and other nucleic acid-based therapeutic agents for CSC treatment. However, a novel discovery of another class of nucleic acid molecules that has replaced the use of antibodies is called aptamers, used for targeted delivery of therapeutic agents in drug-resistant cancer treatment. As we are discussing about nucleic acid aptamers, it is a single-stranded oligonucleotide molecule that forms highly stable 3D structure to bind a wide range of small molecules or even cells with high affinity and specificity. Functionally, it mimics like antibody and modulates several protein functions. The exciting significant future of aptamers is due to its very small size, low systemic toxicity, and non-immunogenicity. These properties make aptamers a novel therapeutic carrier to deliver specific drugs and molecules into diseased cells. Presently, aptamers have been utilized in the designing and formulation of target-specific drug delivery systems in the field of nanobiotechnology. Unnatural aptamers are discovered through an in situ method called SELEX (systematic evolution of ligands by exponential enrichment) with high specificity and stability in a cost-effective manner (Sun et al. 2014b). Based on its application, aptamers can be easily modified with any functional group. Recently, a variety of aptamers are developed with high affinity to target CSC surface markers. Generally, these aptamers are functionalized on the surface of drug-loaded nanoparticles. For example, a CD30-specific RNA aptamer and PEG were functionalized on the surface of hollow gold nanospheres (HAuNS) via covalent S-Au bonds and doxorubicin which were loaded via charge force. The formulated Apt-HAuNS-Dox NPs were highly sensitive to acidic pH. As a result, 80% Dox release was shown in 2 h at pH 5.0. However, gold nanosphere without aptamer conjugation (HAuNs-Dox) released 55% drug at the same pH, which concluded that aptamer conjugation favored pH-induced drug release that selectively killed lymphoma cells and CSCs (Zhao et al. 2013b). Interestingly, PEGylated aptamer against EpCAM (a CSCs marker) showed higher rate of penetration into tumorsphere core after nanocarrier administration and remained in the core for at least 24 h, while EpCAM antibody displayed limited tumor penetration after 4 h incubation. In xenograft tumor, the PEGylated EpCAM aptamers were sustained for 26 h, which was 4.3-fold longer than EpCAM antibody. Also, the accumulation of PEGylated aptamers was 1.67-fold and 6.6-fold higher than antibody in xenograft tumor mice model at 3 h and 24 h after i.v. administration, respectively. Moreover, EpCAM aptamers were detected 200 μm far away from blood vessels in 3 h after i.v. administration than EpCAM antibodies, which were found to be distributed around the blood vessels in xenograft tumors. This study indicated that aptamers are better to antibodies in cancer theranostic application due to their uniform biodistribution, enhanced penetration, and higher retention in tumor sites (Xiang et al. 2015). Furthermore, 19-mer EpCAM RNA aptamer-conjugated PEG-PLGA-Dox nanopolymersomes were developed for targeting EpCAM-overexpressed CSCs isolated from adenocarcinoma cell lines. The in vitro results showed the efficient cell uptake and internalization of nanopolymersomes that exhibited the higher cytotoxicity toward EpCAM+ MCF7 cells (Alibolandi et al. 2015). Also, curcumin-loaded lipid-polymer-lecithin hybrid nanoparticles were synthesized and functionalized with EpCAM RNA Apts for targeting colorectal adenocarcinoma cells. The PEG core of this hybrid nanoparticle was modified with lecithin which is a well-known dispersing agent to improve drug loading capacity and stability. The Apt-CUR-NPs showed enhanced binding affinity to HT-29 cells and increased cellular uptake, due to which cell cytotoxicity was improved and higher curcumin bioavailability was seen over a period of 24 h during in vivo studies after systemic administration of biconjugate nanoparticles (Li et al. 2014). Thereby, several aptamer-conjugated nanodrug formulations were prepared recently, in which most of them are under phase II and III trials. As we have discussed earlier about CD44 expression on CSC surface, a 2'-F-pyrimidine-containing RNA aptamer (Apt1) was designed and conjugated on the surface of PEGylated liposome for targeting CD44+ CSCs. The results showed that Apt1-loaded lipid NPs had higher sensitivity and selectivity compared to blank liposomes. It concluded that Apt1-Lip has promising potency as a specific drug delivery system for CD44+ CSCs (Alshaer et al. 2014). Meanwhile, spherical capsules with alginate-enclosing, chitosan-coated (AEC), iron-saturated bovine lactoferrin, and EpCAM RNA Apt-conjugated calcium phosphate nanoparticles were prepared to reduce the viability of Caco-2 cells. These nanocarriers reduced the expression of CSC markers like CD133+/survivin/CD44+ in xenograft colon cancer models. These nanoparticles induced apoptosis by targeting survivin in drug-resistant cancer and CSCs. During treatment, such nanoparticles maintained iron, zinc, and calcium levels (Kanwar et al. 2015). Next, the application of aptamers has been extended for effective delivery of therapeutic nucleic acids like tumor suppressor gene-carrying plasmid, siRNA, and miRNA. In this context, aptamer-siRNA chimeras were developed to knockdown Plk1 gene in EpCAM+ cancer cells in vitro and in biopsy tissues. The Plk1 EpCAM-AsiCs inhibited EpCAM+ basal and luminal TNBC growth in nude mice and also stopped mammosphere formation in vitro, which provided an efficient approach for treating epithelial cancers (Gilboa-Geffen et al. 2015). Besides drug delivery, the use of aptamers was also extended to photothermal therapy for cancer. Gold nanorods (AuNRs) are well-known nanomaterial that can absorb near-infrared light and generate heat in its surrounding environment for photothermal therapy of cancer. Based on this, two aptamers (Apt CSC1 and Apt CSC13) were functionalized on the surface of gold nanorods to target both prostate cancer and CSCs. A beam of near-IR light was passed on AuNR internalized CSCs and cancer cells by which the temperature increased from 25° to 55 °C in target cells. Such heating caused the destruction of cellular organelles and their membrane which induced apoptosis, while untargeted cells were rarely affected without significant adverse side effects (Wang et al. 2013). All studies discussed here have been summarized in Table 4.1.

As several preclinical and clinical studies are needed to definitively assess how CSCs respond to current therapies, on this basis, the development of effective therapeutic strategies against CSCs is needed to increase the efficacy of conventional cancer treatment. Such potential approaches generally include targeting

Table 4.1 Therapeutic nucleic acid functionalized nanoparticles for CSC therapy

Therapeutic nucleic acids	Chemotherapeutic drugs	Target CSCs and markers	Nanocarrier	Mode of function	References
scL-p53	Temozolomide (TMZ)	GBM CSCs	Cationic liposome 1,2-dioleoyl-3-trimethylammonium propane (DOTAP)/dioleolylphosphatidyl ethanolamine (DOPE)	Upregulating p53 and modulating O^6-methylguanine-DNA methyltransferase enzyme	Kim et al. (2014)
miRNA-34a	–	Lung CSCs, CD44 + B16F10 cells (CSCs like cell population)	Solid-lipid nanoparticles (SLNP)/dimethyldioctadecylammonium bromide (DDAB)	Inhibiting tumor development and tumorigenicity	Shi et al. (2013)
miRNA-200c	Paclitaxel (PTX)	Breast CSCs	Cationic lipid 1,2-dioleoyl-3- trimethylammonium-propane	Reducing the expression of class III β-tubulin (TUBB3) gene	Kouri et al. (2015)
miRNA-182	Tyrosine kinase inhibitors (RTK-is) and TMZ	Bcl2L12, c-Met, and HIF-2α in GBM	Spherical nucleic acid	Suppressing stem cell-associated mRNA signatures	Liu et al. (2016)
siGLUT3	–	GBM CSCs, GLUT-3	PEG-PLA cationic lipid nanoparticles	Knockdown of GLUT3 gene	Xu et al. (2015b)
siMDR1	Paclitaxel	CD133+ human colon CSCs	PEI-Lipid (1:16)	Knockdown of MDR1 gene	Liu et al. (2009)
siPlk1	TGF-β type 1 inhibitor	Breast CSCs derived from MDA-MB-231 cells	Lipid-polymeric nanoparticles	Knockdown of Plk-1 gene	Zuo et al. (2016)
CD30 RNA Apt	Doxorubicin	CD30 overexpressed cancer and CSCs cells	Hollow gold nanosphere	Targeting specific drug cytotoxicity	Zhao et al. (2013b)

2'-F-pyrimidine-containing RNA aptamer (Apt1)	–	CD44+ CSCs	PEGylated liposome	Targeting specific drug carrier	Alshaer et al. (2014)
EpCAM RNA apt	Curcumin	EpCAM+ colon CSCs	Lecithin modified PLGA–PEG NPs	Inducing apoptosis	Li et al. (2014)
EpCAM RNA apt	Iron-saturated bovine lactoferrin-	CD133+/CD44+/survivin colon CSCs	Alginate-enclosed, chitosan-coated, calcium-phosphate (Ca-P) NPs	Inducing apoptosis in colon CSCs	Kanwar et al. (2015)
CSC1 and CSC13 apt	–	Prostate CSCs	Gold nanorods (AuNRs)	Photothermal destruction	Wang et al. (2013)

Abbreviations: *CSCs* cancer stem cells, *CD* cluster of differentiation, *NPs* nanoparticles; *Apt* aptamer, *EpCAM* epithelial cell adhesion molecule, *TGF-β* transforming growth factor β, *GLUT-3* glucose transporters-3, *MDR-1* multidrug resistance-1, *Plk1* polo-like kinase-1, *GBM* glioblastoma multiforme, *HIF-2α* hypoxia inducible factor-2α, *c-met* hepatocyte growth factor receptor, and *Bcl2L12* Bcl-2 like protein-12

CSCs, inhibition of ATP-binding cassette (ABC) transporters, blocking of essential self-renewal and cell survival signaling pathways, or disruption of tumor microenvironment.

4.3.2.2 Chemotherapeutic Drug-Loaded Nanomedicines Against CSCs

The main application of nanomedicine is to deliver hydrophobic drugs that have several serious issues related to its solubility, stability, and toxicity. The use of nanomedicine generally increases the solubility and stability of hydrophobic drugs with reduced side effects. This approach can be useful for such drug candidates that have the aim to eliminate drug-resistant cancer and CSCs. For example, doxorubicin (Dox) is the most commonly used chemo-drug at present time in clinic. To enhance its therapeutic efficacy, a nanodrug formulation based on gold nanoparticles was prepared, called Dox-Hyd-PEG-AuNPs, for targeting BCSCs. This formulation inhibited the mammosphere formation capacity in vitro and further removed all tumor cell subpopulations enriched with CSCs in orthotopic xenograft breast cancer mice model (Sun et al. 2014c). Like Dox, salinomycin (SA) is also a widely used chemo-drug having very low therapeutic efficacy in clinic. Therefore, several nanodrug carriers were used to enhance its clinical use. In an earlier study, SA drug was conjugated to hyaluronic acid (HA)-based nanogels, i.e., made up of cholesterol. These drug-loaded nanogels targeted and killed CD44+ BCSCs in both drug-resistant tumor cell subpopulations and multicellular tumor spheroids (Wei et al. 2013). In addition, paclitaxel (PTX) is the most common drug in the treatment of several solid cancers. The clinical efficacy of this drug was enhanced through the preparation of several nanodrug formulations. For example, PTX-loaded poly(D,L-lactide-coglycolide) nanoparticles were synthesized. Further, these drug-loaded nanoparticles were functionalized with anti-CD133 antibody for targeted drug delivery that resulted into higher cytotoxicity on HepG2 and Huh7 cells in both in vitro and in vivo studies. Also, these actively targeted nanoparticles eliminated CD133+ liver CSCs, which made it a promising candidate for testing in different phase trials (Jin et al. 2014).

The most important advantage of nanocarriers is to deliver multiple drugs at a time. For instance, a combined therapeutic approach was developed using nanoparticle-encapsulated Dox (pluronic F127-chitosan-Dox NPs) and cryoablation technology, which killed almost CD44+ and CD133+ CSCs in a 3D mammosphere model (Rao et al. 2014). In another study, SA drug was used in combination with other therapeutic modalities to eliminate other CSCs. By this approach, a combinational therapy using SA drug with polyelectrolyte-conjugated AuNPs (Au/SA@PDC) was developed and showed a synergistic BCSC inhibition in drug-resistant MCF-7 cells (Xu et al. 2014). Additionally, two different nanodrug formulations like octreotide-conjugated PTX-loaded PEG-b-PCL (Oct-M-PTX) and salinomycin-loaded PEG-b-PCL (M-SA) polymeric micelles were prepared. It was shown that M- SA micelles eliminated the large proportion of CD44+/CD24- BCSCs in a potent manner compared to SA alone. Also, Oct-M-PTX micelles inhibited a large population of MCF-7 cells as compared to M-PTX micelles, while the anticancerous effect of Oct-M-PTX and M-Sal micelle combination was very strong in both in vitro and in vivo system. Hence, this combinational therapy improved the treatment of breast cancer with elimination of BCSCs (Zhang et al. 2012a). Further, both PTX and SA drugs were also conjugated to anti-CD44 functionalized SWCNTs via hydrazine linker. Such pH-responsive nanodrug carriers released both drugs at acidic tumor microenvironment after targeting CD44+ BCSCs and increased the combined therapeutic effect of both drugs in xenograft mice models (Al Faraj et al. 2016). The nanodrug carriers were also used in the codelivery of MDR inhibitors with cytotoxic drugs. These inhibitors increase the antitumor effect of existing drugs after sensitizing CSCs. For example, a combination of ABCG2 mAb- and PTX-conjugated Fe3O4 nanoparticles was tested on multiple myeloma (MM) $CD138^{-}CD34^{-}$CSCs in a xenograft MM NOD/SCID mouse model. The combination induced a strong therapeutic response compared to current conventional regimens for MM patients which is due to instantaneous inhibition of ABC transporters by antibodies and delivery of PTX to CSCs via magnetic nanoparticles (Yang et al. 2015). Nanomedicines were also used to carry such therapeutic modalities that affect CSC's development and maintenance pathways. These agents were reached into different phases of clinical trials but could not be used for a longer time due to their high toxicity, lesser solubility, and nontargeted drug delivery. For example, cyclopamine is a natural Hedgehog (Hh) pathway inhibitor. When this drug was conjugated with HPMA (N-(2-hydroxypropyl)-methacrylamide) copolymer, its systemic toxicity was decreased and aqueous solubility increased. Such polymer-drug conjugate was used as a CSC-selective macromolecular therapy, which removed a large subpopulation of CD133+ CSCs from human prostate cancer epithelial cells (RC-92a/hTERT cells) (Zhou et al. 2012). In another study, HPI-1 is used as a potent antagonist of Hh transcription factor (Gli1) that blocks the downstream signaling events of Smo. A nanoformulation based on HPI-1-encapsulated PLG PEG nanoparticles (NanoHHI) was prepared. The NanoHHI increased the systemic bioavailability of HPI-1 inhibitor and improved its solubility. Such nanoformulations inhibited the growth of Ptch(−/+), Trp53 (−/−) medulloblastoma in xenograft mice models. Especially, the combined therapeutic effect of NanoHHI with gemcitabine significantly inhibited the growth of orthotopic Pa03C pancreatic cancer xenografts (Chenna et al. 2012). Also, NanoHHI decreased the population of CD133+ CSCs in hepatocellular carcinoma (HCC) and potentially inhibited the tumor growth in orthotopic HCC xenografts (Xu et al. 2011). Conclusively, the Gli1 inhibition through NanoHHI displayed antitumor effect in both pancreatic cancer and HCC models. All studies discussed here are summarized in Table 4.2.

4.3.2.3 Targeted Therapy- and Immunotherapy-Based Nanomedicines against CSCs

In few years, the field of drug delivery has become more advanced with the development of actively targeted nanodrug carriers, i.e., basically based on ligand-receptor interactions. The main advantage of such delivery is to increase the therapeutic efficacy of chemo-drugs, small molecular inhibitors and nucleic acid-based therapeutic

Table 4.2 Chemotherapeutic drug loaded different nanodrug carriers for CSC therapy

Nanocarrier	Chemotherapeutic drugs	Target CSCs and markers	Composition	Mode of function	References
Gold nanoparticles (AuNPs)	Doxorubicin	Breast CSCs	PEG- Hyd-Dox, AuNPs	Reducing the no. of tumor cell subpopulations enriched with CSCs	Sun et al. (2014c)
Hyaluronic acid (HA)-nanogel	Salinomycin	CD44+ breast CSCs	HA, cholesterol	Targeting CD44+ CSCs and inhibiting tumor growth	Wei et al. (2013)
Anti-CD133 antibody conjugated poly (D,L-lactide-coglycolide) NPs	Paclitaxel	CD133+ liver CSCs	Anti-CD133 antibody, PLGA NPs	Suppressing CD133+ CSCs population	Jin et al. (2014)
Pluronic F127-chitosan NPs	Doxorubicin	CD44+ CD133+ CSCs	Pluronic F127, chitosan	Combined killing of CD44+ and CD133+ CSCs via Dox and cryoablation technology	Rao et al. (2014)
PDC-AuNPs	Salinomycin	Breast CSCs	PDC, AuNPs	Synergistic inhibition of BCSCs via hyperthermia and SA treatment	Xu et al. (2014)
Octreotide-PEG-b-PCL polymeric micelles	Paclitaxel Salinomycin	Breast CSCs	PEG-b-PCL, octreotide	Eliminating BCSCs to improve breast cancer treatment	Zhang et al. (2012a)
Anti-CD44 antibody functionalized SWCNTs	Paclitaxel Salinomycin	CD44+ breast CSCs	Anti-CD44 antibody, SWCNTs, hydrazine linker	Combined therapeutic effect against BCSCs in xenografts	Al Faraj et al. (2016)
Iron oxide NPs	Paclitaxel	MM $CD138^{-}CD34^{-}$CSCs	ABCG2 mAb, Fe3O4 NPs	Inhibiting ABC transporters by antibody and targeting of CSCs by PTX-loaded magnetic NPs	Yang et al. (2015)
HPMA	Cyclopamine	CD133+ CSCs	HPMA polymer	Removing a large population of CD133+ CSCs from human prostate cancer epithelial cells	Zhou et al. (2012)

PLGA-PEG NPs	HPI-1 gemcitabine	Medulloblastoma, pancreatic cancer	PLGA-PEG copolymer	Inhibiting medulloblastoma growth and Pa03C pancreatic cancer in xenografts	Chenna et al. (2012)
PLGA-PEG NPs	HPI-1	CD133+ CSCs hepatocellular carcinoma (HCC)	PLGA-PEG copolymer	Inhibiting tumor growth in HCC xenografts	Xu et al. (2011)

Abbreviations: *CSCs* cancer stem cells, *CD* cluster of differentiation, *NPs* nanoparticles, *PDC* poly (dimethyldiallylammonium chloride), *PEG* poly ethylene glycol, *PCL* poly-caprolactone, *SWCNT* single-walled carbon nanotube, *HPMA* N-(2-hydroxypropyl) methacrylamide, *HPI* Hedgehog pathway inhibitor, *mAb* monoclonal antibody, and *ABCG* ATP-binding cassette subfamily G

agents without any risky side effects, achieved through the practice of therapeutic antibodies functionalized on nanoparticle surface. Moreover, antibodies have also eliminated CSCs and drug-resistant cancer cells from tumor cell population like other therapeutic modalities. As a result, the patient's survival rate increases during clinical trials (Vinogradov and Wei 2012). For example, an anti-CD133 antibody-decorated SN-38-loaded PEG-PCL nanoparticle-based nanoformulation was prepared to target CD133+ HCT116 cells and showed higher toxicity in colorectal cancer cells compared to nontargeted particles (PEG-PCL-SN38). Further, CD133Ab-PEG-PCL-SN38 NPs significantly reduced the tumor growth in orthotopic HCT116 xenograft mice model compared to CPT-116 and PEG-PCL-SN38 NPs. However, tumor relapse condition was observed in all treatment groups during the off-therapy stage, but no relapse condition was observed in CD133Ab-PEG-PCL-SN38 NP treatment group. Also, the mouse body weight decreased during the treatment stage that indicated better inhibition of tumor growth and then remained constant in the rest of the experimental period as shown in Fig. 4.5. In end, it was concluded that several other targeted nanoparticles can be designed using this cutting-edge study for delaying tumor relapse condition (Ning et al. 2016). Mostly, mAb-functionalized nanodrug carriers are under different clinical phase trials, while there is a concern related to the use of mAb and its origin. In preclinical studies, mAbs generally target human antigens and cannot cross-react with murine antigens, due to which the systemic toxicity and other side effects of mAbs could not be observed. Therefore, there is a need to characterize mAbs in a very efficient manner on different preclinical and clinical platforms. Further, anti-CD44 mAb was developed to target CD44+ CSC population. Based on this, an anti-CD44 mAb-conjugated liposomal nanoparticles loaded with doxorubicin and triple fusion gene were prepared for hepatocellular carcinoma (HCC) treatment. These nanoparticles killed CD44+ CSCs via chemotherapy and gene therapy to reduce the side effects of conventional chemotherapy (Wang et al. 2012b). In addition to therapeutic antibodies, aptamers have also shown their better potential for the development of targeted CSC therapeutics and molecular imaging agents as described earlier in Sect. 4.3.2.1. In another study, two different nanodrug formulations like PLGA-PEG-PTX-CD44 and PLGA-PEG-PTX-CTX were developed to target CD44+ breast and EGFR+ colon cancer cells. These targeted nanoconjugates displayed significantly more therapeutic efficacy in tdTomato+ MCF-7, MDA-MB-231, HCT116, and HCT8 cells. Using this fluorescent CSC model, it was concluded that active targeting sensitized CSCs to PTX treatment (Gener et al. 2015). Furthermore, EGFR and EGFRvIII receptors are also highly expressed on glioblastoma multiforme (GBM) neurospheres and GBM stem-like cells (GSCs). To target this receptor, a nanoformulation of cetuximab-conjugated iron oxide nanoparticles (CTX-IONPs) was formulated that showed a significant antitumor effect with increased apoptosis in EGFRvIII+ GSCs and EGFR+ GBM neurospheres. Further, the survival rate of GBM xenografts was increased with substantial tumor regression after treatment (Kaluzova et al. 2015). Recently, an NTP-conjugated, paclitaxel-loaded biodegradable polyglutamic acid polymer-based nanoformulation was prepared to target NCAM-overexpressed CSCs in Wilms tumor. The results showed the proliferation

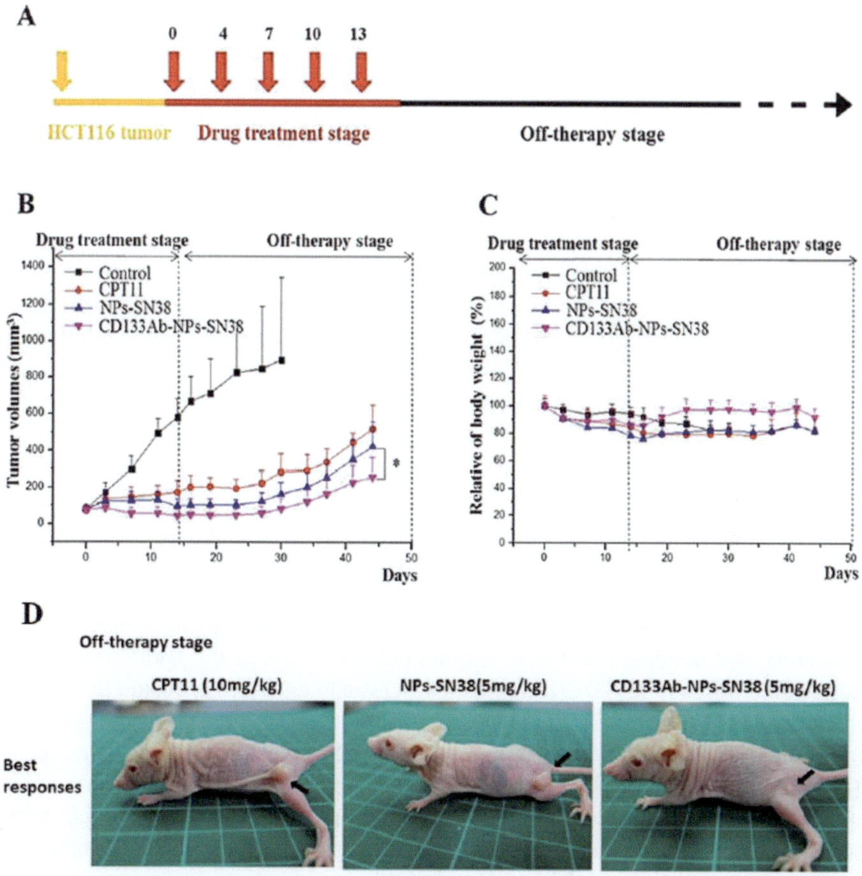

Fig. 4.5 In vivo anticancer efficacy of CD133Ab-NPs-SN-38 in HCT116 tumor xenograft model (**a**) Scheme of the in vivo treatment schedule. (**b**) Tumor volume and (**c**) the changes in body weight of HCT116 tumor-bearing mice treated with control (n = 6), CPT-11 (10 mg/kg, n = 6), NPs-SN-38 (5 mg/kg of SN-38, n = 6), and CD133Ab-NPs-SN-38 (5 mg/kg of SN-38, n = 6). (**d**) Pictures of tumor size in the mice with the best responses in treated groups. * = P < 0.05. (Reprinted with permission from Ning, S. T., Lee, S. Y., Wei, M. F., Peng, C. L., Lin, S. Y. F., Tsai, M. H. et al. (2016). Targeting Colorectal Cancer Stem-Like Cells with Anti-CD133 Antibody-Conjugated SN-38 Nanoparticles. *ACS Applied Materials & Interfaces*, *8*(28), 17,793–17,804. Copyright 2018. American Chemical Society, Ref. no. 82)

and migration inhibition in xenograft-derived WT cells. Also, NTP-PGA-PTX conjugate reduced the tumor size in a patient-derived WT xenografts with dramatic reduction in NCAM-overexpressed CSCs. Further, this nanoconjugate was used in targeting drug-resistant cell population (Markovsky et al. 2017). In fact, these antigens are mainly responsible for tumor development and progression; therefore, the search for novel antigens and the development of new therapeutic antibodies are still going on. The nanodrug carriers have been also utilized in the codelivery of multiple drugs and imaging agents to the drug-resistant cancer for cancer theranostic applications. For instance, bifunctional nanoparticles were designed for tumor imaging and targeted drug delivery in the treatment of hormone refractory prostate cancer. The single-chain prostate-specific antigen (PSA) antibody-conjugated PLGA-SPIO/docetaxel nanoformulation was prepared. These nanoparticles increased antitumor efficacy and improved MRI imaging in vitro via targeted delivery of Dtxl and SPIONS to PC3M cells. In addition, such dual activity comprising of nanoparticles provided a negative MRI contrast enhancement and tumor growth inhibition in PC3M xenograft mice models (Gao et al. 2012). As CSCs overexpress significant biomarkers on their surface for their isolation and targeting, no targeted therapy has the capability to eliminate all CSCs from tumor cell subpopulation. It may be due to the localization of CSCs in tumor necrotic area, where nanodrug carriers are difficult to reach. Thereby, targeted nanodrug carriers can be more suitable in such cancer treatment, where CSCs are freely available like leukemia stem cells (LSCs) in hematological malignancies. All studies pertaining to targeted nanodrug carriers are summarized in Table 4.3.

In recent years, nanoparticle-based delivery has gained a significant attention as potential carriers for cancer vaccine delivery. These nanocarriers have delivered vaccine antigens, adjuvants, and immunomodulatory agents to the specific target sites. In fact, most of the vaccines require additional adjuvants to induce cellular immunity, but nanoparticle-based vaccine delivery reduced the additional use of adjuvants (Shima et al. 2013). As nanoparticles can enter easily inside the cell, it interacts with Toll-like receptors (TLRs) to further improve its efficacy as vaccine adjuvant (Nguyen et al. 2012). The nanocarriers have also increased the antigen-specific cytotoxic T-cell (CD8+) responses that are critical regulators of anticancer immunity. If nanoparticles are manipulated by their surface charge, particle size, particle core hydrophobicity, and surface-bound ligands, then it can easily enter into antigen-presenting cells (APCs) and modulate humoral immune responses to tumor antigens leading to improved anticancer immunity (Cruz et al. 2012). In addition, nanoparticles have also destroyed or decreased the no. of tumor-associated macrophages (TAM) which are responsible for CSC growth in its niche. For example, a siCCR2-encapsulated lipid nanoparticle containing C12–200 lipid, PEG-DMG, cholesterol, and disteroylphosphatidyl choline was formulated. The monocyte-targeted siRNA nanomaterials silenced CCR2 at mRNA, protein, and functional levels in monocyte subsets, preventing their accumulation at inflammation site in both lymphoma-modeled mice and colorectal xenografts. It was already known that inflammatory monocytes differentiate into tumor-associated macrophages under host response. The results showed that siCCR2 decreased the no. of TAMs followed by CSC growth inhibition leading to the reduction in tumor size (Leuschner et al. 2011). Furthermore, several immunotherapeutic molecules were delivered to CSCs through aptamers. For instance, bispecific oligonucleotide aptamer conjugates were used to deliver 4-1BB costimulatory molecules to prostate cancer cells and enhanced T-cell-based antitumor immunity. Nevertheless, the effects of such costimulatory aptamers on CSCs are less understood (Pastor et al. 2011). Also, CART cells were developed for CSC therapy. Up to date, only three animal studies are available on

Table 4.3 Different types of targeted nanodrug carriers for CSC therapy

Nanocarrier	Chemotherapeutic drugs	Ligand/receptor	CSC source and their marker	Mode of function	References
CD133Ab-PEG-PCL-SN38 NPs	SN-38	Anti-CD133 antibody/CD133	Colorectal cancer, HCT-116 cells/CD133+	Inhibit tumor growth in CD133+ HCT116 xenografts	Ning et al. (2016)
Anti-CD44 Ab-liposomal NPs	Doxorubicin	Anti-CD44 antibody/CD44	CD44+ HCC CSCs (HepG2 cells)	Eliminate CCD44+ CSCs and induce apoptosis via chemo and gene therapy	Wang et al. (2012b)
PLGA-PEG-PTX-CD44 PLGA-PEG-PTX-CTX	Paclitaxel	Anti-CD44 antibody/CD44 Cetuximab/EGFR	CD44+ breast CSCs EGFR+ olon CSCs	Inhibit MCF-7, MDA-MB-231, HCT116 and HCT8 tumor cells growth	Gener et al. (2015)
CTX-IONPs	–	Cetuximab/EGFR and EGFRvIII	CD133- EGFR+ GBM neurospheres and EGFRvIII+ GSCs	Inhibit tumor growth and increase survival rate of GBM xenografts	Kaluzova et al. (2015)
NTP-PGA-PTX	Paclitaxel	NCAM targeted peptide(NTP)/NCAM	NCAM + WT CSCs	Reduce the tumor size in a patient derived WT xenografts	Markovsky et al. (2017)
scAb-PLGA-SPIO/Dtxl	Docetaxel	Single-chain PSCA antibody/PSCA	PSCA+ PC3M CSCs	Inhibit tumor growth in PC3M xenograft	Gao et al. (2012)

Abbreviations: *CSCs* Cancer Stem Cells, *CD* Cluster of Differentiation, *NPs* Nanoparticles, *PEG* Poly Ethylene Glycol, *PCL* Poly-Caprolactone, *Ab* Antibody, *PTX* Paclitaxel, *CTX* Cetuximab, *SPIONP* Super Paramagnetic Iron Oxide Nanoparticles, *PLGA* Poly (Lactic-Co-Glycolic) Acid, *PGA* Polyglutamic Acid, *DTX* Docetaxel, *scAb* Single-Chain PSCA Antibody, *EGFR* Epidermal Growth Factor Receptor, *PSCA* Prostate Stem Cell Antigen, *NCAM* Neural Cell Adhesion Molecule, *HCT* Human Colorectal Carcinoma, *HCC* Hepatocellular Carcinoma, *WT* Wilms Tumor, *GBM* Glioblastoma Multiforme, *GSC* Glioblastoma Stem Cells, and *PC* Prostate Cancer

CSC-targeted CART cells in which the first study used an anti-CD133 CART cell that killed patient-derived glioblastoma stem cells both in vitro and in vivo platform (Zhu et al. 2015). In the second study, EpCAM-targeted CART cells showed an antitumor efficacy against prostate cancer in both in vitro and animal models (Deng et al. 2015). Further, in the third study, similar EpCAM CART cells were used to treat peritoneal carcinomatosis in xenograft mice model (Ang et al. 2017). Significantly, a CSC-based dendritic cell vaccine was developed to induce anti-CSC immunity in an effective manner. The DC vaccine displayed a strong antitumor effect in neurospheres compared to glioma xenografts (Ning et al. 2012; Toda 2013). In another study, a CSC-based vaccine prevented the liver metastasis from colon cancer and reduced the tumor size with low incidence in a rat colon carcinoma syngeneic model (Duarte et al. 2013). In few years, cancer immunotherapy has gained a large attention to both the clinicians and scientists because of its use; the survival rate of patients has increased with elimination of relapsed state. Thereby, in future, the combination of cancer immunotherapy with nanotechnology may open novel avenues with several breakthroughs for patient's treatment.

4.3.2.4 Metabolic Target-Based Nanomedicines Against CSCs

The targeting of metabolism in CSCs and drug-resistant cancer cells has always been a challenging task. When CSCs are treated through radiation, then it induces DNA damage through ROS (reactive oxygen species) generation derived from water molecules. Such damages can be seen in long- and short-term consequences. The short-term DNA damage disturbs the DNA metabolism such as DNA replication and RNA transcription. If DNA repair does not work, then it leads to the genomic instability and subsequently tumor development in a long-term manner. In CSC population of different tumors like lung, breast, glioblastoma, and prostate, the DNA repair mechanism is highly active due to the activation of ATR-Chk1 and ATM-Chk2 pathways (Krause et al. 2016). In cellular physiology, ROS is mainly produced during oxygen metabolism leading to the control of different cellular processes like proliferation, differentiation, and survival (Schieber and Chandel 2014). If ROS level is higher inside the cells, then it leads to irreversible oxidative stress and cell death. Thereby, ROS level is maintained through several scavenging molecules like catalase, peroxidase, glutathione, dismutase, and superoxide (Trachootham et al. 2009). In CSCs, the ROS level is generally found to be lower that contributes to the high resistance to genotoxic stress. Furthermore, it was already known that CSC populations reside in hypoxic region of tumors where oxygen level is very low. If tumor oxygenation is carried out, then CSCs can be more susceptible to current treatments (Kobayashi and Suda 2012). In several studies, nanomedicines have been utilized for increased ROS generation that induces necrosis and apoptosis in cancer cells with various morphological and physiological changes, for example, hyperthermia, a noninvasive treatment procedure that usually kills drug-resistant cancer and CSCs via heat shock and tumor reoxygenation. In addition, SPION nanoparticles were developed to generate heat in a localized tumor cell population area under an alternating magnetic field. Such NPs induced magnetic hyperthermia in MDA-MB-231 and A549 cells. Further, several

CSC-associated assays were showing the removal of ALDH+ CSCs in SPION-treated tumor cell subpopulations. During treatment, CSC killing was achieved through higher ROS generation and acute necrosis. In end, these results concluded that magnetic hyperthermia has the ability to eliminate the tumor relapse state compared to conventional cancer treatments (Sadhukha et al. 2013). Likewise, iron oxide magnetic nanoparticles were functionalized with epidermal growth factor for targeting EGFR receptor overexpressed in breast cancer cells. Such actively targeted nanoparticles were found to enter into the lysosomes of MDA-MB-231 cells. Under the effect of alternating magnetic field, these nanoparticles disrupted the lysosomal membrane and killed EGFR+ breast cancer cells with increased ROS production. Hence, the lysosome-mediated cell death pathway is an alternative independent mechanism to kill drug-resistant cancer cells, when apoptosis pathways become resistant (Domenech et al. 2013). In recent studies, breast CSCs were found to be resistant for traditional hyperthermia. Later, this resistance was observed due to higher expression of HSP90 in breast CSCs. Next, the PEG-coated MWCNTs (multiwalled carbon nanotubes) were designed to kill breast CSCs via thermal treatment that was activated through NIR (near infrared) irradiation. In contrast to magnetic hyperthermia, MWCNT-mediated photothermal therapy increased the survival time of mice with complete tumor regression (Burke et al. 2012). Overall, these electromagnetic field-responsive nanoparticles are in their initial stages of development, but such thermal effect will also lead to the development of other novel anti-CSC therapeutics in the future. In another study, the mitochondria-targeted PEGylated liposomes were formulated and encapsulated with daunorubicin and quinacrine drugs. To achieve mitochondrial targeting, the dequalinium regulator was attached on the surface of the liposomes. The results showed the accumulation of such targeted liposomes into the mitochondria which induced the proapoptotic Bax protein activation, reduced the mitochondrial membrane potential, opened the mitochondrial permeability transition pores, released the Cytochrome-C (Cy7C) from mitochondria to cytosol, and activated downstream caspase signaling. Finally, such nanoformulations induced apoptosis in MCF-7 CSCs and reduced the growth of relapsed tumors at large extent arising from MCF-7 CSCs in female NOD/SCID mice after the combined i.v. injection of daunorubicin and quinacrine liposomes as shown in Fig. 4.6 a, b (Zhang et al. 2012b).

4.4 Future Directions in Nanomedicine-Mediated Cancer Stem Cell Therapy

The current state of drug delivery technology clearly suggests the development of novel nanodrug carriers to enhance the therapeutic efficacy of existing anti-CSC therapies. Among all types of drug delivery systems, polymer-based drug delivery vehicles have gained a major attention related to its widespread use. These polymeric nanodrug carriers are designed with few considerations like controlled drug release profile, batch-to-batch reproducibility, and narrow size distribution. Moreover, such nanodrug carriers can also provide synthetic versatility according to

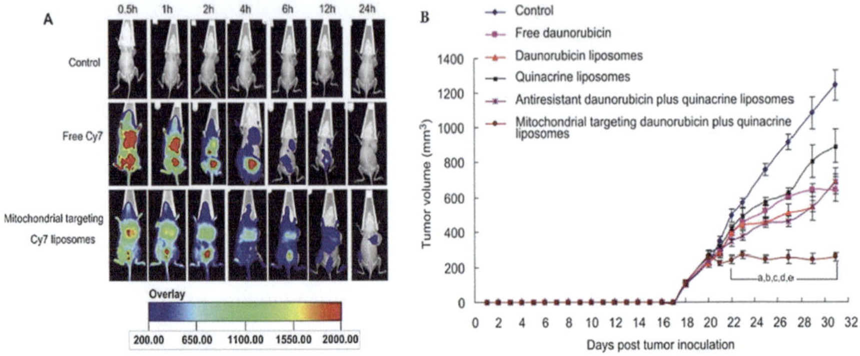

Fig. 4.6 (**a**) In vivo real-time imaging of the relapsed tumors arising from MCF-7 CSCs in female NOD/SCID mice after intravenous injecting PBS (pH 7.4), free Cy7, or mitochondrial targeting Cy7 liposomes. (**b**) Efficacy of mitochondrial targeting daunorubicin plus quinacrine liposomes in treating the relapsed tumors arising from MCF-7 cancer stem cells in female NOD/SCID mice. (Permission obtained from Elsevier press, Ref. no. 107)

the type of application. For example, the surface of polymeric nanodrug carriers can be modified with cancer targeting ligands for site-specific delivery to tumor site. Still, the application of drug delivery technology in CSC-targeted therapies is in an early stage with several unsolved issues. However, mostly nanodrug carriers are under different clinical phase trials and very few of them have been clinically approved. While there is need of more basic and applied research to advance the field of drug delivery for increasing the therapeutic effect of anti-CSC therapy in clinic, we here discuss several less explored issues with more attention to take advantage of such innovative and more effective strategies.

4.4.1 Synthesis of Highly Efficient Targeted Nanoparticles for CSC Therapy

The concept of designing targeted nanoparticles holds their great promises in CSC therapy as these nanoparticles can increase the drug concentration in CSCs for their elimination inside tumor mass. Therefore, the synthesis of such targeted nanoparticles will enhance the therapeutic efficacy of anti-CSC drugs in clinic and reduce the treatment course with patient's better outcome, but still, more attention should be given to the design and optimization of effective nanodrug carriers. If targeting moiety will be modified to achieve higher therapeutic efficacy, then several issues like additional complexity, regulatory barriers, and cost will come into consideration because these problems cannot be ignored. In terms of practice also, many questions were raised related to nanoparticle targeting and drug accumulation in the selected tumor and CSC subpopulation. In this scenario, a fundamental paradox infers that addition of targeting moiety onto the surface of nanoparticles generally compromises with the stealth feature of nanoparticles and can increase their clearance by reticuloendothelial system from host body. For instance, actively and passively targeted doxorubicin-loaded liposomal nanocarriers were used to elucidate the effect of targeting moiety on blood circulation time and tumor-bearing animal survival. Further, the results showed that no difference was observed related to animal survival between both types of targeted nanocarriers. Also, intratumoral doxorubicin concentrations were equal in both treatments (McNeeley et al. 2007). Although high avidity of nanoparticles is another enigma that has been always seen as an advantage, in case of targeted nanoparticles, such effects reduced their infiltration inside the tumor core (Lee et al. 2010). It was also seen that some CSC populations are found in necrotic region of tumors, i.e., very challenging to reach (Keith and Simon 2007). Overall, targeted nanocarriers can be suitable for treating such types of cancer, where CSCs are easily accessible like leukemia diseases.

As most of the CSC markers are used for the development of targeted nanocarriers, they are also expressed on normal stem cells. In result, it can lead to the unwanted toxicities (Xia 2014). Thereby, the search of highly CSC-specific ligands has always been a very challenging task. We also suggest that targeting of CSCs should be more inclusive and circumvent all the discussed downsides to achieve the ultimate objective of enhanced cancer therapeutic efficacy.

4.4.2 Synthesis of Nanoparticles with Deep Penetration Potentials for Effective CSC Therapy

Several evidence suggested that tumors are heterogeneous in nature which contain two different types of cell populations in their microenvironment, i.e., CSCs and non-CSCs. The CSC population are generally found in hypoxic region (low oxygen level responsible for stemness), while non-CSC population can be seen in vascularized region. For example, CD133+ ALDH+ breast CSCs were located in the central region of tumor tissues (Liu et al. 2014). The non-CSCs and CSCs near to vascularized region were killed easily using therapeutic agents, but CSCs enriched in hypoxic region of tumor core could not be easily targeted due to poor penetration of nanodrug carriers or therapeutic drug molecules. Therefore, to achieve such penetration, several nanodrug carriers could be modified on several aspects including particle size, PEG coating, surface charge, and conjugation of tissue-penetrating peptides. By these modifications, the penetration and retention behavior of nanodrug carriers could be enhanced. Based on this, several intelligent and stimuli-sensitive nanoparticles are designed further. These smart nanoparticles contributed the controlled drug release profile and efficient delivery of therapeutic agents in tumor core. For instance, a pH-sensitive, doxorubicin-encapsulated DLC-PEG liposomal dendrimers were prepared for long circulation and better tumor accumulation. However, the drug-loaded dendrimers were released that further penetrated deeply inside the tumor, where doxorubicin was accumulated and killed the MCF-7 cells (Sun et al. 2014d).

4.4.3 Synthesis of Nanoparticles for Better Cellular Internalization for Effective CSC Therapy

In order to eliminate CSCs, the rationally designed nanodrug carriers have the capacity to deliver any types of therapeutic agents to the CSC-enriched target site. Further, these carriers are characterized with advantages like better internalization rate, higher retention time in the blood, and higher accumulation at tumor sites. While these drug delivery systems were not enough to overcome all limitations, therefore intelligent and stimuli-responsive drug delivery systems such as pH, temperature, and tumor microenvironment responsive were designed. Usually, nanoparticles were PEGylated or modified with other hydrophilic polymers to improve their stability and bioavailability, lessen immunogenicity, and prolong circulation time in the blood. Despite such enormous activity, PEGylation decreased the cellular uptake of nanoparticles resulting in the blockage of intracellular trafficking pathways for diminishing the anticancer therapeutic efficacy (Knop et al. 2010; Mishra et al. 2004). Therefore, to overcome this situation, the PEG molecules were conjugated on nanoparticle surface via stimuli-responsive linker. When these nanoparticles entered the cells, then PEG molecules cleaved under specific stimuli. For example, a liposomal formulation having pH-sensitive PEG-coating and cell-penetrating peptide was prepared, in which PEG molecules were conjugated to phosphatidylethanolamine (PE) lipid molecules via hydrazone (HZ) bonds. When such carriers entered into acidified tumor microenvironment, PEG molecules were removed due to cleavage of HZ bonds and showed site-specific delivery to cells due to TATp moieties (Kale and Torchilin 2010). Further, stimuli-sensitive PEG shield is used to coat several types of nanoparticles to enhance their intracellular delivery. However, this is not well studied that such stimuli-sensitive nanodrug carriers may facilitate their interaction with CSCs. Therefore, it can be a worthy area for its exploration.

4.4.4 Development of Nanoparticle-Mediated Genome Engineering for CSC Targeting

As we discussed in Sect. 4.3.1, there are two major therapeutic nucleic acids used in CSC therapy, miRNA and siRNA. Both nucleic acids suppress those genes responsible for CSC survival via RNA interference at mRNA, protein, and functional level. The main drawback of such mechanism is to repress gene expression incompletely. In result, it may lead to the progression of several diseases, where complete ablation of gene functions is required for therapy. Also, RNA interference exhibits off-target effects, which pose a safety issue and reduce the efficacy of gene therapy (Mittal 2004; Jackson and Linsley 2010). However, the recent advances in gene-editing technology like CRISPR-CAS9 system could harness their potential in manipulation or removal of diseased genes with on-target effects. Also, the CRISPR-CAS9 technology has been used with its huge potential to study genomic rearrangements, analyze gene functions, and inactivate deleterious mutations, insertion of therapeutic transgenes, and introduction of protective genetic mutations for treating hereditary disorders (Cong et al. 2013; Ran et al. 2013). In the context of anti-CSC therapy, such RNA-guided CRISPR-CAS9 technology completely inhibited the expression of ABC transporters leading to drug accumulation inside CSCs and their killing (Qi et al. 2013). Also, CRISPR-CAS9 technology introduced the BMP4 gene inside the CD133+ hepatocellular carcinoma CSCs (Zhang et al. 2012c). As a result, CSCs underwent differentiation and lost their self-renewal capacity. The main drawback of gene-editing technology is its delivery as nucleic acids cannot reach easily to tumor core enriched with CSCs. Therefore, there is need of nanodrug delivery vehicles to harbor such gene-editing technology like what cationic lipid nanoparticles have shown (Zuris et al. 2014). It also appeared that CRISPR-CAS9 technology has been used at different preclinical and clinical platforms before entering into therapeutic pipeline. Still, there is concern going on related to its safety, efficacy and specificity. As gene-editing technology provides several possibilities to treat various diseases, therefore the advances in drug delivery technology will further increase its performance against several diseases.

4.5 Conclusions

The efficacy of chemotherapy or other therapeutic modalities is found to be reduced in the relapsed cancer patients. Therefore, the search and development of novel anticancer drugs to circumvent drug resistance and more effective treatment is of utmost importance. In this context, recent developments through nanotechnological advancements toward targeting CSCs along with conventional treatment could be the best strategy to overcome the resistance of anticancer drugs. However, the complexity and very limited understanding of tumor organization hamper the progress of nanotechnological approach in this direction. Therefore, the role of multidisciplinary fields is required to develop multidrug delivery systems that would be essential to improve the clinical translation of anticancer drugs in the near future.

References

Abbaszadegan MR, Bagheri V, Razavi MS, Momtazi AA, Sahebkar A et al (2017) Isolation, identification, and characterization of cancer stem cells: a review. J Cell Physiol 232(8):2008–2018

Al Faraj A, Shaik AS, Ratemi E, Halwani R (2016) Combination of drug-conjugated SWCNT nanocarriers for efficient therapy of cancer stem cells in a breast cancer animal model. J Control Release 225:240–251

Alibolandi M, Ramezani M, Sadeghi F, Abnous K, Hadizadeh F (2015) Epithelial cell adhesion molecule aptamer conjugated PEG–PLGA nanopolymersomes for targeted delivery of doxorubicin to human breast adenocarcinoma cell line in vitro. Int J Pharm 479(1):241–251

Alshaer W, Hillaireau H, Vergnaud J, Ismail S, Fattal E (2014) Functionalizing liposomes with anti-CD44 aptamer for selective targeting of cancer cells. Bioconjug Chem 26(7):1307–1313

Ang WX, Li Z, Chi Z, Du SH, Chen C, Tay JC et al (2017) Intraperitoneal immunotherapy with T cells stably and transiently expressing anti-EpCAM CAR in xenograft models of peritoneal carcinomatosis. Oncotarget 8(8):13545

Atena M, Reza AM, Mehran G (2014) A review on the biology of cancer stem cells. Stem Cell Discovery 4(04):83–89

Barenholz YC (2012) Doxil®—the first FDA-approved nano-drug: lessons learned. J Control Release 160(2):117–134

Burke AR, Singh RN, Carroll DL, Wood JC, D'Agostino RB, Ajayan PM et al (2012) The resistance of breast cancer stem cells to conventional hyperthermia and their sensitivity to nanoparticle-mediated photothermal therapy. Biomaterials 33(10):2961–2970

Burkhardt JK, Hofstetter CP, Santillan A, Shin BJ, Foley CP et al (2012) Orthotopic glioblastoma stem-like cell xenograft model in mice to evaluate intra-arterial delivery of bevacizumab: from bedside to bench. J Clin Neurosci 19(11):1568–1572

Cai J, Fang L, Huang Y, Li R, Xu X, Hu Z et al (2017) Simultaneous overactivation of Wnt/β--catenin and TGFβ signalling by miR-128-3p confers chemoresistance-associated metastasis in NSCLC. Nat Commun 8:15870

Chen ZG (2010) Small-molecule delivery by nanoparticles for anticancer therapy. Trends Mol Med 16(12):594–602

Chen K, Huang YH, Chen JL (2013) Understanding and targeting cancer stem cells: therapeutic implications and challenges. Acta Pharmacol Sin 34(6):732–740

Chenna V, Hu C, Pramanik D, Aftab BT, Karikari C, Campbell NR et al (2012) A polymeric nanoparticle encapsulated small-molecule inhibitor of Hedgehog signaling (NanoHHI) bypasses secondary mutational resistance to smoothened antagonists. Mol Cancer Ther 11(1):165–173

Clement V, Sanchez P, De Tribolet N, Radovanovic I, Altaba AR (2007) HEDGEHOG-GLI1 signaling regulates human glioma growth, cancer stem cell self-renewal, and tumorigenicity. Curr Biol 17(2):165–172

Cong L, Ran FA, Cox D, Lin S, Barretto R, Habib N et al (2013) Multiplex genome engineering using CRISPR/Cas systems. Science 339(6121):819–823

Correa S, Binato R, Du Rocher B, Castelo-Branco MT, Pizzatti L, Abdelhay E (2012) Wnt/β--catenin pathway regulates ABCB1 transcription in chronic myeloid leukemia. BMC Cancer 12(1):303

Cruz LJ, Tacken PJ, Rueda F, Carles Domingo J, Albericio F, Figdor CG (2012) 8 targeting nanoparticles to dendritic cells for immunotherapy. Methods Enzymol 509:143e163

Cui J, Jiang W, Wang S, Wang L, Xie K (2012) Role of Wnt/β-catenin signaling in drug resistance of pancreatic cancer. Curr Pharm Des 18(17):2464–2471

Davis ME, Chen ZG, Shin DM (2008) Nanoparticle therapeutics: an emerging treatment modality for cancer. Nat Rev Drug Discov 7(9):771–782

Dean M, Fojo T, Bates S (2005) Tumour stem cells and drug resistance. Nat Rev Cancer 5(4):275–284

Deng Y, Pu X, Huang M, Xiao J, Zhou J, Lin T, Lin EH (2010) 5-Fluorouracil upregulates the activity of Wnt signaling pathway in CD133-positive colon cancer stem-like cells. Chin J Cancer 29(9):810–815

Deng Z, Wu Y, Ma W, Zhang S, Zhang YQ (2015) Adoptive T-cell therapy of prostate cancer targeting the cancer stem cell antigen EpCAM. BMC Immunol 16(1):1

Domenech M, Marrero-Berrios I, Torres-Lugo M, Rinaldi C (2013) Lysosomal membrane permeabilization by targeted magnetic nanoparticles in alternating magnetic fields. ACS Nano 7(6):5091–5101

Dragu DL, Necula LG, Bleotu C, Diaconu CC, Chivu-Economescu M (2015) Therapies targeting cancer stem cells: current trends and future challenges. World J Stem Cells 7(9):1185–1201

Duarte S, Momier D, Baqué P, Casanova V, Loubat A, Samson M et al (2013) Preventive Cancer stem cell-based vaccination reduces liver metastasis development in a rat Colon carcinoma syngeneic model. Stem Cells 31(3):423–432

Feldmann G, Dhara S, Fendrich V, Bedja D, Beaty R et al (2007) Blockade of hedgehog signaling inhibits pancreatic cancer invasion and metastases: a new paradigm for combination therapy in solid cancers. Cancer Res 67(5):2187–2196

Feldmann G, Fendrich V, McGovern K, Bedja D, Bisht S et al (2008) An orally bioavailable small-molecule inhibitor of hedgehog signaling inhibits tumor initiation and metastasis in pancreatic cancer. Mol Cancer Ther 7(9):2725–2735

Fischer M, Yen WC, Kapoun AM, Wang M, O'Young G et al (2011) Anti-DLL4 inhibits growth and reduces tumor-initiating cell frequency in colorectal tumors with oncogenic KRAS mutations. Cancer Res 71(5):1520–1525

Folkins C, Man S, Xu P, Shaked Y, Hicklin DJ, Kerbel RS (2007) Anticancer therapies combining antiangiogenic and tumor cell cytotoxic effects reduce the tumor stem-like cell fraction in glioma xenograft tumors. Cancer Res 67(8):3560–3564

Ganesh S, Iyer AK, Morrissey DV, Amiji MM (2013) Hyaluronic acid based self- assembling nanosystems for CD44 target mediated siRNA delivery to solid tumors. Biomaterials 34(13):3489–3502

Gao X, Luo Y, Wang Y, Pang J, Liao C, Lu H, Fang Y (2012) Prostate stem cell antigen-targeted nanoparticles with dual functional properties: in vivo imaging and cancer chemotherapy. Int J Nanomedicine 7:4037–4051

Gener P, Gouveia LP, Sabat GR, de Sousa Rafael DF, Fort NB, Arranja A et al (2015) Fluorescent CSC models evidence that targeted nanomedicines improve treatment sensitivity of breast and colon cancer stem cells. Nanomedicine 11(8):1883–1892

Gilboa-Geffen A, Hamar P, Le M, Wheeler LA, Trifonova R, Petrocca F et al (2015) Gene knockdown by EpCAM aptamer-siRNA chimeras suppresses epithelial breast cancers and their tumor-initiating cells. Mol Cancer Ther, Molcanther-0201

Goodison S, Urquidi V, Tarin D (1999) CD44 cell adhesion molecules. Mol Pathol 52(4):189–196

Gupta PB, Chaffer CL, Weinberg RA (2009) Cancer stem cells: mirage or reality? Nat Med 15(9):1010–1012

Huang J, Li C, Wang Y, Lv H, Guo Y et al (2013) Cytokine-induced killer (CIK) cells bound with anti-CD3/anti-CD133 bispecific antibodies target CD133 high cancer stem cells in vitro and in vivo. Clin Immunol 149(1):156–168

Huang X, Kang X, Zhao M (2017) A forecast of targeting leukemia stem cells by Nanomedicine. J Stem Cell Res Ther 7(385):2

Jackson AL, Linsley PS (2010) Recognizing and avoiding siRNA off-target effects for target identification and therapeutic application. Nat Rev Drug Discov 9(1):57–67

Jin C, Yang Z, Yang J, Li H, He Y, An J et al (2014) Paclitaxel-loaded nanoparticles decorated with anti-CD133 antibody: a targeted therapy for liver cancer stem cells. J Nanopart Res 16(1):2157

Kale AA, Torchilin VP (2010) Environment-responsive multifunctional liposomes. Liposomes Methods Protocols 1:213–242

Kaluzova M, Bouras A, Machaidze R, Hadjipanayis CG (2015) Targeted therapy of glioblastoma stem-like cells and tumor non-stem cells using cetuximab-conjugated iron-oxide nanoparticles. Oncotarget 6(11):8788

Kanwar JR, Mahidhara G, Roy K, Sasidharan S, Krishnakumar S, Prasad N et al (2015) Fe-bLf nanoformulation targets survivin to kill colon cancer stem cells and maintains absorption of iron, calcium and zinc. Nanomedicine 10(1):35–55

Keith B, Simon MC (2007) Hypoxia-inducible factors, stem cells, and cancer. Cell 129(3):465–472

Kim SS, Rait A, Kim E, Pirollo KF, Nishida M et al (2014) A nanoparticle carrying the p53 gene targets tumors including cancer stem cells, sensitizes glioblastoma to chemotherapy and improves survival. ACS Nano 8(6):5494–5514

Knop K, Hoogenboom R, Fischer D, Schubert US (2010) Poly (ethylene glycol) in drug delivery: pros and cons as well as potential alternatives. Angew Chem Int Ed 49(36):6288–6308

Kobayashi CI, Suda T (2012) Regulation of reactive oxygen species in stem cells and cancer stem cells. J Cell Physiol 227(2):421–430

Kouri FM, Hurley LA, Daniel WL, Day ES, Hua Y et al (2015) miR-182 integrates apoptosis, growth, and differentiation programs in glioblastoma. Genes Dev 29(7):732–745

Krause M, Dubrovska A, Linge A, Baumann M (2016) Cancer stem cells: Radioresistance, prediction of radiotherapy outcome and specific targets for combined treatments. Adv Drug Deliv Rev:30052–30057

Kwiatkowska-Borowczyk EP, Gąbka-Buszek A, Jankowski J, Mackiewicz A (2015) Immunotargeting of cancer stem cells. Contemp Oncol 19(1A):A52
Lee H, Fonge H, Hoang B, Reilly RM, Allen C (2010) The effects of particle size and molecular targeting on the intratumoral and subcellular distribution of polymeric nanoparticles. Mol Pharm 7(4):1195–1208
Leuschner F, Dutta P, Gorbatov R, Novobrantseva TI, Donahoe JS, Courties G et al (2011) Therapeutic siRNA silencing in inflammatory monocytes in mice. Nat Biotechnol 29(11):1005–1010
Li L, Xiang D, Shigdar S, Yang W, Li Q, Lin J et al (2014) Epithelial cell adhesion molecule aptamer functionalized PLGA-lecithin-curcumin-PEG nanoparticles for targeted drug delivery to human colorectal adenocarcinoma cells. Int J Nanomedicine 9:1083
Lim KJ, Bisht S, Bar EE, Maitra A, Eberhart CG (2011) A polymeric nanoparticle formulation of curcumin inhibits growth, clonogenicity and stem-like fraction in malignant brain tumors. Cancer Biol Therapy 11(5):464–473
Liu C, Zhao G, Liu J, Ma N, Chivukula P et al (2009) Novel biodegradable lipid nano complex for siRNA delivery significantly improving the chemosensitivity of human colon cancer stem cells to paclitaxel. J Control Release 140(3):277–283
Liu T, Cheng W, Lai D, Huang Y, Guo L (2010) Characterization of primary ovarian cancer cells in different culture systems. Oncol Rep 23(5):1277–1284
Liu C, Kelnar K, Liu B, Chen X, Calhoun-Davis T, Li H et al (2011) The microRNA miR-34a inhibits prostate cancer stem cells and metastasis by directly repressing CD44. Nat Med 17(2):211–215
Liu S, Cong Y, Wang D, Sun Y, Deng L, Liu Y et al (2014) Breast cancer stem cells transition between epithelial and mesenchymal states reflective of their normal counterparts. Stem Cell Rep 2(1):78–91
Liu J, Meng T, Yuan M, Wen L, Cheng B et al (2016) MicroRNA-200c delivered by solid lipid nanoparticles enhances the effect of paclitaxel on breast cancer stem cell. Int J Nanomedicine 11:6713–6725
Ma S, Lee TK, Zheng BJ, Chan KW, Guan XY (2008) CD133+ HCC cancer stem cells confer chemoresistance by preferential expression of the Akt/PKB survival pathway. Oncogene 27(12):1749–1758
Malladi S, Macalinao DG, Jin X, He L, Basnet H, Zou Y et al (2016) Metastatic latency and immune evasion through autocrine inhibition of WNT. Cell 165(1):45–60
Marangoni E, Lecomte N, Durand L, De Pinieux G, Decaudin D, Chomienne C et al (2009) CD44 targeting reduces tumour growth and prevents post-chemotherapy relapse of human breast cancers xenografts. Br J Cancer 100(6):918–922
Markovsky E, Vax E, Ben-Shushan D, Eldar-Boock A, Shukrun R, Yeini E et al (2017) Wilms tumor NCAM-expressing Cancer stem cells as potential therapeutic target for polymeric Nanomedicine. Mol Cancer Ther 16(11):2462–2472
Matsui WH (2016) Cancer stem cell signaling pathways. Medicine 95:S8–S19
McGowan PM, Simedrea C, Ribot EJ, Foster PJ, Palmieri D et al (2011) Notch1 inhibition alters the CD44hi/CD24lo population and reduces the formation of brain metastases from breast cancer. Mol Cancer Res 9(7):834–844
McNeeley KM, Annapragada A, Bellamkonda RV (2007) Decreased circulation time offsets increased efficacy of PEGylated nanocarriers targeting folate receptors of glioma. Nanotechnology 18(38):385101
Mishra S, Webster P, Davis ME (2004) PEGylation significantly affects cellular uptake and intracellular trafficking of non-viral gene delivery particles. Eur J Cell Biol 83(3):97–111
Mittal V (2004) Improving the efficiency of RNA interference in mammals. Nat Rev Genet 5:355–365
Nguyen DN, Mahon KP, Chikh G, Kim P, Chung H, Vicari AP et al (2012) Lipid-derived nanoparticles for immunostimulatory RNA adjuvant delivery. Proc Natl Acad Sci 109(14):E797–E803
Niero EL, Rocha-Sales B, Lauand C, Cortez BA, de Souza MM et al (2014) The multiple facets of drug resistance: one history, different approaches. J Exp Clin Cancer Res 33(1):33–37

Ning N, Pan Q, Zheng F, Teitz-Tennenbaum S, Egenti M, Yet J et al (2012) Cancer stem cell vaccination confers significant antitumor immunity. Cancer Res 72(7):1853–1864

Ning ST, Lee SY, Wei MF, Peng CL, Lin SYF, Tsai MH et al (2016) Targeting colorectal Cancer stem-like cells with anti-CD133 antibody-conjugated SN-38 nanoparticles. ACS Appl Mater Interfaces 8(28):17793–17804

Pastor F, Kolonias D, McNamara JO II, Gilboa E (2011) Targeting 4-1BB costimulation to disseminated tumor lesions with bi-specific oligonucleotide aptamers. Mol Ther 19(10):1878–1886

Polakis P (2012) Wnt signaling in cancer. Cold Spring Harb Perspect Biol 4(5):a008052

Qi LS, Larson MH, Gilbert LA, Doudna JA, Weissman JS, Arkin AP et al (2013) Repurposing CRISPR as an RNA-guided platform for sequence-specific control of gene expression. Cell 152(5):1173–1183

Ran FA, Hsu PD, Wright J, Agarwala V, Scott DA, Zhang F (2013) Genome engineering using the CRISPR-Cas9 system. Nat Protoc 8(11):2281–2308

Rao W, Bellotti A, Littrup PJ, Yu J, Lu X, He X (2014) Nanoparticle-encapsulated doxorubicin enhances cryoablation of cancer stem-like cells. Technology 2(01):28–35

Rink JS, Plebanek MP, Tripathy S, Thaxton CS (2013) Update on current and potential nanoparticle cancer therapies. Curr Opin Oncol 25(6):646–651

Sadhukha T, Niu L, Wiedmann TS, Panyam J (2013) Effective elimination of cancer stem cells by magnetic hyperthermia. Mol Pharm 10(4):1432–1441

Schieber M, Chandel NS (2014) ROS function in redox signaling and oxidative stress. Curr Biol 24(10):R453–R462

Shen S, Xia JX, Wang J (2016) Nanomedicine-mediated cancer stem cell therapy. Biomaterials 74:1–18

Shi S, Han L, Gong T, Zhang Z, Sun X (2013) Systemic delivery of microRNA-34a for cancer stem cell therapy. Angew Chem Int Ed 52(14):3901–3905

Shima F, Akagi T, Uto T, Akashi M (2013) Manipulating the antigen-specific immune response by the hydrophobicity of amphiphilic poly (γ-glutamic acid) nanoparticles. Biomaterials 34(37):9709–9716

Singh A, Settleman JEMT (2010) EMT, cancer stem cells and drug resistance: an emerging axis of evil in the war on cancer. Oncogene 29(34):4741–4751

Skubitz AP, Taras EP, Boylan KL, Waldron NN, Oh S et al (2013) Targeting CD133 in an in vivo ovarian cancer model reduces ovarian cancer progression. Gynecol Oncol 130(3):579–587

Smith DC, Eisenberg PD, Manikhas G, Chugh R, Gubens MA, Stagg RJ et al (2014) A phase I dose escalation and expansion study of the anticancer stem cell agent demcizumab (anti-DLL4) in patients with previously treated solid tumors. Clin Cancer Res 20(24):6295–6303

Stratford EW, Bostad M, Castro R, Skarpen E, Berg K, Høgset A et al (2013) Photochemical internalization of CD133-targeting immunotoxins efficiently depletes sarcoma cells with stem-like properties and reduces tumorigenicity. Biochim Biophys Acta (BBA)- Gen Subj 1830(8):4235–4243

Sun T, Zhang YS, Pang B, Hyun DC, Yang M, Xia Y (2014a) Engineered nanoparticles for drug delivery in cancer therapy. Angew Chem Int Ed 53(46):12320–12364

Sun H, Zhu X, Lu PY, Rosato RR, Tan W, Zu Y (2014b) Oligonucleotide aptamers: new tools for targeted cancer therapy. Mol Ther-Nucleic Acids 3:e182

Sun TM, Wang YC, Wang F, Du JZ, Mao CQ, Sun CY et al (2014c) Cancer stem cell therapy using doxorubicin conjugated to gold nanoparticles via hydrazone bonds. Biomaterials 35(2):836–845

Sun Q, Sun X, Ma X, Zhou Z, Jin E, Zhang B et al (2014d) Integration of nanoassembly functions for an effective delivery cascade for cancer drugs. Adv Mater 26(45):7615–7621

Sun R, Liu Y, Li SY, Shen S, Du XJ, Xu CF et al (2015) Co-delivery of all-trans retinoic acid and doxorubicin for cancer therapy with synergistic inhibition of cancer stem cells. Biomaterials 37:405–414

Swaminathan SK, Roger E, Toti U, Niu L, Ohlfest JR, Panyam J (2013) CD133- targeted paclitaxel delivery inhibits local tumor recurrence in a mouse model of breast cancer. J Control Release 171(3):280–287

Takebe N, Miele L, Harris PJ, Jeong W, Bando H, Kahn et al (2015) Targeting Notch, Hedgehog, and Wnt pathways in cancer stem cells: clinical update. *Nature Reviews Clinical Oncology 12*(8):445–464

Toda M (2013) Glioma stem cells and immunotherapy for the treatment of malignant gliomas. ISRN Oncology, 2013. 673793

Trachootham D, Alexandre J, Huang P (2009) Targeting cancer cells by ROS- mediated mechanisms: a radical therapeutic approach? Nat Rev Drug Discov 8(7):579–591

Vermeulen L, Felipe De Sousa EM, Van Der Heijden M, Cameron K, De Jong JH et al (2010) Wnt activity defines colon cancer stem cells and is regulated by the microenvironment. Nat Cell Biol 12(5):468–476

Villegas VE, Rondon-Lagos M, Annaratone L, Castellano I, Grismaldo A, Sapino A et al (2016) Tamoxifen treatment of breast cancer cells: impact on hedgehog/GLI1 signaling. Int J Mol Sci 17(3):308

Vinogradov S, Wei X (2012) Cancer stem cells and drug resistance: the potential of nanomedicine. Nanomedicine 7(4):597–615

Wang CH, Chiou SH, Chou CP, Chen YC, Huang YJ, Peng CA (2011) Photothermolysis of glioblastoma stem-like cells targeted by carbon nanotubes conjugated with CD133 monoclonal antibody. Nanomedicine 7(1):69–79

Wang J, Sullenger BA, Rich JN (2012a) Notch signaling in cancer stem cells. Springer, (Chapter 13)

Wang L, Su W, Liu Z, Zhou M, Chen S, Chen Y et al (2012b) CD44 antibody-targeted liposomal nanoparticles for molecular imaging and therapy of hepatocellular carcinoma. Biomaterials 33(20):5107–5114

Wang J, Sefah K, Altman MB, Chen T, You M, Zhao Z et al (2013) Aptamer-conjugated Nanorods for targeted Photothermal therapy of prostate Cancer stem cells. Chem Asian J 8(10):2417–2422

Wang X, Jung YS, Jun S, Lee S, Wang W, Schneider A et al (2016) PAF-Wnt signaling-induced cell plasticity is required for maintenance of breast cancer cell stemness. Nat Commun 7:10633

Wei X, Senanayake TH, Warren G, Vinogradov SV (2013) Hyaluronic acid-based nanogel–drug conjugates with enhanced anticancer activity designed for the targeting of CD44- positive and drug-resistant tumors. Bioconjug Chem 24(4):658–668

Wu Y, Crawford M, Yu B, Mao Y, Nana-Sinkam SP, Lee LJ (2011) MicroRNA delivery by cationic lipoplexes for lung cancer therapy. Mol Pharm 8(4):1381–1389

Xia P (2014) Surface markers of cancer stem cells in solid tumors. Curr Stem Cell Res Ther 9(2):102–111

Xiang D, Zheng C, Zhou SF, Qiao S, Tran PHL, Pu C et al (2015) Superior performance of aptamer in tumor penetration over antibody: implication of aptamer-based theranostics in solid tumors. Theranostics 5(10):1083

Xu Y, Chenna V, Hu C, Sun HX, Khan M, Bai H et al (2011) A polymeric nanoparticle encapsulated hedgehog pathway inhibitor HPI-1 (NanoHHI) inhibits systemic metastases in an orthotopic model of human hepatocellular carcinoma. Clin Cancer Res, clincanres-0950

Xu Y, Wang J, Li X, Liu Y, Dai L, Wu X, Chen C (2014) Selective inhibition of breast cancer stem cells by gold nanorods mediated plasmonic hyperthermia. Biomaterials 35(16):4667–4677

Xu ZY, Tang JN, Xie HX, Du YA, Huang L, Yu PF, Cheng XD (2015a) 5-Fluorouracil chemotherapy of gastric cancer generates residual cells with properties of cancer stem cells. Int J Biol Sci 11(3):284–294

Xu CF, Liu Y, Shen S, Zhu YH, Wang J (2015b) Targeting glucose uptake with siRNA-based nanomedicine for cancer therapy. Biomaterials 51:1–11

Yang C, Xiong F, Dou J, Xue J, Zhan X, Shi F et al (2015) Target therapy of multiple myeloma by PTX-NPs and ABCG2 antibody in a mouse xenograft model. Oncotarget 6(29):27714

Zhang Y, Zhang H, Wang X, Wang J, Zhang X, Zhang Q (2012a) The eradication of breast cancer and cancer stem cells using octreotide modified paclitaxel active targeting micelles and salinomycin passive targeting micelles. Biomaterials 33(2):679–691

Zhang L, Yao HJ, Yu Y, Zhang Y, Li RJ, Ju RJ et al (2012b) Mitochondrial targeting liposomes incorporating daunorubicin and quinacrine for treatment of relapsed breast cancer arising from cancer stem cells. Biomaterials 33(2):565–582

Zhang L, Sun H, Zhao F, Lu P, Ge C, Li H et al (2012c) BMP4 administration induces differentiation of CD133+ hepatic cancer stem cells, blocking their contributions to hepatocellular carcinoma. Cancer Res 72(16):4276–4285

Zhao C, Chen A, Jamieson CH, Fereshteh M, Abrahamsson A et al (2009) Hedgehog signalling is essential for maintenance of cancer stem cells in myeloid leukaemia. Nature 458(7239):776–779

Zhao Y, Alakhova DY, Kabanov AV (2013a) Can nanomedicines kill cancer stem cells? Adv Drug Deliv Rev 65(13):1763–1783

Zhao N, You J, Zeng Z, Li C, Zu Y (2013b) An ultra pH-sensitive and Aptamer- equipped Nanoscale drug-delivery system for selective killing of tumor cells. Small 9(20):3477–3484

Zhou Y, Yang J, Kopeček J (2012) Selective inhibitory effect of HPMA copolymer- cyclopamine conjugate on prostate cancer stem cells. Biomaterials 33(6):1863–1872

Zhu X, Prasad S, Gaedicke S, Hettich M, Firat E, Niedermann G (2015) Patient- derived glioblastoma stem cells are killed by CD133-specific CAR T cells but induce the T cell aging marker CD57. Oncotarget 6(1):171–184

Zuo ZQ, Chen KG, Yu XY, Zhao G, Shen S et al (2016) Promoting tumor penetration of nanoparticles for cancer stem cell therapy by TGF-β signaling pathway inhibition. Biomaterials 82:48–59

Zuris ZA, Thompson DB, Shu Y, Guilinger JP, Bessen JL, Hu JH et al (2014) Cationic lipid-mediated delivery of proteins enables efficient protein-based genome editing *in vitro* and *in vivo*. Nat Biotechnol 33:73–80

Anticancerous Activity of Transition Metal Oxide Nanoparticles

5

Debasree Ghosh, Sudip Majumder, and Pratibha Sharma

Abstract

Transition metal oxide (TMO) nanoparticles are becoming a major thrust area of research due to their potential anticancer activity. Their variable d-shell configurations, limited size, and a high density of corner or edge surface sites render them to exhibit unique physical and chemical properties. The TMO nanoparticles have shown excellence in various fields including catalysis, biomedical, solar cell, and lithium-ion batteries. Nanoparticles of copper oxide, iron oxide, nickel oxide, and zinc oxide have been shown to exhibit potential cytotoxic effects against some human cancer cell lines. They have been also found to exhibit reactive oxygen species-mediated cell death in some of these cell lines.

Keywords

Transition metal oxide nanoparticles · Cytotoxicity · Reactive oxygen species · Anticancer activity · Apoptosis · Surface modification

5.1 Introduction: Basics of Transition Metal Oxide Nanoparticle

Transition metals possess an important place in medicinal chemistry. They occupy the groups 3–12 in the periodic table. These metals do possess incomplete electronic configuration in their outermost d shells both in their ground state and in first excited state; by virtue of this property, transition metals form coordination complexes with higher coordination number. History and development of metal-based drugs are primarily based on transition metal systems, considering their unique

D. Ghosh · S. Majumder (✉) · P. Sharma
Department of Chemistry, Amity School of Applied Sciences, Amity University Haryana, Gurugram, Haryana, India

S. K. Saxena, S. M. P. Khurana (eds.), *NanoBioMedicine*,
https://doi.org/10.1007/978-981-32-9898-9_5

therapeutic potential and ability to act as lead pharmacophore (Rafique et al. 2010). Among the transition metals, transition metal oxides are playing an important role in various fields of natural sciences including physics, materials science, chemistry, and nano-science. They have varying electronic configurations due to which they can exhibit metallic, semiconductor, or insulator character. In industrial application, oxides are used as fabricating material, sensors, piezoelectric devices, fuel cells, coatings for the passivation of surface against corrosion, and a catalyst (Mishra et al. 2017).

Nanotechnology is a rapidly developing field, and nanoparticles (NPs having dimensions less than 100 nm with their unique physiochemical properties) (Panariti et al. 2012) are becoming progressively relevant in various sectors like biomedical (Saptarshi et al. 2013), electronic, industrial, cosmetic, paints, food additives, etc. (Arivalagan et al. 2011). Metal oxide (MeOx) nanoparticles are extremely important for commercial purpose as they constitute a huge role in modern-day lab-on-chip technology that includes uses in diagnostic devices, robotics, nanochips and nanomotors, bacterial sensors, navigation instruments, photovoltaic and solar cells, etc. (Rodríguez and Fernández-García 2007; Fernández-García et al. 2004). The scope of functionalization and fabrication of these transition metal oxide nanoparticles render an opportunity to use them for specific applications like targeted gene therapy and improved bioavailability of pharmaceutical drugs. Transition metal oxide nanoparticles (TMO) are very useful in imaging devices as they are capable of forming excellent contrast and sensitivity (Falcaro et al. 2015). TMOs present novel physicochemical properties, especially due to their relatively small size and appreciably large surface area. Three important material properties significantly depend on individual particle size. First of them is the lattice symmetry and cell parameters (Ayyub et al. 1995). Bulk metal oxides usually possess robust structure with well-defined cell dimensional parameters. However, a number of physical parameters like surface energy and molecular stress vary with decrease in particle size, and variation in the thermodynamic stability of a nanoparticle can cause significant structural transformation to its cell parameters. In order to satisfy the varying demands of surface energy with decrease in particle size, certain unstable phases in bulk material can acquire significant stability in their nano-form. This structural phenomenon has been detected in TiO_2, VOx, Al_2O3, or MoOx oxides (Millot et al. 2003; McHale et al. 1997; Zhang and Bandfield 1998).

In nanotechnology, there is a goal to make nanostructures or nanoarrays with special properties with respect to those of bulk or single particle species. Due to their size and edge surface area, metal oxide nanoparticles can display novel physicochemical properties. Thus there are oxides with metallic conduction properties like RuO_2, ReO_3, and $LaNiO_3$; similarly there are oxides that have insulating behavior like $BaTiO_3$. Also some metal oxides vary their regimes with the change of temperature, pressure, or composition like V_2O_3 and $La_{(1-x)}$ $Sr_{(x)}$ VO_3. Transition metal oxide nanoparticles are the emerging nanomaterials that have shown excellence in various fields including catalysis, biomedical, solar cell, and lithium-ion battery. TMOs present a potent class of futuristic functional materials due to their relative environmentally friendly properties, cost-effectiveness, and robust physiochemical entity.

The science and developmental application of nanodimensional material is often regarded as nanotechnology. These groups of materials are extensively studied in recent years due to their unique variation in physicochemical properties with decreasing size, which renders larger surface area and higher adsorption quality (Huang et al. 2010).

5.2 Applications of Transition Metal Oxide Nanoparticle

Transition metal oxide is a compound having an oxygen bounded to transition metals. In transition metals, titanium dioxide is used in plastics, and transition metal oxides are also used in paints as pigments.

They have a variety of surface structures which affect the surface energy of these compounds and influence their chemical properties. Transition metal oxides have various properties like optical, mechanical, transport, and chemical, due to which they are mostly preferable in various fields.

Transition metal oxide nanoparticles have a number of therapeutic applications. Wanz et al. have shown effective antimicrobial activities of electrosprayed NiO, ZnO, and CuO nanoparticles against *E. coli* (Wang et al. 2010). Both cuprous oxide and cupric oxide nanoparticles have shown good antimicrobial and fungicidal activities. Copper oxides have been used against a number of hospital infections. CuO nanoparticles have shown good antimicrobial activities against both gram-negative (*E. coli* and *P. aeruginosa*) and gram-positive bacteria (*B. subtilis* and *S. aureus*) (Azam et al. 2012) Awady et al. have showed the antimicrobial activities of TiO_2 nanoparticles toward *C. reinhardtii* and *S. cerevisiae* (Al-Awady et al. 2015). They are used as anti-inflammatory agents, anti-infective agents, antidiabetic agents, and neurological drugs. They are being used nowadays as effective anticancer agents.

5.3 Synthesis of Transition Metal Oxide Nanoparticle

A wide variety of methods have been reported for synthesis of transition metal oxide nanoparticles, including coprecipitation (Lien et al. 2007), sol-gel method (Xu et al. 2007), flow injection (Alvarez et al. 2006), electrochemical (Cabrera et al. 2008), solvothermal (Liu and Kim 2009), hydrothermal (Zhang et al. 2008a), microwave-assisted, sonochemical, etc. (Hong et al. 2006). Among all these methods, coprecipitation technique using precursor salts and precipitating agents is extremely popular due to ease of synthesis, high yield, short time span, and cost-effectiveness. However the challenge with the synthesis remains in controlling the size distribution of nanoparticles without which it is impossible to get a homogenized population. This problem is even more pronounced for transition metal oxides as they are mostly magnetic in nature and initiate spontaneous agglomeration. To overcome this problem, scientists have used physiochemical techniques like calcination or annealing at higher temperature, prolonged time of sonication, etc. However the most common method is use of capping agents or stabilizers which prevents agglomeration and helps to synthesize the nanoparticle in homogeneous population with a narrow size distribution. Most common capping agents are citrate (Naeimi and Nazifi 2013), tannic acid (Arora 2017), polyethylene glycol, (Cahyana et al. 2017) etc. With the advent of technology, every day new techniques are coming for better yield of higher-quality nanoparticles. Recently self-propagating room temperature (SPRT) method has been used to synthesize nanosized cerium oxide powder as reported by (Arsalani 2010) (Fig. 5.1).

However, recently green synthesis of nanoparticles has become extremely popular due to its eco-friendly nature. In this method, different plant extracts are utilized as a reducing agent to synthesize desired nanoparticle by bioreduction of precursor salts. The mechanism of these reductions is yet to be fully understood, although it has been speculated that the nonspecific proteins and polyphenols in plant extracts might play an important part in this process. Synthesis of transition metal oxide nanoparticles using plant extracts has also gained a serious momentum in the last decade.

Sangeetha et al. (Sangeetha et al. 2012) reported synthesis of monodispersed and highly stable cupric oxide (CuO) nanoparticles from *Aloe vera* extract. This method was found to be both eco-friendly and inexpensive. Naika et al. have reported the synthesis of CuO from the leaf extract of *Gloriosa superba* (Naika et al. 2015). Very recently Gultekin et al. have reported biosynthesis of CuO nanoparticle from *Vitis vinifera* (cimin grape) extract (Gultekin et al. 2017). Behra et al. have reported synthesis of cuprous oxide (Cu_2O) nanoparticle in the presence of *Calotropis gigantea*

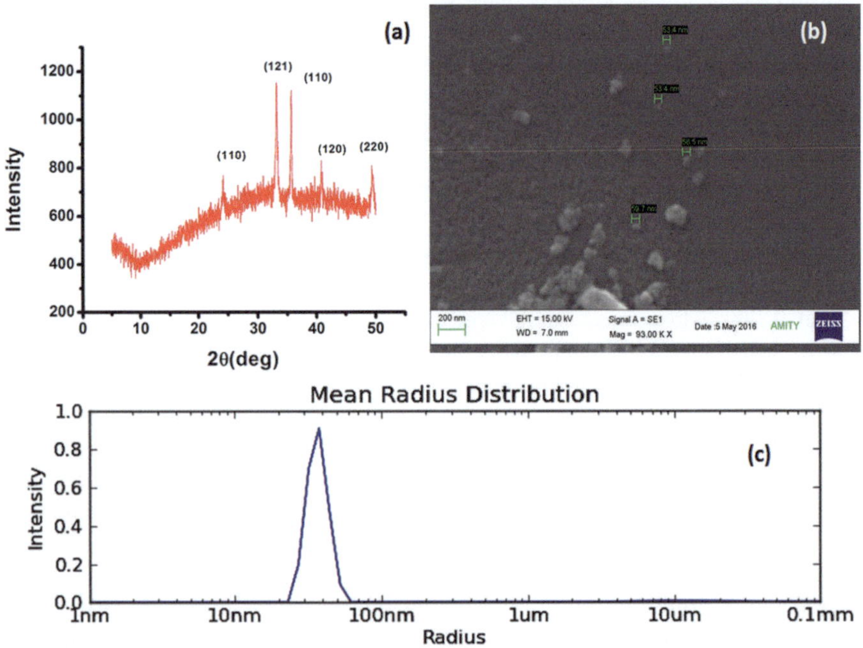

Fig. 5.1 (**a**) XRD, (**b**) SEM micrograph, and (**c**) DLS pattern of Fe_2O_3 nanoparticle

plant extract as a biosurfactant (Behera and Giri 2014). Recently Kaur et al. have reported a green synthesis of Cu_2O nanoparticle from $CuSO_4$ using peel extract of *Punica granatum* as natural reducing agent (Kaur et al. 2016). Muthukumar and Matheswaran (Narayanan and Sakthivel 2010) obtained magnetite (Fe_3O_4) nanoparticles using *Amaranthus spinosus* leaf aqueous extracts. These nanoparticles were having definite spherical shape, monodispersed population, less aggregation tendency, and smaller size compared to the Fe_3O_4 nanoparticles synthesized via chemical route using sodium borohydride (Muthukumar and Matheswaran 2015). In another study, Fe_3O_4 nanoparticles were synthesized by hydrothermal method using *Aloe vera* plant extract (Phumying et al. 2013). Senthil and Ramesh (Senthil and Ramesh 2012) synthesized Fe_3O_4 nanoparticles by reducing precursor ferric chloride solution with *Tridax procumbens* leaf extract where the aldehydic carbohydrate components of leaf extract probably served as the reducing agent. The nanoparticles exhibited good antibacterial activity against *Pseudomonas aeruginosa*. Mukherjee et al. have reported green synthesis of ά-Fe_2O_3 nanoparticles from *Aloe vera* plant extract (Mukherjee and Ghosh 2014). Similarly in a recent report, Khalil et al. have described the facile synthesis of ά-Fe_2O_3 nanoparticles from aqueous extracts of *Sageretia thea* (Osbeck) (Khalil et al. 2017a). Herrera-Becerra et al. (Herrera-Becerra et al. 2010) reported successful use of polyphenolic tannin compound as a stabilizer and anti-aggregation agent for the synthesis of Fe_2O_3 nanoparticles with less than 10 nm size.

Helan et al. (Helana et al. 2016) have reported a green synthesis of NiO nanoparticle mediated by the leaf extract of *Azadirachta indica.* Ezhilarasi et al. (Ezhilarasi et al. 2016a) have reported the synthesis of NiO nanoparticle using *Moringa oleifera* extract; they had also reported its cytotoxic potential against HT-29 cancer cells. Nasseri et al. (Nasseri et al. 2016) have reported *a* green synthesis of NiO nanoparticles using aqueous extract of *Tamarix serotina* (Fig. 5.2).

Santhoshkumar et al. (2017) have described effective synthesis *of* ZnO nanoparticle from *Passiflora caerulea* leaf extract; these nanoparticles showed good results against urinary tract infection. Shah et al. (2015) have reported a green route of

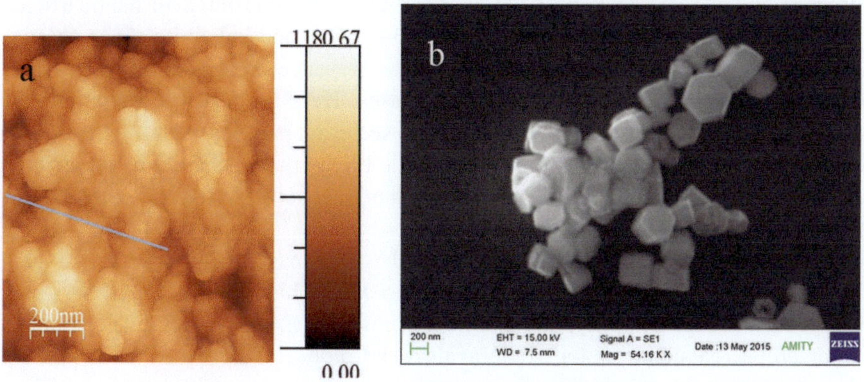

Fig. 5.2 (**a**) AFM picture and (**b**) SEM micrograph of ZnO nanoparticle

synthesizing ZnO nanoparticles using the leaf extract of *Camellia sinensis*. Ramesh et al. (2014) have reported a green synthesis method for obtaining zinc oxide nanoparticles using flower extract of *Cassia auriculata.* Santoshkumar et al. (2014) have reported the green synthesis of TiO_2 nanoparticles from *Psidium guajava* extract having excellent antioxidant activity and antibacterial activity against *Staphylococcus aureus* and *Escherichia coli.* Aqueous leaf extracts of *Trigonella foenum-graecum* have been used for the synthesis of TiO_2 nanoparticles by Subhapriya and Gomathipriya (2018). *Aloe vera* leaf extracts have been used by Rao et al. for synthesizing TiO_2 nanoparticles (Rao et al. 2015). Zahir et al. (2015) have synthesized TiO_2 nanoparticles from *Euphorbia prostrata* leaf extract. *Curcuma longa* plant extracts have been utilized in the synthesis of TiO_2 nanoparticles by Raghad et al. (2016). TiO_2 nanoparticles have been synthesized by Chatterjee et al. (2017) from the legumes of *Vigna radiata* which have been found to have effective cytotoxic activity against Mg-63 osteosarcoma cell lines. At IC50 value of 200 ug/ml, the osteosarcoma cell proliferation was inhibited at a significant rate.

Arumugam et al., Kannan et al., Priya et al., Thovhogi et al., and Maqbool et al. have reported green synthesis method for obtaining CeO_2 nanoparticles using *Gloriosa superba*, *Acalypha indica*, *Aloe barbadensis*, *Hibiscus sabdariffa*, and *Olea europaea*, respectively (Charbgoo et al. 2017). Leaf extracts of *Prosopis juliflora* have also been used as a reducing and stabilizing agent in the ultrasound-assisted biosynthesis of CeO_2 nanoparticles by Arunachalam et al. (2017). Debanjan et al. have reported the green synthesis of CeO_2 nanoparticles from *Aloe vera* extracts which have been found to be effective antioxidant against oxidative stress-induced damage in rat neuroblastoma cells (Dutta et al. 2016).

Apart from plant sources, bacteria, fungi, biomass, etc. have also been used for synthesizing nanoparticles. The biogenic synthesis of copper oxides was performed using *Penicillium aurantiogriseum*, *P. citrinum*, and *P. waksmanii* isolated from soil (Honary et al. 2012). The effect of pH and soil composition on the size of synthesized nanoparticle was also examined by the researchers who reported the production of spherical shaped CuO nanoparticles (Singh et al. 2010). Sing et al. used *E. coli* for synthesizing CuO nanoparticles which resulted in nanoparticles with an inhomogeneous size distribution (10–40 nm) and indefinite shape (partial aggregation). The results indicated possible presence of both monovalent and bivalent copper oxide (Cu_2O and CuO) phases. Here the possible reducing and stabilizing agents were probably the bacterial secretion proteins (ranging from 22 to 52 kDa). Usha et al. (2010) demonstrated the biosynthesis of CuO nanoparticles by a *Streptomyces* sp. that interacted efficiently against *E. coli*, *S. aureus*, and *Aspergillus niger* after 48 h of incubation.

Yaaghoobi et al. (2012a) reported biosynthesis of magnetic iron oxide nanoparticles ($\leq$104 nm) from *Acinetobacter radioresistens*. This group of researchers also analyzed a comparative toxicity profile of biosynthesized magnetite (Fe_3O_4) and commercial magnetite nanoparticle against red blood cells . Byrne et al. (2011) described the production of Fe_3O_4 nanoparticles by *Geobacter sulfurreducens*. They found modulation of active biomass at the start of synthesis significantly affects the goal of achieved nanoparticles with higher monodispersity and smaller size population. This finding supports the hypothesis that adjusting reaction parameters overall influences the physical properties of biosynthesized nanoparticles. It has been found that the mycelia of acidophilic fungi, *Verticillium* sp., and *Fusarium oxysporum* extracellularly form Fe_3O_4 nanoparticles when they are exposed to an aqueous solution of $K_3[Fe(CN)_6]$ and $K_4[Fe(CN)_6]$ (Bharde et al. 2006).

Salvadori and coworkers (2014) synthesized NiO NPs by using nickel chloride as precursor and dead, dried, and living biomass of filamentous fungus *Aspergillus aculeatus* as reducing agent. Among the three types of fungal biomass used in this experiment, maximum adsorption capacity and hence maximum resistance to metal toxicity were exhibited by dead biomass. The same group of researchers synthesized NiO nanoparticles by the biosorption of Ni(II) ions on the *Hypocrea lixii* fungus living, dried, and dead biomass. The Ni(II) retention capacity of Ni for dead biomass was highest for dead biomass which indicated that dried and live biomass were susceptible to toxicity generated by high concentration of metal. Ullah and coworkers (2014) synthesized NiO nanoparticles by using nickel nitrate hexahydrate as precursor and *Rhizopus nigricans* fungus as reducing and stabilizing agent. The fungus was obtained from bread, and its fine pieces were added into precursor solution $NiSO_4{\cdot}6H_2O$ for the subsequent precipitation.

Prasad and Jha (2009) reported the use of probiotic microbes *Lactobacillus sporogenes* for synthesizing homogeneous, small-sized ZnO nanoparticles with potential ability to act as H_2S decontaminator. Jayaseelan et al. (2012) have reported an eco-friendly route for synthesizing ZnO nanoparticle using the reproducible bacteria *Aeromonas hydrophila*. Rajan et al. (2016) have used *Aspergillus fumigatus* fungus for synthesizing ZnO nanoparticles.

Tarafdar et al. (2013) have reported the green synthesis of TiO_2 nanoparticles using *Aspergillus tubingensis*. Green synthesis of TiO_2 nanoparticles having good antimicrobial properties against *Bacillus subtilis* has also been reported by Malarkodi et al. using *Planomicrobium* sp. (Malarkodi et al. 2013). *Aspergillus flavus* has been used for the synthesis of spherical nanoparticles of TiO_2 by Rajakumar et al. (2012). Aenishanslins et al. (2014) have reported the biosynthesis of TiO_2 nanoparticles from the gram-positive bacterium *Bacillus mycoides*. *Lactobacillus*-mediated biosynthesis of TiO_2 nanoparticles has been reported by Jha and Prasad (2010). Subramanyam and Siva (2016) have reported the biosynthesis of TiO_2 nanoparticles by *Fusarium oxysporum*. Recently Venkatesh et al. (2016) have reported a biosynthesis of CeO_2 using the plant pathogenic fungus *F. solani*. Munusamy et al. (2014) have reported synthesis of CeO_2 nanoparticle from the culture of *Curvularia lunata* fungus.

5.4 Role as Potent Anticancer Agent

Metals like arsenic, antimony, bismuth, vanadium, and gold and their oxides have been used for therapeutic purpose from very early stages of human civilization. However transition metal oxides are gaining increasing importance as potential anticancer drugs over the last decade or so. Current conventional drugs include metabolites, biological agents, and alkylating agents, all of which have significant side effects because of their inability to differentiate between normal and cancer cells (Sak 2012). Implementation of TMOs is expected to give better results with appreciably less side effects. A wide range of transition metal oxides have been investigated for their anticancer efficacy (Desoize 2004). The most important transition metal oxides with potential anticancer activity are categorized below.

5.4.1 Copper Oxide (CuO and Cu_2O)

Copper is used in medicinal inorganic chemistry from several years particularly in antibacterial and anticancer agents, because of its characteristic natural bioavailability, role in angiogenesis, and increased uptake in cancerous tissues (Deo et al. 2016). Among all the different transition metal oxides, copper(I) oxide (Cu_2O) and copper(II) oxide (CuO) have become extremely popular to nanotechnologists in the course of the most recent decade for their promising cytotoxic potential against malignancy cell lines. The thrust behind using copper oxide nanoparticles is fundamental because of their relatively less toxicity toward biosystems and slightly higher solubility in water compared to other transition metal oxides which facilitate cellular uptake. Additionally they are very cost-effective compared to Au and Ag NPs and have a longer stability period.

Both monovalent and divalent copper oxides, namely, cuprous oxide (Cu_2O) and cupric oxide (CuO), have shown cytotoxic potentials against different cell lines.

However till now moderately substantial higher percentile of researches have been performed with CuO. In spite of the fact, there are a couple of literary works to set up the anticancerous activity of Cu_2O also. Different metal nanoparticles (NPs) have been accounted for their anticancer properties like gold (Geetha et al. 2013), silver (Sreekanth et al. 2016), cobalt (Khan et al. 2015), and cerium (Pešić et al. 2015); however, copper nanoparticles have turned into a favored destination+ among researchers (Laha et al. 2014; Dipranjan et al. 2012; Pramanik et al. 2015; Nagajyothia et al. 2017; Wang et al. 2012; Guo et al. 2010).

CuO nanoparticles synthesized from different plant sources like *Ficus religiosa* (Sankar et al. 2014) or *Acalypha indica* (Sivaraj et al. 2014a) have shown cytotoxicity against A549 human lung cancer cells and MCF-7 breast cancer cells, respectively. The component of cytotoxicity was demonstrated to be through the acceptance of apoptosis with enhanced ROS generation (Narayanan and Sakthivel 2010).

Cuprous oxide nanoparticles (CONPs) can cause apoptosis to malignant cells via a mitochondrion-linked pathway; therefore, they are reckoned as suitable potential drugs for treatment of melanoma and other cancers. CONPs are reported to restrain the development and metastasis of melanoma in a mice model with low toxicity (Wang et al. 2013a). CONPs were found to prompt cytotoxicity in a human liver carcinoma cell line (HepG2) in a dosage subordinate way, which was presumably intervened through ROS generation and oxidative pressure (Siddiqui et al. 2013a).

CONPs demonstrate high anticancer potential, the IC_{50} value of CuO nanoparticles against human skin cancer cell line is significantly lower (1.71 μg/ml) than many other anticancer drugs, and this fixation had no cytotoxic effect on typical white blood cells. The NPs caused disintegration of cell membrane, degradation of DNA, arrest of cell cycle at G2/M stage, chromosomal condensation, mitochondrial layer depolarization, and apoptotic cell death. Cell apoptosis occurred in the caspase-9-intervened intrinsic pathway (Chakraborty and Basu 2017). The cytotoxic impact of CuO NPs against human breast cancer cell line (MCF-7) was remarkable with 50% mortality at 62.5 μg/ml (Jeronsia et al. 2016). Being a transition element, inherently they have a high cytotoxic impact on malignancy cells even at low dosages, for the most part due to oxidative pressure it causes to the cytoplasm of the cells.

The role of CONPs in cancer treatment is so important that same of copper-based molecules have even progressed to clinical phase III trial (Szymanski et al. 2012). However CuO nanoparticles also do possess significantly higher toxicity compared to other transition metal oxide nanoparticles which involve DNA damage (Karlsson et al. 2008). However, Pandey et al. have shown the CuO nanoparticles are extremely efficient in killing A549 lung cancer cell lines under hypoxic condition and their cytotoxicity includes intracellular ROS generation, where they are much less toxic against normoxic cells (Pandey et al. 2016). Laha et al. (2014) synthesized CuO nanoparticles by biophysical methods and reported its induced autophagy against a human breast cancer cell line (MCF-7) in a period and dosage subordinate way. Siddiqui et al. (2013b) reported that CuO nanoparticles (average size 22 nm) actuated cytotoxicity in human hepatocellular carcinoma (HepG2) cells in a dose-dependent manner (2–50 mg/mL). Their investigation also reports that there is an upregulation of tumor suppressor p53 gene and apoptotic caspase 3 gene when the cells are exposed to CuO nanoparticles.

Sun et al. (2012) found significant cytotoxicity of CuO nanoparticles against the A549, H1650, and CNE-2Z cell line viability. An observation from the study was that treatment of A549 cells with CuO nanoparticles leads to a fundamental expansion of LC3-II biomarker. This suggests that CuO nanoparticle allows a better utilization of autophagy inhibitors such as wortmannin and 3-methyladenine that improve cell survival (Sun et al. 2012). These results indicate that the cytotoxicity of CuO nanoparticles may include the autophagic pathway in A549 cells. These outcomes support the results reported by Laha et al. (2014) in which cancer cells were treated with CuO nanoparticles. Research work on both CuO and Cu_2O nanoparticles shows their high potential to work as future anticancer drugs, though significant risk assessment studies are yet to be undergone (Fig. 5.3).

5.4.2 Iron Oxide (Fe_2O_3 and Fe_3O_4)

The anticancer activity of iron oxide nanoparticles can be classified under two categories: direct and indirect. Iron oxide nanoparticles are capable of binding to tumor sites through covalent bond formation due to their high particulate nature.

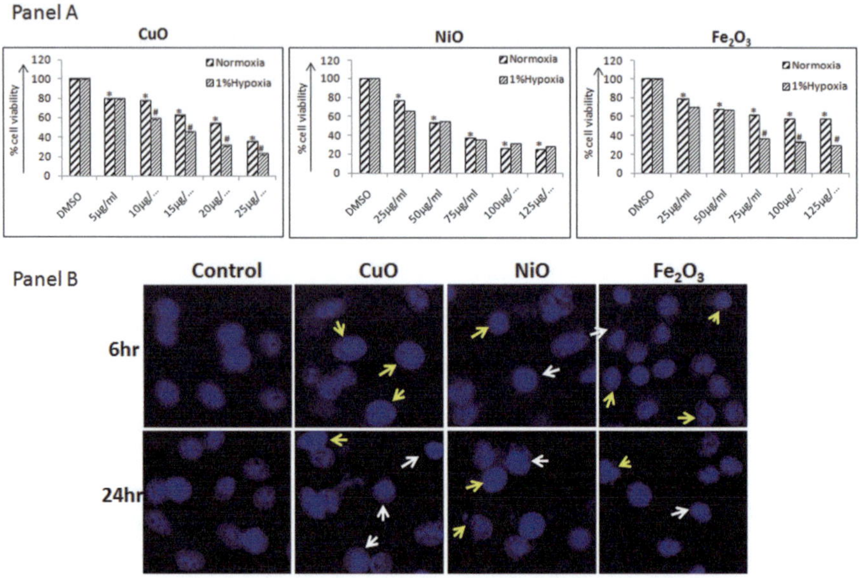

Fig. 5.3 (**a**) In vitro cell cytotoxicity (MTT assay) exhibited by metal oxide nanoparticles in normoxia and hypoxia. The error bars indicate standard deviations ($* = p < 0.05$ vs DMSO control, $\# = p < 0.05$ vs normoxia). (**b**) Representative microphotograph of DAPI staining at 40X magnification where apoptotic features (yellow arrows) and necrosis (white arrows) are shown. (Adapted from Pandey et al. 2016)

Additionally, iron oxide nanoparticles can potentially convert radiant energy into reactive oxygen species which eventually lowers the possible damage to health tissues and cells (Rao et al. 2016). Magnetic iron oxide nanoparticles play important roles as drug carriers and MRI agent and in tissue repair and have potential anticancer activity (Indira and Laksmi 2010; Teja and Pei-Yoong 2009; Jain et al. 2005). Surface modifications of these nanoparticles are done with drugs, protein, and genetic materials (Neuberger et al. 2005; Kohler et al. 2005). The expanded surface zone of these nanoparticles decreases the amount of the drug to the least and also reduces the adverse effect of the drug on normal cells (Tartaj et al. 2003; Jurgons et al. 2006; Ai et al. 2005; Gonzales and Krishnan 2005).

Hyperthermia is a technique of killing the cancer cells by raising their temperature to 41–45 °C. Living cells have the capability to repair themselves; however, tumor cells once damaged cannot revive themselves (Kim et al. 2006; Green 2005; Olsvik et al. 1994). The iron oxide nanoparticles (IONP) with small particle size and uniformity have increased hyperthermia capability. Surface modifications of these nanoparticles are done using organic polymers and bioactive molecules to increase their biocompatibility. These modifications are actually responsible for the slow release of the drugs.

Near-infrared (NIR) or oscillating magnetic fields (MF) are the nontoxic wavelength radiations which can be absorbed by the iron oxide nanoparticles and can act as stimuli for the production of reactive oxygen species (ROS) or hyperthermia which are responsible for killing of the cancer cells (Laurent et al. 2011). These nanoparticles can be directed to the tumor site by attaching targeting moieties and by guiding them by the external use of nearby magnetic fields. High biodegradability, relatively low toxic effect, and most importantly prolonged sustained release with longer retention time have been observed for both magnetite and hematite (Fe_3O_4 and Fe_2O_3) nanoparticles as evident from in vitro and in vivo reports (Pandey et al. 2016; Laurent et al. 2011). These parameters have made them an important tool in biomedical applications. Iron oxide nanoparticles have a large surface area which are utilized for attaching a large number of tumor-targeting ligands like monoclonal antibodies, peptides, or small molecules for diagnostic imaging or delivery of therapeutic agents. In this way these nanoparticles can specifically kill cancer cells via ROS generation or by generation of heat energy. This diminishes the damage to healthy tissues, the main side effects of conventional anticancer treatments.

Nanoparticles of iron oxide coupled with the antitumor drug doxorubicin have been observed to increase the antitumor efficacy as compared with conventional doxorubicin alone. This is because applying an external magnetic field can actively target the drug-conjugated nanoparticle at the tumor site effectively. The anticancer drug activity depends on the activation of the hydroxyl radicals, which breaks mitochondria, lipids, proteins, DNA, and different structures in tumor cells and finally causes apoptosis or necrosis.

Spherical iron oxide nanoparticles have been used as a medical device for magnetic tumor hyperthermia in brain (Van Landeghem et al. 2009; Silva et al. 2011; Johannsen et al. 2010; Maier-Hauff et al. 2011) and prostate cancer, along with radiotherapy or chemotherapy. PEG-covered iron oxide nanocubes have been made to generate heat which have been found to disrupt the extracellular framework and matrix of a tumor under the influence of a magnetic stimulus (Kolosnjaj-Tabi et al. 2014). The different nanostructures used in anticancer hyperthermia treatment (Bhattacharyya et al. 2011) absorb radiation energy from some external source and transform it into heat (Danhier et al. 2010). Removal of the magnetic field leads to the loss of magnetization of the superparamagnetic iron oxide nanoparticles (Hilger and Kaiser 2012). Magnetization leads to nanoparticle aggregation which is required during the treatment but is extremely hazardous in posttreatment conditions.

It has been observed that cancer cells exposed to iron oxide nanoparticles are killed by necrosis instead of apoptosis. In addition, it has been found that iron oxide nanoparticles barely affect normal human fibroblast cells. This result showed that nanoparticles can be used to kill the tumor cells selectively without damaging the normal cells.

Iron oxide magnetic nanoparticles loaded with high dosages of water-insoluble anticancer agents have shown good antiproliferative effects in breast and prostate cancer lines in a dose-dependent manner. Anticancer drugs such methotrexate or malignant cell-targeting agents like chlorotoxin have been chemically conjugated to iron oxide nanoparticle for ensuring better targeted delivery (Laurent et al. 2014).

The in vitro and in vivo efficacy of iron oxide nanoparticles conjugated with methotrexate (therapeutic agent) and chlorotoxin (targeting ligand) has been investigated by Sun et al. by MRI. The results indicated higher selective toxicity to tumor cells, which suggests these functionalized nanoparticles possibly have a wide scope for therapeutic and diagnostic purpose (theranostic application) (Sun et al. 2008). In another examination, C595 monoclonal immunizer (mAb)-conjugated SPIONs have been accounted to detect mucin 1 (MUC1)-expressing ovarian cancers utilizing MRI. Appended C595-mAbs tie to MUC1 to explicitly distinguish ovarian cancer cells. The anticancer effects and MRI parameters of this nanoformulation were investigated. This conjugate was able to detect and recognize the tumor cells without in vivo toxicity to other cells and showed good antiovarian malignancy action (Shahbazi-Gahrouei and Abdolahi 2013). Ma et al. developed and investigated the efficacy of a novel SPION-based MRI contrast agent which is highly specific to cancer cells (Ma et al. 2014). They coupled folic acid to the surface of the modified SPION stacked to polymeric nanoparticles. These nanoparticles were observed to be cytotoxic to the MCF-7 and SPC-A-1 cells. The r2/r1 (transverse versus longitudinal proton relaxation rate) estimation of SPION was around 40, which was higher than that of Resovist® (a commercial MRI contrast agent involving IONPs covered with carboxydextran), demonstrating that SPION had a stronger T2 shortening effect, bringing about high MRI viability (Ma et al. 2014). Iron oxide nanoparticles have been used to synthesize bifunctional probes that can be used for focal imaging and targeted drug carrier for hepatocellular carcinoma (Pilapong et al. 2014). The nanoprobe joined the particular capacity of a cancer-specific molecule [DNA-based epithelial cell attachment molecule (EpCAM) aptamer] and imaging ability of magnetic IONP (Pilapong et al. 2014). IONPs have been investigated not just for the conveyance of low molecular weight drugs; they are found to be reasonably useful for macromolecular drug delivery that includes protein, peptide, and DNA-based drugs. To serve this purpose, cationic polymers like polyethylenimine (PEI), dextran, and chitosan are used for surface decoration and coating of iron oxide nanoparticles. In an experimental investigation, Lin et al. revealed polycation/iron oxide nanocomposite as MRI-visible small interfering (si)RNA bearers to overcome multidrug resistance (MDR) through hushing of target mRNA, therefore bypassing p-glycoprotein (P-gp). Nanoparticles prompted a successful hushing impact that was practically identical to that of a commercial transfection agent, Lipofectamine 2000. In this experiment the cells that were transfected with nanoparticles show significant improvement compared to non-transfected cells (Lin et al. 2014). Moreover, Li et al. additionally created redox-activated nanoparticles including a SPIO inner core and a disulfide-containing polyethylenimine external layer for the release of genes. The nanoparticles also delivered siRNA around the transcriptase genes in HepG2 cells, causing their apoptosis and development hindrance. Moreover, nanoparticles were connected as T2-negative contrast agents for MRI of a tumor xenograft (Li et al. 2014). As mentioned above, iron oxide nanoparticles have already implemented a significant contribution in the field of multimodal imaging, the newly developing paradigm for cancer therapy. Zhou et al. revealed polyethylene glycol (PEG)ylated Fe@Fe3O4 nanoparticles having triple functional properties: targeting, PTT, and imaging. These particles indicated high magnetization and transverse relaxivity. In tumors, the intensity of the MRI signal was around triple, and expanded temperature was roughly twofold that was accomplished without magnetic targeting, demonstrating great magnetic targeting capacity (Zhou et al. 2014). Strangely, inferable from the high photothermal change effectiveness and choosing magnetic targeting, these nanoparticles showed synergistic anticancer action both in vitro and in vivo. A multipurpose novel nanohybrid system capable of cancer diagnosis and therapeutic uses is produced by finishing IONPs onto fullerene (C60) and utilizing FA as dynamic targeting ligand. C60-IONPPEG-FA indicated solid photosensitizing and photothermal ablation impact and selectively killed cancer cells through dynamic cancer cell targeting (Shi et al. 2014). Wang et al. created imaging contrast agents utilizing bovine serum albumin-capped SPION for the determination of pancreatic cancer. The r2:r1 proportion of 13.3 guaranteed the use of these SPIONs as T2-weighted MRI contrast agents. Coupling of near-infrared (NIR) fluorescent dye and mAbs has expanded their capacity for the particular bimodal imaging of pancreatic cancer (Wang et al. 2014). The examinations talked about above uncovered the capability of IONPs in the finding and treatment of cancer and recommend that these have potential as therapeutic agents for malignancy treatment.

The cytotoxicity of magnetites prepared by coprecipitation and magnetosomes isolated from MSR-1 on L929 cells were compared in an experimental study (Han et al. 2007a). Chemically synthesized nanoparticles showed a size distribution of 7–18 nm, while the magnetosomes were having a size range estimate of 10–60 nm. Both types of nanoparticles influenced the activity of L929 cells in a dose and time subordinate way (with a fixation scope of 0.5–1.0 mg/mL and an incubation time of 24–72 h). Engineered iron oxide nanoparticles induced 85% cytotoxicity, while 90% was observed with the biogenic magnetite; the two exposures happened at 1.0 mg/mL and with 72 h of incubation (Han et al. 2007b). It has been perceived that the surface lipid coating of the magnetosomes enhances biocompatibility of the nanomaterial in comparison to their chemically synthesized counterparts (Han et al. 2007b). Toxicity studies with magnetite nanoparticles obtained using the commercial and bacterial sources (*Acinetobacter radioresistens*) were done on fringe blood cells by observing hemagglutination, hemolysis, and morphological changes (Yaaghoobi et al. 2012b). Low nanoparticle concentrations induced lysis and extreme hemagglutination in the samples treated with commercial nanoparticles (50 μg/mL). No morphological change was observed in the peripheral blood cells on incubation with the biogenic iron oxide nanoparticles (Yaaghoobi et al. 2012c). These results suggest the relatively lower toxicity of iron oxide nanoparticles synthesized from bacterial source compared to those synthesized by chemical route.

5.4.3 Zinc Oxide (ZnO)

ZnO nanoparticles have the capability to meet most of the therapeutic indices commonly utilized by chemotherapeutic specialists and show high level of cancer cell selectivity as evident from ex vivo studies (Hanley et al. 2008; Wang et al. 2009a). In biomedical applications, compatibility issues with living tissues are a major problem which needs to be considered before applying any type of nanomaterial in the body. This limits the type of nanomaterial to be used. The FDA considers ZnO nanoparticles to be a "GRAS" (generally recognized as safe). The preferential cytotoxicity of ZnO nanoparticles against cancer cells has made it an effective anticancer agent (Hanley et al. 2008; Wang et al. 2009a).

It has been observed that high specific targeting of these nanoparticles to the cancer cells without affecting the normal cells can be achieved by improving the design of these nanoparticles. This is because extremely high concentrations of ZnO nanoparticles with size range of 4–20 nm have been seen to show harmful effects to normal body cells as well (Hanley et al. 2009). In this regard, functionalization of these nanoparticles with targeting proteins or chemical groups enhances their targeting ability and killing efficiency to cancer cells while rendering inert to normal cells. Size is an important parameter in increasing the cytotoxicity of these ZnO nanoparticles as smaller nanoparticles show greater toxicity (Hanley et al. 2009; Guo et al. 2008; Nair et al. 2009).

In vitro cytotoxicity studies in cancers including glioma, breast, bone, colon, and leukemias and lymphomas have shown better cytotoxicity with nanophase ZnO compared to ZnO of higher dimension (Hanley et al. 2008; Guo et al. 2008; Wang et al. 2009b; Reddy et al. 2007). ZnO nanoparticles were ~ 28–35 times more cytotoxic to the cancer cells of lymphocytic lineage as compared with their normal counterparts (Hanley et al. 2008, 2009; Wang et al. 2009a). The ex vivo therapeutic index of ZnO is even higher than some standard chemotherapeutic drugs like doxorubicin and carboplatin against a variety of leukemia, lymphoma, and tumors. This observation suggests that ZnO nanoparticle is extremely selective while killing cancer cells. The rapidly dividing cells have been found to be most susceptible to the preferential cytotoxicity of these nanoparticles (Hanley et al. 2008, 2009). Studies have shown that the generation of ROS is proposed as a key cytotoxic mechanism of ZnO nanoparticles (Hanley et al. 2009; Jeng and Swanson 2006; Moos et al. 2010) leading to the process of cell death by apoptosis. According to the photodynamic therapy concept, photoactivation of ZnO nanoparticles leads to greater levels of ROS generation, which have the capability of killing cancerous cells if targeted effectively. ZnO nanoparticles conjugated with porphyrin are found to exhibit synergistic cytotoxic effect against ovarian cancer cell lines, which was found to be enhanced with UV radiation (Zhang et al. 2008b). Comparative investigations have shown that simultaneous administration of ZnO nanoparticles and daunorubicin also displays similar synergistic cytotoxic effects on leukemic cancer cells (Guo et al. 2008). It seems from the reports that ZnO nanoparticle conjugated to tumors can be targetedly photoactivated for specifically destroying cancer cells. Direct drug conjugation or encapsulation of the drug within the ZnO nanoparticles in the future is expected to improve further anticancer efficacy of these nanoparticles.

Zinc oxide nanoparticles can exert cytotoxicity alone or in combination with other drugs (Vinardell and Mitjans 2015). Combination of the anticancer drugs to nanoparticles has been adopted as an important strategy nowadays for reducing the side effects of the drugs. Zno nanoparticles combined with standard drugs like paclitaxel, daunorubicin, and cisplatin have been shown to have effective chemotherapeutic effect in cell lines in vitro (Hackenberg et al. 2012). Effective drug targeting and accumulation of the drug daunorubicin with differently sized ZnO NPs in leukemia cancer cells have shown that these nanoparticles are efficient agents for drug delivery (Guo et al. 2008). Doxorubicin-loaded ZnO nanoparticles exhibit higher cellular uptake and more selectivity and lower toxicity against normal cells. Drug-loaded ZnO nanoparticles show excellent cytotoxicity with minimized side effects, which is really beneficial for cancer management.

Zinc oxide nanoparticles have been discovered very effective against T98G cancer cells. ZnO NPs are most effective on T98G cancer cells, moderately effective on KB cells, and mildly toxic on human HEK cells. These outcomes have shown that treatment with ZnO NPs sensitizes T98G cells by increasing cytogenetic damage and causing apoptotic death. The ZnO NPs act as genotoxic drugs, since they bring about micronucleus formation in cells (Wahab et al. 2013a). The intracellular generation of ROS correlated with apoptosis was also measured with melanoma cancer cells with varying ZnO NP doses (Wahab et al. 2013b). Zinc oxide NPs were found to exhibit activity against HepG2 cancer cells at very low concentrations and MCF-7 cancer cells in a dosage subordinate way. MTT assay demonstrated a dose-dependent decrease in the cell viability. It can be clearly stated from the results of these cellular antiproliferation experiments that nanoparticles do sensitize cancer cells. Apoptosis has been observed to increase with an increasing concentration of NPs, resulting in cell demise in both cancer cell lines (Wahab et al. 2014).

Zinc oxide can have different surface charges under acidic and basic conditions. Cancer cells have a negative surface charge. Modification of the surface charge of these nanoparticles enables their uptake in the malignant cells as per the requirements. The photodynamic property of ZnO nanoparticles generates a lot of ROS which results in cell apoptosis (Ryter et al. 2007). Zinc oxide NPs are thus an important drug delivery agent.

Recent studies suggest than ZnO nanoparticles can trigger tumor cell killing through NADPH-dependent oxidative pathway that leads to apoptosis; four representative ZnONP samples of different sizes and specific surface areas showed a remarkable impact on cell toxicity and DNA breakage in macrophages of mice in a p47phox- and Nrf2-independent manner. ZnONPs induced necrosis and apoptosis in these macrophages because of their essential part in the control of immune responses during inflammation and clearance of inhaled particulates. ZnONP enhanced a rapid induction of nuclear condensation, DNA fragmentation, and the arrangement of hypodiploid DNA-containing nuclei and apoptotic bodies (Han et al. 2007a).

In an experimental report, ZnO nanoparticles containing doxorubicin exhibited synergistic effects in cancer cells. ZnO nanoparticles containing doxorubicin showed higher anticancer efficacy than doxorubicin or ZnO nanoparticles used alone due to higher cellular internalization and retention (Sharma et al. 2014). The imaging ability of ZnO has additionally been accounted by several researchers. Hong et al. detailed the utilization of ZnO nanowires for imaging along with drug delivery. Optical imaging of integrin $\alpha v\beta 3$ on U87MG human glioblastoma cells was done by RGD peptide-conjugated green fluorescent ZnO nanowires (Hong et al. 2011). ZnO quantum dots conjugated with transferrin have found application in targeted delivery to cancer cells. The selective binding and internalization of these quantum dots are due to receptor-mediated endocytosis. These nanoprobes were found to be resistant to photobleaching from time-lapsed photobleaching studies. This property is utilized for long-term imaging application (Sudhagar et al. 2011). Recently, surface-modified fluorescent ZnO nanoparticles have been developed by Hong et al., and they conjugated 64Cu and TRC105 (a chimeric mAb against CD105 for cell selective delivery) to the surface of ZnO nanoparticles. This development can be considered a new breakthrough in multimodal imaging as these nanoparticles can have potent use in PET and other fluorescent imaging techniques used for diagnosis of cancer and other tumors (Hong et al. 2015). Several studies suggest that ZnO nanoparticles might have promising potential applications as anticancer agents, especially considering their relatively low toxicity and high biological activity.

In a study by Darroudi et al. (2014a), ZnO nanoparticles (1.5–100 μg/ml) synthesized from gelatin showed cell toxicity on neuro2A cells (a fast-growing mouse neuroblastoma cell line) after incubation for 24 h. Cell death was increased with a gradual increase in nanoparticle dosage above 2 μg/mL (Darroudi et al. 2014b). ZnO nanoparticles synthesized from *Tabernaemontana divaricata* leaf extract (Sivaraj et al. 2014b) exhibited significant toxicity against MCF-7 breast cancer cell lines after a mere period of 24 h treatment. The IC_{50} value was recorded to be 30.65 μg/mL (Sivaraj et al. 2014b). Interestingly, the IC_{50} values for the biosynthesized ZnO nanoparticles were lower than that of biosynthesized copper oxide nanoparticles synthesized from identical plant extract (Sivaraj et al. 2014c).

5.4.4 Nickel Oxide (NiO)

NiO nanoparticles have intense cytotoxic effects in different cells like human airway epithelial (HEp-2) cells, human breast cancer (MCF-7) cells, and many more. NiO nanoparticles have been reported to induce apoptosis in human lung epithelial cells (Ahamed et al. 2008). Effective cytotoxic activity has been observed with NiO nanoparticles against HT-29 (colon cancer cell lines); the nanoparticles were prepared using *Moringa oleifera* plant extract (Ezhilarasi et al. 2016b). Human blood cells like RBCs and macrophages did not show any cytotoxic effects with NiO nanoparticles (Khalil et al. 2017b). The IC50 value (μg/ml) for NiO nanoparticles on A549 (human lung cancer cells) was 28.39, whereas it was 11.46 on MCF-7 (human breast cancer cells), showing significant activity of these nanoparticles on human breast cancer cells (Raj et al. 2014). Distorted morphology of the cancerous cells is indicative of the cytotoxic activity of the NiO nanoparticles (Mariam et al. 2014; Chen et al. 2013). The biocompatibility of NiO NPs topped with biomolecules like, glucose, is well established, and these are utilized as biosensors and heat mediators for cancer hyperthermia (Vaseem et al. 2017). Nickel (Ni) ions on binding to DNA cause strong DNA damage by H_2O_2 (Kawanishi et al. 1989). The other mechanism is circuitous oxidative DNA damage because of aggravation. Essential wellsprings of endogenous oxygen radicals are phagocytic cells, for example, neutrophils and macrophages (Grisham et al. 2000). Presence and aggravation of reactive oxygen species like NO can cause damage in cellular DNA of animal tissues (Ducrocq et al. 1999; Chazotte-Aubert et al. 1999; Eiserich et al. 1998). On the basis of this proposition, it is believed that NiO exerts its cytotoxicity through a bi-pronged mechanism, involving direct inflammation and ROS generation via H_2O_2 formation. This twofold activity may clarify the moderately high cancer-causing hazard related with Ni_3S_2.

NiO NPs have been found to induce cell toxicity in refined HeLa cells (Ada et al. 2010). NiO NPs caused cell cytotoxicity in both a dosage subordinate and a period subordinate way. Apoptosis prompted by mutagenic cancer-causing agents is a novel kind of programmed cell death. NiO NPs generate ROS and impart a severe oxidative stress on cell membrane leading to the breakdown of membrane lipid molecules. The intracellular calcium homeostasis is disturbed, leading to adjustments in metabolic pathways and to apoptosis (Clutton 1997; Knaapen et al. 2004).

5.5 Cerium Oxide (CeO_2)

Cerium oxide NPs (CNPs) are unique and extremely fascinating for radiation therapy, having the ability to specifically initiate the killing of radiation-exposed cancer cells (Wason et al. 2013) while shielding the encompassing tissue from radiation-prompted harm and ROS-induced damage. CNPs have been found to have radio-enduring capability, in addition to being radiosensitizing. In illuminated cells, CNPs cause ROS-induced cell damage leading to programmed cell death in illuminated cancer cells while ensuring normal tissues (Colon et al. 2010; Tarnuzzer et al. 2005a). It has been speculated that at acidic (pH 4.3) environments, the catalase activity of CNPs is blocked leading to the accumulation of H_2O_2 which kills the malignant cells by oxidative damage (Wason et al. 2013). The theory depends on the presumption that the intracellular environment of cancer cells is marginally more alkaline (pH > 7.4) than normal cells, while the extracellular matrix is somewhat acidic because of the Warburg impact, and the pH of the normal tissues decreases from 7.1 to 6.7 (Neri and Supuran 2011). The differential toxicity of CNPs on cancer cells is also due to their SOD-like activity. The SOD enzyme acts as a radiosensitizing agent by deferring the G2 to M transition and permitting DNA repair (Ali et al. 2014). Experiments have demonstrated that CNPs are nearly as lethal as radiotherapy to the pancreatic cancer cells L3.6pl, yet indicated practically zero toxic impact on the normal cells hTERT-HPNE. It is thus evident from this experiment that these NPs can be used for pancreatic cancer treatment (Wason et al. 2013). A report analyzed the relative toxicity of CNPs to both malignant and normal cell lines, and findings suggested that CNPs were specifically toxic to cancer cells. This advantage of the CNPs in normal human cell lines shows that they might be more secure for human utilization in industrial and medicinal applications (Pešić et al. 2015). Different experiments have demonstrated that redox-dynamic CNPs exhibit toxic impacts on several cancer cells and are responsible for making the tumor cells susceptible to radiation while ensuring the normal cells in the stroma encompassing a tumor (Sack et al. 2014a). In melanoma and squamous cell carcinoma of the skin, CNPs have shown pro-apoptotic effects. It has been accounted for that CNPs show either pro-oxidant or antioxidant redox activity. Interestingly, CNPs induce ROS-mediated apoptosis in tumor cells, but simultaneously they also develop antioxidative defense mechanism in normal cells (Alili et al. 2011a). CNPs thus may have therapeutic effects against skin cancer therapy and might also be used with traditional cancer drugs, for example, doxorubicin, to ensure against doxorubicin-actuated cell toxicity. Genetic toxicity with these nanoparticles is dependent on nanoparticle size (De Marzi et al. 2013a). CNPs of 5 nm size are found to be non-genotoxic as against the nanoparticles with 16–22 nm size, where profound DNA damaging impacts were found in cancer cells (Sack et al. 2014b), showing that the toxic activity of these nanoparticles is controlled by the size as well as the shape of the affected cells. The use of traditional chemotherapies with CNPs improves the antitumor activity and brings down the harmful side effects (De Marzi et al. 2013b). In vivo xenograft with immune-deficient nude mice demonstrated lessening of tumor weight and volume after treatment with CNPs. The pro-oxidant and antioxidant properties of CNPs were responsible for its anticancer effects. This examination was the first to demonstrate that CNPs avoid tumor development in vivo (Alili et al. 2013a). The antitumor activity of cerium oxide nanoparticles is enormously subject to their size and shape, albeit both small- and large-sized nanoparticles incite DNA damage in tumor cell lines (Wason et al. 2013).

On chemical modifications, nanomedicines can selectively kill tumor cells by generating ROS in tumor cells (Colon et al. 2009, 2010; Das et al. 2012; Lord et al. 2013a).

As for CNPs, an examination has exhibited their toxicity to malignant cells and prevention of metastasis (Alili et al. 2011b). These nanoparticles have been found to cause cytochrome discharge leading to caspase-3 and caspase-9 activation, promoting apoptosis of cancer cells by the mitochondrian pathway (Wang et al. 2013b). Polymer covering of CNPs increments hydrophilicity, but the redox activities are not affected (Alili et al. 2011b; Asati et al. 2009). Myofibroblasts generally intercede epithelial/stromal flagging. They have an important role in the outflow of extracellular matrix components, including alpha-smooth muscle actin and collagen, to encourage cancerous progression and angiogenesis (Desmoulière et al. 2004). CNPs have the capability of balancing the production of myofibroblast from fibroblast by TGF-β1-induced ROS subordinate articulation of smooth muscle actin. Pretreatment with CNPs induced myofibroblast formation and TGF-β1-actuated alpha-smooth muscle actin articulation in fibroblasts (Alili et al. 2011b; Wason and Zhao 2013a). CNP treatment lessened the capacity of myofibroblasts to initiate invasion by squamous tumor cells. Thus, these outcomes show the immediate negative impacts of CNPs on cancer cells, and also their capacity to adjust the tumor condition and in a roundabout way hinders tumor cell intrusion. These information additionally recommend CNPs as an important anticancer nanomedicine (Alili et al. 2011b).

CNPs have been found to adjust the intracellular ROS levels. Nanoceria surface coating with heparin decides the internalization and ROS-producing capacity of these particles. Heparin nanoceria was viable in diminishing endothelial cell multiplication, demonstrating their role in controlling angiogenesis. Specifically, CNPs induce angiogenesis by balancing the cellular oxygen condition and stabilizing hypoxia, actuating factor 1α endogenously, and CNP surface properties have been found to be interlinked with angiogenesis. Expanded Ce^{3+}/Ce^{4+} proportion and high surface territory are responsible for making the CNPs more active in controlling intracellular oxygen, which prompts the induction of angiogenesis. Thus, it has been found that the reactive surface of CNPs and easy oxygen transport advance angiogenesis (Lord et al. 2013b).

Although experimental investigations have shown that CNPs have been successful in overcoming the multidrug resistance in chemotherapy and are functioning well as chemosensitizers, more research in this field is required for designating these nanoparticles as successful chemotherapeutic agents in these respects.

Furthermore, experiments have demonstrated CNPs to be lethal to cultured bronchial epithelial lung fibroblasts; however, they are not toxic to mammary epithelial cells, macrophages, immortalized keratinocytes, or immortalized pancreatic epithelial cells (Wason and Zhao 2013b).

CNPs can likewise create ROS and peroxidize the lipids of the liposomal membrane, subsequently controlling numerous signaling pathways and affecting the important movements of cells (Lin et al. 2006a). Negatively charged CNPs have the capability to accumulate inside acidic lysosomes, bringing about expanded toxicity in tumor cells.

CNPs, also called nanoceria, of less than 10 nm diameter have been shown to have effective therapeutic activity against neurodegenerative sicknesses as they could pass the blood-brain barrier of rodents. These nanoparticles have shown good antioxidant activity in light of their capacity to switch between the two redox states III and IV (Naz et al. 2017; Arya et al. 2016; Heckman et al. 2013). In the most recent years, an expanding number of literature have shown nanoceria to be an effective antioxidant which has the capability to quench ROS in various cells and tissues in vitro (Walkey et al. 2015). CNPs have shown anti-angiogenic effect in ovarian cancer (Giri et al. 2013). A perfect cancer therapy should include efficient and targeted killing of cancer cells bypassing processes like intrusion and neoangiogenesis, with minimum effect to healthy normal cells. In that specific situation, CNPs might be a decent possibility for a compelling treatment with less side effects because of their specific toxicity to the cancer cells. Favorable position of the utilization of CNP over an unadulterated anti-angiogenic therapeutic approach is that neoangiogenesis is targeted by these nanoparticles and tumor cells are killed through pro-oxidative mechanisms prompting apoptosis (Pešić et al. 2015; Alili et al. 2013b). Additionally, CNP can be combined with, for instance, other anti-angiogenic compounds. In that specific circumstance, volume reduction of the tumor with diminished blood supply has been observed in a glioblastoma mouse model (de Groot and Mandel 2014). Moreover, CNP treatment and radiation therapy sensitize the breast carcinoma cell line MCF-7 to radiation and additionally shield normal breast CRL8798 epithelial cells from radiation-instigated damage (Tarnuzzer et al. 2005b). Further on, we could indicate prior that a mix of doxorubicin with CNP exhibits positive synergistic effect against human melanoma cells, without inducing DNA damage and even providing a shielding effect to human dermal fibroblasts from doxorubicin-prompted cellular toxicity (Sack et al. 2014c; von Montfort et al. 2015). CNPs with less than 10 nm diameter might be a significant and proficient tool to assault and execute malignant glioma cells and to secure normal (healthy) cells from the negative impact of tumor cells.

Some researchers concluded that 20 nm CeO_2 nanoparticles lessen cell suitability in human bronchoalveolar carcinoma-derived cells; 3.5–23.3 μg/ml of CeO_2 has been responsible for bringing about serious oxidative damage by decreasing GSH and α-tocopherol levels and increasing MDA and LDH levels, the markers of lipid peroxidation and membrane damage, respectively (Lin et al. 2006b). The cytotoxic effects of these nanoparticles have been found to be both dose and time dependent.

5.6 Titanium Dioxide (TiO_2)

Nanoparticles of TiO_2 have been reported to induce ROS generation leading to the phenomenon of oxidative stress. Shukla et al. have reported an increase in lipid peroxidation, numerous oxidative DNA damages, and micronuclei in human epidermoid carcinoma cells (A431 cell line) on incubation of these cells with TiO_2. Cytotoxic effects have also been observed by Vamanu et al., human leukemic monocyte lymphoma cells (U937 cell line). UV-excited TiO_2 nanoparticle toxicity has been reported in human hepatocarcinoma cells (SMMC-7721 cell line) by Zhang et al. An oxidative stress-induced DNA double breaks followed by increase in H2AX histone phosphorylation resulted in generating toxicity to human A549 cells (Bogdan et al. 2017). TiO_2 nanoparticles induce oxidative stress and produce inflammatory cytokines in vivo also in a subcutaneous tumor model induced by inoculating CT26 cells in BALB/c mice, as reported by Fujiwara et al. The cytotoxic effects were enhanced by exposure of these mice to UV radiation (Fujiwara et al. 2015). Combined effects of TiO_2 nanoparticles with ultrasound were found to have good cytotoxic effects on human ovarian carcinoma cell line A2780 as reported by Bernard and Mornstein (2016). Combination therapy of TiO_2 nanoparticles with sonodynamic therapy and photodynamic therapy has been found to be very effective in an in vivo model of human prostate cancer, as reported by Miyoshi et al. In this study 5-aminolevulinic acid solution was mixed with the TiO_2 nanoparticles (Miyoshi et al. 2011). TiO_2 nanoparticles have shown dose- and time-dependent inhibition of MCF-7 breast cancer cell line as reported by Lotfian and Nemati (2018). Zegota et al. have reported TiO_2 nanoparticles to have concentration-dependent increase in DNA damage in colon cancer patients (Kurzawa-Zegota et al. 2017). Ag-doped TiO_2 nanoparticles have been shown to induce cell toxicity and oxidative stress in human lung (A549), breast cancer (MCF-7), and liver cancer cells (HepG2) as reported by Ahamed et al. (2017).

5.6.1 Limitations

Primarily the main limitation for transition metal oxides is their cytotoxicity to normal cell lines also. The risk assessment related to toxicity needs to be investigated thoroughly before their possible paradigm shift of being used as anticancer medicines. Toxicity of these nanoparticles toward normal cell line is still under investigation, and it needs to be addressed fully before possible translational conversion of transition metal oxide nanoparticles to therapeutic anticancer drug. Currently, low solubility of transition metal oxide nanoparticles in water is also a big challenge for their utilization in in vitro and in vivo conditions. Less solubility of these transition metal oxides in water creates lots of difficulty in their cellular uptake. However this difficulty can somewhat be alleviated by use of surface compatible capping agents like citrate, starch, etc.

5.6.2 Future Perspectives

The future of transition metal oxide (TMO) nanoparticles is extremely promising for anticancer treatment. These TMOs can be useful both in single dosage and in combinatorial therapy. Their cost-effectiveness can be a big advantage over current expensive anticancer drugs. Use of them can possibly reduce the possible side effects of painful chemotherapeutic agents. Combination of these metallic oxides with other inorganic therapeutic agents can develop new mechanisms for killing cancer cells.

References

Ada K, Turk M, Oguztuzun S, Kilic M, Demirel M, Tandogan N, Ersayar E, Latif O (2010) Cytotoxicity and apoptotic effects of nickel oxide nanoparticles in cultured HeLa cells. Folia Histochem Cytobiol 48(4):524–529

Ahamed M, Karns M, Goodson M, Rowe J, Hussain S, Schlager J, Hong Y (2008) DNA damage response to different surface chemistry of silver nanoparticles in mammalian cells. Toxicol Appl Pharm 233:404–410

Ahamed M, Khan MAM, Akhtar MJ, Alhadlaq HA, Alshamsan A (2017) Ag-doping regulates the cytotoxicity of TiO_2 nanoparticles via oxidative stress in human cancer cells. Sci Rep 7:17662

Ai H, Flask C, Weinberg B, Shuai XT, Pagel MD, Farrell D, Duerk J, Gao J (2005) Magnetite-loaded polymeric micelles as ultrasensitive magnetic resonance probes. Adv Mat 17:1949–1952

Al-Awady MJ, Greenway GM, Paunov VN (2015) Nanotoxicity of polyelectrolyte-functionalized titania nanoparticles towards microalgae and yeast: role of the particle concentration, size and surface charge. RSC Adv 5:37044–37059

Ali D, Alarifi S, Alkahtani S, AlKahtane AA, Almalik A (2014) Cerium oxide nanoparticles induce oxidative stress and genotoxicity in human skin melanoma cells. Cell Biochem Biophys 71:1643–1651

Alili L, Sack M, Karakoti AS, Teuber S, Puschmann K, Hirst SM, Reilly CM, Zanger K, Stahl W, Das S et al (2011a) Combined cytotoxic and anti-invasive properties of redox-active nanoparticles in tumor-stroma interactions. Biomaterials 32:2918–2929

Alili L, Sack M, Karakoti AS et al (2011b) Combined cytotoxic and antiinvasive properties of redox-active nanoparticles in tumor-stroma interactions. Biomaterials 32:2918–2929
Alili L, Sack M, von Montfort C, Giri S, Das S, Carroll KS, Zanger K, Seal S, Brenneisen P (2013a) Downregulation of tumor growth and invasion by redox-active nanoparticles. Antioxid Redox Signal 19:765–778
Alili L, Sack M, von Montfort C, Giri S, Das S et al (2013b) Downregulation of tumor growth invasion by redox-active nanoparticles. Antioxid Redox Signal 19:765–778
Alvarez S, Muhammed M, Zagorodni A (2006) Novel flow injection synthesis of Iron oxide nanoparticles with narrow size distribution. Chem Eng Sci 61:4625
Arivalagan K, Ravichandran S, Rangasamy K, Karthikeyan E (2011) Nanomaterials and its potential applications. Int J Chem Tech Res 3:534–538
Arora O (2017) World J Eng Res 2:81
Arsalani N (2010) Fattahi H & Nazarpoor M. Polym Lett 4:329
Arunachalam T, Karpagasundaram M, Rajarathinam N (2017) Ultrasound assisted green synthesis of cerium oxide nanoparticles using Prosopis juliflora leaf extract and their structural, optical and antibacterial properties. Mat Sci Poland 35:791–798
Arya A, Gangwar A, Singh SK, Roy M, Das M et al (2016) Cerium oxide nanoparticles promote neurogenesis abrogate hypoxia-induced memory impairment through AMPK-PKC-CBP signaling cascade. Int J Nanomedicine 11:1159–1173
Asati A, Santra S, Kaittanis C, Nath S, Perez JM (2009) Oxidase-like activity of polymer-coated cerium oxide nanoparticles. Angew Chem Int Ed Engl 48:2308–2312
Ayyub P, Palkar VR, Chattopadhyay S, Multani M (1995) Phys Rev B 51:6135
Azam A, Ahmed AS, Oves M et al (2012) Size-dependent antimicrobial properties of CuO nanoparticles against gram-positive and -negative bacterial strains. Int J Nanomedicine 7:3527–3535
Behera M, Giri G (2014) Green synthesis and characterization of cuprous oxide nanoparticles in presence of a bio-surfactant. Mat Sci Poland 32:702–708
Bernard V, Mornstein V (2016) The viability of ovarian carcinoma cells a2780 affected by titanium dioxide nanoparticles and low ultrasound intensity. Lékař a technika 46:21–24
Bharde A, Rautaray D, Bansal V, Ahmad A, Sarkar I, Yusuf SM, Sanyal M, Sastry M (2006) Extracellular biosynthesis of magnetite using fungi. Small 2:135–141
Bhattacharyya S, Kudgus RA, Bhattacharya R, Mukherjee P (2011) Inorganic nanoparticles in cancer therapy. Pharm Res 28:237–259
Bogdan J, Czarnak JP, Zarzyńska J (2017) Nanoparticles of titanium and zinc oxides as novel agents in tumor treatment: a review. Nanoscale Res Lett 12:225
Byrne, J.M.; Telling, N.D.; Coker, V.S.; Pattrick, R.A.D.; van der Lann, G.; Arenholz, E.; Tuna, F.; Lloyd, J.R. (2011) Control of nanoparticle size, reactivity and magnetic properties during the bioproduction of magnetite by Geobacter sulfurreducens. Nanotechnology 2011, 22, 455709
Cabrera L, Gutierrez S, Menendez N, Morales P, Herrasti P (2008) Magnetite nanoparticles: electrochemical synthesis and characterization. Electrochim Acta 53:3436
Cahyana AH, Pratiwi D, Ardiansah B (2017) Int Conf on Recent Trends in Physics IOP Conf. Series: Mat Sci and Eng 188: 1
Chakraborty R, Basu T (2017) Metallic copper nanoparticles induce apoptosis in a human skin melanoma A-375 cell line. Nanotechnology 28(10):105101
Charbgoo F, Ahmad MB, Darroudi M (2017) Cerium oxide nanoparticles: green synthesis and biological applications. Int J Nanomedicine 12:1401–1413
Chatterjee A, Ajantha M, Talekar A, Revathy N, Abraham J (2017) Biosynthesis, antimicrobial and cytotoxic effects of titanium dioxide nanoparticles using Vigna unguiculata seeds. Int J Pharmacognosy Phytochem Res 9(1):95–99
Chazotte-Aubert L, Oikawa S, Gilibert I, Bianchini F, Kawanishi S, Ohshima H (1999) Cytotoxicity and site-specific DNA damage induced by nitroxyl anion (NO-) in the presence of hydrogen peroxide. Implications for various pathophysiological conditions. J Biol Chem 274:20909–20915
Chen M, Zhang Y, Huangetal B (2013) Evaluationoftheantitumor activity by Ni nanoparticles with verbascoside. J Nanomater,2013: 623497,6

Clutton S (1997) The importance of oxidative stress in apoptosis. Br Med Bull 53:662–668

Colon J, Herrera L, Smith J et al (2009) Protection from radiation-induced pneumonitis using cerium oxide nanoparticles. Nanomedicine 5:225–231

Colon J, Hsieh N, Ferguson A, Kupelian P, Seal S, Jenkins DW, Baker CH (2010) Cerium oxide nanoparticles protect gastrointestinal epithelium from radiation-induced damage by reduction of reactive oxygen species and upregulation of superoxide dismutase 2. Nanomedicine 6:698–705

Danhier F, Feron O, Preat V (2010) To exploit the tumor microenvironment: passive and active tumor targeting of nanocarriers for anti-cancer drug delivery. J Control Release 148:135–146

Darroudi M, Sabouri Z, Oskuee RK, Zak AK, Kargar H, Hamid MHNA (2014a) Green chemistry approach for the synthesis of ZnO nanopowders and their cytotoxic effects. Ceram Int 40:4827–4831

Darroudi M, Sabouri Z, Oskuee RK, Zak AK, Kargar H, Hamid MHNA (2014b) Green chemistry approach for the synthesis of ZnO nanopowders and their cytotoxic effects. Ceram Int 40:4827–4831

Das S, Singh S, Dowding JM et al (2012) The induction of angiogenesis by cerium oxide nanoparticles through the modulation of oxygen in intracellular environments. Biomaterials 33:7746–7755

de Groot JF, Mandel JJ (2014) Update on anti-angiogenic treatment for malignant gliomas. Curr Oncol Rep 16:380

De Marzi L, Monaco A, de Lapuente J, Ramos D, Borras M, di Gioacchino M, Santucci S, Poma A (2013a) Cytotoxicity and genotoxicity of ceria nanoparticles on different cell lines *in vitro*. Int J Mol Sci 14:3065–3077

De Marzi L, Monaco A, de Lapuente J, Ramos D, Borras M, di Gioacchino M, Santucci S, Poma A (2013b) Cytotoxicity and genotoxicity of ceria nanoparticles on different cell lines in vitro. Int J Mol Sci 14:3065–3077

Deo KM, Pages BJ, Ang DL, Gordon CP, Aldrich-Wright JR (2016) Transition metal intercalators as anticancer agents-recent advances. Int J Mol Sci 17:1–17

Desmoulière A, Guyot C, Gabbiani G (2004) The stroma reaction myofibroblast: a key player in the control of tumor cell behavior. Int J Dev Biol 48:509–517

Desoize B (2004) Metals and metal compounds in cancer treatment. Anticancer Res 24:1529–1544

Dipranjan L, Debalina B, Arindam P, Chitta Ranjan S, Panchanan P et al (2012) Evaluation of copper iodide and copper phosphate nanoparticles for their potential cytotoxic effect. Toxicol Res 1:131136

Ducrocq C, Blanchard B, Pignatelli B, Ohshima H (1999) Peroxynitrite: an endogenous oxidizing and nitrating agent. Cell Mol Life Sci 55:1068–1077

Dutta D, Mukherjee R, Patra M, Banik M, Dasgupta R, Mukherjee M, Basu T (2016) Green synthesized cerium oxide nanoparticle: a prospective drug against oxidative harm. Colloids Surf B: Biointerfaces 147:45–53

Eiserich JP, Hristova M, Cross CE, Jones AD, Freeman BA, Halliwell B, van der Vliet A (1998) Formation of nitric oxide derived inflammatory oxidants by myeloperoxidase in neutrophils. Nature 391:393–397

Ezhilarasi AA, Vijaya JJ, Kaviyarasu K, Maaza M, Ayeshamariam A, Kennedy LJ (2016a) Green synthesis of NiO nanoparticles using Moringa oleifera extract and their biomedical applications: cytotoxicity effect of nanoparticles against HT-29 cancer cells. J Photochem Photobiol B 164:352–360

Ezhilarasi AA, Vijaya JJ, Kaviyarasu K, Maaza M, Ayeshamariame A, Kennedy LJ (2016b) Green synthesis of NiO nanoparticles using Moringa Oleifera extract and their biomedical applications: Cytotoxicity effect of nanoparticles against HT-29 cancer cells https://doi.org/10.1016/j.jphotobiol.2016.10.003

Falcaro P, Ricco R, Yazdi A et al (2015) Application of metal and metal oxide nanoparticles @ MOFs. Coord Chem Rev 307

Fernández-García M, Martínez-Arias A, Hanson JC, Rodríguez JA (2004) Nanostructured oxides in chemistry: characterization and properties. Chem Rev 104:4063

Fujiwara R, Luo Y, Sasaki T, Fujii K, Ohmori H, Kuniyasu H (2015) Cancer therapeutic effects of titanium dioxide nanoparticles are associated with oxidative stress and cytokine induction. Pathobiology 82:243–251

Geetha R, Ashokkumar T, Tamilselvan S, Govindaraju K, Sadiq M et al (2013) Green synthesis of gold nanoparticles and their anticancer activity. Cancer Nanotechnol 4(4–5):91–98

Giri S, Karakoti A, Graham RP, Maguire JL, Reilly CM et al (2013) Nanoceria: a rare-earth nanoparticle as a novel anti-angiogenic therapeutic agent in ovarian cancer. PLoS One 8:e54578

Gonzales M, Krishnan KM (2005) Synthesis of magnetoliposomes with monodisperse iron oxide nanocrystal cores for hyperthermia. J Magnet Mag Mat 293:265–270

Green M (2005) Organometallic based strategies for metal nanocrystal synthesis. Chem Commun (24):3002–3011

Grisham MB, Jourd'heuil D, Wink DA (2000) Chronic inflammation and reactive oxygen and nitrogen metabolism-implications in DNAdamage and mutagenesis [review]. Aliment Pharmacol Ther 14(suppl 1):3–9

Gultekin DD, Nadaroglu H, Gungor AA, Kishali NH (2017) Biosynthesis and characterization of copper oxide nanoparticles using Cimin grape (Vitis vinifera cv) extract. Int J Sec Metabolite 4:77–84

Guo D, Wu C, Jiang H et al (2008) Synergistic cytotoxic effect of different sized ZnO nanoparticles and daunorubicin against leukemia cancer cells under UV irradiation. J Photochem Photobiol B 93:119–126

Guo Y, Zhang J, Yang L, Wang H, Wang F et al (2010) Synthesis of amorphous and crystalline cupric sulfide nanoparticles and study on the specific activities on different cells. Chem Commun (Camb) 46(20):3493–3495

Hackenberg S, Scherzed A, Harnisch W, Froelich K, Ginzkey C, Koehler C, Hagen R, Kleinsasser N (2012) Antitumor activity of photo-stimulated zinc oxide nanoparticles combined with paclitaxel or cisplatin in HNSCC cell lines. J Photochem Photobiol B 114:87–93

Han L, Li S, Yang Y, Zhao F, Huang J, Chang J (2007a) Comparison of magnetite nanocrystals formed by biomineralization and chemosynthesis. J Mag Mag Mat 313:236–242

Han L, Li S, Yang Y, Zhao F, Huang J, Chang J (2007b) Comparison of magnetite nanocrystals formed by biomineralization and chemosynthesis. J Mag Mag Mat 313:236–242

Hanley C, Layne J, Punnoose A et al (2008) Preferential killing of cancer cells and activated human T cells using zinc oxide nanoparticles. Nanotechnology 19:295103–295113

Hanley C, Thurber A, Hanna C et al (2009) The influences of cell type and ZnO nanoparticle size and immune cell cytotoxicity and cytokine induction. Nanoscale Res Lett 4:1409–1420

Heckman KL, DeCoteau W, Estevez A, Reed KJ, Costanzo W et al (2013) Custom cerium oxide nanoparticles protect against a free radical mediated autoimmune degenerative disease in the brain. ACS Nano 7:10582–10596

Helana V, Princea J, Al-Dhabib NA, Arasub MV, Ayeshamariam A, Madhumitha G, Roopan SM, Jayachandrane M (2016) Neem leaves mediated preparation of NiO nanoparticles and its magnetization, coercivity and antibacterial analysis. Results Phys 6:712–718

Herrera-Becerra R, Rius JL, Zorrilla C (2010) Tannin biosynthesis of iron oxide nanoparticles. Appl Phys A Mater Sci Process 100:453–459

Hilger I, Kaiser WA (2012) Iron oxide-based nanostructures for MRI and magnetic hyperthermia. Nanomedicine 7:1443–1459

Honary S, Barabadi H, Gharaeifathabad E, Naghibi F (2012) Green synthesis of copper oxide nanoparticles using Penicillium aurantiogriseum, Penicillium citrinum and Penicillium waksmanii. Digest J Nanomat Biostruct 7:999–1005

Hong RY, Pan TT, Li HZ (2006) Microwave synthesis of magnetic Fe3O4 nanoparticles used as a precursor of nanocomposites and ferrofluids. J Magn Magn Mater 303(1):60–68

Hong H et al (2011) Cancer-targeted optical imaging with fluorescent zinc oxide nanowires. Nano Lett 11:3744–3750

Hong H et al (2015) Red fluorescent zinc oxide nanoparticle: a novel platform for cancer targeting. ACS Appl Mater Interfaces 7:3373–3381

Huang YW, Wu C, Aronstam RS (2010) Toxicity of transition metal oxide nanoparticles: recent insights from in vitro studies. Materials 3:4842–4859

Indira TK, Laksmi PK (2010) Magnetic nanoparticles—a review. Int J pharm Sci. Nanotechnol 3:1035–1042

Jain TK, Morales MA, Sahoo SK, Leslie-Pelecky DL, Labhasetwar V (2005) Iron oxide nanoparticles for sustained delivery of anticancer agents. Mol Pharm 2(3):194–205

Jayaseelan C, Rahuman AA, Kirthi AV, Marimuthu S, Kumar TS, Bagavan A, Gaurav K, Karthik L, BhaskaraRao KV (2012) Novel microbial route to synthesize ZnO nanoparticles using Aeromonas hydrophila and their activity against pathogenic bacteria and fungi. Spectrochim Acta A Mol Biomol Spectrosc 90:78–84

Jeng HA, Swanson J (2006) Toxicity of metal oxide nanoparticles in mammalian cells. J Environ Sci Health A Tox Hazard Subst Environ Eng 41:2699–2611

Jeronsia JE, Raj DJV, Joseph LA, Rubini K, Das SJ (2016) In vitro antibacterial and anticancer activity of copper oxide nanostructures in human breast Cancer Michigan Cancer Foundation-7 cells. J Med Sci 36(4):145–151

Jha AK, Prasad K (2010) Biosynthesis of metal and oxide nanoparticles using lactobacilli from yoghurt and probiotic spore tablets. Biotechnol J 5:285–291

Johannsen M, Thiesen B, Wust P, Jordan A (2010) Magnetic nanoparticle hyperthermia for prostate cancer. Int J Hyperth 26:790–795

Jurgons R, Seliger C, Hilpert A, Trahms L, Oden-bach S, Alexiou C (2006) Drug loaded magnetic nanoparticles for cancer therapy. J Phys: Cond Mat 18:S2893–S2902

Karlsson HL, Cronholm P, Gustafsson J, Möller L (2008) Copper oxide nanoparticles are highly toxic: a comparison between metal oxide nanoparticles and carbon nanotubes. Chem Res Toxicol 21:1726–1732

Kaur P, Thakur R, Chaudhury A (2016) Biogenesis of copper nanoparticles using peel extract of Punica granatum and their antimicrobial activity against opportunistic pathogens. Green Chem Lett Rev 9:33–38

Kawanishi S, Inoue S, Yamamoto K (1989) Site-specific DNAdamage induced by nickel(II) ion and hydrogen peroxide. Carcinogenesis 10:2231–2235

Khalil AT, Ovais M, Ullah I, Ali M, Shinwari ZK, Maaza M (2017a) Biosynthesis of iron oxide (Fe_2O_3) nanoparticles via aqueous extracts of *Sageretia thea* (Osbeck.) and their pharmacognostic properties. Green Chem Lett Rev 10:186–201

Khalil AT, Ovais M, Ullah I, Ali M, Shinwari ZK, Hassan D, Maaza M (2017b) Sageretia thea (Osbeck.) modulated biosynthesis of NiO nanoparticles and their in vitro pharmacognostic, antioxidant and cytotoxic potential. Artif Cells Nanomedicine Biotechnol. https://doi.org/10.1080/21691401.2017.1345928

Khan S, Ansari AA, Khan AA, Ahmad R, Al-Obaid O, Kattan A-W (2015) In vitro evaluation of anticancer and antibacterial activities of cobalt oxide nanoparticles. J Biol Inorg Chem 20(8):1319–1326

Kim J, Park S, Lee JE, Jin SM, Lee JH, Lee IS, Yang I, Kim JS, Kim SK, Cho MH, Hyeon T (2006) Designed fabrication of multi functional magnetic gold nano shells and their application to magnetic resonance imaging and photothermal therapy. Angew Chem Int Ed 45:7754–7758

Knaapen AM, Borm PJ, Albrecht C, Schins RPL (2004) Inhaled particles and lung cancer. Part a. mechanisms. Int J Cancer 109:799–809

Kohler N, Sun C, Wang J, Zhang M (2005) Methotrexate-modified superparamagnetic nanoparticles and their intracellular uptake into human cancer cells. Langmuir 21:8858–8864

Kolosnjaj-Tabi J, di Corato R, Lartigue L, Marango I, Guardia P, Silva AK, Luciani N, Clément O, Flaud P, Singh JV et al (2014) Heat-generating iron oxide nanocubes: subtle "destructurators" of the tumoral microenvironment. ACS Nano 8:4268–4283

Kumar JS, Kumar SV, Kumar SR (2017) Synthesis of zinc oxide nanoparticles using plant leaf extract against urinary tract infection pathogen. Resour-Efficient Technols 3:459–465

Kurzawa-Zegota M, Sharma V, Najafzadeh M, Reynolds PD, Davies JP, Shukla RK, Alok D, Anderson D (2017) Titanium dioxide nanoparticles induce DNA damage in peripheral blood

lymphocytes from *Polyposis coli*, Colon Cancer patients and healthy individuals: an *Ex Vivo/In Vitro* study. J Nanosci Nanotechnol 17(12):9274–9285
Laha D, Pramanik A, Maity J, Mukherjee A, Pramanik P et al (2014) Interplay between autophagy and apoptosis mediated by copper oxide nanoparticles in human breast cancer cells MCF7. Biochim Biophys Acta 1840(1):1–9
Laurent S, Dutz S, Häfeli UO, Mahmoudi M (2011) Magnetic fluid hyperthermia: focus on superparamagnetic iron oxide nanoparticles. Adv Colloid Interf Sci 166:8–23
Laurent S, Saei AA, Behzadi S, Panahifar A, Mahmoudi M (2014) Superparamagnetic iron oxide nanoparticles for delivery of therapeutic agents: opportunities and challenges. Expert Opin Drug Deliv 11(9):1449–1470
Li D et al (2014) Theranostic nanoparticles based on bioreducible polyethylenimine-coated iron oxide for reduction-responsive gene delivery and magnetic resonance imaging. Int J Nanomedicine 9:3347–3361
Lien HL, Jhuo YS, Chen LH (2007) Effect of heavy metals on dechlorination of carbon tetrachloride by iron nanoparticles. Environ Eng Sci 24:21
Lin W, Huang YW, Zhou XD, Ma Y (2006a) Toxicity of cerium oxide nanoparticles in human lung cancer cells. Int J Toxicol 25:451–457
Lin W, Huang YW, Zhou XD, Ma Y (2006b) Toxicity of cerium oxide nanoparticles in human lung Cancer cells. Int J Toxicol 25:451–457
Lin G et al (2014) Delivery of siRNA by MRI-visible nanovehicles to overcome drug resistance in MCF-7/ADR human breast cancer cells. Biomaterials 35:9495–9507
Liu XM, Kim JK (2009) Solvothermal synthesis and magnetic properties of magnetite nanoplatelets. Mater Lett 63:428
Lord MS, Tsoi B, Gunawan C, Teoh WY, Amal R, Whitelock JM (2013a) Anti-angiogenic activity of heparin functionalised cerium oxide nanoparticles. Biomaterials 34:8808–8818
Lord MS, Tsoi B, Gunawan C, Teoh WY, Amal R, Whitelock JM (2013b) Anti-angiogenic activity of heparin functionalised cerium oxide nanoparticles. Biomaterials 34:8808–8818
Lotfian H, Nemati F (2018) Cytotoxic effect of Tio_2 nanoparticles on breast Cancer cell line (MCF-7). IIOABJ 7:219–224
Ma X et al (2014) Exploring a new SPION-based MRI contrast agent with excellent water-dispersibility, high specificity to cancer cells and strong MR imaging efficacy. Colloids Surf B Bio Interfaces 126C:44–49
Maier-Hauff K, Ulrich F, Nestler D, Niehoff H, Wust P, Thiesen B, Orawa H, Budach V, Jordan A (2011) Efficacy and safety of intratumoral thermotherapy using magnetic iron-oxide nanoparticles combined with external beam radiotherapy on patients with recurrent glioblastoma multiforme. J Neuro-Oncol 103:317–324
Malarkodi C, Chitra K, Rajeshkumar S, Gnanajobitha G, Paulkumar K, Vanaja M, Annadurai G (2013) Novel eco-friendly synthesis of titanium oxide nanoparticles by using Planomicrobium sp. and its antimicrobial evaluation. Pelagia Res Library 4:59–66
Mariam AA, Kashif M, Arokiyarajetal S (2014) Bio-synthesisof NiO and Ni nanoparticles and their characterization. Dig J Nanomater Biostruct 9:1007–1019
McHale JM, Auroux A, Perrota AJ, Navrotsky A (1997) Science 277:788
Millot N, Aymes D, Bernard F, Niepce JC, Traverse A, Bouree F, Cheng BL, Perriat P (2003) J Phys Chem B 107:5740
Mishra RK, Thomas S, Kalarikkal N (eds) (2017) Micro and nano fibrillar composites (MFCs and NFCs) from polymer blends. Woodhead, Elsevier
Miyoshi N, Kume K, Tsutumi K, Fukunaga Y, Ito S, Imamurad Y, Bibin AB (2011) Application of Titanium Dioxide (TiO2) nanoparticles in photodynamic therapy (PDT) of an experimental tumor. AIP Confer Proc 1415:21
Moos PJ, Chung K, Woessner D et al (2010) ZnO particulate matter requires cell contact for toxicity in human colon cancer cells. Chem Res Toxicol 23:733–739
Mukherjee D, Ghosh S. (2014). Green synthesis of ά-Fe2O3 nanoparticles from Aloe vera plant extract and its application in Arsenic (V) remediation. The First International Conference on

Emerging Materials: Characterization & Application EMCA-2014 December 4–6, 2014,, At Kolkata, INDIA

Munusamy S, Bhakyaraj K, Vijayalakshmi L, Stephen A, Narayanan V (2014) Synthesis and characterization of xerium oxide nanoparticles using *Curvularia lunata* and their antibacterial properties. pp 2347–3207

Muthukumar H, Matheswaran M (2015) Amaranthus spinosus leaf extract mediated FeO nanoparticles: physicochemical traits, photocatalytic and antioxidant activity. ACS Sustain Chem Eng 3:3149–3159

Naeimi H, Nazifi ZS (2013) J Nanopart Res 15:2026

Nagajyothia PC, Muthuramanb P, Sreekanthc TVM, Kim DH, Shima J (2017) Green synthesis: in-vitro anticancer activity of copper oxide nanoparticles against human cervical carcinoma cells. Arab J Chem 10(2):215–225

Naika HR, Lingaraju K, Manjunath K, Kumar D, Nagaraju G, Suresh D, Nagabhushanae H (2015) Green synthesis of CuO nanoparticles using *Gloriosa superba* L. extract and their antibacterial activity. J Taibah Univ Sci 9:7–12

Nair S, Sasidharan A, Divya Rani V et al (2009) Role of size scale of ZnO nanoparticles and microparticles on toxicity toward bacteria and osteoblast cancer cells. J Mater Sci Mater Med 20:235–241

Narayanan KB, Sakthivel N (2010) Biological synthesis of metal nanoparticles by microbes. Adv Colloid Interf Sci 156:1–13

Nasseri MA, Ahrari F, Zakerinasab B (2016) A green biosynthesis of NiO nanoparticles using aqueous extract of Tamarix serotina and their characterization and application. Appl Organometal Chem 30:978–984

Naz S, Beach J, Heckert B, Tummala T, Pashchenko O et al (2017) Cerium oxide nanoparticles: a 'radical' approach to neurodegenerative disease treatment. Nanomedicine 12:545–553

Neri D, Supuran CT (2011) Interfering with pH regulation in tumors as a therapeutic strategy. Nat Rev Drug Discov 10:767–777

Neuberger T, Schopf B, Hofmann H, Hofmann M, Rechenberg B (2005) Superparamagnetic nanoparticles for biomedical applications: possibilities and limitations of a new drug delivery system. J Magnet Mag Mat 293:483–496

Olsvik O, PopovicT SE, Cudjoe KS, Hornes E, Ugelstad J, Uhlen M (1994) Magnetic separation techniques in diagnostic microbiology. Clin Microbiol Rev 7:43–54

Órdenes-Aenishanslins NA, Saona LA, Durán-Toro VM, Monrás JP, Bravo DM, Pérez-Donoso JM (2014) Use of titanium dioxide nanoparticles biosynthesized by Bacillus mycoides in quantum dot sensitized solar cells. Microb Cell Factories 13:90

Panariti A, Misserocchi G, Rivolta I (2012) The effect of nanoparticle uptake on cellular behaviour: disrupting or enabling functions? Nanotechnol Sci Appl 5:87–100

Pandey N, Dhiman S, Srivastava T, S M (2016) Transition metal oxide nanoparticles are effective in inhibiting lung cancer cell survival in the hypoxic tumor microenvironment. Chem Biol Interact 254:221–230

Pešić M, Podolski R-A, Stojković S, Matović B, Zmejkoski D et al (2015) Anti-cancer effects of cerium oxide nanoparticles and its intracellular redox activity. Chem Biol Interact 5(232):85–93. however copper nanoparticles have turned into a favoured decision among researchers [5-10]

Phumying S, Labuayai S, Thomas C, Amorkitbamrung V, Swatsitang E, Maensiri S (2013) Aloe-vera plant extracted solution hydrothermal synthesis and magnetic properties of magnetite (Fe_3O_4) nanoparticles. Appl Phys A 111(4)

Pilapong C et al (2014) Smart magnetic nanoparticle-aptamer probe for targeted imaging and treatment of hepatocellular carcinoma. Int J Pharm 473:469–474

Pramanik A, Laha D, Chattopadhyay S, Dash SK, Roy S et al (2015) Targeted delivery of "copper carbonate" nanoparticles to cancer cell in-vivo. Toxicol Res 4:1604–1612

Prasad K, Jha AK (2009) ZnO nanoparticles: synthesis and adsorption study. Nat Sci 1:129–135

Rafique S, Idrees M, Nasim A et al (2010) Transition metal complexes as potential therapeutic agents. Biotechnol Mol Biol Rev 5:38–45

Raghad DH, Abdul J, Rasha SN, Ahmed NA (2016) Biological synthesis of titanium dioxide nanoparticles by Curcuma longa plant extract and study its biological properties. WSN 49(2):204–222

Raj KP, Sivakarthik P, Uthirakumar AP (2014) Thangaraj V cytotoxicity assessment of synthesized nickel oxide nanoparticles on MCF-7 and A-549 cancer cell lines. J Chem Pharma Sci 4:269–271

Rajakumar G, Rahuman A, Roopan SM, Khanna G, Elango G, Kamaraj C, Zahir AA, Kanayairam V (2012) Fungus mediated biosynthesis and characterization of TiO2 nanoparticles and their activity against pathogenic bacteria. Spectrochim Acta A Mol Biomol Spectrosc 90:78–84

Rajan A, Cherian E, Baskar G (2016) International. J Modern Sci Technol 1:52–57

Ramesh P, Rajendrana A, Meenakshisundaram M (2014) Green synthesis of zinc oxide nanoparticles using flower extract Cassia Auriculata. J Nanosci Nanotechnol 2:41–45

Rao KG, Ashok CH, Rao KV, Chakra CS, Tambur P (2015) Green synthesis of TiO_2 nanoparticles using Aloe Vera extract. Int J Adv Res Phy Sci 2:28–34

Rao PV, Nallappan D, Madhavi K, Rahman S, Wei LJ, Gan SH (2016) Phytochemicals and Biogenic Metallic Nanoparticles as Anticancer Agents. Hindawi Publishing Corporation Oxidative Medicine and Cellular Longevity Volume 2016, Article ID 3685671, 15 pages https://doi.org/10.1155/2016/3685671

Reddy KM, Feris K, Bell J et al (2007) Selective toxicity of zinc oxide nanoparticles to prokaryotic and eukaryotic systems. Appl Phys Lett 90:213902–213903

Rodríguez JA, Fernández-García M (eds) (2007) Synthesis, properties and applications of oxide nanoparticles. Wiley, New Jersey

Ryter SW, Kim HP, Hoetzel A, Park JW, Nakahira K, Wang X, Choi AM (2007) Mechanisms of cell death in oxidative stress. Antioxid Redox Signal 9:49–89

Sack M, Alili L, Karaman E, Das S, Gupta A, Seal S, Brenneisen P (2014a) Combination of conventional chemotherapeutics with redox-active cerium oxide nanoparticles—a novel aspect in cancer therapy. Mol Cancer Ther 13:1740–1749

Sack M, Alili L, Karaman E, Das S, Gupta A, Seal S, Brenneisen P (2014b) Combination of conventional chemotherapeutics with redox-active cerium oxide nanoparticles—a novel aspect in cancer therapy. Mol Cancer Ther 13:1740–1749

Sack M, Alili L, Karaman E, Das S, Gupta A et al (2014c) Combination of conventional chemotherapeutics with redox-active cerium oxide nanoparticles-a novel aspect in cancer therapy. Mol Cancer Ther 13:1740–1749

Sak K (2012) Chemotherapy and dietary phytochemical agents. Chemother Res Pract 2012:1–11

Salvadori MR, Nascimento CAO, Corr^ea B (2014) Nickel oxidenanoparticlesfilmproducedbydeadbiomassoffilamentousfungus. Sci Rep 4:Article 6404

Sangeetha G, Rajeshwari S, Rajendran V (2012) Aloe barbadensis miller mediated green synthesis of mono-disperse copper oxide nanoparticles: optical properties. Spectrochim Acta Part A 97:1140–1144

Sankar R, Maheswari R, Karthik S, Shivashangari KS, Ravikumar V (2014) Anticancer activity of Ficus religiosa engineered copper oxide nanoparticles. Mater Sci Eng C 44:234–239

Santhoshkumar T, Rahuman AA, Jayaseelan C, Rajakumar G, Marimuthu S, Kirthi AV, Velayutham K, Thomas J, Venkatesan J, Kim S (2014) Green synthesis of titanium dioxide nanoparticles using Psidium guajava extract and its antibacterial and antioxidant properties. Asian Pac J Trop Med 7:968–976

Saptarshi SR, Dschl A, Lopata AL (2013) Interaction of nanoparticles with proteins: relation to bioreactivity of the nanoparticle. J Nanobiotechnol 11:26

Senthil M, Ramesh C (2012) Biogenic synthesis of Fe_3O_4 nanoparticles using tridax procumbens leaf extract and its antibacterial activity on Pseudomonas Aeruginosa. Dig J Nanomater Biostruct 7:1655–1660

Shah RK, Boruah F, Parween N (2015) Synthesis and characterization of ZnO nanoparticles using leaf extract of Camellia sinensis and evaluation of their antimicrobial efficacy. Int J Curr Microbiol App Sci 4:444–450

Shahbazi-Gahrouei D, Abdolahi M (2013) Detection of MUC1-expressing ovarian cancer by C595 monoclonal antibody-conjugated SPIONs using MR imaging. Sci World J 2013:609151

Sharma, H. et al (2014) Development and characterization of metal oxide nanoparticles for the delivery of anticancer drug. Artif Cells Nanomed Biotechnol. Published online November 19, 2014

Shi J et al (2014) A fullerene-based multi-functional nanoplatform for cancer theranostic applications. Biomaterials 35:5771–5784

Siddiqui MA, Alhadlaq HA, Ahmad J, Al-Khedhairy AA, Musarrat J, Ahamed M (2013a) Copper oxide nanoparticles induced mitochondria mediated apoptosis in human hepatocarcinoma cells. PLoS ONE 8, https://doi.org/10.1371/journal.pone.0069534

Siddiqui MA, Alhadlaq HA, Ahmad J, Al-Khedhairy AA, Musarrat J, Ahamed M (2013b) Copper oxide nanoparticles induced mitochondria mediated apoptosis in human hepatocarcinoma cells. PLoS One 8:e69534

Silva AC, Oliveira TR, Mamani JB, Malheiros SM, Malavolta L, Pavon LF, Sibov TT, Amaro E Jr, Tann_us A, Vidoto EL (2011) Application of hyperthermia induced by superparamagnetic iron oxide nanoparticles in glioma treatment. Int J Nanomedicine 6:591–603

Singh AV, Patil R, Anand A, Milani P, Gade WN (2010) Biological synthesis of copper oxide nano particles using Escherichia coli. Curr Nanosci 6:365–369

Sivaraj R, Rahman PK, Rajiv P, Narendhran S, Venckatesh R (2014a) Biosynthesis and characterization of Acalypha indica mediated copper oxide nanoparticles and evaluation of its antimicrobial and anticancer activity. Spectrochim Acta A 129:255–258

Sivaraj R, Rahman PKSM, Rajiv P, Venckatesh HASR (2014b) Biogenic copper oxide nanoparticles synthesis using Tabernaemontana divaricata leaf extract and its antibacterial activity against urinary tract pathogen. Spectrochim Acta Part A Mol Biomol Spec 133:178–181

Sivaraj R, Rahman PKSM, Rajiv GP, Venckatesh R (2014c) Biogenic zinc oxide nanoparticles synthesis using Tabernaemontana Divaricata leaf extract and its anticancer activity against MCF-7 breast cancer cell lines. Int Conf Advan Agric Biol Environ Sci:83–85

Sreekanth TVM, Pandurangan M, Kim DH, Lee YR (2016) Green synthesis: in-vitro anticancer activity of silver nanoparticles on human cervical Cancer cells. J Clust Sci 27(2):671681

Subhapriya S, Gomathipriya P (2018) Green synthesis of titanium dioxide (TiO_2) nanoparticles by Trigonella foenum-graecum extract and its antimicrobial properties. Microb Pathog 116:215–220

Subramanyam SG, Siva K (2016) Biosynthesis, Characterization and application of titanium oxide nanoparticles by Fusarium oxysporum. 4: 69–75

Sudhagar S et al (2011) Targeting and sensing cancer cells with ZnO nanoprobes in vitro. Biotechnol Lett 33:1891–1896

Sun C et al (2008) Tumor-targeted drug delivery and MRI contrast enhancement by chlorotoxin-conjugated iron oxide nanoparticles. Nanomedicine 3:495–505

Sun T, Yan Y, Zhao Y, Guo F, Jiang C (2012) Copper oxide nanoparticles induce autophagic cell death in A549 cells. PLoS One 7:e43442

Szymanski P, Frczek T, Markowicz M, Mikiciuk- Olasik E (2012) Development of copper based drugs, radiopharmaceuticals and medical materials. Biometals 25:1089–1112

Tarafdar A, Raliya R, Wang WN, Biswas P, Tarafdar JC (2013) Green synthesis of TiO_2 Nanoparticle using Aspergillus tubingensis. Adv Sci Eng Med 5:1–7

Tarnuzzer RW, Colon J, Patil S, Seal S (2005a) Vacancy engineered ceria nanostructures for protection from radiation-induced cellular damage. Nano Lett 5:2573–2577

Tarnuzzer RW, Colon J, Patil S, Seal S (2005b) Vacancy engineered ceria nanostructures for protection from radiation-induced cellular damage. Nano Lett 5:2573–2577

Tartaj P, Morales MD, Veintemillas-Verdaguer S, Gonzalez-Carreno T, Serna CJ (2003) The preparation of magnetic nanoparticles for applications in biomedicine. J Phys D: App Phys 36:182–197

Teja AS, Pei-Yoong K (2009) Synthesis, properties, and applications of magnetic iron oxide nanoparticles. Prog Cry Grow Charac Mat 55:22–45

Ullah M, Naz A, Mahmood T, Siddiq M, Bano A (2014) Biochemical synthesis of nickel & cobalt oxide nano-particles by using biomass waste. Int J Enhanc Res Sci Technol Eng 3:415–422
Usha R, Prabu E, Palaniswamy M, Venil CK, Rajendran KR (2010) Synthesis of metal oxide nano particles by Streptomyces sp. for development of antimicrobial textiles. Global J Biotechnol Biochem 5:153–160
Van Landeghem FK, Maier-Hauff K, Jordan A, Hoffmann K-T, Gneveckow U, Scholz R, Thiesen B, Brück W, von Deimling A (2009) Post-mortem studies in glioblastoma patients treated with thermotherapy using magnetic nanoparticles. Biomaterials 30:52–57
Vaseem M, Tripathy N, Khang G, Hahn Y-B (2017) Green chemistry of glucose-capped ferromagnetic hcp-nickel nanoparticlesandtheirreducedtoxicity. RSC Adv 3(25):9698–9704
Venkatesh KS, Gopinath K, Palani NS, Arumugam A, Jose SP, Bahadur SA, Ilangovan R (2016) Plant pathogenic fungus F. solani mediated biosynthesis of nanoceria: antibacterial and antibiofilm activity. RSC Adv 6:4272
Vinardell MP, Mitjans M (2015) Antitumor activities of metal oxide nanoparticles. Nano 5:1004–1021
von Montfort C, Alili L, Teuber-Hanselmann S, Brenneisen P (2015) Redox active cerium oxide nanoparticles protect human dermal fibroblasts from PQinduced damage. Redox Biol 4:1–5
Wahab R, Dwivedi S, Umar A, Singh S, Hwang IH, Shin HS, Musarrat J, Al-Khedhairy AA, Kim YS (2013a) ZnO nanoparticles induce oxidative stress in Cloudman S91 melanoma cancer cells. J Biomed Nanotechnol 9:441–449
Wahab R, Kaushik NK, Kaushik N, Choi EH, Umar A, Dwivedi S, Musarrat J, Al-Khedhairy AA (2013b) ZnO nanoparticles induces cell death in malignant human T98G gliomas, KB and non-malignant HEK cells. J Biomed Nanotechnol 9:1181–1189
Wahab R, Siddiqui MA, Saquib Q, Dwivedi S, Ahmad J, Musarrat J, Al-Khedhairy AA, Shin HS (2014) ZnO nanoparticles induced oxidative stress and apoptosis in HepG2 and MCF-7 cancer cells and their antibacterial activity. Colloids Surf B Biointerfaces 117:267–276
Walkey C, Das S, Seal S, Erlichman J, Heckman K et al (2015) Catalytic properties biomedical applications of cerium oxide nanoparticles. Environ Sci Nano 2:33–53
Wang H, Wingett D, Engelhard MH et al (2009a) Fluorescent dye encapsulated ZnO particles with cell-specific toxicity for potential use in biomedical applications. J Mater Sci Mater Med 20:11–22
Wang H, Wingett D, Engelhard MH et al (2009b) Fluorescent dye encapsulated ZnO particles with cell-specific toxicity for potential use in biomedical applications. J Mater Sci Mater Med 20:11–22
Wang Z, Lee H, Wu B et al (2010) Anti-microbial activities of aerosolized transition metal oxide nanoparticles. Chemosphere 80:525–529
Wang Y, Zi XY, Su J, Zhang HX, Zhang XR et al (2012) Cuprous oxide nanoparticles selectively induce apoptosis of tumor cells. Int J Nanomedicine 7:2641–2652
Wang Y, Yang F, Zhang H-X, Zi XY, Pan XH, Chem F, Luo WD, Li JX, Zhu HY, Hu YP (2013a) Cuprous oxide nanoparticles inhibit the growth and metastasis of melanoma by targeting mitochondria. Cell Death Dis 4. https://doi.org/10.1038/cddis.2013.314
Wang Y, Yang F, Zhang HX et al (2013b) Cuprous oxide nanoparticles inhibit the growth and metastasis of melanoma by targeting mitochondria. Cell Death Dis 4:e783
Wang X et al (2014) Surface engineered antifouling optomagnetic SPIONs for bimodal targeted imaging of pancreatic cancer cells. Int J Nanomedicine 9:1601–1615
Wason MS, Zhao J (2013a) Cerium oxide nanoparticles: potential applications for cancer and other diseases. Am J Transl Res 5:126–131
Wason MS, Zhao J (2013b) Cerium oxide nanoparticles: potential applications for cancer and other diseases. Am J Transl Res 5:126–131
Wason MS, Colon J, Das S, Seal S, Turkson J, Zhao J, Baker CH (2013) Sensitization of pancreatic cancer cells to radiation by cerium oxide nanoparticle-induced ROS production. Nanomedicine 9:558–569
Xu J, Yang H, Fu W, Du K, Sui Y, Chen J, Zeng Y, Li M, Zou G (2007) Preparation and magnetic properties of magnetite nanoparticles by sol–gel method. J Magn Mater 309:307–311

Yaaghoobi M, Emtiazi G, Roghanian R (2012a) A novel approach for aerobic construction of iron oxide nanoparticles by Acinetobacter radioresistens and their effects on red blood cells. Curr Nanosci 8:286–291

Yaaghoobi M, Emtiazi G, Roghanian R (2012b) A novel approach for aerobic construction of iron oxide nanoparticles by Acinetobacter radioresistens and their effects on red blood cells. Curr Nanosci 8:286–291

Yaaghoobi M, Emtiazi G, Roghanian R (2012c) A novel approach for aerobic construction of iron oxide nanoparticles by Acinetobacter radioresistens and their effects on red blood cells. Curr Nanosci 8:286–291

Zahir AA, Chauhan IS, Bagavan A, Kamaraj C, Elango G, Shankar J, Arjaria N, Roopan SM, Rahuman AA, Singh N (2015) Green synthesis of silver and titanium dioxide nanoparticles using Euphorbia prostrata extract shows shift from apoptosis to G0/G1 arrest followed by necrotic cell death in Leishmania donovani. Antimicrob Agents Chemother 59:4782–4799

Zhang H, Bandfield JF (1998) J Mater Chem 8:2073

Zhang ZJ, Chen XY, Wang BN, Shi CW (2008a) Hydrothermal synthesis and self-assembly of magnetite (Fe_3O_4) nanoparticles with the magnetic and electrochemical properties. J Cryst Growth 310(24):5453–5457

Zhang Y, Chen W, Wang SP et al (2008b) Phototoxicity of zinc oxide nanoparticle conjugates in human ovarian cancer. J Biomed Nanotechnol 4:432–438

Zhou Z et al (2014) Iron/iron oxide core/shell nanoparticles for magnetic targeting MRI and near-infrared photothermal therapy. Biomaterials 35:7470–7478

Latest Tools in Fight Against Cancer: Nanomedicines

6

Mala Trivedi, Parul Johri, Aditi Singh, Rachana Singh, and Rajesh K. Tiwari

Abstract

The application of nanotechnology in cancer management is being studied for specifically targeting cancer cells and destroying them with minimum damage to healthy tissues or using the nanoscale devices to detect cancer cells before they have formed tumors. Since the nanoparticles are much smaller than human cells, they easily move in and out of most cells just like large biomolecules of our body and can easily interact with other molecules on the surface as well as inside of the cells. Though the technology is not more than four decades old, it has produced substantial number of nanodiagnostic and nanotherapeutic agents with higher efficiency and safety. Nanotechnology has given a new insight for cancer treatment because of its potential to overcome the side effects of chemotherapeutic agents. An array of nanovehicle platforms can be designed which can specifically target the cancerous tissues, have high drug-loading capacities, and are favorable for endocytic intracellular uptake.

Keywords

Nanomedicine · Cancer · Nanoparticles · Diagnostics · Treatment · Nanomaterial

6.1 Introduction

Cancer, also termed neoplasm or malignant tumor, is characterized by the uncontrolled growth of body cells which can initiate at any part of the body, may spread to other organs or parts, and have various molecular subtypes. According to World

M. Trivedi · P. Johri · A. Singh · R. Singh · R. K. Tiwari (✉)
Amity Institute of Biotechnology, Amity University Uttar Pradesh, Lucknow, Uttar Pradesh, India
e-mail: rktiwari@lko.amity.edu

S. K. Saxena, S. M. P. Khurana (eds.), *NanoBioMedicine*, https://doi.org/10.1007/978-981-32-9898-9_6

Fig. 6.1 Various toxicological assays for testing nanomaterials safety

Characterization of nano materials
↓
Exposure of nano materials
↓
Single and repeated dosage toxicity testing
↓
In vitro genotoxicity
↓
Pharmacokinetics
↓
Teratogenicity / Oncogenicity / Chronic Toxicity
↓
Specialized Studies of the subject

Health Organization, cancer is the second leading cause of mortality worldwide and had accounted for almost eight and half million deaths in 2015 only (http://www.who.int/cancer/en/). The annual economic burden of cancer on the globe was over one trillion US dollar in 2010 itself. It is also reiterated that as per WHO's current evidence, around 30–50% of cancer deaths could have been prevented by effective early diagnosis, screening, treatment, and palliative care as each type of neoplasm requires specific management strategies. Nanotechnology is one such emerging interdisciplinary field of biomedical science that is being explored extensively for cancer management including precise imaging, screening, diagnostic, and therapeutics. Many such examples have been reported, e.g., Davis et al. (2010) have successfully used the targeted nanodelivery system to administer siRNA to induce RNAi mechanism. This gene silencing technique has been done in phase I clinical trials on patients with solid cancers. There is a wide variety of organic and inorganic nanomaterials used in cancer therapeutics and diagnostic applications. The key treatment for the cancer tumors located deep into our body are chemotherapy and radiotherapy, which suffers from their non-specific mode of actions and a wide variety of side effects (Fig. 6.1).

Moreover many multifunctional theragnostic nanosystems have been developed that can carry and deliver the specific drugs for cancer and can monitor the tumor lesions in the body as response of the drug delivered. In the past few years, nanoscale deliverance vehicles for RNAi (RNA interference) have been designed and used effectively in numerous cancer therapeutics (Tharushi et al. 2017). There is a wide variety of bioinformatics databases developed to support and aid the various latest therapeutics for cancers and tumors (Table 6.1).

6.2 Diagnosis

Nanoparticles are being studied for a safe and efficient delivery of therapeutic and diagnostic agent, and many such nanotechnologies have been either commercialized or have reached clinical trial stages. Few examples to cite are Doxil® and Abraxane®. Also in 2016 the polymeric nanoparticle BIND-014 carrying docletaxel

Table 6.1 Nanotechnology databases

S. no.	Name of the database	URL of the database
1.	Nanoparticle Information Library	http://nanoparticlelibrary.net/
2.	National Center for Biomedical Ontology Bioportal	http://www.bioontology.org/
3.	Nanomaterial Biological Interactions Knowledgebase	http://www.nbi.oregonstate.edu/
4.	Toxicology Literature Online	http://toxnet.nlm.nih.gov/cgi-bin/sis/htmlgen?TOXLINE
5.	OECD	http://webnet.oecd.org/NANOMATERIALS/Pagelet/Front/Default.aspx
6.	InterNano	http://www.internano.org/
7.	caNanoLab	https://cananolab.nci.nih.gov/caNanoLab/
8.	Collaboratory for Structural Nanobiology	http://csn.ncifcrf.gov/Advanced_Structure_Analysis/HOME.html
9.	Toxicology Data Network	http://toxnet.nlm.nih.gov/
10.	Nanomaterials Registry	https://www.nanomaterialregistry.org/

has successfully cleared first in human phase I trial in advanced solid tumors. The nanoparticle was well tolerated without any unexpected toxicity (Von Hoff et al. 2016). With the advent of therapeutic approach towards personalized medicines, the opportunities and possibilities for nanotechnology in cancer therapeutics are increasing. Also there are many added advantages of nanotechnology over conventional methods, such as delivery of poorly soluble drugs, modifying tissue distribution and limiting the drugs' cytotoxicity. One such example is the decreased cardiac toxicity in liposome-encapsulated doxorubicin (Mamot et al. 2012; Manzoor et al. 2012) and higher tolerance in albumin-stabilized paclitaxel (nab-PTX) (Golden et al. 1998).

Diagnostic methods are essential for the early detection of diseases to enable their prompt treatment, minimizing possible damage to the rest of the organism. The importance of imaging methods to diagnose, treat, and follow up cancer, cardiovascular, and neurological patients is well-known.

Cancer is one of the main causes of mortality worldwide and accounted for **7.6 million deaths** (around 13% of all deaths) in 2008.

6.2.1 Nanomedicine for Early Diagnosis of Cancers

Cancer biomarkers are indicators produced by tumor cells spreading in the body and are commonly used in cancer detection. However they are present in too low concentrations to be efficiently detected in early phases. However the targeted delivery of specific nanoparticles into the tumor can induce a local interaction with cancer cells and forces them to significantly increase the production of these biomarkers. Biomarkers detection becomes thus much easier and can provide an **earlier diagnosis to doctors than biopsies**. Early detections of cancers allow early and **less burdensome treatments,** increasing also the chances of recovery.

6.2.2 Cancer Detection

Nanoparticles have been explored as synthetic scaffolds for imaging probes used in the detection and monitoring of cancer and its treatment, because of their unique physical and chemical properties. Nanoparticles provide an excellent platform for drug delivery, like tunable surface properties, and also make them a good platform for diagnostic imaging. Noticeable progress has been made in this area using several different nanotechnology-based imaging modalities.

6.2.3 Quantum Dots

Quantum dots are semiconductor nanocrystals. They are one of the most coveted luminescence probes, with tremendous potential in many biological and biomedical applications. Because of their unique features, viz., small size (2–10 nm), size-variable optical properties, tunable surface properties, and extraordinary photostability, quantum dots are being used as exclusive probes for optical imaging (Bruchez Jr et al. 1998; Chan and Nie 1998; Akerman et al. 2002; Jaiswal et al. 2003; Alivisatos 2004). The most common quantum dot formulations used in biological applications are cadmium selenide (CdSe), indium phosphide (InP), cadmium telluride (CdTe), and indium arsenide (InAs). According to their size, quantum dots can reemit absorbed light of different specific wavelengths (400 to 1350 nm). The main drawback of quantum dots is to make a stable dispersed formulation under aqueous/physiological conditions; as soon as the dots come in contact with an aqueous solution, fluorescence is quenched. Most part of this problem has been overcome by the use of different coating materials. Some coating material increases the fluorescent yield of quantum dots. Moreover, surface coating reduces the adverse toxic effects that can be produced by toxic materials such as Cd, Se, and As. These composite nanomaterials have been found to be extremely efficient agents for cancer diagnosis in vivo. The small size of the quantum dots permits for unhampered access to the systemic circulation and conjugation of targeting molecules (Wu et al. 2003; Gao et al. 2004; Bharali et al. 2005; Michalet et al. 2005).

Cancer targeting and imaging using quantum dots (QDs) was first reported in vivo using live animals (Gao et al. 2004). Subcutaneous injection of QD-conjugated prostate cancer cells followed by systemic injection of multifunctional QD probes enabled multicolor fluorescence imaging of cancer cells with high sensitivity.

Chemical conjugation of tetrac to PEGylated QDs was demonstrated by Bharali et al. 2009. Notably, nano formulated tetrac retained its antiproliferative activity. This result highlights the incredible potential of QDs for site specific drug delivery and detection of cancer cells.

The use of QDs for NIR imaging (700–1000 nm wavelength range) has been reported by Frangioni (2003), Bhushan et al. (2008), and Gao et al. (2010) in different time periods. NIR QDs can enhance the depth of tissue penetration which helps in correct and sensitive detection of photons in vivo. With the help of NIR QDs, the difficulty of auto-fluorescence that is associated with the optical imaging of naturally occurring compounds in animal tissue can also be overcome. These QDs have great potential in terms of in vivo imaging. Use of this technology for lymphatic mapping in animal models has been validated successfully by Parungo et al. (2005). In a study by Allen et al. (2010) it is demonstrated the feasibility of using NIR QDs with an InAs (ZnCdS) core shell for biological imaging. These NIR QDs surface coated with PEG enabled imaging of the tumor vasculature as deep as 200 μm, whereas visible QDs generated images with very poor vascular contrast.

In short, semiconductor QDs have emerged as perfect nanodevices because of their distinguishing properties including high levels of light emission.

6.2.4 Magnetic Resonance Imaging

6.2.4.1 Iron Oxide Nanoparticles

From a clinical point of view, magnetic resonance imaging (MRI) is one of the most predominant and robust noninvasive imaging tools for detection and monitoring of disease. In recent years, noticeable progress has been seen in the development of nanoparticle systems for upgraded cancer imaging and diagnosis by MRI (Sun et al. 2008; Shubayev et al. 2009; Veiseh et al. 2010). Magnetic nanoparticles, an elite class of nanomedicines, have the potential to strongly change currently available clinical diagnostic and therapeutic paradigms. Magnetic nanoparticles used in biomedical applications contain an inorganic nanoparticle core and a surface coating that deliberates stability in aqueous dispersions (Zhang et al. 2002; Kohler et al. 2004; Cheng et al. 2008; Ho and Li 2008). Appropriate coatings may enable targeting of these nanoparticles and real-time monitoring or both. The success of magnetic nanoparticles in biomedical applications is mainly due to their exceptional improved proton relaxation capabilities (Josephson et al. 1988). Super magnetic iron oxide (SPIO) nanoparticles are used extensively with various trade names as bowel contrast agents (Lumerin, Gastromark) and for spleen/liver imaging (Endorem, Feridex) (Wang et al. 2001; Sun et al. 2008).

In recent years, great effort has been made to develop magnetic nanoparticles for targeted delivery of chemotherapeutic agents and various drugs. Strategies developed are magnetic drug targeting and conjugation of targeting moieties to the nanoparticle. Most commonly used nanoparticles for these types of applications are iron oxide nanoparticles. Certain degree of success has been noticed with these particles in preclinical trials. But the low efficacy of iron oxide nanoparticles to reach up to the desired sites or depths and their comparatively low efficiency have restricted their use. Developments in the synthesis of next-generation magnetic nanoparticle-based MRI contrasting agents have significant potential to change the imaging paradigm. These next-generation MRI contrasting agents consist of various core materials, such as iron oxide, coated with suitable materials and conjugated to tumor-specific moieties for improved efficacy and tumor targeting capabilities.

SPIO nanoparticles were first used in the clinical setting for imaging liver tumors (Reimer and Tombach 1998). These nanoparticles were readily taken up by Kupffer cells (hepatic macrophages located in the hepatic parenchyma). Since the most part of liver tumors are without macrophages, macrophage specific uptake of SPIOs increases the contrast between healthy and diseased tissue. With this technique, small liver tumors or metastases of 2–3 mm could be detected.

Another most widely used application of super paramagnetic nanoparticles is the diagnosis and treatment of tumors of the central nervous system (CNS). Specifically SPIOs are major agents for this type of application because they can be used as intravascular contrast agents and also for cellular imaging. Ferumoxtran-10 and its derivative ferumoxytol are examples of SPIOs that have been shown to be effective as MRI contrast agents for improved imaging of tumors in animal models (Varallyay et al. 2002; Neuwelt et al. 2004; Manninger et al. 2005).

6.2.4.2 Gadolinium-Incorporated Nanoparticles

Gadolinium-based complexes are the most frequently used contrast agents for clinical MRI. Gd complexes enhance contrast by selectively reducing the water molecule near the complex. Though this methodology has been used widely in MRI, it has gained comparatively low sensitivity. With the advancement in this area of nanotechnology, significant improvements have been achieved to overcome the low sensitivity and other issues related with Gd.

Xu et al. (2007) synthesized a G2 PAMAM dendrimer incorporating Gd(III)-1B 4M DTPA for targeting ovarian cancer in vivo. The dendrimer was fluorescently labeled with rhodamine green for MRI and optical fluorescence. When the nanoformulation was administered to mice bearing ovarian tumor xenografts, it effectively targeted the tumor tissue and delivered sufficient amounts of chelated Gd(III) and fluorophore to the tumor to produce visible changes in the tumor tissue by MRI and fluorescence imaging (Xu et al. 2007).

The major concerns in treatment of malignant gliomas are inefficient drug delivery due to the blood brain barrier and failure to image drug permeation throughout tumor tissue. Sarin et al. in 2009, detailed a unique method to overcome these problems, in which different-sized PAMAM dendrimers were synthesized with Gd-DTPA functionalized surfaces. In his work, the physiologic upper limit of nanoparticle size was less than 11.7–11.9 nm in diameter for traversing the blood brain/tumor barrier. These dendrimer nanoformulations also exhibited prolonged blood half-lives. Due to noninvasive nature and higher patient compliance, positron emission tomography (PET) is one of the most commonly used medical imaging tools for diagnosis and monitoring of number of diseases, including various types of cancer and heart disease. Nanoparticulate carrier systems labeled with diagnostic radionucleotides can additionally enhance the sensitivity of PET, chiefly for cancer detection when the nanoparticles are engineered to target the cancer sites by passive or active targeting.

Similarly, biodistribution studies of poly (butyl 2-cyanoacrylate) nanoparticles labeled with a 99mTc–dextrancomplex showed significant RES accumulation (~60% of the injected dose) within 2 min of injection. Antibodies against the $\alpha V\beta 3$ integrin, which is instrumental in tumor angiogenesis, have been conjugated to lipid nanoparticles to selectively deliver radionuclides to tumor sites (Li et al. 2002;

Guccione et al. 2004). Significantly higher levels of anti-αVβ3 integrin-conjugated DTPA derivatized nanoparticles radiolabeled with 111In accumulated in tumors after 72 h as compared to unconjugated nanoparticles (3%).

PET and bioluminescence were used to quantify the in vivo impact, including biodistribution and efficacy of tumor targeted small inhibitory siRNA nanoparticles. For PET imaging, 1,4,7,10-tetraazacyclododecane-1,4,7,10 tetraacetic acid was conjugated to the 5′ end of the siRNA molecule, allowing 64Cu-labeling for PET imaging. For targeted delivery, transferrin was also conjugated to the siRNA nanoparticle. Mice bearing luciferase-expressing Neuro2A subcutaneous tumors were analyzed by bioluminescence imaging and PET before and after injection to correlate functional efficacy with bio distribution data. Transferrin targeted siRNA nanoparticles reduced tumor luciferase activity by ~50% relative to nontargeted siRNA nanoparticles one day after injection. These results highlight on major important differences between intracellular and extracellular tumor spaces of targeted and non-targeted entities (Bartlett et al. 2007).

Synthesis of biocompatible nanoparticles for multimodal fusion imaging for cancer diagnosis and monitoring was reported by Nahrendorf et al. (2010). Mice were injected with a fluorophore derivatized RGD peptide, which targeted the αVβ3 integrin (Integrisense), a protease sensor (Prosense), and a nanoparticulate PET agent (64Cu-CLIO-VT680). Using multichannel FMT/PET-CT, the investigators were able to measure tumor proteases, macrophage content and integrin expression simultaneously, as well as distinct tumor localization of the probe. Therefore, multichannel FMT/PET-CT fusion can seamlessly integrate and visualize a number of important therapeutic parameters. This type of advancement will accelerate the progress of next-generation PET and optical molecular imaging agents.

Nanobiotechnology has incredible potential in terms of clinical imaging which has made possible convergence of different imaging modalities (e.g., PET and optical imaging). Combination of dual or multimodal imaging moieties in the same nanoparticulate carrier system can advance our understanding of biological systems by providing corresponding information. For example, PET imaging can deliver useful information based on whole body imaging while optical imaging can provide valuable molecular information about the cancer microenvironment.

On the road to recovery, the successful clinical translation of nanomedicines may be achieved by following the underneath steps:

1. Understand the interaction between the pathophysiology of tumor and the behavior of nanomedicine on that tumor.
2. Implementation of patient-focused nanomedicines, for a specific patient population (rendition from formulation-focused research to disease-driven development strategies).
3. More clinically relevant animal models should be used in order to bridge the gap between research and treatment (Edelman et al. 2010).
4. In the clinical trials (phase II and phase III), only the right patients should be selected and treated (use companion diagnostics for patient selection).

Table 6.2 Anticancer nanomedicines and their fate in clinical trials

S. no	Type of nanomedicine	Drug name	Indication (type of cancer)	Phase of testing
1.	Liposomes	Vincristine	Brain metastases Glioma	Phase II
2.	Polymeric nanoparticles	Docetaxel + prostate specific Membrane antigen	Cervical cancer Bladder cancer Head and neck cancer	Phase II
3.	Liposomes	Oxaliplatin	Gastrointestinal adenocarcinoma	Phase II
4.	Polymeric conjugates	Asparaginase	Acute lymphoblastic leukemia	Approved
5.	Polymeric micelles	DACH-platin	Non-small cell lung cancer Bladder cancer	Phase III
6.	Polymeric conjugates	Docetaxel	Solid tumors	Phase I
7.	Polymeric micelles	Paclitaxel	Ovarian cancer Breast cancer	Approved

A number of nanomedicines have been designed and tested for various trials (Table 6.2).

6.3 Treatment

Traditional System of medicines always played a vital role in meeting the global healthcare needs in past, which is continuing at present, and shall also play key role in the future. India is well-known for its rich, centuries-old heritage of traditional medicinal systems. Ancient Vedas and other scriptures point out the practice of traditional medicines in India.

6.3.1 Homeopathy and Cancer Therapy

There are various systems for treatment of diseases that include traditional and modern system. Traditional system, including homeopathy, Ayurvedic, Unani, and allopathic, comes under modern system. Germany is a place of origin of this system of medicine; from there it reaches to India in the early eighteenth century. Homeopathy system was not only well accepted in Indian traditional system of medicines but enriched it [Prasad et al. 2003]. Homeopathy is a system that was designed on the theory of "treating like with like." For treating any sickness, a homeopath prescribes small doses of drug, as higher doses are harmful to patients. Homeopathic medicines are prepared from extracts of plants and animal substances and also from minerals. They are generally diluted with water to prepare medicines.

Treatment Preferences of Homeopathy People prefer homeopathic medicines for treatment of ailments as they don't find any side effects of them. They can also be used along with other types of treatment. The mode of treatment in homeopathy is based on symptoms and conditions. Mode of treatment in homeopathy is to relax patient and control side effects, anxiety and depression; basically it's a placebo effect where people feel better after taking medicines prescribed by homeopath. Various clinical trials were conducted to test effects of homeopathic drugs in treating illness; however none confirm their effect in treating cancer. Research is still going on to test the role of these medicines in reducing cancers, enhancing immunity, etc. National Health and Medical Research Council (NHMRC), Australia, conducted a comparative study of homeopathic medicines on more than 150 people as a test and control with no medicines. There was no significant difference between the two groups. On the basis of the report of NHMRC, it was suggested that these medicines should not be recommended for serious or chronic diseases or diseases that could become serious in the future.

6.3.2 Ayurveda and Cancer Therapy

This is one of the oldest and indigenous systems of Indian medicine, based on plants. Ayurveda was used since ancient time in preventing or suppressing tumorous growth. According to Ayurveda, there are three systems for normal body functioning. These are "Vata," "Pitta," and "Kapha," i.e., nervous system, venous system, and arterial systems, respectively. Cancer can be defined as inflammatory/non-inflammatory swellings; this definition is from "Charak" and "Sushrut samhita" (Bhishagratha 1991). Malignancy is mentioned as "tridoshas." In malignant tumors, coordination between abovementioned three systems gets out of control leading to proliferation of cells (Balachandran and Govindarajan 2005). Ayurvedic therapy works on the principle of "finding cause of illness." There are four types of therapy in this system of medicine: (1) maintenance of health, (2) restoration of normal function, (3) cure, and (4) spiritual approach (Thatte and Dhahanukar 1991).

In Ayurveda various types of herbal preparations and also crude herbs are used for treatment. The advantage of this therapy is that there are no side effects and complications (Smith et al. 1995). Various anticancerous compounds have been isolated from plants, viz., vincristine, vinblastine, and taxol. Nine plant derived compounds have shown anticancer potential (Cravotto et al. 2010; Patel et al. 2010).

6.3.3 Modern Therapy

The modern therapy for cancer treatment includes radiotherapy and chemotherapy, to completely cure the ailments. However, this leads to many side effects that are sometimes toxic too.

6.3.3.1 Cancer Treatment

Scientists across the world are working day and night to find out cure for cancer. Drugs that were developed during this process were either ineffective or very toxic to patients. Drugs used for chemotherapy are insoluble in water; therefore chemical solvents are used that are toxic to our body [Kwon 2003].

To overcome the problem, hydrophobic nanocarriers were incorporated to anticancer drugs, viz., Doxil and Abraxane, to improve solubility of lipophilic compounds. Both of these medicines are available on the market for cancer treatment. Abraxane is used in treatment of breast cancer [Sparreboom et al. 2005; Gradishar et al. 2005; Moreno-Aspitia, A., & Perez, E. A. 2005]. This is an albumin-bound nanoparticle-based medicine. One more drug, doxorubicin, commercial name Doxil, is liposome-based medicine, effective against cancer [Martin 1998; Nishiyama, N. & Kataoka, K. 2006; Park 2002].

For most effective cancer treatment, targeted delivery of drugs to cancerous tumor is the most effective. The idea of liposomal-based nanomedicines is helpful in direct delivery of nanomedicine to target cancer cells. To target cancerous cells of such liposomal-based medicines, there is also a need of targeting ligand; for this purpose a conjugated nanoparticle can be designed for solubilization or functionalization requirement [Gao et al. 2005]. Recently the focus of the treatment of cancer is on synergistic drug treatment, where multiple compounds are integrated in a single particle. The particle starts delivering compounds to tumor in a sequential method. The chemotherapeutic agent doxorubicin is used in combination with antiangiogenesis agent, combretastatin [Sengupta et al. 2005].

Nanomedicines are small in size therefore they can easily target malignant cell in comparison to traditional medicine.

6.3.3.2 Nanotechnology in Drug Delivery Systems

The limitations encountered with delivery of drugs to tumor cells are poor solubility, poor distribution in tissues, combination of multiple drugs with distinct pharmacokinetic and pharmacodynamic properties, drug resistance at cellular or non-cellular levels, and distribution and clearance of drugs. Tumors show resistance to therapeutic agents by poor vascularization of tumors and thus reducing access of drug to tumor. Number of biochemical alterations, like altered enzyme activity, altered apoptosis, or multidrug resistance-associated protein (Brigger et al. 2012; Links and Brown 1999; Krishna and Mayer 2000), at cellular level may induce resistance to therapeutic agents. Lastly, the high toxicity of anticancer drugs on normal cells along with tumors limits their use for the strong side effects. The field of nanotherapeutics has been exploring possibilities to overcome these limitations. According to official registered records of clinicaltrials.gov, more than 1300 nanomedicine formulations for cancer therapy have been registered. Nanomaterials have been synthesized from organic, inorganic, lipid, glycan, proteins metrials as well as from synthetic polymers (Wicki et al. 2015). Following are the strategies for application of nanoparticle in drug delivery.

Passive Targeting

First-generation nanomedicine has been synthesized by modulating pharmacokinetic and biodistribution properties of the compound. Two such examples of first-generation nanomedicine based on passive targeting which have been commercialized are Doxil®, a doxorubicin HCl liposome injection indicated for ovarian cancer, multiple myeloma, and Kaposi's sarcoma, and Abraxane®, an albumin protein-bound paclitaxel injectable suspension recommended for metastatic breast cancer and non-small cell lung cancer (NSCLC). The phenomenon of enhanced permeability and retention (EPR) effect has been used in accumulation of nanodrugs in tumor. A study on macromolecule and nanoparticle transport via gaps in blood vessels has led to the understanding of transvascular pathway, transport across it, and microenvironment, thus concluding that along with other factors permeability to a molecule is not dependent on pore size as long as the molecule is much less in size (Hobbs et al. 1998). But passive targeting does not prevent accumulation of nanocarriers in other fenestrated endothelial organs (Wicki et al. 2012).

Active Targeting

To overcome limitations of passive targeting, second generation nanomedicines with improved functionalities and increased efficacy are being developed. In active targeting, a high-affinity ligand is attached to the surface of nanoparticle, which binds selectively to target cell receptor with high specificity. Various small and macromolecules have been used for this purpose, such as folic acid, carbohydrates, proteins, antibodies, etc. The important points to take into consideration are that ligand must be specific to binding with cancer cells with minimum binding with normal healthy cells; they should be stable enough to avoid premature degradation or cleavage and must not initiate unwarranted initiation of immune system. The approach also requires the optimal density of ligand on nanocarrier so that high level of target efficiency and optimal internalization are obtained (Bareford and Swaan 2007). Bhattacharyya et al. (2010; 2012) have shown different uptake mechanisms of gold nanoparticles with variable density of target antibodies.

In some cases, the nanocarriers are not always internalized, but deposition of the small drug molecules in vicinity of target tumors is sufficient, e.g., small molecules of doxorubicin can cross cell membrane by passive diffusion. Curative properties of noninternalized antibody-drug conjugates have been successfully studied in experimental animals (Pereira et al. 2014).

External Stimuli-Trigger Release

The release of drugs from nanocarriers in response to physical, chemical, or biological triggers has been tried with good results. The drug is released after a stimulus is generated by the neoplastic tissue. The internal triggers may be pH, redox state, ionic strength, or stress in target tissue (Jhaveri et al. 2014). pH of endosomes or lysosomes is different from that of blood or cytoplasm, and this change can be used as an internal stimuli. A pH-responsive liposome-containing glutamic acid-based zwitterionic lipids were synthesized and found to be having higher efficacy against breast cancer xenograft, as a result of better intracellular drug delivery and blood persistence as compared to conventional drug delivery (Obata et al. 2010). Another advantage of drug release by tumor microenvironment response is that it decreases drug side effect and of the many strategies is the enzyme sensitive release. Many biochemical molecules and enzymes are exposed differently in normal and cancer cells, like proteases, glucuronidases, lipases, oxidoreductases, etc., and this concept has been successfully employed for designing drug delivery system triggered by biocatalytic action (De La Rica et al. 2012). Also by decreasing tumor collagen intensity or decreasing internal fluid pressure by normalizing tumor microvessels, the accumulation and uptake of nanoparticles could be enhanced in tumors (Tong and Langer 2015). Disulfide-containing nanocarriers are another such strategy which is based on trigger release by hypoxic condition and poor nutrient levels in microenvironment (Fleige et al. 2012).

External stimuli like local hyperthermia have been used to release drugs from nanocarriers which are temperature responsive. ThermoDox, thermosensitive liposomes, are used for intravascular release of doxorubicin after mild hyperthermia (Manzoor et al. 2012). Similarly, ultraviolet and infrared light or ultrasound waves are some other external stimuli being tested (Fomina et al. 2010; Bardhan et al. 2011; Neofytou et al. 2012; Shim and Kwon 2012).

6.3.4 Multifunctional Nanomedicine

Multifunctional nanocarriers are another promising and latest methodology to combine more than one function in parallel, like delivery of multi-targeted drugs or combination therapies, such as doxorubicin and DNA-loaded nanoparticles (Liu et al. 2013) and polymersomes of doxorubicin and paclitaxel (Ahmed et al. 2006). More emerging multifunctional nanocompounds are theranostics, where diagnostics and treatment are carried out with same nanoformulations (Liong et al. 2008). Being used as personalized medicine as per individual patient need is another important achievement of theranostics. Though not much clinical trials have been done, significant results are obtained in a number of preclinical studies, such as miRNA theragnostic with wide application in targeted delivery of personalized medicine in multiple myeloma patients (Ahmad et al. 2014).

6.3.5 Nanomedicines

The major types of nanocompounds being used in cancer treatment are viral vectors, drug conjugates, lipid based, etc. Many of these nanomedicines have cleared phase I clinical trial in solid tumors, whereas some have been approved for clinical cancer care (Table 6.1).

6.3.5.1 Natural Compound Nanocarriers

There are wide varieties of natural compounds for drug delivery, ranging from lipids, proteins to glycans.

Lipid-Based Nanocarriers

Lipids have been extensively used as nanocarriers due to their advantage of self-assembly (Mura et al. 2015). Liposomes or phospholipids bilayers and micelles are the most commonly used lipids. They have been found to have much higher drug carrying capacity per molecule as compared to other forms. Doxil®, a liposomal doxorubicin, is the first US FDA-approved nanocarrier (Barenholz 2012; Harrison et al. 1995). nal-IRI, nanoliposomal irinotecan, is an efficient, high-loading nanomedicine with improved biodistribution and pharmacokinetics and with less systemic toxicity (Ko 2016). Recently, nanoliposomal irinotecan with fluorouracil/leucovorin has cleared phase III randomized clinical trial for treatment in advanced pancreatic cancer (Glassman et al. 2018) and is a US FDA-approved nanomedicine for metastatic pancreatic cancer. Myocet® (non-pegylated liposomal doxorubicin; Chan et al. 2004), DaunoXome® (non-pegylated liposomal daunorubicin; Gill et al. 1996), DepoCyt® (non-pegylated liposomal cytarabine; Gökbuget et al. 2011), Marqibo® (vincristine sulfate liposomes; O'Brien et al. 2013), and Mepact® (liposomal mifamurtide; Frampton 2010) are some of the other FDA-approved lipid-based nanomedicines for cancer treatment. However all these nanomedicines are not target based.

Immunoliposomes (ILs) are liposomes conjugated with antibody for selective targeting of antigen-expressing cells. ILs have been used for improving efficacy and reducing toxicity as well as in immunoassays, immunotherapy, and imaging; however very limited clinical trials and no FDA-approved drug have been reported yet (Wang et al. 2018). Similarly, targeted liposomes for siRNA delivery have resulted in enhanced drug uptake and reduced cytotoxicity in number of xenograft studies (Eloy et al. 2018). Thermosensitive liposome (TSL) is another example of target-specific drug delivery system, wherein target tissue is exposed to localized hyperthermia by an image-guided device and TSLs do precise drug delivery. Recent advancement in the field is intravascular trigger release in which the drug is released within seconds when TSLs pass through heated tissue region, thus enabling 20–30 times higher uptake of drugs (Haemmerich and Motamarry 2018).

Protein- and Peptide-Based Nanocarriers

Protein-based nanocarriers have attracted much attention due to their many advantages like low cytotoxicity, high drug binding capacity, increased uptake into targeted cells, high nutritional value, as well as being GRAS (generally regarded as safe). Also the ease to prepare them from many renewable sources and scale-up in manufacture make them a promising candidate for drug and gene delivery. The various functional groups available in a polypeptide give the flexibility for different three-dimensional networks to be synthesized for providing a protective matrix to molecules and increasing specific targets at the site of action. Proteins from animal as well as plant origins have been tried for nanocarriers. Gelatin, collagen, albumin, silk proteins, and elastin of animal origin are some such examples, whereas zein, gliadins, soy proteins, and lectins are plant-derived proteins. The protein albumin being nontoxic, nonimmunogenic, and biocompatible has emerged as the most successful protein to be used in nanoparticle synthesis. Albumin has a high binding capacity due to higher percentage of charged amino acids and thus presents many binding sites for drugs. Paclitaxel-albumin nanoparticle, Abraxane®, is the only FDA-approved chemotherapeutic nanodrug for breast cancer, non-small cell lung carcinoma, and pancreatic cancer. Abraxane® is a nab-paclitaxel with improved drug solubility but without much difference in overall survival.

Though animal protein, gelatin, has been reported as nanocarriers of drugs for HIV, malaria, analgesics, etc., it has only been tried as paclitaxel-loaded nanoparticle in cancer for longer retention and higher accumulation in tissues (Yeh et al. 2005). Nanomicelles of β-casein containing chemotherapeutic drugs, mitoxantrone, vinblastine, docetaxel, and irinotecan, have been synthesized but not reached clinical trials yet. Silk sericin protein is used in formation of self-assembled micellar nanoparticle carrying hydrophilic FITC-inulin and hydrophobic paclitaxel drugs with promising results in cytotoxicity assay in vitro using breast cancer MCF-7 cells in comparison to free paclitaxel (Mandal and Kundu 2009).

The hydrophobic plant proteins zein and gliadin have been found to have some advantage over hydrophilic animal proteins, by providing sustained drug release and being more cost-effective. They also reduce risk of infections like mad cow disease (spongiform encephalitis; Ezpeleta et al. 1996; Lai and Guo 2011). However, only paclitaxel-loaded poly(lactic-co-glycolic acid) (PLGA) nanoparticles against colon cancer cells have been studied with some success (Wang et al. 2010). Lectins, a diverse class of carbohydrate-binding proteins, have been extensively studied for glycotargeting of cancer drugs and have shown significant autophagy in cancer cells (Bies et al. 2004).

Despite being synthesized in various combinations, protein nanoparticles have not given much satisfactory results due to their heterogeneity, batch-wise variations in synthesis, and also rapid solubilization due to their hydrophilic nature.

Glycan-Based Nanocarriers

Poly lactic-co-glycolic acid (PLGA) is one of the most studied biodegradable polymers of lactic acid and glycolic acid, which has got US FDA and European Medicine Agency approval for parenteral drug delivery system. A number of in vivo and in vitro studies have been done to evaluate PLGA-based anticancer nanodrugs (Danhier et al. 2012). PEGylated PLGA nanoparticles containing doxorubicin were found to have enhanced anti-tumoral efficacy than free drug. However, no approved glycan based nanodrug has yet come in market. Cyclodextrin nanoparticle-based camptothecin drug for advanced solid tumors has cleared phase II clinical trials (Weiss et al. 2013). Also cyclodextrin nanocarriers with methotrexate for melanoma, lonidamine for prostate cancer, exemestane for breast cancer, and vorinostat for lymphoma are some other examples (Gidwani and Vyas 2015). The advantages of cyclodextrin encapsulated nanomedicines are increased drug loading capacity, enhanced drug circulation, reduced toxicity, and targeted sustained release. The ADME properties of cyclodextrin nanodrugs have been described and reviewed by Maeda et al. (2009). Polymer particle of cyclodextrin and PEG (CALAA-01) for siRNA delivery is studied to silence the expression of ribonucleotase reductase (Davis 2009).

6.3.5.2 Virus Nanocarriers

Virus and virus-like particles (VLP) have emerged as important nanocarriers platform due to their various advantages like uniform morphology, biocompatibility, and ability to self-assemble to package viral nucleic acid. This has given advantage of tailoring viruses at genetic level to be used as reagents, catalysts, etc. Pox viruses, like vaccinia virus, replicate specifically in cancer cells, which have been utilized in JX-594 pox virus to destroy tumor cells by activating EGFR-Ras-MAPK signaling (Park et al. 2008). In another study, dose-dependent viral replication and anti-tumor activity by i.v. injection of JX-594 has been shown. In the phase III clinical trial (OPTiM) for stage III and IV melanoma patients, intralesional administration of oncolytic virus talimogene laherparepvec (T-Vec) has improved response and survival as compared to granulocyte-macrophage colony stimulating factor (GM-CSF). Thirty-six percent of patients have shown higher durable response rate, whereas 29.5% of patients had complete response with T-Vec against 0% with GM-CSF (Andtbacka et al. 2016).

The major drawback of these nanomedicines is their biosafety and cytocompatibility concerns, and that is the reason none of the several oncolytic viruses have been approved for cancer therapy, despite clearing clinical trials (Aghi and Martuza 2005; Vähä-Koskela et al. 2007).

Synthetic Polymer Nanocarriers

Chemically versatile synthetic polymers are quite promising nanocarrier tool in cancer therapeutics. Many synthetic polymers are undergoing clinical trials; some such examples are docetaxel-PNP for solid tumors, Lipotecan for liver and renal cancer, and Nanotax and Nanoxel-polymer nanoparticles of paclitaxel for peritoneal neoplasma and advanced breast cancer, respectively (Ghamande et al. 2014; Madaan et al. 2013). Polyethylene glycol and polyaspartate (PEG-PAA) polymeric micelle, NK-105, is a synthetic nanocarrier for paclitaxel and has cleared phase II/III clinical trials for gastric and breast cancer (W-262). Similarly, SP1049C, a doxorubicin polymeric micelle, is undergoing clinical trial III for advanced adenocarcinoma (Valle et al. 2011). NC-6004, a PEG-PAA polymer micelle (Nanoplantin), has been found with less neuro- and ototoxicity and is undergoing phase III trials for pancreatic cancer (Plummer et al. 2011). Docetaxel-encapsulated PLGA-PEG nanoparticle, BIND-014, is the first targeted polymer nanoparticle undergoing clinical trials (Jung et al. 2012; Madaan et al. 2013; Ghamande et al. 2014). A reduction responsive drug delivery nanocarrier is synthesized from a linear polyester with disulfide bonds, conjugated with PEG. The nanoparticle has shown faster payload release in response to intracellular reducing environment and thus generating superior anticancer activity towards PC-3 cells.

Drug Conjugates

Drug conjugates form the most wide and successful nanomedicine group in cancer therapeutics. In contrast to any natural or synthetic nanocarriers where the drugs are usually encapsulated, here the active molecules are conjugated covalently to a target antibody or peptide; so the resultant conjugate formulation is a monomer or oligomer and thus has a minimum effect on drug solubility or biodegradability (Wicki et al. 2012). The three approved antibody drug conjugates (ADC) in the market today are Adcetris®, Kadcyla®, and Zevalin®. The drug conjugate Adcetris® (by Seattle Genetics) contains brentuximab vedotin targeting CD30 in non-Hodgkin lymphoma (NHL) and was approved in 2011 (Younes et al. 2012). Vedotin is much less toxic and more effective when combined with anti-CD30 antibody (brentuximab), which redirects the drug selectively to CD30 expressing cancer cells. Kadcyla® by Roche/ImmunoGen contains trastuzumab emtansine targeting HER-2 in breast cancer and got approval in 2013 (W-73). Similarly, Zevalin® is also approved for NHL in 2002 (Morschhauser et al. 2008). Drug conjugates with polymer are also a vast group of nanodrugs and at least 20 different anticancer conjugates undergoing or completing different phases of clinical trials (Wicki et al. 2012).

N-(2-Hydroxypropyl) methacrylamide (*HPMA*)-based copolymer is a promising carrier with enhanced tumor uptake and anti-tumor activity (Chytil et al. 2018). PK-1 is a HPMA polymer-doxorubicin conjugate, which have significantly low cardiotoxicity and alopecia than free doxorubicin (Seymour et al. 2009), whereas PK-2 is a modified PK-1 with galactosamine residues making it the first drug conjugate for active targeting in hepatocellular carcinoma (Seymour et al. 2002). Many ADC with better properties are expected to clear clinical trials and get approval in future.

Inorganic Nanoparticles

Because of their small size and unique shape, inorganic molecules have found great potential in the biomedical field. Nanoparticles using inorganic substances are being used in imaging, radiotherapy and drug delivery in cancers. Gold, silver, and platinum metal nanoparticles are showing good scope as drug delivery system for cancer therapeutics (Bhattacharyya et al. 2011). To increase tumor targeting and enhanced permeability and retention (EPR) effect for breast cancer treatment, a precise gold nanoparticle (AuNP) system is developed by coadministering iRGD peptide and legumain. The modified AuNPs showed higher penetration and accumulation in breast tumor tissues in vivo (Yang et al. 2018). The most studied metal nanoparticles are iron oxide nanoparticles, mainly for diagnostic purposes. A multifunctional silica nanosphere containing superparamagnetic iron oxide nanocrystals and anticancer drugs has been synthesized for simultaneous functions of enhanced drug delivery, fluorescent imaging, and target delivery (Liong et al. 2008).

NanoTherm® (Magforce®) are superparamagnetic iron oxide nanoparticles in late clinical trial stages, to be used in combination with radiotherapy for glioblastoma patients. NanoTherm® is injected into tumor and heated up by magnetic field, thereby either destroying the tumor cells or sensitizing them for radio- or chemotherapy. The nanodrug is being investigated for brain, pancreatic, prostate, and esophageal cancers also and has got marketing approval in many European countries. Many different iron oxide nanoparticles were approved by European Union, but only three iron oxide nanoparticles, Feraheme®, Feridex® and GastroMARK™, have been approved by FDA, of which two of them were later withdrawn also. Synthesized from Hafnium metal oxide, NBTXR3 (developed by Nanobiotex) is a novel metallic nanoparticle which increases radiotherapy efficacy. It had reached phase II/III clinical trials for soft tissue sarcoma by 2016 (Bobo et al. 2016).

Till date more than 50 nanomedicines have been approved by FDA, whereas at least 75 others are undergoing clinical trials at different levels (Bobo et al. 2016). Lastly, the future of nanomedicine looks very promising, as indicated by hundreds of research papers coming out every year. However, due to increasing concerns of nanomaterial toxicity to human beings and environment, very few cancer nanodrugs have been finally approved for therapeutic usage.

6.4 Opportunities and Challenges

There are many strategies and challenges in the development of anticancer nanomedicines. The formulation of most of the nanomedicines (in clinical and developmental trials) is designed for cytotoxics which can address the issues with tolerability and makes the drug cost-effective and viable. The upcoming opportunities for nanomedicines include

(a) Delivering the next generation of drugs
(b) Molecularly targeted agents, peptides
(c) Toxin-like agents that induce cell death
(d) DNA-/RNA-based therapeutics
(e) Drug combinations, etc.

The significant off-target accretion, thin therapeutic window, crossing the cell membrane, and achieving synergistic drug ratios at the end are the major delivery challenges for the abovementioned agents (Aliosmanoglu and Basaran 2012; Tsigelny and Simberg 2011).

There are several reasons for the lack of success of targeted nanoparticles; few amongst them are

1. Heterogeneity of the tumor.
2. The cell receptors that could be targeted are less in number.
3. Most of the nanoparticles are taken up by the immune organs which affects the pharmacokinetics of the NP.
4. The diffusion and penetration problems of various tumors.

The affinity of nanoparticles for tumor cell receptors can be significantly improved by in silico designing. The science of systems biology is growing rapidly. Various in silico models could be simulated which specifically complement the in vivo animal models, for studying the behavior of an organism at various stages and dose response in cancer therapeutics. The introduction of high-throughput experimental tools, various image and data analysis softwares, and the advent of personalized medicines have paved the path for intervening the disease-perturbed biological systems (Ayyappa and Suresh 2015). There is variety of softwares and resources for cancer systems biology:

(a) Minimum Information for Biological and Biomedical Investigations (MIBBI)
(b) Minimum Information About a Microarray Experiment (MIAME)
(c) Minimum Information About a Proteomics Experiment (MIAPE)
(d) Core Information for Metabolomics Reporting (CIMR)
(e) Minimal Information About a RNAi Experiment (MIARE)

The tremendous growth in computational methods, storage capacities of computers, various databases, data analysis, and retrieval tools has led to a new era for the upcoming fields like nanotechnology. The formal way of representing facts from a living body into a coded form with the help of programming languages is coined as ontology. A detailed list of ontology-related databases is given below (Table 6.3).

The next important aspect for nanomaterials is their curation. The various processes of sample preparation and experimental conditions can very much affect the data measurement; hence, curation is an important step for making the data useful. The Nanomaterial Registry developed a very efficient method to assist the data curation. This web based tool does the programmatically generated data upload as well as the entry of manual data (Fig. 6.2). Other than data curation, text mining (computerized methods for extracting knowledge from existing literature) could be used to predict the various protein-nanoparticle interactions in nanotechnology (Suresh and Sangdun 2014).

Various molecular modelling (in silico structure determination method) tools can be used to model the cancerous protein structure and for studying nanoparticle interactions. There are various softwares for molecular visualization which could be

Table 6.3 Ontology resources related to nanoparticle

S. no	Name	URL	Description
1.	Ontology Lookup Service	http://www.ebi.ac.uk/ontology-lookup/browse.do?ontName=REX	It provides a web service interface to query various ontologies from a single site
2.	BioPortal	http://bioportal.bioontology.org/	It provides right of entry to commonly used biomedical ontologies and tools for their investigation
3.	Cancer Open Biomedical Resource	http://www.bioontology.org/caOBR	A tool for indexing open biomedical resources with cancer nanotechnology informatics knowledge
4.	NanoParticle Ontology (NPO)	http://www.nano-ontology.org/	Developed to provide knowledge regarding the characterization of nanomaterials in cancer nanotechnology research
5.	Chemical Entities of Biological Interest	http://www.geneontology.org/	Ontology of chemical entities of biological interest

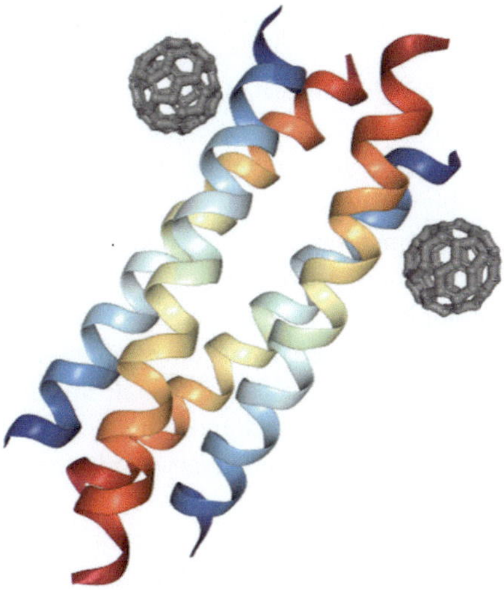

Fig. 6.2 5HKN crystal structure de novo designed fullerene organizing protein complex with fullerene

Table 6.4 Tools for computational designs of bimolecular nanostructure

S. no.	Name	URL	Operating system
1.	Ninithi	http://sourceforge.net/projects/ninithi/files/	Windows
2.	caDNAno	http://cando-dna-origami.org/	Web Accessible
3.	TubeGen	http://turin.nss.udel.edu/research/tubegenonline.html	Web Accessible
4.	Tiamat	http://yanlab.asu.edu/Resources.html	Windows
5.	Wrapping	http://www.photon.t.u-tokyo.ac.jp/~maruyama/wrapping3/wrapping.html	Windows

used for visualization of modelled structure. In addition, the docking and molecular dynamics simulation studies can also assist in studying the interactions. Quantitative Structure-Activity Relationships (QSAR) can predict the toxicity of nanomaterials. High-Throughput Screening Data Analysis Tools (HDAT) offer various normalization methods, clustering, and heat maps and are used for studying the toxicity of engineered nanomaterials. The invention of medical imaging technologies for various biomedical applications, image databases, and image analysis tools can be used to enhance image annotation at nano level. There are many softwares available for nanomaterials designs and DNA origami (Table 6.4).

The data generated by nanomaterials analysis is much more complex than the molecular data or sequence information. The storage, retrieval, analysis, and data interpretation of such complex information cannot be imagined without the aid of computational methods. There is a great need for development of animal models in order to reduce the animal studies for cancer and other life-threatening diseases.

6.5 Conclusions

Nanomedicine is the application of nanotechnology to achieve innovation in healthcare. It uses the properties developed by a material at its nanometric scale 10^{-9} m which often differ in terms of physics, chemistry, or biology from the same material at a bigger scale.

Nanomedinces have the potential to enable early detection and prevention and to drastically improve diagnosis, treatment, and follow-up of many diseases including but not limited to cancer. Overall, nanomedicine has nowadays more than 70 products under clinical trials, covering all major diseases including cardiovascular, neurodegenerative, musculoskeletal, and inflammatory. Enabling technologies in all healthcare areas, nanomedicine is already accounting for 77 marketed products, ranging from nanodelivery and pharmaceutical to imaging, diagnostics, and biomaterial.

As any medical devices or drugs, nanomedicines are strictly regulated and have to follow thorough characterization, toxicity assessment, and multi-stage clinical trials before benefiting patients with their whole potential.

Nanomedicine is understood to be a key enabling instrument for personalized, targeted, and regenerative medicine by delivering the next level of new drugs, treatments, and implantable devices to clinicians and patients, for real breakthroughs in healthcare.

Beyond that, nanomedicine provides important new tools to deal with the grand challenge of an aging population and is thought to be instrumental for improved and cost-effective healthcare, one crucial factor for making medicines and treatments available and affordable to all.

References

Aghi M, Martuza RL (2005) Oncolytic viral therapies — the clinical experience. Oncogene 24:7802–7816. https://doi.org/10.1038/sj.onc.1209037

Ahmad N, Haider S, Jagannathan S, Anaissie E, Driscoll JJ (2014) MicroRNA theragnostics for the clinical management of multiple myeloma. Leukemia 28(4):732–738

Ahmed F, Pakunlu RI, Brannan A, Bates F, Minko T, Discher DE (2006) Biodegradable polymersomes loaded with both paclitaxel and doxorubicin permeate and shrink tumors, inducing apoptosis in proportion to accumulated drug. J Control Release 116:150–158

Akerman ME, Chan WC, Laakkonen P, Bhatia SN, Ruoslahti E (2002) Nanocrystal targeting in vivo. Proc Natl Acad Sci USA 99:12617–12621

Aliosmanoglu A, Basaran I (2012) Nanotechnology in cancer treatment. Nanomedicine Biotherapeutic Discov 2(4):1–3

Alivisatos P (2004) The use of nanocrystals in biological detection. Nat Biotechnol 22:47–52

Allen PM, Liu W, Chauhan VP, Lee J, Ting AY, Fukumura D et al (2010) In As (ZnCdS) quantum dots optimized for biological imaging in the near-infrared. J Am Chem Soc 132:470––471

Andtbacka RH, Agarwala SS, Ollila DW, Hallmeyer S, Milhem M et al (2016) Cutaneous head and neck melanoma in OPTiM, a randomized phase 3 trial of Talimogene laherparepvec versus granulocyte-macrophage colony-stimulating factor for the treatment of unresected stage IIIB/IIIC/IV melanoma. Head Neck 38(12):1752–1758

Ayyappa KSK, Suresh PK (2015) In silico comparative analysis of cancer and stem cell microarray data to demonstrate molecular transitions and the relative involvement of glycolysis and oxidative phosphorylation cost-effective correlation of bioenergetics and differentiation processes. Appli In Drug Develop 7(1):69–81

Balachandran P, Govindarajan R (2005) Cancer: an ayurvedic perspective. Pharmacol Res 51:19–30

Bardhan R, Lal S, Joshi A, Halas NJ (2011) Theranostic nanoshells: from probe design to imaging and treatment of cancer. Acc Chem Res 44:936–946

Bareford LM, Swaan PW (2007) Endocytic mechanisms for targeted drug delivery. Adv Drug Deliv Rev 59:748–758. https://doi.org/10.1016/j.addr.2007.06.008

Barenholz Y (2012) Doxil®—the first FDA-approved nano-drug: lessons learned. J Control Release 160:117–134

Bartlett DW, Su H, Hildebrandt IJ, Weber WA, Davis ME (2007) Impact of tumorspecific targeting on the biodistribution and efficacy of siRNA nanoparticles measured by multimodality in vivo imaging. Proc Natl Acad Sci USA 104:15549–15554

Bharali DJ, Lucey DW, Jayakumar H, Pudavar HE, Prasad PN (2005) Folatereceptor-mediated delivery of InP quantum dots for bioimaging using confocal and two-photon microscopy. J Am Chem Soc 127:11364–11371

Bharali DJ, Khalil M, Gurbuz M, Simone TM, Mousa SA (2009) Nanoparticles and cancer therapy: a concise review with emphasis on dendrimers. Int J Nanomedicine 4:1–7

Bhattacharyya S, Bhattacharya R, Curley S, McNiven MA, Mukherjee P (2010) Nanoconjugation modulates the trafficking and mechanism of antibody induced receptor endocytosis. Proc Natl Acad Sci USA 107:14541–14546

Bhattacharyya S, Kudgus RA, Bhattacharya R, Mukherjee P (2011) Inorganic nanoparticles in cancer therapy. Pharm Res 28:237–259

Bhattacharyya S, Singh RD, Pagano R, Robertson JD, Bhattacharya R, Mukherjee P (2012) Switching the targeting pathways of a therapeutic antibody by nanodesign. Angew. Chem Int Ed Engl 51:1563–1567. https://doi.org/10.1002/anie.201105432

Bhishagratha KL (1991) Sushruta samhita. Sushruta samhita. Choukhamba Orientalia, Varanasi

Bhushan KR, Misra P, Liu F, Mathur S, Lenkinski RE, Frangioni JV (2008) Detection of breast cancer microcalcifications using a dual-modality SPECT/NIR fluorescent probe. J Am Chem Soc 130:17648–17649

Bies C, Lehr CM, Woodley JF (2004) Lectin-mediated drug targeting: history and applications. Adv Drug Deliv Rev 56:425–435

Bobo D, Robinson KJ, Islam J, Thurecht KJ, Corrie SR (2016) Nanoparticle-based medicines: a review of FDA-approved materials and clinical trials to date. Pharm Res 33:2373–2387

Brigger I, Dubernet C, Couvreur P (2012) Nanoparticles in cancer therapy and diagnosis. Adv Drug Deliv Rev 64:24–36

Bruchez M Jr, Moronne M, Gin P, Weiss S, Alivisatos AP (1998) Semiconductor nanocrystals as fluorescent biological labels. Science 281:2013–2016

Chan S, Davidson N, Juozaityte E, Erdkamp F, Pluzanska A et al (2004) Phase III trial of liposomal doxorubicin and cyclophosphamide compared with epirubicin and cyclophosphamide as first-line therapy for metastatic breast cancer. Ann Oncol 15:1527–1534

Chan WC, Nie S (1998) Quantum dot bioconjugates for ultrasensitive nonisotopic detection. Science 281:2016–2201

Cheng Y, Wang J, Rao T, He X, Xu T (2008) Pharmaceutical applications of dendrimers: Promising nanocarriers for drug delivery. Front Biosci 13:1447–1471

Chytil P, Koziolová E, Etrych T, Ulbrich K (2018) HPMA copolymer-drug conjugates with controlled tumor-specific drug release. Macromol Biosci 18(1):1700209. https://doi.org/10.1002/mabi.201700209

Cravotto G, Boffa L, Genzini L, Garella D (2010) Phytotherapeutics: an evaluation of the potential of 1000 plants. J Clin Pharm Ther 35:11–48

Danhier F, Ansorena E, Silva JM, Coco R, Le Breton A et al (2012) PLGA-based nanoparticles: an overview of biomedical applications. J Control Release 161:505–522

Davis ME (2009) The first targeted delivery of siRNA in humans via a self-assembling, cyclodextrin polymer-based nanoparticle: From concept to clinic. Mol Pharm 6:659–668

Davis ME, Zuckerman JE, Choi CHJ, Seligson D, Tolcher AW et al (2010) Evidence of RNAi in humans from systemically administered siRNA via targeted nanoparticles. Nature 464:1067–1070

De La Rica R, Aili D, Stevens MM (2012) Enzyme-responsive nanoparticles for drug release and diagnostics. Adv Drug Deliv Rev 64(11):967–978

Edelman LB, Eddy JA, Price ND (2010) *In silico* models of cancer. Wiley Interdiscip Rev Syst Biol Med 2(4):438–459

Eloy JO, Petrilli R, Raspantini GL, Lee RJ (2018) Targeted liposomes for siRNA delivery to cancer. Curr Pharm Des. https://doi.org/10.2174/1381612824666180807121935

Ezpeleta I, Irache JM, Stainmesse S, Chabenat C, Gueguen J et al (1996) Gliadin nanoparticles for the controlled release of all-transretinoic acid. Int J Pharm 131:191–200

Fleige E, Quadir MA, Haag R (2012) Stimuli-responsive polymeric nanocarriers for the controlled transport of active compounds: concepts and applications. Adv Drug Deliv Rev 64:866–884

Fomina N, McFearin C, Sermsakdi M, Edigin O, Almutairi A (2010) UV and near-IR triggered release from polymeric nanoparticles. J Am Chem Soc 132:9540–9542

Frampton JE (2010) Mifamurtide: a review of its use in the treatment of osteosarcoma. Paediatr Drugs 12:141–153

Frangioni JV (2003) In vivo near-infrared fluorescence imaging. Curr Opin Chem Biol 7:626–634

Gao J, Chen X, Cheng Z (2010) Near-infrared quantum dots as optical probes for tumor imaging. Curr Top Med Chem 1:209–217

Gao X, Cui Y, Levenson RM, Chung LW, Nie S (2004) In vivo cancer targeting and imaging with semiconductor quantum dots. Nat Biotechnol 22:969–976

Gao X, Yang L, Petros JA, Marshall FF, Simons JW, Nie S (2005) In vivo molecular and cellular imaging with quantum dots. Curr Opin Biotechnol 16:63–72

Ghamande S, Lin CC, Cho DC, Shapiro GI, Kwak EL et al (2014) A phase 1 open-label, sequential dose-escalation study investigating the safety, tolerability, and pharmacokinetics of intravenous TLC388 administered to patients with advanced solid tumors. Invest New Drugs 32:445–451. https://doi.org/10.1007/s10637-013-0044-7

Gidwani B, Vyas A (2015) A comprehensive review on cyclodextrin-based carriers for delivery of chemotherapeutic cytotoxic anticancer drugs. BioMed Res Int 2015:198268. https://doi.org/10.1155/2015/198268

Gill PS, Wernz J, Scadden DT, Cohen P, Mukwaya GM et al (1996) Randomized phase III trial of liposomal daunorubicin versus doxorubicin, bleomycin, and vincristine in AIDS-related Kaposi's sarcoma. J Clin Oncol 14:2353–2364

Glassman DC, Palmaira RL, Covington CM, Desai AM, Ku GY et al (2018) Nanoliposomal irinotecan with fluorouracil for the treatment of advanced pancreatic cancer, a single institution experience. BMC Cancer 18(1):693–703

Gökbuget N, Hartog CM, Bassan R, Derigs HG, Dombret H et al (2011) Liposomal cytarabine is effective and tolerable in the treatment of central nervous system relapse of acute lymphoblastic leukemia and very aggressive lymphoma. Haematologica 96:238–244

Golden PL, Huwyler J, Pardridge WM (1998) Treatment of large solid tumors in mice with daunomycin-loaded sterically stabilized liposomes. Drug Deliv 5:207–212. https://doi.org/10.3109/10717549809052036

Gradishar WJ, Tjulandin S, Davidson N, Shaw H, Desai N, Bhar P et al (2005) Phase III trial of nanoparticle albumin- bound paclitaxel compared with polyethylated castor oil-based paclitaxel in women with breast cancer. J Clin Oncol 23:7794–7803

Guccione S, Li KC, Bednarski MD (2004) Vascular-targeted nanoparticles for molecular imaging and therapy. Methods Enzymol 386:219–236

Haemmerich D, Motamarry A (2018) Thermosensitive liposomes for image-guided drug delivery. Adv Cancer Res 139:121–146

Harrison M, Tomlinson D, Stewart S (1995) Liposomal-entrapped doxorubicin: an active agent in AIDS-related Kaposi's sarcoma. J Clin Oncol 13:914–920

Ho KM, Li P (2008) Design and synthesis of novel magnetic core–shell polymeric particles. Langmuir 24:1801–1807

Hobbs SK, Monsky WL, Yuan F, Roberts WG, Griffith L et al (1998) Regulation of transport pathways in tumor vessels: role of tumor type and microenvironment. Proc Natl Acad Sci USA 95:4607–4612

Jaiswal JK, Mattoussi H, Mauro JM, Simon SM (2003) Long-term multiple color imaging of live cells using quantum dot bioconjugates. Nat Biotechnol 21:47–51

Jhaveri A, Deshpande P, Torchilin V (2014) Stimuli-sensitive nanopreparations for combination cancer therapy. J Control Release. https://doi.org/10.1016/j.jconrel.2014.05.002

Josephson L, Lewis J, Jacobs P, Hahn PF, Stark DD (1988) The effects of iron oxides on proton relaxivity. Magn Reson Imaging 6:647–653

Jung J, Park SJ, Chung HK, Kang HW, Lee SW et al (2012) Polymeric nanoparticles containing taxanes enhance chemoradiotherapeutic efficacy in non-small cell lung cancer. Int J Radiat Oncol Biol Phys 84:e77–e83. https://doi.org/10.1016/j.ijrobp.2012.02.030

Ko AH (2016) Nanomedicine developments in the treatment of metastatic pancreatic cancer: focus on nanoliposomal irinotecan. Int J Nanomedicine 11:1225–1235

Kohler N, Fryxell GE, Zhang M (2004) A bifunctional poly(ethylene glycol) silane immobilized on metallic oxide-based nanoparticles for conjugation with cell targeting agents. J Am Chem Soc 126:7206–7211

Krishna R, Mayer LD (2000) Multidrug resistance (MDR) in cancer—mechanisms, reversal using modulators of MDR and the role of MDR modulators in influencing the pharmacokinetics of anticancer drugs. Eur. J Cancer Sci 11:265–283

Kwon GS (2003) Polymeric micelles for delivery of poorly water-soluble compounds. Crit Rev Ther Drug Carrier Syst 20:357–403

Lai LF, Guo HX (2011) Preparation of new 5-fluorouracil-loaded zein nanoparticles for liver targeting. Int J Pharm 404:317–323

Li KC, Guccione S, Bednarski MD (2002) Combined vascular targeted imaging and therapy: a paradigm for personalized treatment. J Cell Biochem Suppl 39:65–71

Links M, Brown R (1999) Clinical relevance of the molecular mechanisms of resistance to anticancer drugs. Expert Rev Mol Med 1:1–21

Liong M, Lu J, Kovochich M, Xia T, Ruehm SJ et al (2008) Multifunctional inorganic nanoparticles for imaging, targeting, and drug delivery. ACS Nano 2(5):889–896

Liu C, Liu F, Feng L, Li M, Zhang J, Zhang N (2013) The targeted co-delivery of DNA and doxorubicin to tumor cells via multifunctional PEI-PEG based nanoparticles. Biomaterials 34:2547–2564

Loo C, Lowery A, Halas N, West J, Drezek R (2005) Immunotargeted nanoshells for integrated cancer imaging and therapy. Nano Lett 5:709–711

Madaan A, Singh P, Awasthi A, Verma R, Singh AT et al (2013) Efficiency and mechanism of intracellular paclitaxel delivery by novel nanopolymer-based tumor-targeted delivery system, Nanoxel(TM). Clin Transl Oncol 15:26–32. https://doi.org/10.1007/s12094-012-0883-2

Maeda H, Bharate GY, Daruwalla J (2009) Polymeric drugs for efficient tumor-targeted drug delivery based on EPR-effect. Eur J Pharma Biopharma 71(3):409–419

Mamot C, Ritschard R, Wicki R, Stehle G, Dieterle T et al (2012) Tolerability, safety, pharmacokinetics, and efficacy of doxorubicin-loaded anti-EGFR immunoliposomes in advanced solid tumours: a phase 1 dose-escalation study. Lancet Oncol 13:1234–1241. https://doi.org/10.1016/S1470-2045(12)

Mandal BB, Kundu SC (2009) Self-assembled silk sericin/poloxamer nanoparticles as nanocarriers of hydrophobic and hydrophilic drugs for targeted delivery. Nanotechnology 20:355101

Manninger SP, Muldoon LL, Nesbit G, Murillo T, Jacobs PM, Neuwelt EA (2005) An exploratory study of ferumoxtran-10 nanoparticles as a blood–brain barrier imaging agent targeting phagocytic cells in CNS inflammatory lesions. AJNR Am J Neuroradiol 26:2290–2300

Manzoor AA, Lindner LH, Landon CD, Park JY, Simnick AJ et al (2012) Overcoming limitations in nanoparticle drug delivery: triggered, intravascular release to improve drug penetration into tumors. Cancer Res 72:5566–5575. https://doi.org/10.1158/0008-5472.CAN-12-1683

Martin FJ (1998) Clinical pharmacology and antitumor efficacy of DOXIL (pegylated liposomal doxorubicin). In: Lasic DD, Papahadjopoulos D (eds) Medical applications of liposomes. Elsevier, New York, pp 635–688
Michalet X, Pinaud FF, Bentolila LA, Tsay JM, Doose S, Li JJ et al (2005) Quantum dots for live cells, in vivo imaging and diagnostics. Science 307:538–544
Moreno-Aspitia A, Perez EA (2005) Nanoparticle albumin-bound paclitaxel (ABI-007): a newer taxane alternative in breast cancer. Future Oncol (London, England) 1:755–762
Morschhauser F, Radford J, Van Hoof A, Vitolo U, Soubeyran P et al (2008) Phase III trial of consolidation therapy with yttrium-90-ibritumomab tiuxetan compared with no additional therapy after first remission in advanced follicular lymphoma. J Clin Oncol 26:5156–5164. https://doi.org/10.1200/JCO.2008.17.2015
Mura M, Hopkins TG, Michael T, Abd-Latip N, Weir J, Aboagye E, Mauri F, Jameson C, Sturge J, Gabra H et al (2015) LARP1 post-transcriptionally regulates mTOR and contributes to cancer progression. Oncogene 34:5025–5036
Nahrendorf M, Keliher E, Marinelli B, Waterman P, Feruglio PF, Fexon L et al (2010) Hybrid PET-optical imaging using targeted probes. Proc Natl Acad Sci USA 107:7910–7915
Neuwelt EA, Varallyay P, Bago AG, Muldoon LL, Nesbit G, Nixon R (2004) Imaging of iron oxide nanoparticles by MR and light microscopy in patients with malignant brain tumours. Neuropathol Appl Neurobiol 30:456–471
Nishiyama N, Kataoka K (2006) Current state achievements, and future prospects of polymeric micelles as nanocarriers for drug and gene delivery. Pharmacol Theory 112:630–648
Obata Y, Tajima S, Takeoka S (2010) Evaluation of pH-responsive liposomes containing amino acid-based zwitterionic lipids for improving intracellular drug delivery in vitro and in vivo. J Control Release 142:267–276
O'Brien S, Schiller G, Lister J, Damon L, Goldberg S et al (2013) High-dose vincristine sulfate liposome injection for advanced, relapsed, and refractory adult Philadelphia chromosome-negative acute lymphoblastic leukemia. J Clin Oncol 31:676–683
Park JW (2002) Liposome-based drug delivery in breast cancer treatment. Breast Cancer Res 4:95–99. 436: 568–572
Park BH, Hwang T, Liu TC, Sze DY, Kim JS et al (2008) Use of a targeted oncolytic poxvirus, JX-594, in patients with refractory primary or metastatic liver cancer: A phase I trial. Lancet Oncol 9:533–542. https://doi.org/10.1016/S1470-2045(08)70107-4
Parungo CP, Ohnishi S, Kim SW, Kim S, Laurence RG, Soltesz EG et al (2005) Intraoperative identification of esophageal sentinel lymph nodes with nearinfrared fluorescence imaging. J Thorac Cardiovasc Surg 129:844–850
Patel B, Das S, Prakash R, Yasir M (2010) Natural bioactive compound with anticancer potential. Int J Advan Pharmaceut Sci 1:32–41
Pereira DIA, Bruggraber F, Poots LK, Tagmount MA, Aslam MF, Frazer DM, Vulpe CD, Anderson GJ, Powell JJ (2014) Nanoparticulate iron(III) oxohydroxide delivers safe iron that is well absorbed and utilised in humans. Nanomedicine 10(8):1877–1886
Perrino E, Steiner M, Krall N, Bernardes GJL, Pretto F et al (2014) Curative properties of non-internalizing antibody-drug conjugates based on maytansinoids. Cancer Res 74:2569–2578
Plummer R, Wilson RH, Calvert H, Boddy AV, Griffin M et al (2011) A Phase I clinical study of cisplatin-incorporated polymeric micelles (NC-6004) in patients with solid tumours. Br J Cancer 104:593–598. https://doi.org/10.1038/bjc.2011.6
Prasad E, Kenneth R, Shang JW, Kose MA (2003) The effects of financial globalization on developing countries: Some empirical evidence. IMF Occasional Paper no. 220. International Monetary Fund, Washington, DC
Reimer P, Tombach B (1998) Hepatic MRI with SPIO: Detection and characterization of focal liver lesions. Eur Radiol 8:1198–1204
Sengupta S, Eavarone D, Capila I, Zhao G, Watson N, Kiziltepe T et al (2005) Temporal targeting of tumour cells and neovasculature with a nanoscale delivery system. Nature 436(7050):568–572
Seymour LW, Ferry DR, Anderson D, Hesslewood S, Julyan PJ et al (2002) Hepatic drug targeting: phase I evaluation of polymer-bound doxorubicin. J Clin Oncol 20:1668–1676

Seymour LW, Ferry DR, Kerr DJ, Rea D, Whitlock M et al (2009) Phase II studies of polymer-doxorubicin (PK1, FCE28068) in the treatment of breast, lung and colorectal cancer. Int J Oncol 34:1629–1636

Shim MS, Kwon YJ (2012) Stimuli-responsive polymers and nanomaterials for gene delivery and imaging applications. Adv Drug Deliv Rev 64:1046–1059

Shubayev VI, Pisanic TR, Jin S (2009) Magnetic nanoparticles for theragnostics. Adv Drug Deliv Rev 61(6):467–477. 1872-8294 (Electronic), 0169409X (Linking)

Smith HF, Woerdenbag HJ, Singh RH, Meulenbeld GJ, Labadie RP, Zwaving JH (1995) Ayurvedic herbal drugs with possible cytostatic activity. J Ethnopharmacol 47:75–84

Sparreboom A, Scripture CD, Trieu V, Williams PJ, De T, Yang A et al (2005) Comparative preclinical and clinical pharmacokinetics of a cremophor-free, nanoparticle albumin- bound paclitaxel (ABI-007) and paclitaxel formulated in Crem- ophor (Taxol). Clin Cancer Res 11:4136–4143

Sun B, Ranganathan B, Feng SS (2008) Multifunctional poly(D, L-lactide-coglycolide)/montmorillonite (PLGA/MMT) nanoparticles decorated by Trastuzumab for targeted chemotherapy of breast cancer. Biomaterials 29:475–486

Suresh P, Sangdun C (2014) Nanoinformatics: emerging databases and available tools. Int J of Mol Sci 15:7158–7182

Tharushi RS, Udawatte C, Ratnaweera CN (2017) An *in-silico* approach to study the binding interaction of coumarin derivatives to aromatase. 4th Ann Cong Drug Dis Design 6(3):34

Thatte U, Dhahanukar S (1991) Ayurveda, the natural alternative. Sci Today 2001:12–18

Tong R, Langer R (2015) Nanomedicines targeting the tumor microenvironment. Cancer J 21(4):314–321

Tsigelny IF, Simberg D (2011) Has the time for in silico design of nanomedicines finally arrived? Nanomed & Biothera Dis 1(2):1–2

Vähä-Koskela MJV, Heikkilä JE, Hinkkanen AE (2007) Oncolytic viruses in cancer therapy. Cancer Lett 254:178–216. https://doi.org/10.1016/j.canlet.2007.02.002

Valle JW, Armstrong A, Newman C, Alakhov V, Pietrzynski G et al (2011) A phase 2 study of SP1049C, doxorubicin in P-glycoprotein-targeting pluronics, in patients with advanced adenocarcinoma of the esophagus and gastroesophageal junction. Invest New Drugs 29:1029–1037. https://doi.org/10.1007/s10637-010-9399-1

Varallyay P, Nesbit G, Muldoon LL, Nixon RR, Delashaw J, Cohen JI et al (2002) Comparison of two superparamagnetic viral-sized iron oxide particles ferumoxides and ferumoxtran-10 with a gadolinium chelate in imaging intracranial tumors. AJNR Am J Neuroradiol 23:510-–519

Veiseh O, Gunn JW, Zhang M (2010) Design and fabrication of magnetic nanoparticles for targeted drug delivery and imaging. Adv Drug Deliv Rev 62:284–304

Von Hoff DD, Mita MM, Ramanathan RK, Weiss GJ, Mita AC et al (2016) Phase I study of PSMA-targeted Docetaxel-containing nanoparticle BIND-014 in patients with advanced solid tumors. Clin Cancer Res 22(13):3157–3163

Wang C, Ho PC, Lim LY (2010) Wheat germ agglutinin-conjugated PLGA nanoparticles for enhanced intracellular delivery of paclitaxel to colon cancer cells. Int J Pharm 400:201–210

Wang D, Sun Y, Liu Y, Meng F, Lee RJ (2018) Clinical translation of immunoliposomes for cancer therapy: Recent perspectives. Expert Opin Drug Deliv 15(9):893–903

Wang YX, Hussain SM, Krestin GP (2001) Superparamagnetic iron oxide contrast agents: Physicochemical characteristics and applications in MR imaging. Eur Radiol 11:2319–2331

Weiss GJ, Chao J, Neidhart JD, Ramanathan RK, Bassett D et al (2013) First-in-human phase 1/2a trial of CRLX101, a cyclodextrin-containing polymercamptothecin nanopharmaceutical in patients with advanced solid tumor malignancies. Invest New Drugs 31:986–1000

Wicki A, Rochlitz C, Orleth A, Ritschard R, Albrecht I, Herrmann R et al (2012) Targeting tumor-associated endothelial cells: anti-VEGFR2 immunoliposomes mediate tumor vessel disruption and inhibit tumor growth. Clin Cancer Res 18:454–464

Wicki A, Witzigmann D, Balasubramanian V, Huwyler J (2015) Nanomedicine in cancer therapy: challenges, opportunities, and clinical applications. J Control Release 200:138–157

Wu X, Liu H, Liu J, Haley KN, Treadway JA, Larson JP et al (2003) Immunofluorescent labeling of cancer marker Her2 and other cellular targets with semiconductor quantum dots. Nat Biotechnol 21:41–46

Xu H, Regino CA, Koyama Y, Hama Y, Gunn AJ, Bernardo M et al (2007) Preparation and preliminary evaluation of a biotin-targeted, lectin-targeted dendrimer-based probe for dual-modality magnetic resonance and fluorescence imaging. Bioconjug Chem 18:1474–1482

Yang Y, Chen Q, Li S, Ma W, Yao G et al (2018) iRGD-mediated and enzyme-induced precise targeting and retention of gold nanoparticles for the enhanced imaging and treatment of breast cancer. J Biomed Nanotechnol 14(8):1396–1408

Yeh TK, Lu Z, Wientjes MG, Au JLS (2005) Formulating paclitaxel in nanoparticles alters its disposition. Pharm Res 22:867–874

Younes A, Gopal AK, Smith SE, Ansell SM, Rosenblatt JD et al (2012) Results of a pivotal phase II study of brentuximab vedotin for patients with relapsed or refractory Hodgkin's lymphoma. J Clin Oncol 30:2183–2189. https://doi.org/10.1200/JCO.2011.38.0410

Zhang Y, Kohler N, Zhang M (2002) Surface modification of superparamagnetic magnetite nanoparticles and their intracellular uptake. Biomaterials 23:1553–1561

Part III

Healthcare Applications of NanoBioMedicine

HIV: Biology to Treatment

7

Ashish Swarup Verma, Vipin Kumar, Malay Kumar Saha,
Shanta Dutta, and Anchal Singh

Abstract

AIDS is one of the most dreaded diseases of the twenty-first century caused by human immunodeficiency virus (HIV). Recently, there are reports which show decline in new infections due to better access to anti-retroviral drugs. Still on a daily basis, ~2356 new HIV infections are being reported globally. New treatments and anti-HIV drugs are being continuously developed with the aim to control and cure AIDS. The anti-HIV drugs that are in use usually target HIV entry and replication inside the host cells. However, these drugs are only partially effective in slowing the rate of HIV replication. Nevertheless, the virus manages to replicate at much slower rates even when anti-retroviral treatment is ongoing. The HIV seropositives who are on anti-retroviral treatment for long periods of time are now developing different kinds of other complications including neuroAIDS. The latest development in HIV therapy is a novel kind of bone marrow transplantation from donors who have a homozygous mutation in CCR5 gene.

Keywords

AIDS · ARD · ART · HAART · HIV · NeuroAIDS

A. S. Verma
Jadavpur University, Kolkata, WB, India

V. Kumar · A. Singh (✉)
Department of Biochemistry, Institute of Science, Banaras Hindu University, Varanasi, UP, India

M. K. Saha · S. Dutta
National Institute of Cholera and Enteric Diseases (NICED), Kolkata, WB, India

S. K. Saxena, S. M. P. Khurana (eds.), *NanoBioMedicine*,
https://doi.org/10.1007/978-981-32-9898-9_7

Abbreviations

AIDS	Acquired immunodeficiency syndrome
AML	Acute myeloid leukemia
ARD	Anti-retroviral drug/s
ART	Anti-retroviral treatment/s
BBB	Blood-brain barrier
CNS	Central nervous system
Env	Envelope
Gag	Group-specific antigen
HIV	Human immunodeficiency virus
InI	Integrase inhibitors
LTR	Long terminal repeat/s
Nef	Negative factor
NNRTI	Non-nucleoside reverse transcriptase inhibitor/s
NRTI	Nucleoside reverse transcriptase inhibitor/s
NtRTI	Nucleotide reverse transcriptase inhibitor/s
PDI	Protein disulfide isomerase
PI	Protease inhibitor/s
PNS	Peripheral nervous system
Pol	Polymerase
Rev	Regulator of viral expression
Tat	Transactivator of transcription
Vif	Viral infectivity factor
Vpr	Viral protein R
Vpu	Viral protein U

7.1 Introduction

A possibility of life without having infections or suffering from infection/s is just beyond imagination in any natural conditions or scenario. Infections affect us irrespective of our habit and habitat. No matter where we live? How we live? No matter in what hygienic conditions we live!!! All of us have to live and to face infections numerous times throughout our life.

If we go through the historical records, history of infections is prehistorical in existence. Infections and diseases caused by infective agents have been recorded even in ancient times, too. Biblical records clearly mentioned about the havoc caused by plague. Although descriptions of infections and infectious diseases in older times were not well documented, infections at the present time have been well documented and recorded. Human life always remains vulnerable to infections throughout his/her life irrespective of age, sex, race, or any other variable that we can imagine. One should not be surprised to know that human life can be infected even before someone is born. The infections before birth are commonly called as prenatal infections, e.g., infection of herpes simplex virus (HSV) is one of the most common prenatally reported infections. We can easily conclude a fact that infections and life are very closely associated with each other. During the last century, medical sciences have been credited with two major discoveries, i.e., vaccines and antibiotics. Vaccines and antibiotics brought dramatic changes reducing the morbidity as well as preventing the mortality caused by some of these infections. The success to control infections with the use of vaccines and antibiotics is so tremendous that these discoveries almost lead us to believe that *we are wining the war against microbes/infections*. Knowing the current status as well as realizing the current scenario about infection and infectious diseases can we claim if the abovementioned statement still holds the water!!! A sudden eruption and reporting of new and more fatal diseases like SARS, HIV, Ebola, swine flu, etc., to name a few if not all. Some of these diseases remained confined within certain geographical boundaries, while some were presented with their sporadic presence. HIV infections have been reported as a new type of infection. During the last quarter of the twentieth century, HIV infections are indiscriminate to age, sex, and race, and soon it was realized that HIV infections do not even know any geographical boundaries. HIV does not show any sporadic presence or any seasonal dependency. As a matter of fact, HIV infections raised eyebrows of policy makers due to the global presence of HIV which soon started taking the shape of a global epidemic. HIV being a global epidemic has received unprecedented attention from all quarters of life may it be the government or nongovernment organizations. HIV even attracted attention of various religious groups due to the homosexual behavior of the initial HIV patients reported with HIV infections. This mysterious infection due to its fatal nature and quantum of infected people has received more than warranted attention from scientists, clinicians, policy makers, media, etc. But what were the reasons behind such a public attention or outcry for HIV infections? Still there are no conclusive answers to it, but there are certain theories which prevail behind it.

7.2 History

History of HIV and AIDS epidemic begins with illness, fear and death, which was faced by the world due to a new and unknown virus.

The history of HIV and AIDS goes back to the year 1981. It was the year 1981, when the initial cases of HIV/AIDS were reported. Ideally these cases were unknowingly "known" cases of HIV and AIDS. These initial cases were reported with a rare lung infection. These patients were suffering from *Pneumocystis carinii pneumonia* (PCP) (Gottlieb et al. 1981). PCP is known as an opportunistic infection. Only severely immune compromised host or patients get affected by PCP which is considered as a fatal disease. In the year 1981, the term AIDS was not even coined, so these patients were diagnosed with PCP not as patients of HIV/AIDS. Later it was recognized that these PCP patients were actually patients of AIDS. This is the reason these cases can be considered as unknowingly "known" patients of AIDS.

The characteristic feature of AIDS is almost a devastated immune system with the assurance of death at the end, and the onset of death in case of HIV/AIDS patient is simply a matter of time. It will not be wrong to say *AIDS is the ultimate and final chapter of life of a HIV patient.* Clinically AIDS is defined as high viral load and low $CD4^+$ T-cell counts (<200 cell/ml of blood) along with the presence of AIDS-defining diseases like HAD, HIV wasting syndrome, AIDS-defining cancers like Kaposi's Sarcoma, etc. (www.who.int). It is commonly believed that HIV is a single virus, though as a matter of fact there are two types of HIV, and those two different types of HIV are HIV-1 and HIV-2. Out of these two different types of HIV, HIV-1 is more prevalent with global presence and with fatal consequences, while in contrast HIV-2 is less prevalent and causes a slow progressing disease. This chapter is focused on HIV-1 and the term HIV in the chapter implies HIV-1.

The first report about PCP among five gay men was published by Dr. Michael S. Gottlieb and his group in 1981 issue of MMWR. MMWR is a publication from Centers for Disease Control and Prevention (CDC), Atlanta, USA. The rationale question is why these patients were classified as PCP patients rather than AIDS patients. The answer is that, at that time the term AIDS was neither a part of general vocabulary nor even a part of medical dictionary. Nonexistence of the AIDS term is the major reason for the existence of this discrepancy in literature. The most fascinating fact about the saga of AIDS is that the world community is indebted to the alertness of a technician Ms. Sandra Ford at CDC where she was responsible for dispensing a rare and controlled drug, pentamidine. She had seen an unusual increase in numbers of requests to dispense this very rare medicine. Due to her alertness, Ms. Ford brought this matter to the attention of her bosses at CDC. This fact was verified in her interview to a famous US magazine *Newsweek*. Her statement is like this:

> *A doctor was treating a gay man in his 20's who had pneumonia, 2 weeks later he (doctor) called to ask for a refill of a rare drug that I handled. This was unusual nobody ever asked for a refill. Patients usually were cured in one 10-days treatment or they died.* By Sandra Ford

Realizing the significance of this new observation, CDC USA formed a "Task Force on Kaposi's Sarcoma and Opportunistic Infections (KSOI)." At this point of time, very limited information was available about this new and deadly disease. The major concerns were about the mode of transmission for this new and fatal disease. Since no official name existed for this new disease, this new disease was initially identified as lymphadenopathy or KSOI.

Majority of cases for this new disease were initially reported only among homosexual gay men so it is obvious that various new names were adopted in literature for reporting this new disease. These new names somehow adopted the gay cultivate/word to evolve new names for the disease. Some of the common names that were used in literature for this new disease were gay compromise syndrome (GCS), gay-related immune deficiency (GRID), acquired immunodeficiency disease (AID), community-acquired immune dysfunction (CAID), and gay cancer. The role of gay lifestyle started giving strength to this new and fatal disease though very soon this assumption turned out to be wrong because by late 1982, a 20-month-old child was also reported with this new and fatal disease. As per the clinical history, the child had received multiple blood transfusions. Later it was realized that the cause of this disease in this 20-month-old child could be the blood, or transmission of new infection could be due to the transfusion of infected blood.

This case of 20-month-old child raised an alarm for the need of safe blood supply from blood banks. Soon cases for this new and fatal disease were getting reported from all walks of life. These observations have been helpful to delink this myth about the connection of new and fatal disease with unique lifestyle followed by gay men. Therefore, lifestyle-related name used for this disease has lost their relevance. To avoid further confusions among clinicians, there was a dire need to coin a new term and appropriate name for this new disease. To address this overgrowing problem, a meeting was convened on July 24, 1982, in Washington, and in this meeting AIDS was coined as a new name for this new and fatal disease.

Since the formation of KSOI and close monitoring of the disease status, cases were reported from all corners of the USA. Soon detection of this new disease in Haitian hemophiliacs was one of the reasons to believe that the disease might have originated from Haiti. By the end of 1982, AIDS cases were reported from various European countries as well as from the countries of African continents. The quantum of this disease can be realized from the spread of this disease with a fact that by year 1983, 3064 cases of AIDS were reported out of which 1292 had already died (https://www.avert.org/professionals/history-hiv-aids/overview). These numbers about AIDS patients had been a motivating factor to realize the sense of urgency regarding this disease among various sections of the society. The concerns and seriousness related to the disease compelled for an urgent need to find out the causative agent as well as effective treatment strategies against the disease.

Initially researches on this unknown disease were started in France and the USA. The causative agent for this new disease was reported by Luc Montagnier's group who was working at Pasteur Institute, Paris, France. In 1983, Montagnier's group named this virus as lympadenopathy-associated virus (LAV). Montagnier's work was published in May 20, 1983, issue of *Science*. This first publication was authored by 12 investigators (Barre-Sinousi et al. 1983). The contribution of Montagnier's work was finally recognized by awarding him the Nobel Prize in Physiology and Medicine in 2008. Unfortunately, Montagnier's publication did not attract the required attention from scientific community for almost a year till the findings of another group from the USA were published in the same journal. In May 4, 1984, issue of *Science,* findings from Robert Gallo's group at National Cancer Institute (NCI), Bethesda, USA, were published (G). This new article was authored by 13 investigators of Gallo's group wherein they have given the name of this virus as human T-cell lymphotropic virus. Later it was decided that both of them (Montagnier and Gallo) should be equally credited for the discovery of this new organism.

In 1985, the first case of HIV transmission from breast milk to infant was reported. In the meantime HTLV-III and LAV names started creating confusion in literature about the name as well as nomenclature for the same virus. Therefore, in 1986, the International Committee on Taxonomy of Viruses decided to drop both the names, i.e., LAV and HTLV-III. The new name for this virus was given as human immunodeficiency virus (HIV). This new name has been in use since the HIV term is very pertinent with the symptoms of the disease caused by HIV.

Year 1987 has been important in the history of HIV and AIDS because of two important developments: (1) the US Food and Drug Administration (USFDA) has approved Azidothymidine (Zidovudine) as the first drug to treat HIV infection, and (2) the World Health Organization (WHO) also launched the Global Programme on AIDS to raise awareness, to generate evidence-based policies, and to provide technical and financial support to countries affected by this disease (www.who.int). WHO also declared December 1st as World AIDS Day in 1991. By the year 1996 the United Nations Organization (UNO) had established a Joint United Nations Programme (UNAIDS) to advocate for global action on HIV/AIDS epidemic and to coordinate the response across the world. The annual report of UNAIDS is considered as one of the most authentic reports on the global status of HIV and AIDS. As per UNAIDS report of 2018, there is a significant decline of new infections (especially among children of age less than 15 years) and deaths due to AIDS, while access to ART and funds to deal with HIV/AIDS epidemics has seen an increasing trend (http://www.unaids.org/en/resources/documents/2018/unaids-data-2018).

7.3 HIV Disease Burden: An Improvement

Since the establishment of office of UNAIDS by United Nations, UNAIDS publishes annual reports about the status of HIV and AIDS at global level. Annual reports published by UNAIDS are considered as the most accurate and authentic document about the epidemiology of HIV and AIDS. The most recent report of UNAIDS was published in July 2018 (http://www.unaids.org/en/resources/documents/2018/unaids-data-2018). This report has provided details about the global status of HIV and AIDS as of the year 2017 (Table 7.1). There is no doubt that the

Table 7.1 HIV epidemiology: a comparison between 2010 & 2017

	2010	2017	
	(in millions)		Change in (%)
HIV Infection (*Total*)	33.7 (28.4–40.0)	36.9 (31.1–43.9)	+9.4%
HIV infection (*new*)	2.1 (1.4–2.7)	1.8 (1.4–2.4)	–14%
(a) Adults	1.8 (1.4–2.4)	1.6 (1.3–2.1)	–11.1%
(b) Child	0.23 (0.15–0.34)	0.18 (0.11–0.26)	–21%
(c) Death	1.2 (0.88–1.7)	0.94 (0.67–1.3)	–21.1%
ART Access	11.4	21.7	+90.3%
Nos.	(10.1–11.9)	(19.1–22.6)	
Percentage	25%	59%	
Resources (*committed*)	18.8%	21.3%	+13.3%

Values present in 3rd column is representing the percent increase or decrease in the year 2017 compared to year 2010. Value present in parenthesis represents the range

data presented in the report is very encouraging with regard to the current status of HIV and AIDS. UNAIDS report is quite comprehensive on various aspects related to HIV and AIDS like new infections, access to anti-retroviral drugs, deaths due to AIDS and AIDS-related diseases, vertical transmission of HIV, and even the financial resources available or committed to deal with HIV and AIDS epidemic (UNAIDS Report 2012).

As per these new estimates presented in UNAIDS report 2017, ~36.9 million (a range of 31.1–43.9) people are living with HIV infections, which is ~9.4% higher compared to the people living with HIV in 2010. An increase in total number of HIV-infected people is due to two confounding factors, i.e., (1) new HIV infection and (2) now HIV-infected persons who are living longer due to improved ART as well as a significant decline in AIDS-related deaths. Out of the 36.9 million infected people, adults account for 35.1 million out of which men are 18.2 million and women are 16.8 million, while children below the age of 15 years are 1.8 millions. The most encouraging fact about HIV infections is consistent decline in new infections since the first report. In 2017, new infections reported are 1.8 million (in the range of 1.4–2.4 million) which is ~14% decline of new HIV infections in the last 7 years, i.e., 2010. Out of these 1.8 million new HIV infections, 1.6 million cases are reported among adults, while only 0.18 million new HIV infections are reported from children below the age of 15 years. Decline of new infections among children is 21% compared to adults where decline of new infections is only 11.1%. This new observation is certainly a very good sign toward a better control over spread of new HIV infections in the present time. A significant decline in vertical transmission of HIV infection must have played a crucial role for this decline among children. Similarly deaths related to AIDS have also been reduced to 0.94 million compared to 1.2 million in 2010 which amounts to ~21% decline in death due to AIDS and AIDS-related diseases in 2017 compared to 2010. The most satisfying part of this data is the better access for ART treatment to HIV-infected patients. At present around 21.4 million people with HIV infection have access to ART compared to only 11.4 million in 2010. This data suggests that in the last 7 years, access to ART has almost doubled and in the near future access to ART is expected to improve further globally. Since 2010, there is a consistent increase in the financial resources available to deal with HIV epidemic. By the end of 2017, more than 21 billion US dollars were available to deal with HIV epidemic.

As per the latest data available to us, ~4931 individuals get new HIV infections, while ~2575 HIV infected people die every day. So on a daily basis, there is an addition of ~2356 HIV-infected persons to the already existing pool of HIV-infected people. If we do the math, it means that on annual basis, we are adding more than 0.8 million HIV-infected people to the already existing pool of HIV-infected individuals, which is almost more than the population of majority of capital cities of European nation, barring few large cities; the majority of HIV seropositives and AIDS patients live in underdeveloped nations as well as developing nations. Such a high number of HIV seropositives in these nations is indicative of multiple factors, viz., poor implementation of surveillance program, poor health care and health-care system, poor living conditions, lack of resources, social factors, etc. There is no doubt that more serious and concerted efforts are warranted at local, national, and international levels to bring a better control of HIV epidemic especially in underdeveloped and developing nations.

A closer look of UNAIDS 2018 report suggests the following important issues:

1. A decline in new HIV infections.
2. Increase in total number of people living with HIV.
3. A decline in AIDS-related deaths.
4. More than 75% people living with HIV know their status about HIV infections.
5. ~59% of people with HIV infection have access to ART.
6. ~ 47% decline in new HIV infections compared to 1996.
7. ~ 51% decline in AIDS-related deaths since 2004.
8. >77.3% millions have been infected with HIV since the start of HIV epidemic.
9. >35.4 million have already died with AIDS-related illnesses since the start of HIV and AIDS epidemic.
10. >2/3 (~65%) of total HIV infections are reported only from the countries of African continent.
11. Least number of cases of HIV and AIDS are reported from the Pacific.

7.4 HIV: Molecular Biology

HIV is a virus which belongs to *Retroviridae* family and its genus is *Lentivirus*. HIV is a small-size virus which is roughly spherical in shape. The diameter of HIV virion is ~120 nm. The characteristic feature of *Retroviridae* family is its genome. The genome of retroviruses is RNA not the DNA. HIV contains a single-stranded RNA (ssRNA), and this ssRNA is a positive sense RNA which acts as genome. This ssRNA passes the genetic information to the next generation during replication. Presence of a unique enzyme, i.e., reverse transcriptase (RT), is a must for retroviruses. The main role of RT is to reverse-transcribe RNA to DNA, and this newly synthesized DNA from RNA is called as "pro-viral DNA." The genome size of HIV is ~9.8 kb. HIV genome consists of nine genes in total which are flanked by long terminal repeats (LTRs) on either side of the genome (Fig. 7.1). These 9 genes of HIV produce 15 different viral proteins during replication of HIV (Frankel and Young 1998). The main role of LTR is to integrate pro-viral DNA to the host cell DNA. Synthesis of new virion starts from LTR because transcription factors from host cells bind to LTR of viral genome for the integration of pro-viral DNA to the genome of host cell. These 15 viral proteins are divided into 3 categories (1) structural proteins, (2) regulatory proteins, and (3) accessory proteins. Gag, Pol, and Env are structural proteins; Tat and Rev are regulatory proteins; and Nef, Vif, Vpr, and Vpu act as accessory proteins of HIV.

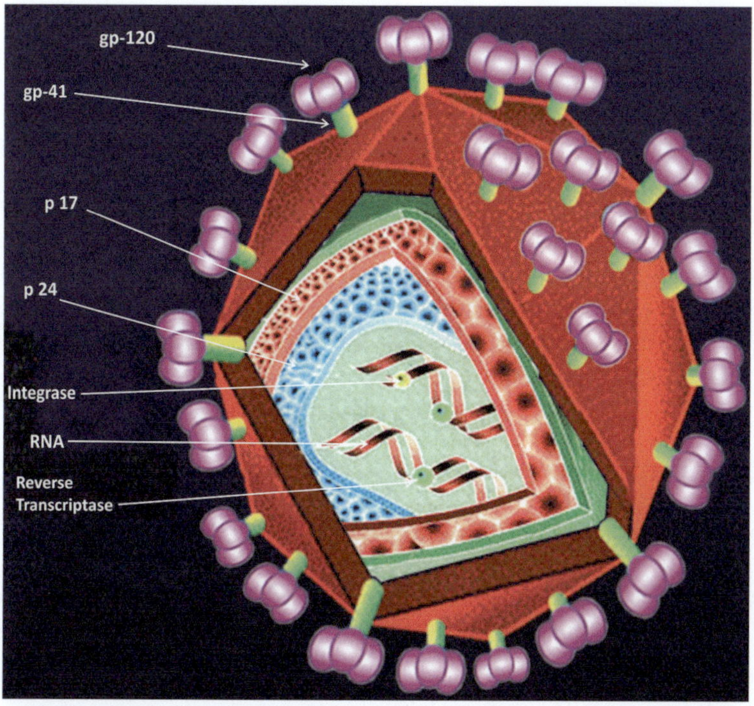

Fig. 7.1 **HIV genome**
Schematic representation of HIV genome. This size of HIV genome is 9.8 kB. HIV genome consists of nine different genes. HIV genes are flanked by long-term repeats (LTR) on either side of full genome. During HIV replication, totally 9 genes of HIV produces 15 different proteins. *env* envelope, *gag* group specific antigen, *LTR* long terminal repeats, *nef* negative factor, *pol* polymerase, *rev* regulator of viral expression, *tat* transactivator of transcription, *vif* viral infectivity factor, *vpr* viral protein R, *vpu* viral protein U

7.4.1 Envelop (Env)

The term Env has been developed from the term "envelope." The size of env gene is 2570 nucleotides which are located between 5571 and 8341. The size of env precursor protein is 160 kDa. Env is a glycoprotein and synthesized as polyprotein precursor. The precursor of Env protein is known as gp160. gp160 is processed or cleaved by cellular proteases which results in the production of two different glycoproteins: (i) gp120 and (2) gp41. Gp120 acts as surface glycoprotein, while gp41 acts as a transmembrane glycoprotein. gp120 interacts with CD4 receptor, while gp41 interacts with chemokine receptors. gp41 consists of three domains: (1) ectodomain, which is responsible for the fusion, (2) transmembrane domain which acts as an anchor, and (3) cytoplasmic tail. The inding of gp120 to its receptors induces conformational changes in gp41. Conformational changes lead to the formation of gp120/gp41 glycoprotein complex which result in the fusion of virus to the host cell.

7.4.2 Group-Specific Antigen (Gag)

The Gag word has been derived from "group-specific antigen." The size of gag gene is 1502 nucleotides and is located between 336 and 1838 nucleotides. Gag gene produces a protein of 55 kDA which is comprised of 550AA. Gag gene encodes for Pr55gag, which acts as a polypeptide precursor. The main role of this polyprotein precursor is to complete the viral assembly at the end of viral replication cycle. Pr55gag cleaves into different proteins, viz., p6, p7, p17, and p24 as well, as two spacer proteins, i.e., p2 and p1. Out of these four proteins, p17 is a matrix protein (MA), p24 is a capsid protein (CA), p7 is a nuclear capsid (NA), and p6 is known as Vpr binding proteins. N-terminal of MA is mainly responsible for targeting and binding to plasma membrane.

7.4.3 Negative Factor (Nef)

The Nef term is coined from "negative factor." The size of Nef gene is 620 nucleotides which is present between 8343 and 8963. Nef is an accessory protein with molecular weight of 27 kDa which is comprised of 206AA. Nef is expressed during early stages of viral replication in host cells. Nef downregulates cell receptors like CD4, CD8, CXCR4, CCR5, etc. Due to the downregulation of receptors for HIV infection, the role of nef is important for HIV pathogenesis. The in vitro studies of nef suggest the following main functions of nef: (1) perturbs endocytosis, (2) modulates signal transduction pathway in HIV-infected cells, (3) supports enhancement of viral infectivity, and (4) supports fusion of HIV to permissive target cells. Nef is also considered to play a significant role in altering CNS functions in HIV-infected individuals. A higher level of nef has been reported to affect the growth of astrocytes. Nef has been reported to alter electrophysiology of neurons.

7.4.4 Polymerase (Pol)

The Pol name has been coined from the word "polymerase." Pol gene consists of 2803 nucleotides, and the location of pol gene is from 1839 to 4642 nucleotides. The molecular weight of pol protein is 112 kDa and it comprises of 935 AA. Pol protein is initially synthesized as a large polyprotein precursor, i.e., Pr160-Gag-Pol. Due to cleavage of Pr160-Gag-Pol by viral protease, it produces viral proteases (PR or p10), reverse transcriptase (RT or p64), and integrase (IN or p32). Out of these proteins, RT is a heterodimer of two subunits, i.e., 64 kDa (p64) and 51 kDa (p51). These two subunits are derived from the same region of Pr160-Gag-Pol precursor protein. Integrase is a single polypeptide with molecular mass of 32 kDa. Integrase promotes the integration of pro-viral DNA (linear but double-stranded) to the genome of host cell. Integration is absolutely a necessary step for viral replication. It has been reported that integrase mutant viruses fail to spread HIV infections.

7.4.5 Regulator of Expression of Viral Proteins (Rev)

Rev is coined from the word "regulator of expression of viral proteins." Rev protein is encoded by two exons. Rev gene comprises 75 to 274 nucleotide sequences. The rev gene is located between 5516–5591 and 7925–8199. The collective length of Rev gene is 349 nucleotides. Rev works as a regulatory protein and is important for the regulation of viral replication. The molecular mass of Rev protein is 191 kDa with 116 AA in total. Rev downregulates splicing of viral RNA post-transcriptionally. Rev contains two domains: (1) arginine-rich domain and (2) leucine-rich domain.

7.4.6 Transactivator of Transcription (Tat)

The term tat is coined from "transactivator of transcription." Tat gene comprises 259 (214 and 45 nucleotides) nucleotides and location of tat gene is between 5377–5591 and 7925–7970. Tat is 86–110 AA-long protein with molecular mass of 16 kDa. Tat is a nonstructural protein which is a product of two exons. It has three main domains: (1) activation domain, (2) basic RNA-binding domain, and (3) overlapping nuclear localization signal. Secreted tat can be taken up by neighboring cells; therefore tat affects both infected and the non-infected cells. Tat has been shown to affect HIV-infected individuals with known CNS pathology. Tat can induce apoptosis in neurons via oxidative stress pathway.

7.4.7 Viral Infectivity Factor (Vif)

The term Vif is coined from the word "viral infectivity factor." Vif gene has conserved sequences among lentiviruses. The size of Vif gene is 578 nucleotides, and its location is between 4537 and 5165. The molecular mass of Vif protein is 23 kDa which is comprised of 192 AA. Vif is expressed during the late stage of viral replication and is present in the cytoplasm of infected cells. Vif inhibits the premature processing of p55Gag in cell cytoplasm. The main role of Vif is to produce competent virions.

7.4.8 Viral Protein R (Vpr)

The term Vpr is coined from "viral protein R." The size of Vpr gene is 296 nucleotides which is localized between 5105 and 5396. Vpr is an accessory protein of HIV and is not essential for HIV replication. The molecular mass of Vpr is 14/15 kDa, and it has 96AA arranged in three domains. Vpr is known to have numerous biological functions like transcription of viral genome, apoptosis, etc. Since Vpr can breach the cell membrane, extracellular Vpr can enter in non-infected cells too (Verma et al. 2012).

7.4.9 Viral Protein U (Vpu)

The term Vpu is coined from "viral protein U." The size of Vpu gene is 248 nucleotides, and location of Vpu gene in HIV genome is between 5608 and 5856. The molecular mass of Vpu is 16 kDa accounting for 62AA. Vpu is a type 1 integral membrane phosphoprotein. Vpu is an accessory protein so it is not essential for HIV replication. Vpu gets expressed during HIV replication. Vpu cannot be detected in virus itself because Vpu does not get packed into viral particle. Vpu consists of two domains: (1) N-terminal and (2) C-terminal. The main function of Vpu is to enhance the release of viral particles and degradation of CD4.

7.4.10 Long Terminal Repeats (LTRs)

LTR term is derived from "long terminal repeats." LTR is present on either side of viral genome. LTR serves as the initiation site for the transcription of viral genome. It harbors cis-acting elements which are needed for RNA synthesis. LTR contains three regions: (1) U3, (2) R, and (3) U5. U3 is known as "unique 3′end," R is known as "repeat," and U5 is known as "unique 5′end." Various elements present in U3 support the direct binding of RNA polymerase II (pol II) to DNA templates. Newly synthesized viral RNA falls under three main categories: (1) unspliced RNA which acts as precursor for Gag and gag-pol polyprotein; (2) partially spliced mRNA (~5 kb) which encodes for Env, Vif, Vpu, and Vpr proteins; and (3) small but multiple spliced mRNAs (1.7–2.0 kb) which encode for Rev, Tat, and Nef. Normally the basal transcriptional activity of LTR is very low, but presence of Tat enhances the transcription rate to many folds.

7.5 HIV Replication

HIV is a virus so it can be easily considered as a bag of nucleoprotein complex. Like any other virus, HIV is also a dead (or non-living) complex of nucleoproteins when HIV is out of its host. Being a virus, HIV has a specific host, i.e., humans. Even in humans there is a specific target cell for the infection of HIV. The main target for HIV infectivity is T-helper cells or $CD4^+$ T-cells. A significant decline of $CD4^+$ cell count is the hallmark for HIV infection. HIV infections are fatal in nature with high morbidity. Even though initial cases of HIV infections were reported in 1981, the term HIV was coined only in the year 1987. Infections with such fatal consequences necessitated the need for development of drugs to find a cure for HIV infections. For the development of new drugs for any infectious agents like HIV, it is of utmost importance to know the drug targets. For the identification of drug targets, it is very useful to have a better understanding about different steps or stages of viral life cycle or viral replication. Therefore serious efforts have been made to delineate the various steps of HIV replication in details which can be used to identify the right targets for the action of anti-retroviral drugs. Various aspects of HIV replication cycle, viz., cell types, cell receptor/s, process and mechanism of receptor-ligand binding, integration of viral genome to the host cell genome, production of new virions, etc., have been studied in great depth. Due to better understanding about HIV replication, it is now possible to develop various groups of anti-retroviral drugs (ARDs). These ARDs have been very effective to control HIV infections, but these ARDs are still unable to cure HIV.

The most characteristic feature of HIV is the genetic material itself. The genetic material of HIV is RNA not the DNA. RNA as a genetic material is a very unique feature of retroviruses. Usually the genetic information is transferred from one generation to the next by DNA, because DNA is known as genetic material for living organisms. The problem about transfer of genetic information from one generation to the next in case of retroviruses has been overcome due to the presence of unique enzyme called reverse transcriptase (RT). RT enzyme can reverse-transcribe RNA into DNA. Reverse transcription is an essential part of replication cycle of HIV. HIV cannot replicate itself in the absence of host/host cells because HIV does not have its own synthetic machinery. Hosts' cells compensate the requirement for the synthetic machinery for HIV replication. Replication of HIV starts only after the hosts' cells get infected with HIV. For HIV replication, it is essential that virion should attach to the specific receptor present on these target cells, and then the replication process continues till the production and release of new virions takes place (Freed 2001). T-cells are the main cells for HIV infections. The process of HIV replication has been divided into five different steps (Fig. 7.2):

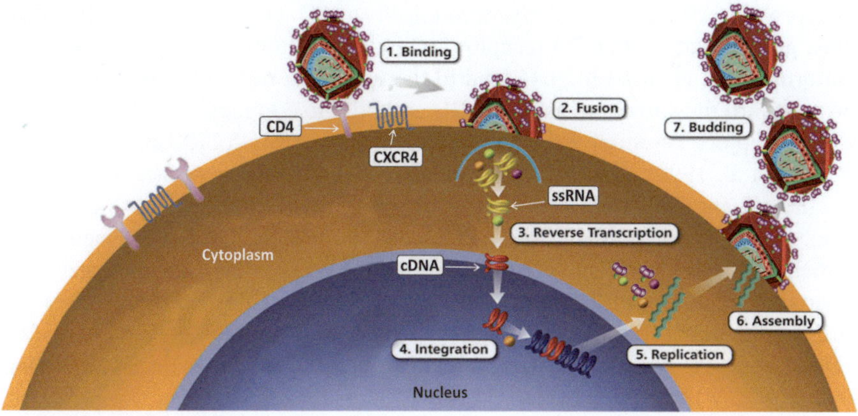

Fig. 7.2 HIV Replication
Schematic representation of different step of HIV replication. During infections, HIV attaches to the CD4 receptor on cell surface of target cells permissive for HIV. After attachment, HIV fuses with the cell membrane of target cells. HIV RNA and other proteins are released inside the target cell after attachment and fusion of HIV with plasma membrane. After activation, reverse transcriptase enzyme produces pro-viral DNA in cytoplasm. Now pro-viral DNA moves to the nucleus and gets integrated with the genome of host cells with the help of integrase enzyme of HIV. After integration to the host cell genome, pro-viral DNA produces mRNA. Newly synthesized viral mRNA translates into different types of viral proteins necessary for the production of new virions. Viral proteases cleave these proteins for the assembly of new virions. These new virions are released into host circulation from the infected cells, Viral RNA; pro-viral DNA

1. Viral attachment and viral entry
2. Reverse transcription
3. Integration of pro-viral DNA
4. Transcription and translation
5. Completion and release

7.5.1 Viral Attachment and Viral Entry (to the Cell)

The first and the essential step of viral (HIV) replication cycle is the attachment of viral particle to the target cells. Interaction of virus to the cells is like a ligand-receptor binding phenomenon. A specific ligand is present on virus, while a specific receptor is present on the target cells. In case of HIV, CD4 is the main receptor, while ligand is gp120. CD4 is the main receptor for the attachment of HIV to the target cells, i.e., T-cells. CD4 is a 58 kDa glycoprotein present at the cell surface of a specific T-cell population which is also known as T-helper cell or $CD4^+$ T-cells. Apart from T-helper cells, CD4 receptors are also present on some other cell types, viz., monocytes, macrophages, dendritic cells, microglial cells, etc. The binding of gp120 to CD4 is the first and most essential step for the entry of HIV into the permissive cells. The binding can occur due to the attachment of HIV to CD4 cells. Attachment occurs when T-helper (permissive) cells and HIV are present in closer vicinity.

The attachment and entry of HIV into T-cells takes place due to the presence of two different envelop proteins present at the surface of HIV. These proteins are glycoprotein 120 (gp120) and glycoprotein 41 (gp41). Out of these two glycoproteins, gp120 binds with CD4 receptor and induces conformational changes. The conformational changes in gp41 lead to the unfolding of gp41 allowing gp41 to move toward the surface of same T-cells and bind to co-receptors. The two types of co-receptors essential for HIV replication are (1) CXCR4 and (2) CCR5 (Benger et al. 1999). Binding of gp120 and gp41 to the cell surface through the specific receptor on cells leads to the fusion of viral envelop with the plasma membrane of the target cells. The fusion of viral envelop with plasma membrane of the target cells is the most crucial step for the entry of HIV to the target cells (Fenouillet et al. 2001). Fusion of viral envelop to the cell membrane paves the path for the delivery of HIV genome and other HIV proteins to the host cells. Delivery of viral genome and viral proteins to the target cells is an essential step to initiate HIV replication cycle.

Binding of virus to the target cells leads to the fusion; therefore the fusion step is one of the interesting targets for the development of new anti-HIV drugs. Although blocking of fusion step of HIV replication cannot be the most effective means to control HIV infections, drugs which can block the entry of HIV to the target cells could also be an effective means to prevent the HIV entry to the new target cells after the production and release of new viruses from infected cells (Ryser and Fluckiger 2005). Realizing the significance of fusion inhibitors, active researches are going on to develop more effective fusion inhibitors. Some of the fusion inhibitors are, e.g., Fuzeon, etc.

7.5.2 Reverse Transcription

Viral genome and proteins are delivered to the target cells after the fusion of HIV with the target cells. Untill this stage RNA is the viral genome, the next step of HIV replication cycle is to convert RNA into DNA. This conversion is accomplished by the activity of reverse transcriptase enzyme of HIV. After release of HIV components into the cytoplasm of target cells, the RT enzyme starts working by making copies of viral RNA into complementary DNA (c-DNA). The new synthesized DNA produced from viral RNA is called "pro-viral DNA." On the one hand, RT is an advantage to HIV, while on the other hand, HIV-RT has been a major disadvantage for drug development because HIV-RT lacks the ability for "proofreading" of DNA. Due to the inability of RT to do DNA proofreading, the process for pro-viral DNA synthesis is highly prone to errors. The inability of proofreading by RT is an advantage for HIV for its survival, but at the same time this has been a great disadvantage for the treatment of HIV infections. Inability to proofread leads to rapid generation of mutations in HIV. These mutations are the major reason for resistance in HIV against ART.

The production of pro-viral DNA due to the activity of RT enzyme has been the most attractive target for the drug development. The simple rationale to target this step by drug is to block or inhibit the production of new pro-viral DNA. If one can stop production of new pro-viral DNA, then HIV infection can be controlled effectively. This approach did not turn out to be as easy as it was conceptualized earlier. The major stumbling block for the development and treatment of HIV infections is none other than the inability of RT to do proofreading.

Majority of anti-retroviral drugs (ARD) have been developed to target RT step of HIV replication. These drugs are grouped as nucleoside reverse transcriptase inhibitors (NRTIs), nucleotide reverse transcriptase inhibitors (NtRTIs), and non-nucleoside reverse transcriptase inhibitors (NNRTIs). Different drugs of these classes are known to inhibit RT activity of HIV with different mechanisms. The advantage of these drugs is that different analogues can be used alternatively to address the concerns of drug resistance developed against any of these drugs. After evolution of resistance to one analogue, another analogue can be used for treatment of HIV infection, e.g., zidovudine, lamivudine, stavudine, etc.

7.5.3 Integration of Pro-viral DNA

After the production or synthesis of pro-viral DNA, it is essential that pro-viral DNA must integrate with target cell genome to produce new progeny of HIV. This step is known as integration. Integration of pro-viral DNA takes place due to the enzymatic activity of integrase (In) enzyme. Integrase is a viral enzyme in origin. HIV genome has nine genes in total which are flanked by long terminal repeat (LTR) on either sides. So LTR of viral genome and enzyme of viral origin are essential for the integration of pro-viral-DNA to the genome of host cell.

Integration step has turned out to be an excellent target for development of new drugs which can stop HIV replication. Rationally, if one can stop the viral integration to the genome of target cell, then further replication steps will not proceed. Integrase inhibitors work in combination with other drugs. Integrase inhibitors alone are not very effective to treat HIV infections. But integrase inhibitors can be an excellent choice of drug to deal with drug resistance against HIV. Some of the drugs of this group were approved by FDA in 2007, e.g., Isentress.

7.5.4 Transcription and Translation

After integration of pro-viral DNA in the genome of target (host) cells, transcription and translation of viral gene starts for the production of new virions. In certain circumstances, pro-viral DNA gets integrated in the genome of target, but integrated pro-viral DNA does not start active transcription and translation. This is the reason these cells do not show signs for active replication of HIV. Cells with integrated pro-viral DNA remain latently infected. This phenomenon is called as latent infection. At some point of time, latently infected cells start active replication of HIV, but still there are no concrete answers for the trigger of active replication. With time, it is getting clearer that latently infected cells are one of the major hurdles to find a cure for HIV.

After integration various transcriptional factors of host/target cells get activated. These host transcription factors support transcription of viral RNA with help from cellular RNA polymerases. Splicing of RNA transcripts takes place in a planned manner. Transcription leads to the production of mRNA to synthesize various viral proteins in cytoplasm of host cells. Tat and Rev are two different proteins that are produced at this stage and support HIV replication in activated T-cells. Env and Gag proteins are produced at a later stage. After the production of gag proteins, new viral RNA molecules (full-length) bind with gag proteins and are packed as new virions.

7.5.5 Assembly and Release

Assembly and release of new virions is the last and final stage of HIV replication cycle. At this stage Env passes through the cytoplasmic reticulum and finally gets transported to Golgi complex. Env proteins undergo proteolytic cleavage in Golgi complex. After proteolytic cleavage, Env produces two different glycoproteins, i.e., gp120 and gp41. Now gp120 and gp41 move toward the plasma membrane of host cell. At this stage gp41 anchors gp120 to the cell membrane of infected cells. Now, gag-pol polyproteins help in the budding of new virions from infected host cells. Proteases from HIV cleave three polyproteins which result into various functional HIV proteins and HIV enzymes. Cleavage of polyproteins completes the maturation and release of new infective virions.

This step is also an interesting target to stop HIV replication. This is the reason that various protease inhibitors (PIs) are used as drugs. If these PIs can prevent the activity of HIV protease that will prevent the completion and maturation of infective HIV, this prevents further infection. This step of HIV replication can be blocked by protease inhibitors, e.g., lopinavir, ritonavir, etc.

7.6 Anti-retroviral Drugs

Nonavailability of any drug to treat HIV/AIDS patient was the major hurdle at the time when initial cases of HIV/AIDS were reported in the year 1981. It took almost 6 years before the first drug to treat HIV/AIDS infection was approved by the US Food and Drug Administration (USFDA or FDA), Maryland, USA. USFDA approved azidothymidine (AZT) as the first drug to treat HIV/AIDS in the year 1987. Drugs to treat HIV infections are also commonly called as anti-retroviral drugs (ARD) or anti-HIV drugs. Some of the common retroviral infections are caused by human T-cell leukemia virus-1 (HTLV-1) and human T-cell leukemia virus-2 (HTLV-2). Interestingly HTLV-1 and HTLV-2 are slow-growing viruses compared to HIV. The incubation period for HTLV-1 and HTLV-2 is very long, almost about 30–40 years.

The enhanced morbidity and mortality associated with HIV infections was really a matter of concern and worry to various sections of society. These serious concerns were the engines to find a cure for HIV by developing new drugs which can treat HIV infections. Thank God! Serious and sincere efforts in this area have been fruitful to develop numerous new drugs to control HIV infections. In the meantime, a better understanding about HIV replication evolved which helped in identification of targets for new drugs which can either block or inhibit HIV replication. Blockage or inhibition of HIV replication can be one of the means to cure HIV, may be in the future, but not at the present time.

Unfortunately, a cure for HIV is still not on the horizon, but a possibility in the future to find a permanent cure for HIV/AIDS cannot be denied. Knowledge about the different crucial steps of HIV replication cycle has been helpful in the development of different groups of medications, and those are (1) nucleoside reverse transcriptase inhibitors (NRTIs), (2) nucleotide reverse transcriptase inhibitors (NtRTIs), (3) non-nucleoside reverse transcriptase inhibitors (NNRTIs), (4) protease inhibitors (PIs), and (5) integrase inhibitors (InIs) (Table 7.2). Apart from these well-established group of ARD, some new types of drugs are under active investigation like fusion inhibitor, etc.

7.6.1 Nucleoside Reverse Transcriptase Inhibitors (NRTIs)

As the name of the drug suggests, these drugs blocks the activity of reverse transcriptase enzyme. Basically these drugs act as “false building block” for HIV replication cycle serving as an alternative substrate for RT enzyme. Mechanistically,

Table 7.2 Anti-retroviral drugs

Class	Generic name	Chemical name
E.I. or F.I	Enfuvirtide	T-20
	Maraviroc	MVC
Int.I	Raltegravir	RGV
NRTI	Abacavir	ABC
	Azidothymidine	AZT
	Didanosine	ddI
	Lamivudine	3TC
	Stavudine	d4T
	Zalcitabine	ddC
	Zidovudine	ZDV
NNRTI	Delavirdine	DLV
	Efavirenz	EFV
	Etravirine	ETR
	Nevirapine	NVP
NtRTI	Tenofovir	TDF
PI	Atazanavir	ATV
	Amprenavir	APV
	Indinavir	IDV
	Lopinavir	LPV
	Nelfinavir	NFV
	Ritonavir	RTV
	Saquinavir	SQV
	Tipranavir	TPV

E.I. Entry Inhibitors, *F.I.* Fusion Inhibitors, *Int. I* Integrase Inhibitors, *NNRTI* Non-Nucleoside Reverse Transcriptase Inhibitors, *NRTI* Nucleoside Reverse Transcriptase Inhibitors, *NtRTI* Nucleotide Reverse Transcriptase Inhibitors, *PI* Protease Inhibitors

NRTIs compete with physiological nucleoside during pro-viral DNA synthesis. NRTIs are designed in such a manner that azido group is added to ribose sugar. The presence of azido group serves as a chain terminator for DNA synthesis. During synthesis of pro-viral DNA, the presence of NRTI terminates DNA synthesis. Termination of DNA synthesis stops further synthesis of new DNA for completion of HIV replication. The first anti-retroviral drug azidothymidine belongs to this group. Apart from thymidine, some other nucleosides have been added in this group of medicine, and those are adenine (A), cytosine©, and guanosine (G). Some of the most common NRTIs are Azt, ddI, ddc, d4T, 3TC, and FTC. Out of these different drugs, Azt and d4T are thymidine analogues, FTC and 3TC are cytosine analogues, ddI is an adenosine analogue, and ABV is a guanosine analogue.

On one hand discovery of NRTIs was a great relief for the treatment of HIV patients, while on the other hand drug resistance turned out to be a major bottleneck for the effective use of NRTI for the treatment of HIV patients. The problem of drug resistance is attributed to the inability to do DNA proofreading by RT of HIV. This feature of RT has been an advantage for HIV to survive under drug pressure. Additionally the development of different analogues for the NRTI group of medicine has been a great advantage to the clinicians to overcome or to deal with the problem of drug resistance up to a greater extent.

7.6.2 Nucleotide Reverse Transcriptase Inhibitors (NtRTIs)

Another version of NRTI is NtRTI group which uses nucleotides instead of nucleosides. NtRTIs have some advantages over NRTI. This NRTI group of drugs has to be activated to show its anti-retroviral and inhibitory activities by converting nucleoside into nucleotide. Therefore, NtRTIs have an added advantage to skip this step of activation to perform inhibitory activities of drug. This is the reason that NtRTIs are more toxic compared to the drugs of NRTI group. The mode of action of NtRTI is similar to NRTI, but NtRTI is known to be presented with more side effects. Tenofovir disoproxil fumarate (TDF) is a NtRTI. TDF is commercially available as Viread. Nowadays NtRTI is available in various combinations, but a combination of TDF with ddI should be avoided.

7.6.3 Non-nucleoside Reverse Transcriptase Inhibitors (NNRTIs)

This is another group of drug which also inhibits the activity of RT enzyme, but the mechanisms of action of NNRTIs are different than the mode of action of NRTI or NtRTI. NNRTI binds directly to RT enzyme. NNRTIs interfere with the active (catalytic) site of RT enzyme, so NNRTIs affect the normal functioning of RT enzyme by binding close to the substrate binding sites (i.e., dNTP binding site). The binding of NNRTI to RT enzyme prevents the nucleoside binding to the catalytic site. Binding of NNRTI to RT enzyme does not require any further activation (Boehringer Ingelheim Pharmaceuticals Incorporated 2005). The example of NNRTI group of drugs is nevirapine, delavirdine, efavirenz, etc.

7.6.4 Integrase Inhibitors (InIs)

Integration of pro-viral DNA into the genome of infected cells occurs due to the activity of integrase enzyme. Integrase enzyme has been studied as one of the potential drug targets to control HIV replication. USFDA has approved a drug in this category in 2007 though InIs alone cannot control HIV infection, so integrase inhibitors have to be given in combination. Raltegravir, the only drug of this group, is available in the market as Isentress.

7.6.5 Protease Inhibitors (PIs)

As the name suggests, this group of medicine inhibits the activity of protease enzyme. PIs inhibit the activity of viral proteases (proteases of HIV). During HIV replication cycle, the role of viral proteases is very important for maturation of viruses. Protease is a homodimer comprised of 99 amino acid monomers. HIV protease cleaves gag and gag-pol polyproteins which are precursors to produce mature and active proteins, and these proteins are protease, reverse transcriptase, RNase H, and integrase. This is the reason viral protease is an important drug target. Protease inhibitors block active site of enzyme so that viral proteases cannot function. A nonfunctional viral protease produces immature and non-infective virions, so a protease inhibitor can reduce or inhibit the production of new virions from infective cells. HIV can still replicate in the presence of PI but will not be able to form a mature virus. PIs are essential components of combination therapy, i.e., HAART. Use of PIs has reduced the problem of drug resistance. Toxicity and side effects are common with PIs. Some of the common PIs are indinavir, lopinavir, nelfinavir, ritonavir, etc. (Saag et al. 2018).

7.6.6 Fusion Inhibitors (FIs)

The first step for HIV infection is the interaction of HIV with receptor and co-receptors present on cells. This interaction takes place through gp120 and gp41 of HIV. Interaction leads to fusion of HIV to plasma membrane of HIV-permissive cells. gp41 plays an important role in initial step involving fusion of HIV with permissive cell envelop. This step is a good target to prevent HIV infections. One drug as fusion inhibitor was approved by USFDA in 2003. The brand name of the drug is Fuzeon which is also called as T-20. Fuzeon is a polypeptide comprising of 36 amino acids. Since Fuzeon is a polypeptide, this drug is not orally bioavailable. Bioavailability is one of the major drawbacks associated with FIs. This medicine can be injected parenterally (subcutaneous). One injection a day does not offer the expected result, but twice a day regimen of FI is useful to demonstrate its effect. FIs are good in combination therapy.

7.7 Anti-retroviral Treatment and HIV Drug Resistance

The first drug to treat HIV infections was azidothymidine (Azt). Azt was a very effective drug, and soon drug resistance was commonly observed among HIV patients with the use of Azt. The main culprit for the development of drug resistance was none other than the inability of RT (HIV) enzyme to perform the function of DNA proofreading. A consistent increase in the cases for drug resistance led to development of new modalities to treat HIV patients. To discuss all possible treatment modalities and to deal with drug resistance are beyond the scope of this chapter; therefore we have discussed three most commonly used treatment modalities for HIV seropositives who are presented with symptoms of drug resistance (Olender et al. 2012).

7.7.1 Highly Active Anti-retroviral Treatment (HAART)

HAART is a new name given to the kind of combination therapy for HIV patients. HAART has became a reality due to the availability of different classes of drugs. Drugs from different classes not only have different mechanism actions, but these drugs also work to block different steps of HIV replication. HAART regimen has been widely accepted since 1996 to treat patients facing the problem of drug resistance. Regimen for HAART was developed and decided by a panel of experts from the National Institutes of Health and other international organizations. Basically HAART uses a combination of three or more anti-HIV drugs, preferably from different classes. HAART has various advantages like the following: (1) HAART reduces the complexity of dose scheduling, (2) HAART reduces the problems related to the complex drug schedules and regimen, (3) HAART is helpful to achieve synergism between drugs used to treat HIV infections, (4) HAART reduces the drug toxicity with the decrease in the doses of different drugs, (5) HAART decreases the possibility for the development of drug resistance, and (6) HAART targets various steps of HIV replication, Therefore HAART can be very effective to control HIV infections. So many advantages associated with HAART have been the reason for effective use of HAART regimen. It is difficult to realize or even to recognize the probable number of combinations that can be used in a HAART regimen. So far ~25 different drugs are licensed to be used as anti-HIV drug.

7.7.2 Salvage Therapy

As the name suggests, salvage therapy is to salvage the patients from the limitations of available medications to treat HIV infections. In certain cases it has been observed that even HAART therapy is unable to give the expected outcome with reference to control over HIV infections. Then salvage therapy is recommended for the treatment. In principle salvage therapy can also be recognized as mega-HAART therapy. In salvage therapy more numbers of drugs are used compared to HAART regimen. Salvage therapy is used among those patients who are either resistant or nonresponder to other anti-HIV treatment strategies (Tang and Shafer 2012). In the case of salvage therapy, up to nine different drugs are given to HIV/AIDS patient in the form of combination therapy. Certainly salvage therapy is an expensive therapy with some serious side effects. Side effects of salvage therapy could either be bearable or unbearable too.

7.7.3 Drug Holiday

As the name suggests, during these treatment protocols, drugs go on a holiday. It means patients are not given any anti-HIV drug for certain period of time before patient starts taking anti-HIV drug again. Drug holiday regimen is evolved to address the problem of drug resistance. The theoretical foundation for this protocol is that withdrawing the drug selection pressure in HIV patients will make HIV or new virion more susceptible to the drug. This can also be called as "drug interruption therapy." Drug holiday regimen has shown promising results.

7.8 How to Test Anti-HIV Drug

Since the discovery of HIV as a causative agent for AIDS, there was a dire need to develop an assay for the testing of anti-retroviral activity of different anti-retroviral drugs or agents. Nonavailability of any suitable animal model for HIV infections is one of the major hurdles to evaluate anti-retroviral drugs in vivo. So, scientists have developed various in vitro assays to evaluate toxicity and anti-retroviral activity of anti-HIV drugs. This test is based on the survival of cells infected with HIV in the presence of anti-retroviral drugs or anti-HIV drugs. These in vitro assays have been adopted in a format to evaluate anti-HIV drugs even on a large scale. The basic principle to evaluate anti-HIV drugs is to assess either cell death or cell viability in the presence of anti-HIV drugs. The advantage of this assay is that one can evaluate two important aspects of anti-HIV drugs, and those aspects are (1) to evaluate toxicity of anti-HIV drugs and (2) to evaluate the anti-HIV effect. These tests are called as MTT assay for the evaluation of anti-HIV drugs.

7.9 MTT Assay for the Evaluation of Anti-HIV Effect of Anti-retroviral Drugs

The basic foundation of this assay is to evaluate the cell death or cell mortality of HIV-permissive cells in the presence of infective HIV as well as drug/s. A 96-well format is a miniature form of the assay so that a large number of samples can be tested with convenience. These two are the characteristic features of MTT assay. This is the reason that MTT assay has been adopted for ART testing.

In this anti-HIV drug assay, MTT dye is used. MTT dye gets metabolized by mitochondrial enzyme to give a blue-colored formazan product. The conversion of formazan from MTT dye is only possible due to the enzymatic activity of mitochondrial enzyme. Mitochondrial enzyme works only in living cells so dead cells do not convert MTT dye into formazan. The intensity of blue color is the end point for the determination of cell viability or cell death. The cell viability in this assay is indicative of toxicity when used for drug toxicity assay and anti-HIV effect when used for the evaluation of anti-HIV effect of drug which is under investigation.

When MTT is added to the cells, the mitochondrial enzyme converts yellow color of MTT dye into blue color crystals of formazan. These crystals are finally dissolved into organic solvent and Optical Density (OD) of sample is monitored. If someone is working in the linear range of assay, then cell viability is directly proportional to the OD of sample. MTT assay is very versatile in nature and can be easily adopted for a 96-well plate format. Adaptation of MTT assay for testing of anti-HIV drug has numerous advantages like the following:

1. Large number of samples can be tested.
2. Number of drugs can be tested at the same time.
3. Various concentrations of drugs can be tested at the same time.
4. The requirement of number of cells needed for assay is very low.
5. A lesser number of HIV virions are needed.
6. It is an efficient and effective method for the testing of large number of samples at any single given time. To evaluate the anti-HIV effect, the following steps are needed: (1) growing of HIV-permissive cells, (2) plating of cells, (3) infection of cells with HIV, (4) addition of anti-HIV drugs, (5) addition of MTT, (6) stopping of MTT reaction, (7) acquisition of data, and (8) analysis of data (Zhu 2005; Peters et al. 2013).

7.10 Rationale of Each Essential Step

7.10.1 Growing or Culturing of HIV-Permissive Cells

For the evaluation of anti-HIV drugs, we need to use cells permissive for HIV. It means cells must be $CD4^+$. The most commonly used cells are T-cells which can be either primary cells like PBMCs or secondary cells or cell lines like MT-4, HUT_{78}, H-9, etc. Cells have to be grown overnight for their use in this assay.

7.10.2 Plating of HIV-Permissive Cells in 96-Well Plate

Overnight grown cells have to be washed and counted. A single cell suspension with the required cell density in required number of wells of the 96-well plate. (Usually 2x concentration of cells is added into the wells, so that drug concentration can also be used 2x. After mixing contents in each well, cells and drug will be diluted in a 1:1 ratio. The final concentration of cells and drug in each well would be 1x or the desired concentration.)

7.10.3 Thawing of HIV Stock

To infect cells in a 96-well plate, one has to thaw an HIV stock of known titer (various expressions of titer are used like MoI, RT, p24, etc.). After thawing, appropriate dilution has to be added into each well in specific volume for HIV stock. HIV-infected cell will show cell death at the end of assay, and effective concentration of anti-HIV drug will prevent cell death caused by HIV.

7.10.4 Preparation of Drug Dilution

One has to use nontoxic concentrations of drug to test anti-HIV activity of the drug under investigation. Initially it is advisable to try five different concentrations or dilutions of drug and to use dilutions of drug/s in log fold (e.g., 2x, 5x, 10x, etc.). Drug dilutions should be prepared, giving due consideration for the volume of drug to be added into specific wells.

7.10.5 Addition of Drug to the Wells

After calculating the drug concentrations, the required volume of drug has to be added to the respective wells. Remember to add the drug to the respective well so that anti-HIV effect of drug can be evaluated at termination of anti-HIV assay.

7.10.6 Incubation of Drug and HIV with Cells

After addition of cells, HIV and candidate drug have to be incubated at 37C with 5% CO_2 and 95% humidity in an incubator. The incubation period has to be empirically determined. It can vary from 24 h to 7 days. Numerous factors like cell counts, HIV titer, and cell death should be taken in account for the determination of incubation period.

7.10.7 Addition of MTT Dye

At the end of incubation, MTT dye has to added to the wells of a 96-well plate to evaluate anti-HIV effect of drug on HIV-permissive cells. After addition of MTT dye to respective wells, MTT has to be mixed properly in each well. Mixing of MTT can be achieved either by plate mixer or by multichannel pipette. After mixing of MTT dye, a 96-well plate is incubated, so that mitochondrial enzyme of live cells can metabolize MTT to produce formazan crystal of blue color.

7.10.8 Stopping the MTT Reaction

After addition of MTT at certain of time, MTT reaction has to be stopped by addition of stopping solution into respective well. Stopping solution has to be mixed in wells to dissolve formazan crystal, and mixing can be done by using either plate mixer or multichannel pipette.

7.10.9 Reading of Plate

After mixing of stopping solution, the OD of each well has to be read at specific wavelength using ELISA plate reader, and data must be stored for analysis. Data sheet must have details of experiment like cell type, HIV type, drug, drug concentration, etc.

7.10.10 Calculating the Data

The OD values of different wells are used to compare the viability of cells in that specific well or group of well with different treatments. On the basis of viability of cells, one can determine the anti-HIV effect as well as the most effective concentration of drug for its anti-HIV effect.

7.11 Important Note

To set up a 96-well plate, it is important to prepare the chart or map for the plate with all the relevant information present on it. Preparation of chart is very helpful to interpret result without any errors or omissions. While setting up an experiment in a 96-well plate, each plate must have the following necessary components:

1. Blank well (only media)
2. Cell control well (only cells, no HIV and no drug)
3. Drug control well (cells and drug, no HIV)
4. HIV control well (cell and HIV, no drug)
5. Positive HIV control well (cell and HIV, no drug)
6. Experimental well (Cell, HIV, and test drug)

7.12 NeuroAIDS

Improvement in the area of anti-retroviral drugs is enormous. The first drug to treat HIV infection was approved by the USFDA in 1987. The first ever drug approved for the treatment of HIV infection was zidovudine, and since then many more drugs have been developed and available to treat HIV infections. These ADTs belong to different groups of drugs, i.e., NRTI, NtRTI, NNRTI, protease inhibitors (PIs), integrase inhibitors (InIs), and even the fusion inhibitors (FIs). Since a variety of drugs are now available which can simultaneously target different stages of HIV replication, at present it can be said that the problems of drug resistance have been addressed very effectively due to availability of different drugs. As a result of the availability of better anti-retroviral drugs for HIV patients, now an HIV seropositive can lead a quality life like a normal person. These drugs have prolonged the asymptomatic stage among HIV seropositives which now extends for 20 years or more.

But no one ever recognized or realized that some of these advantages to treat HIV seropositives may turn out to be a double-edged sword. On the one hand, these improved medications have extended the life span of HIV seropositives, while on the other hand, these advantages have led to new health complications among HIV seropositives. These new complications in HIV seropositives are now grouped as NeuroAIDS. So, what is neuroAIDS? And what are the consequences of neuroAIDS?

NeuroAIDS is really a matter of serious concern due to very high prevalence of neuropsychiatric complications among HIV seropositives. The prevalence of neuropsychiatric complications among the general population is ~10–15%, while prevalence of neuropsychiatric complications has been reported >50% among HIV seropositives. Various neuropsychiatric complications are HIV-associated dementia (HAD), HIV-associated encephalopathy (HIVE), HIV-associated minor cognitive and motor disorder (MCMD), etc. (McArthur et al. 2005). Presently, different types of neuropsychiatric complications in HIV/AIDS patients have been grouped as neuroAIDS (McCombe et al. 2009). It seems like neuroAIDS is another epidemic waiting to happen in the near future (Power et al. 2009). Some of the common neuropsychiatric complications in the case of neuroAIDS have been mentioned in Table 7.3.

Addition of new infections along with the improved survival of HIV seropositives is the main reason for increase of numbers of HIV/AIDS patients among an already existing pool of HIV patients. This new change in demography of HIV/AIDS is a matter of concern for all of us.

These neuropsychiatric complications among HIV seropositives are known to affect both central nervous system (CNS) and peripheral nervous system (PNS). CNS is one of the most well-protected organ systems of the body. CNS is protected by the blood-brain barrier (BBB) which regulates the entry of any biomolecule in the CNS. The reasons or causative factors for the development of neuroAIDS get more intriguing by knowing the fact that most of the cells in CNS are not permissive to HIV and neurons in particular. Most of the cells of CNS are non-permissive to HIV because cells of CNS do not express receptors for HIV infections. Entry of HIV to the brain is also not possible due to the very fine control of the BBB. One of the most probable answers for HIV infections to CNS can be explained on the basis of "Trojan horse hypothesis." As per the assumption of this hypothesis, HIV enters

Table 7.3 NeuroAIDS: neuropsychiatric disorders

Addiction
Anxiety
Depression
Epilepsy
Mania
Mood disorders
Neurocognitive impairment
Neuropathic pain
Physical disability
Seizures

into the CNS through the HIV-infected cells from systemic circulation like monocytes or T-cells. These infected cells may find their entry to the brain due to alteration in the permeability of BBB (Verma et al. 2010). There could be various factors and reasons for alteration in the permeability of BBB. After entry of infected cells to the brain, these cells remain latently infected and remain in the brain for a long period of time. At certain point of time, these latently infected cells may start HIV replication. The trigger for the activation of these latently infected cells is not known or not yet well defined, but active replication of HIV in the brain could lead to the production of various other biomolecules of viral origins. In turn these biomolecules may start or initiate the induction as well as production of different pro-inflammatory and inflammatory cytokines in the local milieu. Certainly these pro-inflammatory and inflammatory cytokines are important in protective immune response, but in CNS these biomolecules could induce neuronal death via apoptosis. Since neurons do not regenerate, therefore inadvertent loss of neurons could lead to permanent damage to CNS or the brain (Olivier et al. 2018).

According to a recent study, only 30% of HIV-infected children living in sub-Saharan Africa are receiving anti-retroviral therapy. The situation is quite worrisome as an estimated 2.3 million children less than 15 years of age are infected with HIV in sub-Saharan Africa. These children may manifest mild to severe neurocognitive disorders with psychiatric symptoms in the long run (Wilmhurst et al. 2018). As per some estimates, NeuroAIDS may have potential to become one of the major health concerns among long-term HIV seropositives who are under proper anti-retroviral treatments. Some of the major concerns for neuroAIDS patients are the following: (1) the total number of HIV/AIDS patients is increasing globally on a daily basis; (2) among the majority of HIV seropositives, neuroAIDS hits them at the prime of their life span, e.g., mostly between the age group of 35 and 45; (3) symptoms of neuroAIDS render these patients least productive in their prime of life; (4) neuroAIDS results in substantial loss of either individual or family income; (5) need of a caretaker causes extra burden to financial resources of the family or individuals; (6) ART is already an expensive treatment, and further additional cost for the treatment of neuroAIDS just exaggerates or worsens the financial situation; etc.

Realizing the potentials and consequences of neuroAIDS at personal level, family level, as well as society level, it is very important to direct financial resources toward the better understanding about neuroAIDS as well as to find a cure for neuroAIDS.

7.13 Berlin Man: An Example for HIV Cure

It has been >35 years since the initial cases of HIV/AIDS have been reported in 1981. Serious, committed, and dedicated research efforts in the area of HIV/AIDS have been quite successful to bring a better control over the HIV/AIDS epidemic. Still a cure for HIV/AIDS is far from reality. Everyone in this world still have a hope that one day we will be able to find a cure for HIV/AIDS. The question is how? and when? Some of the probable answers do come from the nature itself!!! Some clues can be drawn from elite controllers and long-term non-progressors with reference to HIV infections. One and the only example for this approach is none other than the "Berlin man." Berlin man is an HIV-positive patient who has undergone bone marrow transplantation (hematopoietic stem cell transplantation) from a healthy donor. Since the bone marrow transplantation, Berlin man remained free from HIV infections for more than 10 years by now. So what was the reason for bone marrow transplantation in this patient? What are the scientific rationales for the success behind this new and probable cure for HIV?

High awareness about HIV/AIDS was the reason for active participation of people in different HIV/AIDS programs and frequenting HIV/AIDS clinics. Information was collected from certain individuals, who were classified under high-risk category to contract HIV based on their lifestyles. The most interesting observation was that even though these individuals were following a high-risk lifestyle they did not get infected with HIV. The value for this observation was further added with a known fact that these high-risk individuals were never on any type of anti-retroviral treatments. Therefore, it was concluded that there are very good possibilities that these high-risk individuals for HIV/AIDS may have some kind of "natural resistance" against HIV infections. This natural resistance in these individuals prevents HIV infections to them, even though they are following a high-risk lifestyle.

Further studies indicated that the T cell co-receptors may be responsible for the resistance against HIV infections among these high-risk individuals. At this stage it is important to reiterate that the primary receptor of HIV infections is CD4 receptors, but co-receptors are also important for HIV infections along with the requirement of primary receptors. Two different types of co-receptors have been identified for HIV infectivity, and these two co-receptors are CXCR4 and CCR5. On the basis of requirement of co-receptors for HIV infectivity, even HIV strains are divided into two categories. HIV strain which needs CXCR4 co-receptor for its infectivity is called as X-4 strain, and the HIV strain which requires CCR5 for its infectivity is called R5.

The extensive researches have demonstrated that these high-risk individuals which are resistant to HIV infection are presented with mutation in one of the co-receptors, i.e., CCR5. This mutation is a deletion mutation for CCR5 gene. A 32 bp sequence of DNA was found to be deleted in these individuals. Due to deletion of 32 bp, this mutation is known as Δ32 deletion. Story about the mutation of CCR5 co-receptors gets further complicated due to homozygous nature of mutation as well as heterozygous nature of mutation. It was observed that individuals with homozygous mutations for Δ32, i.e., CCR5 Δ32/ Δ32, are fully resistant to HIV infections. On the other hand, individuals with heterozygous mutation were presented with a slow progression of disease even after the exposure to HIV. So far, it has been reported that Δ32 mutation CCR5 exists only among Caucasians. The frequency of CCR5 Δ32 mutation is very low. The frequency of this mutation among Caucasian population is merely ~1–3%. This observation has significant clinical significance toward a possible cure for HIV. Such a low frequency for this mutation could be a major bottleneck for the application of this clinical approach for the treatment of HIV seropositives along with other relevant aspects related to it. A low frequency of this mutation among Caucasians must not discourage us to make this as a possible cure for HIV. There are numerous advancements in the area of gene therapy and stem cell research which may add value to this strategy as one of the probable cures for HIV. So far, Berlin man has remained a success story for the cure of incurable HIV infections.

As mentioned in the previous paragraphs, Berlin man has earned reputation as the first case to testify the significance of CCR5Δ32/Δ32 mutation with a probable potential to cure HIV. Berlin man is a 40-year-old Caucasian male who was HIV positive for >10 years. He has an excellent control over HIV because he was on HAART regimen. But problem started with him when he was diagnosed with acute myeloid leukemia (AML). This patient was not responding to the AML treatment. Under these circumstances except allogeneic bone marrow transplantation, this patient was left with no other choices for the treatment of AML. At this point of time, the surgeon who was treating this patient took extra precautions for allogeneic bone marrow transplantation. The surgeon chose to do the bone marrow transplantation from a donor carrying CCR5Δ32/Δ32 mutation to this patient (recipient of bone marrow transplantation). After a successful bone marrow transplantation with mutated (CCR5Δ32/Δ32) bone marrow, this patient did not show any signs for HIV positivity, even though anti-retroviral treatment was stopped for this patient (Hütter et al. 2009). Clinicians are hopeful that this patient may remain HIV-free, but still the concerns about the possibility for the activation of HIV from HIV reservoir do exist in this patient. The answer to this question or these questions has to wait simply for time.

Acknowledgments Authors are thankful to Mr. Dinesh Kumar, Amity University Uttar Pradesh (AUUP), Noida, for his excellent contribution to develop graphics for this book chapter. Mr. Abhijit Mukherjee is duly acknowledged for his excellent skills and contribution toward the word processing of the manuscript. Research scholars Ms. Ayushi Mishra and Ms. Riya Chaudhary working in the Biochemistry Dept. of BHU are acknowledged for proofreading the manuscript. The role of Ms. Dharashree Rout for the accessory services toward the completion of this manuscript is duly acknowledged.

References

Barre-Sinousi F, Chermann JC, Rey F, Nugeyre MT, Chamaret S, Gruest J, Dauget C, Axler-Blin C, Vezinet-Brun F, Rouzioux C, Rozenbaum W, Montagnier L (1983) Isolation of a T-Lymphotrophic retrovirus from a patient at risk for Acquired Immune Deficiency Syndrome. Science 220:868–871

Berger EA, Murphy PM, Farber JM (1999) Chemokine receptors as HIV-1 co-receptors: roles in viral entry, tropism, and disease. Annu Rev Immunol 17:657–700

Boehringer Ingelheim Pharmaceuticals Incorporated (2005) NNRTI mode of action (animation). www.boehringer-ingelheim.com

Fenouillet E, Barbouche R, Courageot J, Miquelis R (2001) The catalytic activity of Protein Disulfide Isomerase is involved in Human Immunodeficiency Virus envelope-mediated membrane fusion after CD4 cell binding. J Infect Dis 183:744–752

Frankel AD, Young JA (1998) HIV-1: fifteen protein and an RNA. Annu Rev Biochem 67:1–25

Freed EO (2001) HIV-1 replication. Somat Cell Mol Genet 26:13–36

Gallo RC, Salahuddin SZ, Popovic M, Shearer GM, Kaplan M, Haynes BF, Palker TJ, Redfileld R, Oleske J, Safai B, White G, Foster P, Markham PD (1984) Frequent detection and isolation of cytopathic retroviruses (HTLV-III) from patients with AIDS and at risk for AIDS. Science 224:500–503

Gottlieb MS, Schroff R, Schanker HM, Weisman JD, Fan PT, Wolf RA, Saxon A (1981) *Pneumocystis carinii* pneumonia and mucosal candidiasis in previously healthy homosexual men: evidence of a new acquired cellular immunodeficiency. N Engl J Med 305:1425–1431

Hütter G, Nowak D, Mossner M, Ganepota S, Mubig A, Allers K, Schneider T, Hofmann J, Kücherer C, Blau O, Blau IW, Hofmann WK, Thiel E (2009) Long-term control of HIV by CCR5 delta32/delta32 Stem-cell transplantation. N Engl J Med 360:692–698

McArthur JC, Brew BJ, Nath A (2005) Neurological complications of HIV infection. Lancet Neurol 4:543–555

McCombe JA, Noorbakhsh F, Buchholz C, Trew M, Power C (2009) NeuroAIDS: a watershed for mental health and nervous system disorders. J Psychiatry Neurosci 34:83–85

Olender S, Wilkin TJ, Taylor BS, Hammer SM (2012) Advances in anti-retroviral therapy. Top Antivir Med 20:61–68

Olivier IS, Cacabelos R, Naidoo V (2018) Risk factors and pathogenesis of HIV-associated neurocognitive disorder: the role of host genetics. Int J Mol Sci 19(11):E3594

Peters PJ, Richards K, Clapham P (eds) (2013) Human immunodeficiency viruses: propogation, quantitation and storage. In: Current protocols in microbiology, vol 28. Wiley, Hoboken

Power C, Boisse L, Rourke S, John GM (2009) NeuroAIDS: an evolving epidemic. Can J Neurosci 36:285–295

Ryser HJP, Fluckiger R (2005) Progress in targeting HIV-1 entry. Drug Discov Today 10:1085–1094

Saag MS, Benson CA, Gandhi RT, Hoy JF, Landovitz RJ, Mugavero MJ, Sax PE, Smith DM, Thompson MA, Buchbinder SP, Del Rio C, Eron JJ Jr, Fätkenheuer G, Günthard HF, Molina JM, Jacobsen DM, Volberding PA (2018) Antiretroviral drugs for treatment and prevention of HIV infection in adults: 2018 recommendations of the international antiviral society-USA panel. JAMA 320(4):379–396

Tang MW, Shafer RW (2012) HIV-1 Antiretroviral resistance: scientific principles and clinical applications. Drug 80(71, 9):e1–e25

UNAIDS Report (2012) Report on the global AIDS epidemic

Verma AS, Singh UP, Dwivedi PD, Singh A (2010) Contribution of CNS cells in neuroAIDS. J Pharm Bioallied Sci 2:300–306

Verma AS, Singh UP, Mallick P, Dwivedi PD, Singh A (2012) NeuroAIDS and omics of HIV *vpr*. In: Barh D, Blum K, Madigan MA (eds) Omics biomedical perspectives and applications. CRC Press, Boca Raton, pp 477–511

Wilmshurst JM, Hammond CK, Donald K, Hoare J, Cohen K, Eley B (2018) NeuroAIDS in children. Handb Clin Neurol 152:99–116

Zhu T (ed) (2005) Method in molecular biology vol. 304: human retrovirus protocols: virology and molecular biology. Humana Press, Totowa, New Jersey.

Further Reading

Alfred M, Behrens G, Braun P and Bredeek UF (2011) HIV-2011. www.hivbook.com

Barnett T, Whiteside A (2006) AIDS in the twenty-first century: disease and globalization. Palgrave Macmillan, New York

Lever AML (1996) The molecular biology of HIV/AIDS. Wiley, Hoboken
Levy JA (1998) HIV and the pathogenesis of AIDS. American Society for Microbiology, Washington, DC
Sax PE, Cohen CJ, Kuritzkes DR (2012) HIV essentials. Jones and Bartlett Learning, Burlington
Stine G (2011) AIDS update 2012/ edition 21st. McGraw Hill Companies, Inc., New York
Weeks BS, Alcamo IE (2010) AIDS: the biological basis. Jones and Bartlett Learning, Burlington

Modern Approaches in Nanomedicine for NeuroAIDS and CNS Drug Delivery

8

Shailendra K. Saxena, Vimal K. Maurya, Swatantra Kumar, and Madan L. B. Bhatt

Abstract

NeuroAIDS is a neuropsychiatric disorder primarily caused by chronic human immunodeficiency virus (HIV). The most common neurological complications during NeuroAIDS are HAD (HIV-associated dementia) and HAND (HIV-associated neurocognitive disorder). HAART (highly active antiretroviral therapy) is a group of ARV (antiretroviral) drugs used to inhibit the key proteins involved in HIV replication. ARV drugs are extensively used to decrease the viral load and risk of opportunistic infections and to prolong survival of HIV-infected individuals. Significant advancement in antiretroviral (ARV) drugs has been achieved. However, the elimination of HIV from the CNS remains a difficult task. The complex structure of blood-brain barrier (BBB), high plasma protein binding, low patient compliance, and poor pharmacokinetics and biodistribution are the major limitations which restrict the effective concentration of drug to the CNS resulting in low efficacy of HAART in the CNS. Development of novel nanotechnology-based drug delivery methods for ARV drugs can increase the effectiveness of regimen with fewer side effects and better patient compliance which reduces financial load of healthcare system. The present article discusses about the nanotechnology-based approaches that can improve the delivery of ARV, called nano-ART. Nano-ART has excellent BBB permeability and improved pharmacokinetics and biodistribution that have significant clinical advantages to cure the CNS infection of HIV.

Keywords

HIV · NeuroAIDS · HAART · HAD · BBB · HAND · Nanogels · MNPs · Nanoemulsions · Liposomes

S. K. Saxena (✉) · V. K. Maurya · S. Kumar · M. L. B. Bhatt
Centre for Advanced Research, Faculty of Medicine,
King George's Medical University (KGMU), Lucknow, India
e-mail: shailen@kgmcindia.edu

S. K. Saxena, S. M. P. Khurana (eds.), *NanoBioMedicine*,
https://doi.org/10.1007/978-981-32-9898-9_8

Abbreviations

HIV	Human immunodeficiency virus
HAND	HIV-associated neurocognitive disorders
CNS	Central nervous system
ARV	Antiretroviral
ART	Antiretroviral therapy
BBB	Blood-brain barrier
HAD	HIV-associated dementia
HAART	Highly active antiretroviral therapy
SLN	Solid lipid nanoparticles
MNP	Magnetic nanoparticles

8.1 Introduction

NeuroAIDS is a neuropsychiatric disorder of the central nervous system (CNS) primarily caused by chronic human immunodeficiency virus (HIV) (McCombe et al. 2009). The neuropathology is associated with the early infiltration of HIV-1 into the CNS through infected immune cells such as $CD4^+$ T lymphocytes, monocytes/macrophages, and dendritic cells. These infected cells act as cellular reservoirs for HIV-1 during infection. Because of its neurovirulent nature, HIV reaches the brain through the infected monocytes/macrophages ("Trojan horse" approach) during the early stage of infection and causes HIV encephalitis (HIVE) (Saxena et al. 2012a, b). The infected macrophages/monocytes then start to interrupt the brain parenchyma which results in slow neurodegeneration of prefrontal cortex, hippocampus, white matter, and basal ganglia (Kaul et al. 2005). Furthermore, HIV-associated dementia (HAD), neurosyphilis, chronic meningitis, peripheral neuropathies, vacuolar myelopathy, and escalating multifocal leukoencephalopathy and central nervous system lymphomas are the major clinical conditions associated with NeuroAIDS (Schouten et al. 2011). Development of asymptomatic neurocognitive impairments such as HIV-associated dementia (HAD) and HIV-associative neurocognitive disorder (HAND) manifests as a clinical syndrome of cognitive, motor, and behavioral dysfunction during the CNS infection of HIV. HAND-related neurodegeneration seems to be enhanced in drug-abusing HIV-1-positive individuals (Sanmarti et al. 2014). The various proposed mechanisms for HIV-1 invasion into the CNS are histone modifications, presence of Tat-activated elongation factor (P-TEFb) and restriction factors via cellular pathways, etc. (Mehla et al. 2010, Nicoli et al. 2018). Clinical symptoms and neuropsychological assessment are the primary diagnostic criteria for the patients with mild motor and cognitive disorders (Lindl et al. 2010). Early identification of NeuroAIDS biomarkers such as blood biomarkers, cerebrospinal fluid biomarkers, and neuroimaging biomarkers in infected individuals is expected to assist in the better management of HIV and lower their susceptibility to neurodegenerative diseases (Rahimian and He 2017) (Fig. 8.1). Although there is significant advancement in antiretroviral (ARV) drugs,

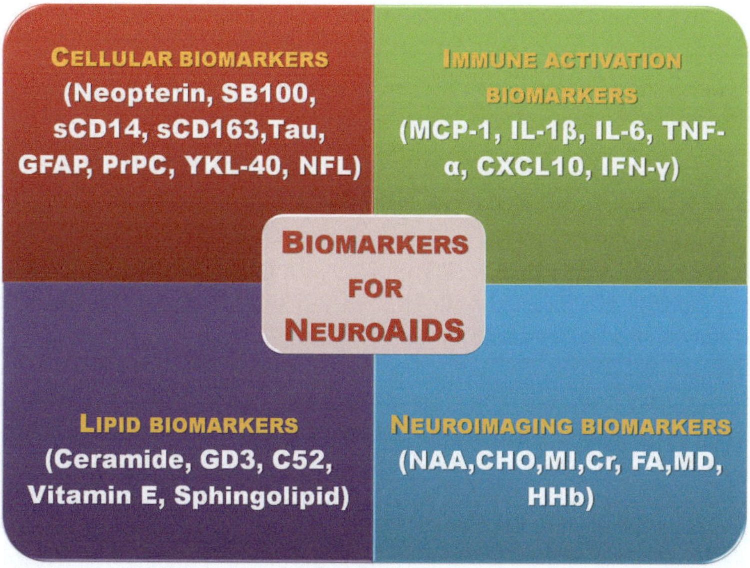

Fig. 8.1 NeuroAIDS biomarkers: *S100B* calcium binding protein B, *sCD14* soluble CD14, *sCD163* soluble CD163, *GFAP* glial fibrillary acidic protein, *PrPC* protease-resistant protein, cellular isoform, *YKL-40* human cartilage glycoprotein 39, *NFL* Neurofilament proteins, *MCP-1* monocyte chemoattractant protein-1, *IL-1β* interleukin-1β, *IL-6* interleukin-6, *TNF-α* tumor necrosis factor α, *CXCL-10* interferon gamma-induced protein 10, *IFN-γ* interferon γ, *NAA* N-acetylaspartate, *CHO* choline, *Cr* creatine, *FA* fractional anisotropy, *MD* mean diffusion, *HHb* deoxyhemoglobin, *GD3* ganglioside

eradication of HIV from the CNS still remains a challenging task. After the systemic administration, the inability of ARV drugs to cross the BBB makes the brain one of the leading sources of HIV. Hence monitoring, screening, and eradication to the virus source in the CNS anticipate a therapeutically challenging task along with vital achievements. So implementation and exploration of nanomedicines has potential against management of NeuroAIDS (Varghese et al. 2018).

8.2 Epidemiology of NeuroAIDS

The subtypes of HIV-1 are divided phylogenetically into different groups (M, N, O, and P) and in subtypes/clades (A–K) as well as many recombinant forms (~89). Subtypes of A–D are extremely widespread, while others with small geographical distributions have low incidence. There are two (CRF002 AG and CRF01 AE) recombinant forms circulating in Southeast Asia and West Africa (Hemelaar 2012). Most HIV research in Western Europe, Australia, and America has been performed for subtype B of the HIV-1 and covers only 12% of worldwide HIV infection. In comparison with HIV-1 subtype B, less studies were performed for HIV-1 subtype C, which accounts for about 50% of worldwide HIV cases predominantly in India and Africa (Geretti et al. 2009). Antiretroviral therapy (ART) was developed against

HIV-1 subtype B and found to be unsuccessful against HIV-1 subtype C (Häggblom et al. 2016, Tauber et al. 2016). HIV-1C has been underdiagnosed in sub-Saharan Africa and India mainly due to the social stigma and shorter life span of HIV-infected people (Xu et al. 2004).

8.3 Challenges for the Treatment

The World Health Organization (WHO) indicated that over 35 million individuals residing globally are infected with HIV. In 2013, 12.9 million patients (37% of 35 million) had access to ART, which further increased to 15 million by 2015 (Kaushik et al. 2016). The UNAIDS recommended 90-90-90 goal for 2020. According to this proposal, 90% of patients diagnosed with HIV will have access to ART, 90% of individuals infected with HIV will understand their status, and 90% of individuals on ART will be virally abolished (Bain et al. 2017). Although the removal of HIV from the CNS still remains a challenge, the use of ART has considerably reduced the viral load among HIV-infected individuals. After systemic administration, most of the ARV drugs bind to the plasma proteins which restrict the effective concentration of the drug to the CNS leading to HAART being less effective in lowering virus replication in the CNS than in the blood (Maurya et al. 2018). Similarly, the complex structure of BBB limits the transmigration of ARV drugs into the brain and reduces the efficacy of ARV drugs, which results in the development of resistant viral strain against HAART treatment. The principle behind the ARV drug resistance during the treatment for NeuroAIDS can be described as genetic variation of HIV and their continuous mutation during infection (Yilmaz et al. 2012). Other major drawbacks associated with ARV drugs are poor pharmacokinetics coupled with poor patient compliance which also influences the drugs' bioavailability. Therefore, maintaining the effective drug concentration and avoiding the side effects are challenging tasks in the management of NeuroAIDS (Wong et al. 2010). In summary, the key challenges in drug development for NeuroAIDS are (1) knowledge about the pathways for virus internalization, replication, glial activation, and reservoirs, in order to determine the novel targets for treatment, and (2) development of selective pharmaceutical agent and its appropriate delivery method within the CNS (Alam et al. 2010).

8.4 Nanotechnology-Based Approaches for the Management of NeuroAIDS

Nanotechnology offers various strategies to combat all the problems associated with targeted delivery of ARV drugs to the CNS for the management of NeuroAIDS (Kaushik et al. 2018). The failure of HAART treatment in NeuroAIDS patients is a global concern. Significant research has been done for the development of novel drug delivery systems to decrease the dosing incidences, increase the CNS penetration, enhance the bioavailability, inhibit the CNS efflux, and enhance the site-specific delivery without any toxic effects of ARV drugs (Kumar et al. 2018). Nanoformulations of ARV drugs reported to have good BBB penetration, high specificity, and less toxicity compared to the conventional HAART regimen (Saxena et al. 2012a, b). Development of novel nanotechnology-based drug delivery methods for ARV drugs can increase the effectiveness of regimen with fewer side effects and better patient compliance which reduces financial load of healthcare system (Nair et al. 2016). The main objective to design a drug delivery system is to optimize the therapeutic index of a drug by restricting its pharmacological activity on the targeted organ or site of action which reduces the drug dose and increases treatment efficacy (Sagar et al. 2014).The present article discusses the various nanotechnology-based approaches that can enhance the delivery of antiretroviral drugs, called nano-ART, across the BBB and increase thebio distribution and effectiveness in the CNS (Fig. 8.2).

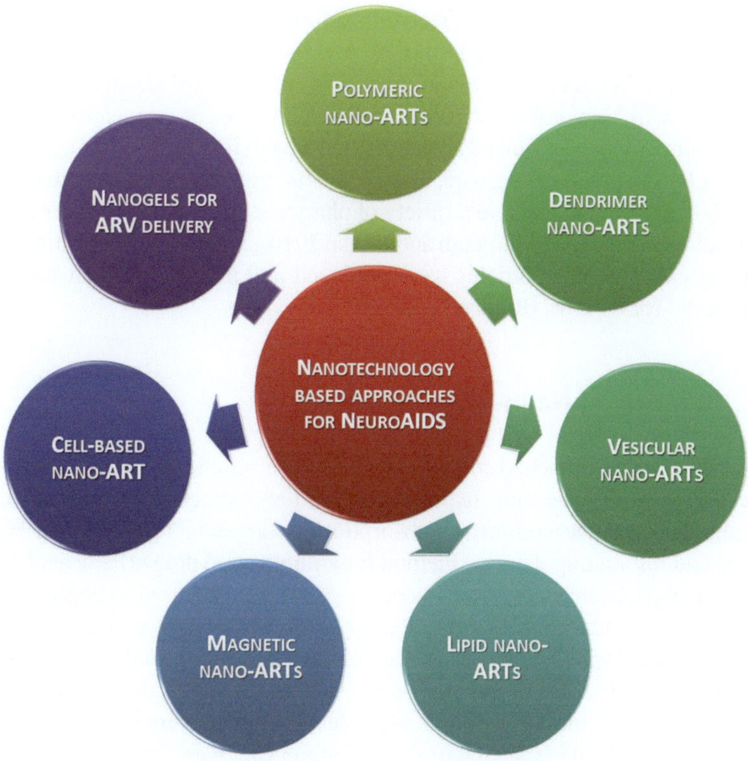

Fig. 8.2 Nanotechnology-based approaches for the management of NeuroAIDS

8.4.1 Polymeric Nano-ARTs

They have shown to be a promising approach for delivery of ARV drugs due to their encapsulation and brain targeting capacity. Polymeric NCs possess various properties like biodegradability and biocompatibility, high encapsulation efficiency, stability in biological system, excellent drug release profile, and permeability. These properties altogether promote the drug entry to the brain (Nowacek and Gendelman 2009). For the effective drug delivery to the brain, various poly(lactic-coglycolic acid) (PLGA) or polylactide (PLA)-, polyethylene glycol (PEG)-, and poly(methacrylic acid)-based polymeric nanoparticles have been developed where both the hydrophobic and hydrophilic drugs can be encapsulated and used for the controlled release for a longer time (Zhang et al. 2009). Alginate embedding, agglutinin coating, and PEGylation are the other surface modification approaches that have been suggested for effective delivery of ARV drugs to the brain. Destache et al. have reported that PLGA nanoparticles containing ritonavir, efavirenz, and lopinavir maintained a sustained peak of about 4 weeks in the mouse model (Destache et al. 2010). In the same way, Rao et al. also reported that PLGA nanoparticles with ritonavir in conjugation with Tat peptides increase the level of ritonavir 800-fold higher as compared with drugs delivered in solution in a mouse brain at 14 days of post-administration (Rao et al. 2008). Polymers such as poly(butyl cyanoacryalate) (PBCA) and methylmethacrylate-sulfopropylmethacrylate (MMA-SPM) have been investigated with zidovudine, stavudine, and lamivudine for permeation across the BBB both in in vitro and in vivo models of NeuroAIDS (Kuo 2005). Cellulose, chitosan, gliadin, pullulan, alginate, and gelatin are the natural biodegradable polymers widely used to encapsulate a variety of pharmaceutical drugs for their effective delivery across the BBB (Mallipeddi and Rohan 2010). Tween 80-coated lamivudine-containing chitosan nanoparticles were produced and proved excellent stability in mouse model for 60 days (Ramana et al. 2014).

8.4.2 Dendrimer Nano-ARTs

Dendrimers are the branched macromolecules with unique topological and structural characteristics. The vital features such as low molecular weight, small size (100 nm), and relative comfort of absorption of targeting ligands make them a potential and fascinating delivery method for antiretroviral drugs (Peng et al. 2013). Dendrimers have been widely used to deliver gene and antiretroviral peptides across the BBB for the inhibition of HIV. Recently, it is investigated that lamivudine-loaded mannose-capped poly(propylene imine) dendrimers increase the cellular uptake of lamivudine and decrease its toxicity with respect to the free drug (Dutta and Jain 2007). The main drawbacks associated with dendrimers are polycationic features and complicated manufacturing technique which are appropriate for delivery of drugs across negatively charged cell membranes.

8.4.3 Vesicular Nano-ARTs

Liposomes is a fabrication of phospholipid bilayers. They are unilamellar or multilamellar in character having colloidal spherical vesicles which are biocompatible and biodegradable. Due to their lipophilic nature, liposomes can be used as a remarkable carrier for the delivery of ARV drugs that have limited CNS permeability (Pant et al. 2012). Nowadays liposomes are the most promising method to deliver the drugs across the BBB with various advantages including site specificity, increased stability of loaded drug, and low immunogenicity that makes them more effective and safe. Zidovudine, zalcitabine, and didanosine are the hydrophobic anti-HIV drugs with poor absorption, bioavailability, and BBB penetration. Liposomal formulations of these drugs have shown to increase pharmacokinetics and biodistribution profile which ultimately enhances the efficacy of the therapy (Kabanov and Gendelman 2007). Kim et al. reported that the delivery of zalcitabine-loaded multivesicular liposomes to the Sprague Dawley rat cerebrospinal fluid increases the half-life of lipozalcitabine in the brain of rats by up to 23 h compared to 1.1 h for non-encapsulated drugs (Kim et al. 1990). Niosomes are biocompatible nanoformulations and considered as a promising method for the delivery of antiretroviral drugs to the brain. They are mainly composed of tyloxapol, cholesterol, and biological surfactant. In niosomes, various concentrations of cholesterol have been used to encapsulate the antiretroviral drugs (nevirapine). It is reported that niosome-based nanoformulation of nevirapine increases its therapeutic index and decreases side effects (Mehta and Jindal 2015).

8.4.4 Lipid Nano-ARTs

The lipophilic nature of lipid nanoparticles facilitates their entry across the BBB by endocytosis. Solid lipid NPs (SLNs) are mainly used for incorporation of various ARV drugs for the treatment of NeuroAIDS. Recent studies have shown that SLNs show less toxicity on nonspecific cells than PLGA nanoparticles which are considered as a standard for biocompatible materials (Bondì et al. 2012). Zidovudine palmitate-loaded SLNs are the first reported antiretroviral SLNs which are synthesized by Heiati et al. by using the trilaurin as a lipid core (Heiati et al. 1997). Recently, SLNs incorporated with saquinavir, stavudine, and delavirdine are individually investigated for drug entrapment efficiency, lipophilicity, and in vitro BBB penetration using human brain microvascular endothelial cells (BMVECs). The results of this study reported that SLNs loaded with saquinavir demonstrated the maximum entrapment efficiency with excellent lipophilicity and indicated that SLNs are appropriate nanocarriers for lipophilic drugs. Similarly, SLNs, encapsulation has been found to improve (4–11 times) BBB permeability (Kuo and Su 2007). The targeted drug delivery of SLNs has been enhanced by using surface charge modification approaches. In this context, positively charged SLNs have been used to deliver the larger amount, whereas negatively charged SLNs for the lower amount to the brain.

8.4.5 Magnetic Nano-ARTs

Magnetic nanoparticle (MNP)-based nanoformulation has been widely investigated for drug delivery, target specificity, drug release, and bioavailability for the treatment of NeuroAIDS. Magnetic materials such as maghemite (γ-Fe2O3) and magnetite (Fe3O4) are widely used magnetic nanoparticles in medicine (Nair et al. 2013). MNP-based drug delivery has shown to be superior over other methods of drug delivery such as liposomes, polymeric nanoparticles, and micelles. Magnetoliposomes also known as hybrid nanoparticles are a combination of liposomes and MNPs and can be used for monocyte/macrophage-based drug delivery to diverse pathological cases including brain carcinomas and inflammations (Saiyed et al. 2010). The magnetic nano-ART can reduce drug deposition in the reticuloendothelial system (RES) along with clearance and metabolism. In recent years researchers have designed magnetically guided layer-by-layer (LbL) combined nanocarrier co-encapsulation of tenofovir and vorinostat which demonstrated excellent BBB penetration with in vitro antiviral efficacy against HIV infection in primary human astrocytes over a period of 5 days (Jayant et al. 2015).

8.4.6 Cell-Based Nano-ARTs

Natural migrant properties of inflammatory response cells such as bone marrow-derived mesenchymal stromal cells, monocytes/macrophages, neuronal stem cells, dendritic cells, neutrophils, and lymphocytes in the direction of neuroinflammation can be accomplished for the site-specific delivery (Das et al. 2016). The immunocyte-based drug delivery has several benefits than traditional delivery methods like high permeability across the BBB, and it spontaneously targets the site of inflammation, injury, and tumors which reduces the cytotoxicity and immunogenicity of the loaded pharmaceutical agents. The drug-loaded nanomedicines such as nanoparticles, magnetoliposomes, and liposomes can be easily immersed by the immunocytes for targeted delivery at the site of action. Various cell surface receptors (mannose and Fc receptors) mediate the entry of drug-loaded nanomedicines into the immunocytes (Gorantla et al. 2006). Cell-mediated delivery of nanoformulated drugs has significant application in the management of neurological disorders such as Parkinson's, Alzheimer's, epilepsy, brain cancer. etc. The application of cell-mediated delivery has shown promising results in HIV-related neuropathogenesis (Batrakova et al. 2011). Recently, Dou and co-workers reported that macrophage-based nanocarrier can be used for the delivery of ARV drugs to the brain. Recently, indinavir was suspended in lipid nanocrystals and enclosed into ex vivo cultivated bone marrow-derived macrophages and administered intravenously into severely combined immunodeficient HIV-1 encephalitis (HIVE) mice. This nanoformulation has shown to inhibit the HIV replication in the HIVE brain with prolonged drug release in various parts of the brain for 14 days (Dou et al. 2009).

8.4.7 Nanogels for ARV Delivery

Nanogels are nanoscalar polymer networks and considered excellent delivery method for pharmaceutical drugs which have low stability in biological system. Nanogel-based drug delivery enables the delivery of both hydrophobic and hydrophilic drugs with highly selective uptake of the drug into targeted cells (Vinogradov et al. 2004). The developments in nanotechnology have come up with the advancement of nanoformulation for encapsulation of the antiretroviral drugs in order to increase the efficacy for long-term therapy. Nanogels contain several important properties like swelling property in aqueous media, higher drug loading capacity, colloidal stability, non-immunologic response, particle size (20–200 nm), and electromobility that make them robust strategy for delivery drugs in various emerging diseases like autoimmune diseases, diabetes, cancer, and neurodegenerative disorders (Senanayake et al. 2013). Senanayake and co-workers reported that the nanogel conjugates of abacavir, lamivudine, and zidovudine have demonstrated tenfold inhibition of reverse transcriptase activity in HIV-infected macrophages, and among all nanoformulated nanogel conjugates, lamivudine was the most effective single drug (Senanayake et al. 2015) (Table 8.1).

Table 8.1 Summary of preclinical nanocarrier/nanomaterial-based CNS delivery of ARV drugs for prevention/treatment of NeuroAIDS

S.no.	Antiretroviral therapy (ART)	Nanomaterials
1	Zidovudine, stavudine, didanosine, zalcitabine, foscarnet, indinavir	Liposome
2	Stavudine	Liposome-laden macrophages
3	Zidovudine	Mannose- and galactose-targeted liposome, Mannose-targeted liposome
4	Lamivudine	Mannose-targeted dendrimer
5	Efavirenz	Tuftsin dendrimers
6	Ritonavir, lopinavir, zidovudine, stavudine	PLGA-poly(D,L-lactide-co-glycolide) nanoparticles
7	Nevirapine, ritonavir, efavirenz	Tat-conjugated nanoparticle
8	Ritonavir, darunavir, atazanavir	Poly(ε-caprolactone) nanoparticles
9	Dapivirine, saquinavir	RMP-7/MMA-SPM nanoparticles
10	Saquinavir, stavudine, delavirdine, atazanavir	Solid lipid nanoparticles (SLN)
11	Tenofovir, enfuvirtide, AZTTP	Magnetic nanoparticles
12	3′-azido-2′,3′-dideoxy thymidine-5′-triphosphate (AZTTP)	Magnetoelectric nanoparticles
13	Efavirenz, lamivudine	Cyclodextrins
14	Rilpivirine	Poloxamer 338/TPGS 1000
15	Zidovudine, lamivudine, efavirenz, nelfinavir, ritonavir	Micelles P85 (Pluronics)

(continued)

Table 8.1 (continued)

S.no.	Antiretroviral therapy (ART)	Nanomaterials
16	Zidovudine, lamivudine, efavirenz, indinavir, ritonavir, atazanavir	Monocyte-derived macrophages/nanoparticles
17	Stavudine, zidovudine, lamivudine, delavirdine	PBCA, poly(butylcyanoacryalate); MMA-SPM (methylmethacrylate-sulfopropylmethacrylate)
18	Amprenavir, saquinavir	Transferrin (Tf)-conjugated quantum dots
19	Lamivudine	Carbon nanotubes

8.5 Conclusions

NeuroAIDS is still an important challenge for people with AIDS. The treatment scheme is ineffective for patients because of patients' inability to continue lifelong medication and poor BBB penetration profile of ARV which eventually reduces the effectiveness of the regimen and may develop resistant viral strain against HAART treatment. This can be overcome by improving the delivery methods of ARV drugs. The development of approaches based on nanotechnologies allowed us to design novel drug delivery methods for the transport of drugs across the BBB for the better management of NeuroAIDS-related neuropathology. For encapsulation of ARV drugs, various advanced approaches such as polymeric nanocarriers, liposome and magnetic nanoparticle-based nanomedicines, and nanogels are developed for targeted delivery to the brain.

8.6 Future Perspectives

The CNS infection of HIV is still challenging and needs extensive exploration. There is urgent need to develop various in vitro models to study the neuropathogenesis of HIV and design a novel drug candidate for effective management with minor side effects. In addition, local, epidemiological, socioeconomic, and genetic factors should be considered in order to determine the rates of HAD/HAND in the neurocognitive evaluation. The treatment strategy of NeuroAIDS can further be enhanced by improving the drug delivery methods of antiretroviral medicines. These newer methods should be investigated in vivo and ex vivo to determine the effectiveness and neurotoxicity in patients with HAD/HAND. The development of prognostic markers for clade-specific HIV which is based on signature sequences helps in the early diagnosis of NeuroAIDS.

Acknowledgments The authors are grateful to the Vice Chancellor, King George's Medical University (KGMU), Lucknow, for the encouragement and support for this work. SK Saxena is also supported by CCRH, Government of India, and US NIH grants: R37DA025576 and R01MH085259. The authors have no other relevant affiliations or financial involvement with any organization or entity with a financial interest in or financial conflict with the subject matter or materials discussed in the manuscript apart from those disclosed.

References

Alam MI, Beg S, Samad A, Baboota S, Kohli K, Ali J, Ahuja A, Akbar M (2010) Strategy for effective brain drug delivery. Eur J Pharm Sci 40(5):385–403

Bain LE, Nkoke C, Noubiap JJN (2017) UNAIDS 90-90-90 targets to end the AIDS epidemic by 2020 are not realistic: comment on "Can the UNAIDS 90-90-90 target be achieved? A systematic analysis of national HIV treatment cascades". BMJ Glob Health 2(2):e000227

Batrakova EV, Gendelman HE, Kabanov AV (2011) Cell-mediated drug delivery. Expert Opin Drug Deliv 8:415–433

Bondì ML, Di Gesù R, Craparo EF (2012) Lipid nanoparticles for drug targeting to the brain. Methods Enzymol 508:229–251

Das MK, Sarma A, Chakraborty T (2016) Nano-ART and NeuroAIDS. Drug Deliv Transl Res 6(5):452–472

Destache CJ, Belgum T, Goede M, Shibata A, Belshan MA (2010) Antiretroviral release from poly (DL-lactide-co-glycolide) nanoparticles in mice. J Antimicrob Chemother 65(10):2183–2187

Dou H, Grotepas CB, McMillan JM, Destache CJ, Chaubal M, Werling J (2009) Macrophage delivery of nanoformulated anti-retroviral drug to the brain in a murine model of Neuro-AIDS. J Immunol 183(1):661–669

Dutta T, Jain NK (2007) Targeting potential and anti-HIV activity of lamivudine loaded mannosylated poly (propyleneimine) dendrimer. BiophysActa 1770:681–686

Geretti AM, Harrison L, Green H, Sabin C, Hill T, Fearnhill E, Pillay D, Dunn D, UK Collaborative Group on HIV Drug Resistance (2009) Effect of HIV-1 subtype on virologic and immunologic response to starting highly active antiretroviral therapy. Clin Infect Dis 48(9):1296–1305

Gorantla S, Dou H, Boska M (2006) Quantitative magnetic resonance and SPECT imaging for macrophage tissue migration and nanoformulated drug delivery. J Leukoc Biol 80(5):1165–1174

Häggblom A, Svedhem V, Singh K, Sönnerborg A, Neogi U (2016) Virological failure in patients with HIV-1 subtype C receiving antiretroviral therapy: an analysis of a prospective national cohort in Sweden. Lancet HIV 3(4):e166–e174

Heiati H, Tawashi R, Shivers RR, Phillips NC (1997) Solid lipid nanoparticles as drug carriers. I. Incorporation and retention of the lipophilic prodrug 3′-azido-3′-deoxythymidine palmitate. Int J Pharm 146(1):123–131

Hemelaar J (2012) The origin and diversity of the HIV-1 pandemic. Trends Mol Med 18(3):182–192

Jayant RD, Atluri VSR, Agudelo M, Sagar V, Kaushik A, Nair M (2015) Sustained-release Nano-ART formulation for the treatment of Neuro-AIDS. Int J Nanomedicine 10:1077–1093

Kabanov AV, Gendelman HE (2007) Nanomedicine in the diagnosis and therapy of neurodegenerative disorders. Prog Polym Sci 32(8–9):1054–1082

Kaul M, Zheng J, Okamoto S, Gendelman HE, Lipton SA (2005) HIV-1 infection and AIDS: consequences for the central nervous system. Cell Death Differ 12:878–892

Kaushik A, Jayant RD, Nair M (2016) Advancements in nano-enabled therapeutics for neuroHIV management. Int J Nanomedicine 11:4317–4325

Kaushik A, Jayant RD, Nair M (2018) Nanomedicine for neuroHIV/AIDS management. Nanomedicine (Lond) 13(7):669–673

Kim S, Scheerer S, Geyer MA, Howell SB (1990) Direct cerebrospinal fluid delivery of an antiretroviral agent using multivesicular liposomes. J Infect Dis 162:750–752

Kumar S, Maurya VK, Dandu HR, Bhatt ML, Saxena SK (2018) Global perspective of novel therapeutic strategies for the management of NeuroAIDS. Biomol Concepts 9(1):33–42

Kuo YC (2005) Loading efficiency of stavudine on polybutylcyanoacrylate and methylmethacrylate- sulfopropylmethacrylate copolymer nanoparticles. Int J Pharm 290:161–172

Kuo Y-C, Su F-L (2007) Transport of stavudine, delavirdine, and saquinavir across the blood–brain barrier by polybutylcyanoacrylate, methylmethacrylate- sulfopropylmethacrylate, and solid lipid nanoparticles. Int J Pharm 340(1–2):143–152

Lindl KA, Marks DR, Kolson DL, Jordan-Sciutto KL (2010) HIV-associated neurocognitive disorder: pathogenesis and therapeutic opportunities. J Neuroimmune Pharmacol 5(3):294–309

Mallipeddi R, Rohan LC (2010) Progress in antiretroviral drug delivery using nanotechnology. Int J Nanomedicine 5:533–547

Maurya VK, Kumar S, Saxena SK (2018) Molecular mechanism of neuropathogenesis and current therapeutics for neuroAIDS. Res Rep 1:e1–e8

McCombe JA, Noorbakhsh F, Buchholz C, Trew M, Power C (2009) NeuroAIDS: a watershed for mental health and nervous system disorders. J Psychiatry Neurosci 34(2):83–85

Mehla R, Bivalkar-Mehla S, Zhang R, Handy I, Albrecht H, Giri S, Nagarkatti P, Nagarkatti M, Chauhan A (2010) Bryostatin modulates latent HIV-1 infection via PKC and AMPK signaling but inhibits acute infection in a receptor independent manner. PLoS One 5(6):e11160

Mehta SK, Jindal N (2015) Tyloxapol niosomes as prospective drug delivery module for antiretroviral drug nevirapine. AAPS PharmSciTech 16(1):67–75

Nair M, Guduru R, Liang P, Hong J, Sagar V, Khizroev S (2013) Externally controlled on-demand release of anti-HIV drug using magneto-electric nanoparticles as carriers. Nat Commun 4:1707

Nair M, Jayant RD, Kaushik A, Sagar V (2016) Getting into the brain: potential of nanotechnology in the management of NeuroAIDS. Adv Drug Deliv Rev 103:202–217

Nicoli F, Gallerani E, Sforza F, Finessi V, Chachage M, Geldmacher C, Cafaro A, Ensoli B, Caputo A, Gavioli R (2018) The HIV-1 tat protein affects human CD4+ T-cell programming and activation, and favors the differentiation of naïve CD4+ T cells. AIDS 32(5):575–581

Nowacek A, Gendelman HE (2009) NanoART, neuroAIDS and CNS drug delivery. Nanomedicine (Lond) 4(5):557–574

Pant M, Garg P, Seth P (2012) Central nervous system infection by HIV-1: special emphasis to NeuroAIDS in India. Proc Natl Acad Sci India Sect B 82(1):81–94

Peng J, Wu Z, Qi X, Chen Y, Li X (2013) Dendrimers as potential therapeutic tools in HIV inhibition. Molecules 18:7912–7929

Rahimian P, He JJ (2017) HIV/neuroAIDS biomarkers. Prog Neurobiol 157:117–132

Ramana LN, Sharma S, Sethuraman S, Ranga U, Krishnan UM (2014) Evaluation of chitosan nanoformulations as potent anti-HIV therapeutic systems. Biochim Biophys Acta 1840:476–484

Rao KS, Reddy MK, Horning JL, Labhasetwar V (2008) TAT-conjugated nanoparticles for the CNS delivery of anti-HIV drugs. Biomaterials 29(33):4429–4438

Sagar V, Pilakka-Kanthikeel S, Pottathil R, Saxena SK, Nair M (2014) Towards nanomedicines for neuroAIDS. Rev Med Virol 24(2):103–124

Saiyed ZM, Gandhi NH, Nair MP (2010) Magnetic nanoformulation of azidothymidine 5′-triphosphate for targeted delivery across the blood-brain barrier. Int J Nanomedicine 5:157–166

Sanmarti M, Ibáñez L, Huertas S, Badenes D, Dalmau D, Slevin M, Krupinski J, Popa-Wagner A, Jaen A (2014) HIV-associated neurocognitive disorders. J Mol Psychiatry 2(1):2

Saxena SK, Tiwari S, Nair MP (2012a) A global perspective on HIV/AIDS. Science 337(6096):798

Saxena SK, Tiwari S, Nair MP (2012b) Nanotherapeutics: emerging competent technology in neuroAIDS and CNS drug delivery. Nanomedicine (Lond) 7(7):941–944

Schouten J, Cinque P, Gisslen M, Reiss P, Portegies P (2011) HIV-1 infection and cognitive impairment in the cART era: a review. AIDS 25(5):561–575

Senanayake TH, Warren G, Wei X, Vinogradov SV (2013) Application of activated nucleoside analogs for the treatment of drug-resistant tumors by oral delivery of nanogel-drug conjugates. J Control Release 167:200–209

Senanayake TH, Gorantla S, Makarov E, Lu Y, Warren G, Vinogradov SV (2015) Nanogel-conjugated reverse transcriptase inhibitors and their combinations as novel antiviral agents with increased efficacy against HIV-1 infection. Mol Pharm 12(12):4226–4236

Tauber SC, Staszewski O, Prinz M, Weis J, Nolte K, Bunkowski S, Brück W, Nau R (2016) HIV encephalopathy: glial activation and hippocampal neuronal apoptosis, but limited neural repair. HIV Med 17(2):143–151

Varghese NM, Senthil V, Saxena SK (2018) Nanocarriers for brain specific delivery of anti-retro viral drugs: challenges and achievements. J Drug Target 26(3):195–207

Vinogradov SV, Batrakova EV, Kabanov AV (2004) Nanogels for oligonucleotide delivery to the brain. Bioconjug Chem 15:50–60

Wong HL, Chattopadhyay N, Wu XY, Bendayan R (2010) Nanotechnology applications for improved delivery of antiretroviral drugs to the brain. Adv Drug Deliv Rev 18(62(4–5)):503–517

Xu Y, Kulkosky J, Acheampong E, Nunnari G, Sullivan J, Pomerantz RJ (2004) HIV-1-mediated apoptosis of neuronal cells: proximal molecular mechanisms of HIV-1-induced encephalopathy. Proc Natl Acad Sci U S A 101(18):7070–7075

Yilmaz A, Price RW, Gisslén M (2012) Antiretroviral drug treatment of CNSHIV-1 infection. J Antimicrob Chemother 67(2):299–311

Zhang J, Li S, Li X (2009) Polymeric nano-assemblies as emerging delivery carriers for therapeutic applications: a review of recent patents. Recent Patents Nanotechnol 3:225–231

Biomedical Applications of Viral Nanoparticles in Vaccine Therapy

9

Sangita Venkataraman, Vijay S. Reddy, and S. M. Paul Khurana

Abstract

Viral nanoparticles (VNPs) are versatile systems for the delivery of vaccines and other therapeutic agents for the treatment of diseases such as cancer or those related to the immune-system, degenerative diseases, and infections caused due to agents like viruses, bacteria, and fungi. Additionally, the VNPs are also used in molecular diagnostics, in the development of films and arrays for applications in electronics and tissue engineering, the design of data storage devices, and devices for tissue-specific imaging and therapy. The ability of viruses and bacteriophages to invade and infect different kinds of host cells empower them as suitable nanocarriers wherein they are able to cross biological barriers that obstruct drug delivery. VNPs are generally produced by genetic or chemical engineering by inserting heterologous sequences or ligands of interest into surface-exposed loops of the capsid protein (CPs). The CPs of viruses are the protein-building blocks that can self-assemble and are inherently biodegradable. The capsids that are generated artificially using the modified version of the CPs are known as virus-like particles (VLPs) to which ligands, peptides, and other agents are conjugated to generate VNPs. VNPs can self-assemble either as discrete structures or may organize themselves into films and arrays. Specific formulations of VNPs loaded with biomolecules such as antibodies and aptamers or biomimetics significantly improve efficacy and function by modulating tissue localization. The potential of VNP-based vaccine therapy is yet to be fully

S. Venkataraman (✉)
Department of Biotechnology, Acharya Nagarjuna University, Guntur, Andhra Pradesh, India

V. S. Reddy
Department of Integrative Structural and Computational Biology,
The Scripps Research Institute, La Jolla, CA, USA

S. M. P. Khurana
Amity Institute of Biotechnology, Amity University Haryana, Gurgaon, Haryana, India

S. K. Saxena, S. M. P. Khurana (eds.), *NanoBioMedicine*,
https://doi.org/10.1007/978-981-32-9898-9_9

explored and it is expected that it will continue to evolve with more effective design and applicability for the treatment of various diseases.

Keywords
Virus-like particles · Coat protein · Biomimetic · Antibiotic · Vaccine

9.1 Introduction

Viral nanoparticles (VNPs) are self-assembling virus-like particles (VLPs) that have potential applications in medicine, biotechnology, tissue engineering, bioimaging, data storage, and energy. Viruses are nucleo-protein entities that feature a proteinaceous capsid protecting a nucleic acid genome. The capsid itself is made up of many repeating units of one or a few types of proteins that are known as coat proteins (CP). The CPs are arranged in a highly symmetric fashion to produce a closed capsid shell within which the genome is encapsidated during the process of virus particle assembly. The capsids are built on either helical or spherical symmetry (Crick and Watson 1956). The symmetry of viral scaffolds gives a powerful advantage in the design of VNPs as the subunits may be arranged into different shapes and sizes and offer the convenience of easy and reproducible modifications that result in avidity effects (Wen and Steinmetz 2016).

More than a century of research has gone into virus research with the discovery of about 5000 different kinds of viruses that vary in shape, sizes, host and transmissibility, nature of genome, etc. Towards the beginning of the century, researchers around the world started to explore the potential of using viruses in nanotechnology and nano-materials by using them for encapsulating drugs and other materials within the capsid or using capsid proteins to tag antibodies or aptamers for delivery into different kinds of cells. The VNPs provide specific, safe, and efficacious delivery of molecules of interest to target cells compared to organic or inorganic nanoparticles. Furthermore, the chemical agents are also saddled by the limited bioavailability, solubility, and undesirable side effects (Allen and Cullis 2004). These challenges are readily overcome in VNPs. Additionally, the VNPs self-assemble to form a uniform and robust capsid that are relatively easy to produce in large quantities. A variety of viruses are being used as VNPs that include the bacterial, plant, and animal viruses, which offer distinct advantages (Van Hest et al. 2014). While the animal viruses display better tissue tropism and drug delivery, the plant and phage-based VNPs are non-infectious to humans and are easier to manufacture through fermentation and "molecular farming" (Marsian and Lomonossoff 2016). The VNPs are generated through modifications of the viruses utilizing the methods of genetic engineering, bioconjugation, biomineralization, and encapsulation. Use of VNPs in vaccine therapy primarily involves the usage of VNPs as epitope-display platforms. VNPs display multiple copies of the antigen of interest that result in avidity effects associated with the multi-valency (Koudelka et al. 2016). Engineered VLPs have been used successfully in vaccine therapy that include GenHevac B for hepatitis B virus (HBV, (Souli et al. 1991)), Ceravix for human papillomavirus (HPV) (Agnandji et al. 2012), RTS, S (Phase I), and PEV3 (Phase II) for malaria (El-attar et al. 2009). Genetically modified herpes virus is available under the brand name of imlygic cancer-hunting virus 41, developed by BioVex company in the USA, which is FDA approved for treating cancer (Bell and McFadden 2015). Such VNPs are used to target various immune cells including B lymphocytes, macrophages, and dendritic cells (DCs) to elicit humoral or cellular responses.

VNPs have also found their use in the field of bioimaging where synthetic dyes, quantum dots, and green fluorescent proteins (GFPs) are regularly used for studying the characteristics of organs and internal structures (Yoo et al. 2014). Cowpea mosaic virus (CPMV) has been successfully used in intravital vascular imaging by serving as carriers of various dyes such as Fluorescein dextran, A488 dextran, and Rhodamine (Lewis et al. 2006). VNPs based on red clover necrotic mosaic virus (RCNMV) have been successfully used as carriers for abamectin plant parasitic nematode control (Cao et al. 2015; Zhang et al. 2017). P22 bacteriophage encapsulating hydrogenase enzyme is incubated with protons (H+) and electrons (e-) to generate molecules of H_2 within the viral capsid and is being considered as carbon-free hydrogen fuel (Jordan et al. 2016). However, these suffer from low sensitivity and photostability issues.

Many challenges such as those associated with the purity, scalability, and cost-effectiveness exist in manufacturing VNPs (Brune et al. 2018). However, the efforts are ongoing to address these issues and expand the applicability, efficacy, and safety aspects of VNPs. The present chapter gives an overview of biomedical applications of VNPs in vaccine therapy focusing particularly on the different viruses being used, display of antigens of interest, the extent of immune responses generated, the diseases being treated, and production methodologies.

9.2 VNPs in Vaccine Therapy

According to the definition of NCI, "Vaccine therapy is a type of treatment that uses a substance or group of substances to stimulate the immune system to destroy a tumor or infectious microorganisms such as bacteria or viruses". Traditionally, the vaccines comprise of attenuated or dead bacteria, viruses, or other pathogens that elicit an immune response in the host. Cancer vaccines mimic a cancer cell by using specific tumour-associated antigens (TAAs) and aid the immune system to eventually eliminate them (Bell and McFadden 2015). Many such cancer vaccines are in various phases of clinical trials. The cancer vaccine Sipuleucel-T (Provenge®) is one such vaccine that is used for the treatment of prostate cancer that is asymptomatic and metastatic (Gardner et al. 2012). The vaccine uses the antigen prostatic acid phosphatase (PAP) that features on prostate cancer cells for generating an immune response against active prostate cancer cells and improving the survivability. VNPs hold tremendous potential for use in vaccine therapy. The following sections describe the various aspects of the use of VNPs in vaccine therapy.

9.2.1 Criteria for Virus Selection

Viruses exhibit enormous diversity in their host range, genome composition, structures, and transmissibility. Based on the nature of hosts, the viruses are classified as bacteriophages (those infecting bacteria, e.g., *Caudovirales*), plant viruses (e.g., Potyviruses), animal viruses (e.g., Poxviruses), protozoan viruses (e.g., Mimiviruses), etc. According to the nature of genome and the strategy used for replication, viruses are classified into seven Baltimore classes, namely, those with double-stranded (ds) DNA genome (Group I, e.g., adenoviruses), single-stranded (ss) DNA genome (Group II, e.g., geminiviruses), ds RNA genome (Group III, e.g., reoviruses), positive sense (+) ss RNA genome (Group IV, e.g., picornaviruses), negative sense (-) ss RNA genome (Group V, e.g., orthomyxoviruses), (+) ss RNA genome with dsDNA intermediate (Group VI, e.g., Retroviruses), and ds DNA viruses with (+) ss RNA intermediate (Group VII, e.g., Caulimoviruses). Viruses are transmitted from one host to another either through mechanical means (e.g., tobacco mosaic virus, TMV), through vectors such as mosquitoes (e.g., Dengue virus), or through contact with contaminated substances (e.g., noroviruses, adenoviruses). Viruses range in structure from the spherical ones with icosahedral symmetry, for example, Sesbania mosaic virus (SeMV), helical rod-shaped viruses (e.g., TMV), or filamentous (e.g., Pepper vein binding virus, PVBV) (Fig. 9.1a), to the spherical or helical viruses covered with an additional layer of lipid, known as enveloped viruses (e.g., influenza viruses), and complex viruses (e.g., Vaccinia virus). Icosahedral viruses display icosahedral point group symmetry of 532 comprising a shell made of pentamers alone or with pentamers and hexamers (Venkataraman et al. 2018). The triangulation number also called as the T number defines the multiple of 60 CPs, which is used for building the capsid (Fig. 9.1a). As an example, a $T = 1$ virus is made of 60 CPs, while a $T = 3$ virus has 180 CPs (Venkataraman et al. 2018). In terms of using viruses as VNPs, the spherical viruses provide a higher surface area for a given mass and hence are effective for the infusion and encapsulation of assorted cargo. They are also efficient in improving cell targeting specificities, while the elongated viral particles fare better in avoiding immune clearance by phagocytosis (Fig. 9.1b) (Champion and Mitragotri 2010). Champion and Mitragotri have argued that due to minimized phagocytosis, the elongated viral particles are not only efficient in the delivery of payloads but also have significantly improved circulation time leading to the requirement of much lower doses and reduced side effects. In terms of the sizes of the viruses, VNPs based on larger viruses have larger surface areas resulting in greater accumulation in the cells. On the contrary, VNPs based on smaller viruses upon entry induce significantly weaker hydrodynamics and shear forces that leads to deeper penetration than the larger ones (Steinmetz and Manchester 2011). However, the size of VNPs is seen to play a much lesser role when compared to shape or morphology during internalization (Koudelka et al. 2016). It has been known that the pharmacological features of the synthetic and biological carriers have significant differences.

The positively charged non-biological carriers that include the dendrimers and polymers are known to induce toxicity by causing disturbances in the membrane

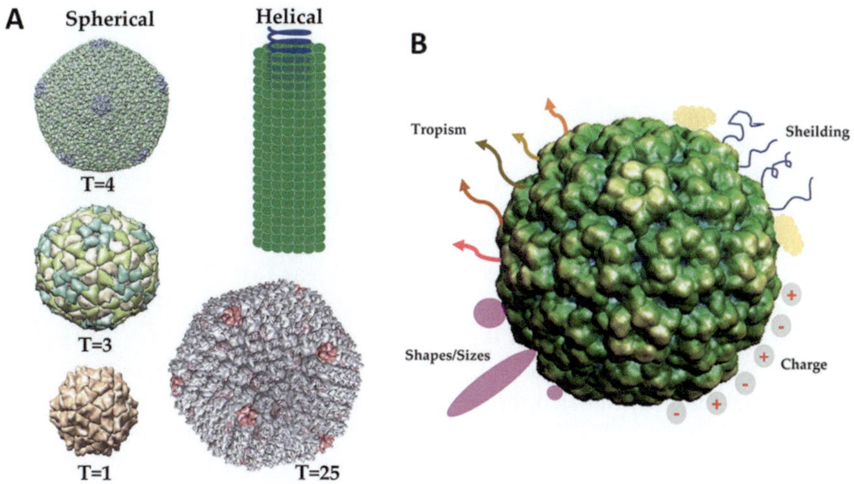

Fig. 9.1 (**a**) Variation in capsid morphologies exhibited by the plant viruses that includes spherical and helical architectures. The spherical viruses may further show the difference in size and symmetry depending on the differences in T numbers. The images of spherical capsids were downloaded from www.rcsb.org (Berman et al. 2000): T=4 (Phage T4, PDB ID: 5vf3) (Chen et al. 2017), T=3 (Seneca valley virus, PDB ID: 3cji) (Venkataraman et al. 2009), and T=1 (SeMV, PDB ID: 1vak) (Sangita et al. 2005b). Picture of TMV is reproduced from https://commons.wikimedia.org/wiki/File:Non-enveloped_helical_virus.svg (**b**) The various parameters that are involved in engineering the VNPs including shape, size, charge (positive or negative), shielding (using polymers or peptides for masking), and targeting (using ligands for different receptors). The suitability of individual virus for use as VNP is governed by its intended application. The picture was created using UCSF Chimera (Pettersen et al. 2004) (CCMV, PDB ID: 1f15) (Smith et al. 2000)

structure (Fig. 9.1b) (MC and Kallinteri 2006; Hong et al. 2012). However, they interact favourably with mammalian cells due to enhanced capacity for tumour transport and longer retention times (Wen et al. 2013). VNPs such as bacteriophage Qβ that are positively charged are seen to have longer half-lives of circulation (more than 3 h), which are notably higher than those of virus particles such as CPMV and Cowpea chlorotic mottle virus (CCMV) that are negatively charged (half-lives <15 min) (Kaiser et al. 2007; Singh et al. 2010; Jeevanandam et al. 2018). Investigations of CPMV functionalized with Gd-DOTA in mouse models and its bio-distribution show rapid plasma clearance (within 20 min) and accumulation of the majority of VNPs in the liver and spleen (post 30 mins) (Singh et al. 2010). Substitution of a single glutamic acid residue with lysine in bacteriophage λ was seen to result in more than 1000-fold increased circulation time (Vitiello and Adhya 2005), indicating the significance of charge of CPs in the efficacy of VNPs. The presence of proteoglycan in the cell membrane of mammalian cells confers a negative charge thereby aiding in the binding of positively charged VNPs, preventing their aggregation and improving their penetration capacity (Wu et al. 2013; Shukla et al. 2014a).

9.2.2 Types of Viruses Used in VNPs

Many viruses are used as VNPs for generating new vaccines. These include the icosahedral plant viruses such as CMV (Zeng et al. 2013), CPMV (Singh et al. 2010), CCMV (Brumfield et al. 2004), RCNMV (Loo et al. 2007), Hibiscus chlorotic ringspot virus (HCRSV) (Ren et al. 2007), Johnson grass chlorotic stripe mosaic virus (JgCSMV) (Alemzadeh et al. 2017), Physalis mottle virus (PhMV) (Masarapu et al. 2017), Alfalfa mosaic virus (AlMV) (Yusibov et al. 1997), and SeMV (Abraham et al. 2016) and helical viruses such as TMV (Staczek et al. 2000), Potato virus X (PVX) (Lee et al. 2017), and Papaya mosaic virus (PapMV) (Denis et al. 2008). In the last decade of the twentieth century, Douglas and Young from Montana State University, Bozeman, MT, USA, considered using viral capsids as nanomaterial and that initiated the use of viruses in nanotechnology applications (Douglas and Young 1998). They used CCMV as a nanocarrier encapsulating polyanethole sulfonic acid in exchange of their genome through structural transitions that are pH- or metal ion-dependent. This, in turn, led to the success of CCMV as VNPs for many other applications. A research pioneered by a team led by Mann (University of Bristol, UK) used TMV particles as templates to successfully fabricate metalized nanotube structures through mineralization techniques (Shenton et al. 1999; Nam et al. 2006; Lee et al. 2009; Pires et al. 2016).

9.3 Production of VNPs

Chimeric VLPs are routinely produced using heterologous expression systems. Such VLPs are used for encapsulating medical cargos such as drugs and other agents. VNPs are produced by many methods that include fermentation, molecular farming, and cell culture using bacterial and insect cells, yeast, plant cells, and other cell-free systems.

9.3.1 Bacterial Expression

The simplest, cost-effective, and widely used platform to express the VLPs is through the expression of CP genes in *Escherichia coli*. Many viral CPs are commonly expressed in *E. coli* and are seen to self-assemble into elegant VLPs (Venkataraman et al. 2018). VLPs of plant viruses such as Sesbania mosaic virus (SeMV) (Sangita et al. 2005a, b), TMV (Hwang et al. 1994), bacteriophages such as Qβ (Bessa et al. 2008) and MS2 (Pickett and Peabody 1993), and Mammalian hepatitis B virus (HBV) core particles (Birnbaum and Nassal 1990) were successfully expressed in *E. coli*. Such VLPs lack their genome but are not empty as they may encapsidate the cellular nucleic acids during assembly (Sangita et al. 2004, 2005a, b). In SeMV VLPs, a significant amount of 23s ribosomal RNA from the host was detected and similarly, ~25% of the mass of Qβ VLPs consisted of *E. coli* RNA (Jennings and Bachmann 2009). Hence, routine measures are undertaken to get rid of the packaged nucleic acids. In those viruses where the over-expression of VLPs in bacterial system leads to insolubility issues and accumulation of protein in the inclusion bodies, optimization of factors such as the use of *E. coli* strain that are resistant to chloramphenicol aids in inhibiting protein progression to an insoluble state or the use of lower temperature or a lower concentration of isopropyl-β-D-1-thiogalactopyranoside (IPTG) for inducing protein expression helps greatly in circumventing solubility issues (Wena and Steinmetza 2016). In vitro assembly of VLPs has been successfully demonstrated for mammalian viruses such as HPV (Rose et al. 1998; Wang and Roden 2013) and Human immunodeficiency virus (HIV) (Morikawa et al. 2004; Kessans et al. 2013), plant viruses like PVX (Lomonossoff and Commandeur 2015; Lee et al. 2017), and CCMV (Díaz-Valle et al. 2015; Hassani-Mehraban et al. 2015) and bacteriophages P22 (Patterson et al. 2008), and PP7 (Caldeira and Peabody 2007; Tumban et al. 2011), using bacterial expression system.

9.3.2 Yeast and Baculovirus Expression

The eukaryotic-based expression systems are preferred to achieve post-translational modifications including glycosylation, lipidation, disulphide bond formation, and proper proteolytic processing. These include expression systems such as yeast and plant-based or baculovirus-based systems as they are scalable like bacterial systems using fermentation technologies (Lebel et al. 2015). Yeast-based systems that routinely include *Saccharomyces cerevisiae* and *Pichia pastoris* (Kim and Kim 2017) have been successfully employed for the production of VLPs of Qβ (Freivalds et al. 2006; Pokorski et al. 2012), CCMV (Brumfield et al. 2004; Hassani-Mehraban et al. 2015), and HPV (Wang and Roden 2013). The latter is commercially sold by Merck in the name of the vaccine Gardasil. Expression system based on baculovirus are usually cultured in *Spodoptera frugiperda* (Sf) lines that are popular insect cell lines (Molinar et al. 2008; Kueh et al. 2016) and *Trichoplusia ni* moth cells and are very useful in expressing multiple genes together (Krammer et al. 2015). Such systems are used to produce viruses such as flock house virus (FHV) that infect insects or plant viruses like CPMV and tomato bushy stunt virus (TBSV), or mammalian viruses like canine parvovirus (CPV) and HPV (Commercially produced Ceravix) (Schneemann et al. 1993; Shanks and Lomonossoff 2000; Singh et al. 2010; Wang and Roden 2013).

9.3.3 Plant-Based Expression

In the early 1980s, plant viruses began to be used as expression vectors for producing many proteins of pharmaceutical importance as they were devoid of contamination, had low production costs, and high expression levels. These include a variety of therapeutic proteins including antibodies that were produced using viruses like CPMV, TMV, and PVX (Steinmetz and Manchester 2011; Lebel et al. 2015;

Koudelka et al. 2016). VNPs and VLPs, which are produced using plant-based expression systems, include RCNMV, Brome mosaic virus (BMV), CCMV, CPMV, PVX, and TMV (Chen and Lai 2013; AR and II 2016). The plant-based expression is generally achieved via mechanical inoculation of viruses purified from infected leaf samples, cDNA, or *in vitro* transfer of the viral RNA transcripts using gentle abrasion of the leaves of the given plant (Marsian and Lomonossoff 2016). Agrobacterium tumefacien-based agroinfiltration involves injecting a bacterial suspension into leaves to express a modified viral genome (Fischer et al. 1999). Recombinant viruses generated through such techniques are used for the production of VLP-based vaccines for diseases such as influenza, rabies, and Ebola. Another advantage of such plant-based VLPs is the significantly reduced cost of administration as the VLPs can be directly taken in using ingestible plants and plant parts (Wen and Steinmetz 2016). Tissue culture–based methods primarily involve the use of suspension cultures for the rapid and continuous production of recombinant VLPs such as rPVP (recombinant poliovirus peptide) using transgenic plant cell lines (Angel and Lim 2007; Lebel et al. 2015).

9.4 Antigen Expression by Capsid Modification

The internal and external capsid surface of viruses could be chemically modified for either the encapsulation of sensitive compounds or precise display of target compounds. The versatility of the techniques is such that it is possible to incorporate multiple functional groups simultaneously within a single virus-based particle. Techniques that are routinely used for such modifications include genetic engineering, infusion, bioconjugation, and biomineralization (Fig. 9.2).

In the genetic engineering approach, epitopes or target sequences of interest are displayed on the capsid surface by modifying the viral genome encoding the CPs (Yan et al. 2015). Through genetic engineering, genetically modified CP genes could be self-assembled into VLPs displaying or encapsulating a chemical moiety (Fig. 9.2). In this technique, unnatural amino acids or residues such as cysteine are inserted or replaced to add functional groups (Miller et al. 2007; Geiger et al. 2013; Rohovie et al. 2017).

VNPs have demonstrated their usefulness in biomineralized-based virus shell engineering (BVSE) targeting Coxsackie adenovirus receptor (CAR) (Wang et al. 2012) and bioconjugation by using azide-alkyne cycloaddition (Smith et al. 2013) (Fig. 9.2). Epitopes of vaccines, protein domains, antibodies, or other short peptides have been genetically tagged to the coat protein and expressed as part of capsid (Staczek et al. 2000; Marusic et al. 2001; Takahashi et al. 2008; Luckanagul et al. 2012; Shukla et al. 2014b; Lebel et al. 2015). Bioconjugation strategies have emerged as strong players in the engineering of VNPs in recent years (Pokorski and Steinmetz 2011; Rohovie et al. 2017). The amino acid residues such as cysteine, lysine, tyrosine, and aspartic/glutamic acid residues offer themselves as ideal candidates for functional bioconjugation reactions involving carbodiimide activation, N-hydroxysuccinimide ester conjugation, azo coupling chemistries, and Michael

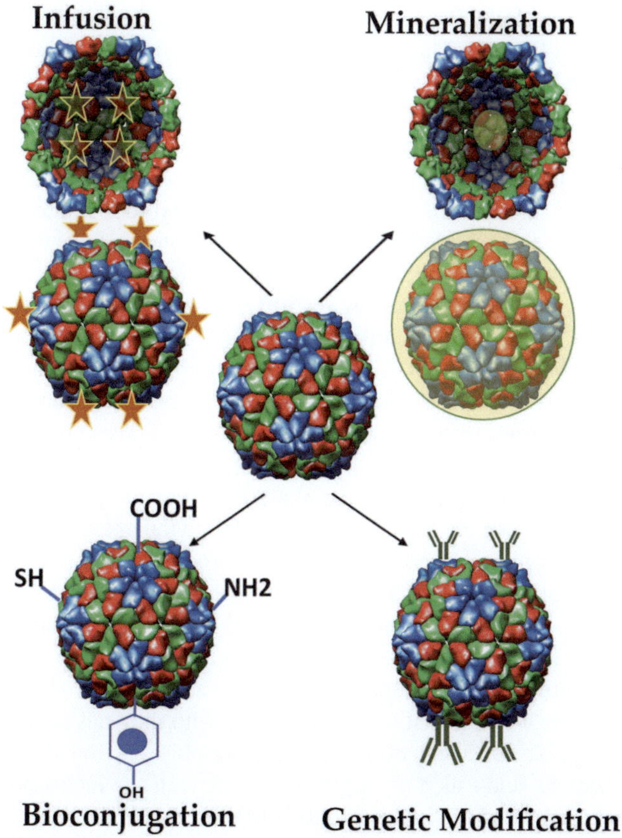

Fig. 9.2 The modifications of VLPs for better antigen presentation involving techniques such as infusion, mineralization, bioconjugation, and genetic modifications. Picture generated using Chimera (Pettersen et al. 2004) and Inkscape with recombinant SeMV capsid (PDB ID: 1×33).

addition (Koudelka et al. 2016). A deviation of this method involves replacing methionine residues with azidohomoalanine (AHA) or homopropargyl glycine (HPG) leading to the addition of alkyne or azide functionalities, respectively (Strable et al. 2009). Unnatural amino acids such as p-acetylphenylalanine, 3-(2-naphthyl) alanine, O-methyltyrosine, p-aminophenylalanine, and pazidophenylalanine could be incorporated using mutant tRNA synthetases leading to the incorporation of azide or amide groups (Strable et al. 2009). Such groups aid in copper (I)-catalysed azide-alkyne cycloaddition (Prasuhn et al. 2007; Bruckman et al. 2008; Singh et al. 2010; Steinmetz et al. 2010; Uchida et al. 2012). Ketone and aldehyde groups can be incorporated via pyridoxal 5′ phosphate at the amino-terminus that can be used for oxime or hydrazine condensation reactions (Schlick et al. 2005; Dirksen and Dawson 2008; Venter et al. 2011).

Modifications of viral CP through polymer hybrid conjugates involved the use of polyethylene glycol (PEG) as a means of shielding biological interactions that help to improve the efficiency and gain more control over polydispersity (Fig. 9.2). In yet another method, referred to as Atom-transfer radical polymerization (ATRP), small initiators are coupled to the CPs first and these are in turn assembled as capsids aiding in ease of purification prior to polymerization (Lucon et al. 2012; Pokorski et al. 2012; Hovlid et al. 2014). This method is popularly used for the attachment of glycoproteins, siRNA, and chemotherapeutics. Other strategies include the use of supramolecular chemistry in which moieties like β-cyclodextrin are tagged to the viral CPs and used as displaying platforms for PEG, drugs, and ligands (Pokorski and Steinmetz 2011). The process of infusion involves the entrapment of foreign cargo inside the viral interior (Fig. 9.2). Electrostatic or affinity-based interactions or polymer conjugates prevent the diffusion of small molecules, drugs, and other moieties outside the capsid (Prasuhn et al. 2007; Hovlid et al. 2014; Wen et al. 2015).

Using various gating techniques involving changes in pH and metal ions, the structural transitions in viruses could be altered to allow the entry of moieties and later prevent escape from the capsid (Loo et al. 2008; Honarbakhsh et al. 2013; Cao et al. 2015). Metal precursors can be introduced through infusion into the capsid to serve for the mineralization of capsid interior of both icosahedral (Douglas et al. 2002; Aljabali et al. 2011; Wen et al. 2013) and rod-shaped viruses (Knez et al. 2003; Lim et al. 2010; Aljabali et al. 2011). Such techniques play a crucial role in the specific identification of peptide nucleators and binders in phage display approaches that are used against varied substrates with minerals such as GaAs and ZnS (Giannini et al. 1995; SR et al., 2000). Assembly of viral particles through addition on nucleic acids such as oligonucleotides with the inclusion of different cargos has been achieved in viruses like RCNMV (Loo et al. 2007).

An interesting study with CCMV demonstrated that two different pathways for the assembly of hybrid capsids were employed that comprised either of 90 or 30 dimers of CP fused with elastin-like polypeptides, depending on the concentration of NaCl (van Eldijk et al. 2013). CCMV CPs have been used to demonstrate enzyme facilitation in which sortase A was covalently attached to the N-terminus of the CP via a C-terminal LPETG tag for protein encapsulation (Schoonen et al. 2018). Hydrophilic poly(2-methyl-2-oxazoline, PMeOx) shielding was shown to protect the TMV-based VNPs from recognition by the immune system (Bludau et al. 2017).

9.5 Role of VNPs in Immune Response

VLPs are known to elicit strong immune responses including both the humoral and the cell-based responses (Van Hest et al. 2014) when introduced into the host system mainly owing to their particulate nature and their ability to exhibit the epitopes in a dense repeating array (i.e., avidity effects). The use of plant VNPs as immunomodulators and their use in conjunction with adjuvants have provided successful treatment strategies for viral vaccines. Table 9.1 summarizes the kind of immune response elicited by plant-based VNPs. VNPs are generally recognized by toll-like

Table 9.1 Immune response induced by plant-based VNPs such as cowpea mosaic virus (CPMV), potato virus X (PVX), tobacco mosaic virus (TMV), cucumber mosaic virus (CMV), alfalfa mosaic virus (AlMV), papaya mosaic virus (PapMV), bamboo mosaic virus (BaMV), tomato bushy stunt virus (TBSV), and plum pox potyvirus (PPX)

Response Type	VLPs
Humoral	CPMV, PVX, TMV, CMV, AlMV, PapMV, BaMV, TBSV, PPX
Cellular	CPMV, PVX, TMV, CMV, AlMV, PapMV, BaMV
Immunomodulation	CPMV, PVX, TMV, PapMV
Adjuvant	PapMV

receptors (TLRs) that are pattern recognition receptors (PRRs) of host cells and entrapped by antigen processing cells (APCs) (Fakruddin et al. 2012; Moghadam et al. 2015). The processing of antigens is then achieved via either the MHC class I pathway for activation of CD8+ T cells or through the MHC class II pathway. They are also known to promote maturation and migration of DCs that is pivotal for the activation of the innate immune response (Grgacic and Anderson 2006; Ponterio et al. 2013; Yan et al. 2015; Liu et al. 2016).

9.5.1 Humoral Response

The plant-based VNPs with a greater payload carrying capacity have an inherent tropism for APCs making them effective as vaccine carriers (Shukla et al. 2017). They are known to elicit a humoral response upon administration that aids in protecting the individuals against infection or disease (Lebel et al. 2015). The humoral response is associated with the production of antibodies, such as IgGs and IgAs. While the production of IgGs is triggered following intraperitoneal, intra-nasal, and subcutaneous injections, IgAs are detected following intra-nasal and oral administration the in mucosa (Marusic et al. 2001; Grgacic and Anderson 2006; Chattopadhyay et al. 2017; Zilker et al. 2017). Systemic and mucosal antibodies are produced when the VLPs carrying foreign epitopes are introduced by various routes. Epitopes with optimal localization and folding are known to elicit a massive humoral response (Grgacic and Anderson 2006; Chattopadhyay et al. 2017). For example, the CMV viruses that were designed to express the HCV R9 peptide were recognized by antibodies from HCV-infected patients (Piazzolla et al. 2005). The higher surface area provided by VLPs provide a better immune response than peptide-conjugates (Skwarczynski and Toth 2016).

The VLP-based vaccines are generally administered along with adjuvants that are less toxic like RIBI, QS-21, and saponin-based QuilA (McInerney et al. 1999; Palmer et al. 2006), or without any adjuvant (Lebel et al. 2015). Plant-based VNPs are highly efficient in inducing antibody response at lower concentrations than commercially available vaccines or peptides that were tagged with keyhole limpet haemocyanin (Mallajosyula et al. 2014; Lebel et al. 2015). Experiments involving immunization of animal models with various plant-based VLPs such as bamboo mosaic virus (BaMV) (Yang et al. 2007; Chen et al. 2012), CPMV (Dalsgaard et al.

1997; Langeveld et al. 2001; Khor et al. 2002), TMV (Jiang et al. 2006; Palmer et al. 2006; Mallajosyula et al. 2014), or PVX (Massa et al. 2008; Jobsri et al. 2015) tagged to different antigenic epitopes were successful in mounting humoral response, generating specific antibodies, and serving as effective vaccines especially for weakly immunogenic candidates. An effective humoral response was demonstrated when the HCV R9 peptide was displayed on PVX (Uhde-Holzem et al. 2010), or the peptides 10 and 18 derived from the outer membrane protein F of the gram negative bacteria *Pseudomonas aeruginosa* were exhibited on the outer surface of CPMV VNPs (Brennan et al. 1999b), thereby eliciting a strong antibody response (Kaltgrad et al. 2007). Cross-linking of B-cell receptor with the VLPs and presenting multiple epitopes in close proximity were successful in eliciting rapid proliferation of not only B lymphocytes but also T cells and the rapid differentiation of B cells into plasma cells (Batista and Harwood 2009). In a recent study, the efficacy of icosahedral and filamentous VNPs was assessed for their abilities of delivering the HER2 epitopes for the treatment and prophylaxis of tumours associated with HER2+ (Shukla et al. 2017). The results showed that CPMV-based VNPs had particularly improved lymph node transmission and retention, and increased uptake by APCs over filamentous PVX particles.

9.5.2 Cellular Immune Response

Plant-based VLPs are also known to elicit cellular immune response towards the epitope carried by the VLPs. A combined humoral and cellular response is more effective towards antigens presented on the VLPs (Lebel et al. 2015). Cellular response was observed to be mounted when VLPs of PapMV (Leclerc et al. 2007; Hanafi et al. 2010), rPVPs, such as TMV (McCormick et al. 2006a, 2006b; Jobsri et al. 2015), CMV (Piazzolla et al. 2005), CPMV (McInerney et al. 1999), and PVX (Massa et al. 2008; Lico et al. 2009) fused to different epitopes were able to elicit activation of CD8+ displaying T lymphocytes. The production of IFN-γ is associated with the mounting of cellular response. Glycan conjugated with CPMV were successfully used for the treatment of tumours by inducing a strong T cell–dependent immune response (Miermont et al. 2008). Potent activation of CD8+ was observed with discernible production of IFN-γ when peripheral blood mononuclear cells (PBMCs) were incubated with AlMV tagged to respiratory syncytial virus (RSV) epitope in patients suffering from HCV (Yusibov et al. 2005; Yusibov and Rabindran 2008). Such a response was also evident for CMV-based VNPs that displayed the R9 epitope of HCV (Piazzolla et al. 2005; Nuzzaci et al. 2007). Vaccination with cancer cell lines such as B16-OVA and Eg.7-OVA (McCormick et al. 2006b), foot and mouth disease virus (FMDV) (Joelson et al. 1997), and lymphocytic choriomeningitis virus (Lacasse et al. 2008) correlated with cellular immune responses. A study involving plant-made VLP vaccines tagged to H1 or H5 hemagglutinin demonstrated that the VNPs rapidly accessed the draining lymph nodes and interacted with activated B lymphocytes, macrophages, and DCs (Hendin et al. 2017). The work emphasized the balance in humoral and cell-mediated responses subsequent to plant-based VNP vaccination of older animals and the young mice. In a recent study, plant-made monovalent VLP vaccines presenting influenza hemagglutinin proteins H1 or H5 induced the presence of long-term cross-reactive memory CD4+ T cells 6 months after immunization in healthy adults (Pillet et al. 2018). The findings corroborated with yet another study where intranasal vaccination with VLPs bearing multiple ectodomains of matrix protein 2 of influenza virus induced both humoral and cellular immune responses and conferred cross-immunity against different subtypes of influenza (Lee et al. 2018).

9.5.3 Immunomodulators and Adjuvants

As mentioned in previous sections, plant-based VLPs are very suitable in eliciting either humoral or cell-mediated immune response. However, to make the response effective, APCs must be activated as they are capable of presenting the antigens to the cells belonging to the adaptive immune system (Storni et al. 2002). VLPs based on rPVPs, PVX (Marusic et al. 2001; Jobsri et al. 2015), TMV (McCormick et al. 2006b; Kemnade et al. 2014), PapMV (Acosta-Ramirez et al. 2008; Lacasse et al. 2008; Lebel et al. 2014), and CPMV (Gonzalez et al. 2009) were capable of activating DCs that in turn led to the upregulation of a variety of co-stimulatory molecules such as MHCI, MHCII, CD40, CD80, CD86, and CCR7 (Lebel et al. 2014; Jobsri et al. 2015) and the production of pro-inflammatory cytokines like IL-6, IL-12, IFN-α, and TNF-α (Acosta-Ramirez et al. 2008; Savard et al. 2011; Lebel et al. 2014; Jobsri et al. 2015). Apart from the DCs, cells such as B cells, NK cells, and macrophages are also known to stimulate the activation markers post rPVP encounter (Acosta-Ramirez et al. 2008; Gonzalez et al. 2009). Due to their ability for ready uptake by the APCs, VLPs such as rPVPs and PapMV have been successfully used as adjuvants together with various types of vaccines. Administration of PapMV along with DCs displaying OVA elicited strong cell-based immune responses in mice against listeria monocytogenes-OVA (Lebel et al. 2014). Administration of PapMV as an adjuvant in mice together with the seasonal trivalent influenza vaccine (TIV) also elicited a similar immune response with enhanced production of antibodies (Acosta-Ramirez et al. 2008; Denis et al. 2008; Savard et al. 2011). Further, the use of PapMV not only primes the immune system but also induces a wide range of immune response targeted towards TIV antigens, thereby providing cross-protection against other strains of influenza absent in the vaccine (Savard et al. 2011; Mathieu et al. 2013). The mechanism by which PapMV particles are able to activate APCs are via their ssRNA which is recognized by TLR7 present in the endosomes of APCs, triggering the production of IFN-α (Lebel et al. 2014; Jobsri et al. 2015). This is further backed by the fact that monomers of PapMV that lack the ssRNA are unable to activate murine splenocytes upon administration suggesting the role of RNA as the major immunomodulator in PapMV (Lebel et al. 2014).

9.6 Diseases Targeted Using Plant VNPs

An early study involving CPMV VNPs bearing short epitopes from the CP VP2 of mink enteritis virus (MEV) demonstrated that it imparted complete protection against clinical disease in mink with a small dose of 1 mg (Dalsgaard et al. 1997). Henceforth, plant-based VLPs have proven to be extremely versatile in the treatment of various diseases. Among the PVX VLP–based vaccine formulations are those displaying HIV-1 epitopes (Marusic et al. 2001), *Staphylococcus aureus* D2 FnBP (Brennan et al. 1999a), hepatitis C virus (HCV) epitopes (Uhde-Holzem et al. 2010), FMDV epitopes (Andrianova et al. 2011), alternanthera mosaic virus (AltMV) CP (Tyulkina et al. 2011), and human papillomavirus (HPV) epitopes (Erovska et al. 2012; Vaculik et al. 2015). In novel approaches involving tagging VNPs with photosensitizers that are used for photodynamic therapy applications, VLPs based on CCMV were successfully encapsulated with chelated Gd3+ and Zn2+ phthalocyanine (Allen et al. 2005; Luo et al. 2016). TMV-based VNPs have been successfully employed for the targeted delivery of thrombolytic therapies to improve clot localization and minimize bleeding risk (Pitek et al. 2018). CPMV- and PVX-based VNPs that display the immunodominant lipopeptide from lipocalin were used to effectively diagnose Sjögren's syndrome (Tinazzi et al. 2015). In a recent study, the preventive potential of major mugwort pollen allergen Art v 1 that was targeted to both the inner and outer surface of the envelope of VNP was determined in humanized mouse models with pollen allergy. The study presented the efficacy of VNPs in serving as effective platforms for the delivery of allergens *in vivo* and specifically target T lymphocytes to prevent allergies (Kratzer et al. 2018).

9.6.1 Vaccines for Infectious Diseases

Plant-based VNP vaccines that are developed against human papillomavirus (HPV) are known to generate immune response against the virus and protect against carcinomas and cancers induced by HPV (Chan and Berek 2007; Koudelka et al. 2016). CPMV-based VNPs carrying foreign epitopes of *Pseudomonas aeruginosa* elicit a humoral immune response and protect the host against bacterial infections. Inactivated recombinant CPMV VNPs conjugated with VP2 CP of canine parvovirus (CPV) that protected dogs from a fatal challenge with CPV (Phelps et al. 2008). CPMV is known to particularly exhibit distinct specificity in its interactions with surface vimentins that are associated with endothelial, tumour, and inflammatory cells. This aspect has found pivotal application in utilizing CPMV-based VNPs for delivery into a number of cancer cells, intravital imaging of vasculature in cancer cells, study of atherosclerotic lesions, and angiogenesis (Wena and Steinmetza 2016). In yet another application of CPMV-based VNPs, the tropism of the virus to APCs was used for fighting infectious diseases caused by viruses such as arenavirus lymphocytic choriomeningitis virus (Wen et al. 2015).

9.6.2 Vaccines for Cancer

Cancer vaccines work by training the immune system to identify TAAs and provide long-term protection against tumour metastases and relapse (Yaddanapudi et al. 2013; Bell and McFadden 2015; Koudelka et al. 2016; Hefferon 2018). The use of VNPs for displaying the cancer-epitopes is effective as the epitopes are presented in a non-native environment thereby overcoming the problem of self-tolerance (Shukla et al. 2014c). CPMV particles have been successfully used for the display of Tn antigen which is a glycoprotein that is exceedingly expressed on cancer cells of breast, colon, or prostate (Miermont et al. 2008). CPMV, RCNMV, and TMV VLPs carrying the cancer markers as antigen were able to elicit an effective humoral response as measured by the high titres of Tn-specific antibodies including IgG and IgM (Miermont et al. 2008; Yin et al. 2012). CPMV and hibiscus chlorotic ringspot virus (HCRSV)–based VLPs were successfully targeted to cancer cells by conjugating them with up to 300 and 950 doxorubicin molecules, respectively (Ren et al. 2007). A preclinical test involving VNP based on CMV modified with folic acid had a better *in vivo* efficacy and reduced cardiotoxicity while targeting ovarian cancer cells (Zeng et al. 2013). A more recent study involving dual bioconjugated truncated hepatitis B core antigen VNPs showed increased accrual and uptake of doxorubicin in the cancerous cells of human cervical and colorectal origin as opposed to free doxorubicin, thereby enhancing the cytotoxicity of the drug towards the tumour cells (Biabanikhankahdani et al. 2018). CPMV Fluorescein labelled and VEGFR1-conjugated CPMV have been particularly targeted to HT-29 cancer cells in mice (Brunel et al. 2014; Shukla and Steinmetz 2016). The use of TMV for the delivery of cationic drugs such as phenanthriplatin into triple negative breast cancer xenograft models of mouse showed greater effectiveness with 4-fold smaller tumour growth as opposed to clinically used cisplatins or free drug usage (Czapar et al. 2016; Patel et al. 2018). Physalis mottle VLPs were surface modified with fluorophores employing esters of reactive lysine-N-hydroxysuccinimide or through cysteine-maleimide chemistries in a range of cancer cells and found to be effective for the delivery of bioimaging agents and drugs (Masarapu et al. 2017).

9.6.3 Vaccines for Neurological Diseases and Addiction

Conjugation of nicotine or amyloid beta (Aβ) tagged to Qβ showed an increased production of drug-specific IgGs in vaccinated mice (Lemere 2013; Lee and Steinmetz 2017). The antibody binding to nicotine was demonstrated to result in up to 90% reduction in nicotine levels in the brain in individual mice. In a recent study, chimeric infectious bursal disease (IBD) VLPs served as effective vaccines for the eradication of HPV-16 E7–dependent cancers (Gonza et al. 2012; Choi 2013). A VLP-based vaccine was developed for the control of hypertension that fared well in both the preclinical studies and phase II clinical trials with good safety levels and efficacies (Ambühl et al. 2007). The conjugation of angiotensin-II peptide with the VNPs of Qβ (AngQβ) showed high immunogenicity in animal models as well as humans. In the phase II clinical studies with the AngQβ VNPs, the blood pressure levels dropped to values that are comparable to those obtained with the use of angiotensin-converting enzyme blockers additionally with lesser side effects (Steinmetz and Manchester 2011; Pan et al. 2017). Qβ particles tagged with deoxyoligonucleotides comprising the CpG-motifs are known to stimulate the innate immune system via toll receptor 9 and are useful in the treatment of allergic rhinitis and asthma (Storni et al. 2018).

9.7 Future Perspectives

Plant-based VNPs have been widely used in the past 30 years for a variety of biological applications in the diagnosis and treatment of diseases including vaccine therapy, bioimaging, data storage, nanoreactors, and sensing devices owing to their symmetric design. Their use in vaccine therapy has revolutionized the delivery of drugs and display of antigenic epitopes for eliciting effective immune responses. Such VNPs have demonstrated their capabilities in inducing humoral and cellular immune responses in experimental mice, non-human primates, and humans. VNPs are versatile in targeted drug delivery as they reduce drug toxicity and degradation, and increase specificity in drug targeting, drug bioavailability, and circulation. Plant-based VNPs as therapeutic vaccines are fast emerging as the future vaccine candidates for cancer, drug addiction, and other infectious diseases. Modifications involving both the interior and exterior of the VNPs, their hierarchical assembly, and immunogenic properties have led to the success of VNPs as vaccines. However, further research is mandatory to exploit the full potential of plant-based VNPs in vaccine therapy.

Acknowledgement SV acknowledges financial support from SERB, DST (ECR/2016/000242).

References

Abraham A et al (2016) Intracellular delivery of antibodies by chimeric Sesbania mosaic virus (SeMV) virus like particles. Nat Publ Group 6:21803. https://doi.org/10.1038/srep21803

Acosta-Ramirez E et al (2008) Translating innate response into long-lasting antibody response by the intrinsic antigen-adjuvant properties of papaya mosaic virus. Immunology 124:186–197

Agnandji S et al (2012) A phase 3 trial of RTS,S/AS01 malaria vaccine in African infants. New Engl J Med Orig 367:2284–2295. https://doi.org/10.1056/NEJMoa1208394

Alemzadeh E, Izadpanah K, Ahmadi F (2017) Generation of recombinant protein shells of Johnson grass chlorotic stripe mosaic virus in tobacco plants and their use as drug carrier. J Virol Methods 248:148–153

Aljabali AAA, Lomonossoff GP, Evans DJ (2011) CPMV-polyelectrolyte-Templated gold nanoparticles. Biomacromolecules 12(7):2723–2728

Allen TM, Cullis PR (2004) Drug delivery systems: entering the mainstream. Science 303:1818–1822. https://doi.org/10.1126/science.1095833

Allen M et al (2005) Paramagnetic viral nanoparticles as potential high-relaxivity magnetic resonance contrast agents. Magn Reson Med 54(4):807–812
Ambühl P et al (2007) A vaccine for hypertension based on virus-like particles: preclinical efficacy and phase I safety and immunogenicity. J Hypertens 25(1):63–72
Andrianova E et al (2011) Foot and mouth disease virus polyepitope protein produced in bacteria and plants induces protective immunity in Guinea pigs. Biochemistry 76:339–346
Angel M, Lim G (2007) Transgenic plants in therapeutically valuable protein production. Trans Plant J 1(2):256–266
AR v Z, II H (2016) Purification of virus-like particles (VLPs) from plants. Method Mol Biol 1404:569–579
Batista FD, Harwood NE (2009) The who, how and where of antigen presentation to B cells. Nat Rev Immunol 9:15–27
Bell J, McFadden G (2015) Viruses for tumor therapy. Cell Host Microbe 15(3):260–265. https://doi.org/10.1016/j.chom.2014.01.002.Viruses.
Berman H et al (2000) The protein data Bank. Nucleic Acids Res 28(1):235–242. https://doi.org/10.1093/nar/28.1.235
Bessa J et al (2008) Efficient induction of mucosal and systemic immune responses by virus-like particles administered intranasally: implications for vaccine design. Eur J Immunol 38(1):114–126
Biabanikhankahdani R et al (2018) A dual bioconjugated virus-like nanoparticle as a drug delivery system and comparison with a pH-responsive delivery system. Nano 8:236. https://doi.org/10.3390/nano8040236
Birnbaum F, Nassal M (1990) Hepatitis B virus nucleocapsid assembly: primary structure requirements in the core protein. J Virol 64(7):3319–3330
Bludau H et al (2017) POxylation as an alternative stealth coating for biomedical applications Herdis. Eur J Immunol 88:679–688. https://doi.org/10.1016/j.eurpolymj.2016.10.041. POxylation.
Brennan F et al (1999a) Immunogenicity of peptides derived from a fibronectin-binding protein of S. aureus expressed on two different plant viruses. Vaccine 17(15–16):1846–1857
Brennan FR et al (1999b) Pseudomonas aeruginosa outer-membrane protein F epitopes are highly immunogenic in mice when expressed on a plant virus. Microbiology 145:211–220
Bruckman M et al (2008) Surface modification of tobacco mosaic virus with "click" chemistry. Chembiochem 9(4):519–523
Brumfield S et al (2004) Heterologous expression of the modified coat protein of cowpea chlorotic mottle bromovirus results in the assembly of protein cages with altered architectures and function. J Gen Virol 85:1049–1053. https://doi.org/10.1099/vir.0.19688-0
Brune KD, Howarth M, Howarth M (2018) New routes and opportunities for modular construction of particulate vaccines: stick, click, and glue. Front Immunol 9:1050–1074. https://doi.org/10.3389/fimmu.2018.01432
Brunel FM et al (2014) A hydrazone ligation strategy to assemble multifunctional viral nanoparticles for cell imaging and tumor targeting. Nano Lett 10(3):1093–1097. https://doi.org/10.1021/nl1002526.A.
Caldeira JC, Peabody DS (2007) Stability and assembly in vitro of bacteriophage PP7 virus-like particles. J Nanobiotechnol 10:1–10. https://doi.org/10.1186/1477-3155-5-10.
Cao J et al (2015) Development of Abamectin loaded plant virus nanoparticles for efficacious plant parasitic nematode control. ACS Appl Mater Interfaces 7(18):9546–9553
Champion JA, Mitragotri S (2010) Shape induced inhibition of phagocytosis of polymer particles. Pharm Res 26(1):244–249. https://doi.org/10.1007/s11095-008-9626-z.
Chan J, Berek J (2007) Impact of the human papilloma vaccine on cervical cancer. J Clin Oncol 25:2975–2982
Chattopadhyay S et al (2017) Nanoparticle vaccines adopting virus-like features for enhanced immune potentiation. Nano 1(3):244–260
Chen Q, Lai H (2013) Plant-derived virus-like particles as vaccines. Hum Vaccin Immunother 9(1):26–49

Chen TH et al (2012) Induction of protective immunity in chickens immunized with plant-made chimeric bamboo mosaic virus particles expressing very virulent infectious bursal disease virus antigen. Virus Res 166:109–115

Chen Z et al (2017) Cryo-EM structure of the bacteriophage T4 isometric head at 3. 3-Å resolution and its relevance to the assembly of icosahedral viruses. Proc Natl Acad Sci:E8184–E8193. https://doi.org/10.1073/pnas.1708483114

Choi K (2013) Review article a virus-like particle vaccine against infectious bursal disease virus: potential uses and applications. Br J Virol 3(1):1–5

Crick FH, Watson JD (1956) Structure of small viruses. Nature 177:473–475

Czapar AE et al (2016) Tobacco mosaic virus delivery of Phenanthriplatin for Cancer therapy. ACS Nano 10(4):4119–4126. https://doi.org/10.1021/acsnano.5b07360.Tobacco.

Dalsgaard K et al (1997) Plant-derived vaccine protects target animals against a viral disease. Nat Biotechnol 15:248–252

Denis J et al (2008) Development of a universal influenza a vaccine based on the M2e peptide fused to the papaya mosaic virus (PapMV) vaccine platform. Vaccine 26:3395–3403

Díaz-Valle A et al (2015) Highly efficient strategy for the heterologous expression and purification of soluble cowpea chlorotic mottle virus capsid protein and in vitro pH-dependent assembly of virus-like particles. J Virol Methods 225:23–29

Dirksen A, Dawson PE (2008) Rapid Oxime and Hydrazone ligations with aromatic aldehydes for biomolecular labeling. Bioconjug Chem 19(12):2543–2548

Douglas T, Young M (1998) Host–guest encapsulation of materials by assembled virus protein cages. Nature 393:152–155

Douglas T et al (2002) Protein engineering of a viral cage for constrained nanomaterials synthesis. Adv Mater 14(6):415–418

El-attar L et al (2009) Comparison of the efficacy of rotavirus VLP vaccines to a live homologous rotavirus vaccine in a pig model of rotavirus disease. Vaccine 27:3201–3208. https://doi.org/10.1016/j.vaccine.2009.03.043

Erovska NOC et al (2012) Transient expression of human papillomavirus type 16 L2 epitope fused to N- and C-terminus of coat protein of potato virus X in plants. J Biosci 37(January):125–133. https://doi.org/10.1007/s12038-011-9177-z.

Fakruddin M, Hossain Z, Afroz H (2012) Prospects and applications of nanobiotechnology: a medical perspective. J Nanobiotechnol 10(1):1–8

Fischer R et al (1999) Towards molecular farming in the future: moving from diagnostic protein and antibody production in microbes to plants. Biotechnol Appl Biochem 30(2):101–108

Freivalds J et al (2006) Assembly of bacteriophage Qβ virus-like particles in yeast Saccharomyces cerevisiae and Pichia pastoris. J Biotechnol 123(3):297–303

Gardner T, Elzey B, Hahn N (2012) Sipuleucel-T (Provenge) autologous vaccine approved for treatment of men with asymptomatic or minimally symptomatic castrate-resistant metastatic prostate cancer. Hum Vaccin Immunother 8(4):534–539

Geiger F et al (2013) TMV nanorods with programmed longitudinal domains of differently addressable coat proteins. Nanoscale 5(9):3808–3816

Giannini C et al (1995) Structural analysis of ZnS/GaAs heterostructures grown by hydrogen transport vapor-phase epitaxy. J Appl Phys 77:2429

Gonza L et al (2012) Chimeric infectious bursal disease virus-like particles as potent vaccines for eradication of established HPV-16 E7 – dependent tumors. PLoS One 7(12):1–12. https://doi.org/10.1371/journal.pone.0052976

Gonzalez MJ et al (2009) Interaction of cowpea mosaic virus (CPMV) nanoparticles with antigen presenting cells in vitro and in vivo. PLoS One 4:E7981

Grgacic E, Anderson D (2006) Virus-like particles: passport to immune recognition. Methods 40(1):60–65

Hanafi LA et al (2010) Two distinct chimeric potexviruses share antigenic cross-presentation properties of mhc class I epitopes. Vaccine 28:5617–5626

Hassani-Mehraban A et al (2015) Feasibility of cowpea chlorotic mottle virus-like particles as scaffold for epitope presentations. BMC Biotechnol 15:80

Hefferon KL (2018) Repurposing plant virus nanoparticles. Vaccine 6:11. https://doi.org/10.3390/vaccines6010011
Hendin H et al (2017) Plant-made virus-like particle vaccines bearing the hemagglutinin of either seasonal (H1) or avian (H5) influenza have distinct patterns of interaction with human immune cells in vitro. Vaccine 35(19):2592–2599
Honarbakhsh S et al (2013) Polymeric systems incorporating plant viral nanoparticles for tailored release of therapeutics. Adv Healthc Mater 2(7):1001–1007
Hong V et al (2012) Analysis and optimization of copper-catalyzed Azide–alkyne cycloaddition for bioconjugation. Angew Chem Int Ed 48(52):9879–9883. https://doi.org/10.1002/anie.200905087.Analysis.
Hovlid ML et al (2014) Terms of use Encapsidated atom-transfer radical polymerization in Q β virus-like nanoparticles. ACS Nano 8(8):8003–8014
Hwang D, Roberts I, Wilson T (1994) Expression of tobacco mosaic virus coat protein and assembly of pseudovirus particles in Escherichia coli. Proc Natl Acad Sci 91(19):9067–9071
Jeevanandam J et al (2018) Review on nanoparticles and nanostructured materials : history, sources, toxicity and regulations. Beilstein J Nanotechnol 9:1050–1074. https://doi.org/10.3762/bjnano.9.98
Jennings G, Bachmann M (2009) Immunodrugs: therapeutic VLP-based vaccines for chronic diseases. Annu Rev Pharmacol Toxicol 49:303–326
Jiang L et al (2006) A modified tmv-based vector facilitates the expression of longer foreign epitopes in tobacco. Vaccine 24:109–115
Jobsri J et al (2015) Plant virus particles carrying tumour antigen activate TLR7 and induce high levels of protective antibody. PLoS One 10:e0118096
Joelson T et al (1997) Presentation of a foreign peptide on the surface of tomato bushy stunt virus. J Gen Virol 78:1213–1217
Jordan PC et al (2016) Self-assembling biomolecular catalysts for hydrogen production. Nat Chem 8:179–185
Kaiser CR et al (2007) Biodistribution studies of protein cage nanoparticles demonstrate broad tissue distribution and rapid clearance in vivo.', *Journal of.* Nanomedicine 2:715–733
Kaltgrad E et al (2007) Anti-carbohydrate antibodies elicited by polyvalent display on a viral scaffold. Chembiochem 8:1455–1462
Kemnade JO et al (2014) Tobacco mosaic virus efficiently targets DC uptake, activation and antigen-specific T cell responses in vivo. Vaccine 32:4228–4233
Kessans SA et al (2013) Biological and biochemical characterization of HIV-1 gag/dgp41 virus-like particles expressed in Nicotiana benthamiana. Plant Biotechnol J 11(6):681–690
Khor IW et al (2002) Novel strategy for inhibiting viral entry by use of a cellular receptor-plant virus chimera. J Virol 76:4412–4419
Kim H, Kim H (2017) Yeast as an expression system for producing virus-like particles: what factors do we need to consider? Lett Appl Microbiol 64(2):111–123
Knez M et al (2003) Biotemplate synthesis of 3-nm nickel and cobalt nanowires. Nano Lett 3(8):1079–1082
Koudelka KJ et al (2016) Virus-based nanoparticles as versatile Nanomachines. Annual Review of Virology 2(1):379–401. https://doi.org/10.1146/annurev-virology-100114-055141. Virus-Based.
Krammer F et al (2015) Europe PMC Funders Group Trichoplusia ni cells (High Five ™) are highly efficient for the production of influenza A virus-like particles: a comparison of two insect cell lines as production platforms for influenza vaccines. 45(3):226–234. https://doi.org/10.1007/s12033-010-9268-3.Trichoplusia.
Kratzer B et al (2018) Running title: shielding of allergens inside virus-like nanoparticles. Exp Allergy Immunol. https://doi.org/10.1111/all.13573
Kueh CL et al (2016) Virus-like particle of Macrobrachium rosenbergii nodavirus produced in Spodoptera frugiperda (Sf9) cells is distinctive from that produced in Escherichia coli. Biotechnol Prog 33(2):549–557

Lacasse P et al (2008) Novel plant virus-based vaccine cell, induces protective cytotoxic T-lymphocyte-mediated antiviral immunity through dendritic maturation. J Virol 82:785–794
Langeveld JP et al (2001) Inactivated recombinant plant virus protects dogs from a lethal challenge with canine parvovirus. Vaccine 19:3661–3670
Lebel ME et al (2014) Nanoparticle adjuvant sensing by TLR7 enhances CD8+ T cell-mediated protection from listeria monocytogenes infection. Immunology 192:1071–1078
Lebel M-è et al (2015) Plant viruses as nanoparticle-based vaccines and adjuvants. Vaccine 3:620–637. https://doi.org/10.3390/vaccines3030620
Leclerc D et al (2007) Proteasome-independent major histocompatibility complex class I cross-presentation mediated by papaya mosaic virus-like particles leads to expansion of specific human T cells. J Virol 81:1319–1326
Lee KL, Steinmetz N (2017) Virus-based nanoparticles as platform technologies for modern vaccines. Wiley Interdiscip Rev Nanomed Nanobiotechnol 8(4):554–578. https://doi.org/10.1002/wnan.1383.Virus-based
Lee C et al (2009) Production of FMDV virus-like particles by a SUMO fusion protein approach in Escherichia coli. 7:1–7. https://doi.org/10.1186/1423-0127-16-69.
Lee KL et al (2017) Genetic engineering and chemical conjugation of potato virus X. Methods Mol Biol 1108:3–21. https://doi.org/10.1007/978-1-62703-751-8.
Lee Y et al (2018) Intranasal vaccination with M2e5x virus-like particles induces humoral and cellular immune responses conferring cross- protection against heterosubtypic influenza viruses. PLoS One 13(1):e0190868
Lemere CA (2013) MER5101, a novel Aβ1-15:DT conjugate vaccine, generates a robust anti-Aβ antibody response and attenuates Aβ pathology and cognitive deficits in APPswe/PS1ΔE9 transgenic mice. J Neurosci 33(16):7027–7037. https://doi.org/10.1523/JNEUROSCI.5924-12.2013.MER5101
Lewis JD et al (2006) Viral nanoparticles as tools for intravital vascular imaging. Nat Med 12(3):354–360. https://doi.org/10.1038/nm1368.Viral.
Lico C et al (2009) Plant-produced potato virus X chimeric particles displaying an influenza virus-derived peptide activate specific CD8+ T cells in mice. Vaccine 27:5069–5076
Lim J et al (2010) Formation of au/Pd alloy nanoparticles on TMV. J Nanomater 2010:620505. https://doi.org/10.1155/2010/620505
Liu J et al (2016) Virus like particle-based vaccines against emerging infectious disease viruses. Virol Sin 31(4):279–287. https://doi.org/10.1007/s12250-016-3756-y
Lomonossoff GP, Commandeur U (2015) Production of hybrid chimeric PVX Particles Using a combination of TMV and PVX-Based expression Vectors. Front Bioeng Biotechnol 3(November):1–12. https://doi.org/10.3389/fbioe.2015.00189
Loo L et al (2007) Encapsidation of nanoparticles by red clover necrotic mosaic virus. J Am Chem Soc 129(36):11111–11117
Loo L et al (2008) Infusion of dye molecules into red clover necrotic mosaic virus. Chem Commun 1:88–90
Luckanagul J et al (2012) Porous alginate hydrogel functionalized with virus as three-dimensional scaffolds for bone differentiation. Biomacromolecules 13(12):3949–3958
Lucon J et al (2012) Polymerization with high-density cargo loading. Nat Chem 4:781–788. https://doi.org/10.1038/NCHEM.1442.
Luo D, Carter KA, Lovell JF (2016) Nanomedical engineering : shaping future nanomedicines. Wiley Interdiscip Rev Nanomed Nanobiotechnol 7(2):169–188. https://doi.org/10.1002/wnan.1315.Nanomedical
Mallajosyula JK et al (2014) Single-dose monomeric HA subunit vaccine generates full protection from influenza challenge. Hum Vaccin Immunother 10:586–595
Marsian J, Lomonossoff GP (2016) Molecular pharming—VLPs made in plants. Curr Opin Biotechnol 37:201–206
Marusic C et al (2001) Chimeric Plant Virus particles as immunogens for inducing murine and human immune responses against Human Immunodeficiency Virus Type 1. 75(18):8434–8439. https://doi.org/10.1128/JVI.75.18.8434

Masarapu H et al (2017) Physalis mottle virus-like particles as Nanocarriers for imaging reagents and drugs. Biomacromolecules 18(12):4141–4153

Massa S et al (2008) Antitumor activity of DNA vaccines based on the human papillomavirus-16 E7 protein genetically fused to a plant virus coat protein. Hum Gene Ther 19:354–364

Mathieu C et al (2013) Induction of innate immunity in lungs with Pneumoniae, virus-like nanoparticles leads to protection against influenza and Streptococcus challenge.', *Nanomedicine*. Nanotechnology 9:839–848

MC G, Kallinteri P (2006) Nanomedicines and nanotoxicology: some physiological principles. Occup Med 56(5):307–311

McCormick A, Corbo T, Wykoff-Clary S, Palmer K et al (2006a) Chemical conjugate tmv-peptide bivalent fusion vaccines improve cellular immunity and tumor protection. Bioconjug Chem 17:1330–1338

McCormick A, Corbo T, Wykoff-Clary S, Nguyen L et al (2006b) Tmv-peptide fusion vaccines induce cell-mediated immune responses and tumor protection in two murine models. Vaccine 24:6414–6423

McInerney TL et al (1999) Analysis of the ability of five adjuvants to enhance immune responses to a chimeric plant virus displaying an HIV-1 peptide. Vaccine 17:1359–1368

Miermont A et al (2008) Cowpea mosaic virus capsid: a promising carrier for the development of carbohydrate based antitumor vaccines. Chemistry 14(16):4939–4947

Miller RA, Presley AD, Francis MB (2007) Self-assembling light-harvesting systems from synthetically modified tobacco mosaic virus coat proteins. J Am Chem Soc 129(11):3104–3109

Moghadam MM, Motalleb G, Fakheri A (2015) Nanoparticles and plant biotechnology. Research in Biotechnology 6(5):1–15

Molinar P, Peralta A, Taboga O (2008) Production of rotavirus-like particles in Spodoptera frugiperda larvae. Method J Virol 147(2):364–367

Morikawa Y, Goto T, Momose F (2004) Human immunodeficiency virus type 1 gag assembly through assembly intermediates. J Biol Chem 279:31964–31972

Nam K et al (2006) Virus-enabled synthesis and assembly of nanowires for lithium ion battery electrodes. Science 312(5775):885–888

Nuzzaci M et al (2007) Cucumber mosaic virus as a presentation system for a double hepatitis C virus-derived epitope. Arch Virol 152:915–928

Palmer KE et al (2006) Protection of rabbits against CUTANEOUS papillomavirus infection using recombinant tobacco mosaic virus containing l2 capsid epitopes. Vaccine 24:5516–5525

Pan Y et al (2017) A Therapeutic Peptide Vaccine Against PCSK9', Scientific Reports. Springer US, 7, p 12534. https://doi.org/10.1038/s41598-017-13069-w

Patel R et al (2018) Radiation therapy combined with Cowpea Mosaic Virus nanoparticle in Situ Vaccination initiates immune-mediated tumor regression. 6–11. https://doi.org/10.1021/acsomega.8b00227.

Patterson D, Schwarz B, Western B, Terra M, Krugler P et al (2008) Engineering virus like particles towards directing immunologic responses. Faculty Posters. Book 9. http://hdl.handle.net/10950/362

Pettersen EF et al (2004) UCSF chimera—a visualization system for exploratory research and analysis. J Comput Chem 25:1605–1612. https://doi.org/10.1002/jcc.20084

Phelps JP, Dang N, Rasochova L (2008) Inactivation and purification of cowpea mosaic virus-like particles displaying peptide antigens from Bacillus anthracis. J Virol Methods 141(2):146–153

Piazzolla G et al (2005) Immunogenic properties of a chimeric plant virus expressing a hepatitis C virus (HCV)-derived epitope: new prospects for an HCV vaccine. J Clin Immunol 25(2):142–152

Pickett G, Peabody D (1993) Encapsidation of heterologous RNAs by bacteriophage MS2 coat protein. Nucleic Acids Res 21(19):4621–4626

Pillet S et al (2018) Humoral and cell-mediated immune responses to H5N1 plant- made virus-like particle vaccine are differentially impacted by alum and GLA-SE adjuvants in a Phase 2 clinical trial, npj Vaccines. Springer US, 3, p. 3. doi:https://doi.org/10.1038/s41541-017-0043-3.

Pires DP et al (2016) Genetically Engineered Phages: a review of advances over the last. 80(3):523–543. https://doi.org/10.1128/MMBR.00069-15.Address

Pitek AS et al (2018) Elongated plant virus-based nanoparticles for enhanced delivery of thrombolytic therapies. Mol Pharmacol 14(11):3815–3823. https://doi.org/10.1021/acs.molpharmaceut.7b00559.Model.

Pokorski JK, Steinmetz NF (2011) The art of engineering viral nanoparticles. Mol Pharmacol 8(1):29–43. https://doi.org/10.1021/mp100225y.The.

Pokorski JK, Breitenkamp K, Finn MG (2012) Functional virus-based polymer-protein nanoparticles by atom transfer radical polymerization. J Am Chem Soc 133(24):9242–9245. https://doi.org/10.1021/ja203286n.Functional

Ponterio E et al (2013) Pattern of activation of human antigen presenting cells by genotype GII. 4 norovirus virus-like particles. J Transl Med 11:8–10

Prasuhn DJ et al (2007) Viral MRI contrast agents: coordination of Gd by native virions and attachment of Gd complexes by azide-alkyne cycloaddition. Chem Commun (12):1269–1271

Ren Y, Wong S, Lim L (2007) Folic acid-conjugated protein cages of a plant virus: a novel delivery platform for doxorubicin. Bioconjug Chem 18:836–843

Rohovie MJ, Nagasawa M, Swartz JR (2017) Virus-like particles: next-generation nanoparticles for targeted therapeutic delivery. (August 2016), pp 43–57. https://doi.org/10.1002/btm2.10049

Rose R et al (1998) Human papillomavirus type 11 recombinant L1 capsomeres induce virus-neutralizing antibodies. J Virol 72(7):6151–6154

Sangita V et al (2004) T=1 capsid structures of Sesbania mosaic virus coat protein mutants : determinants of T=3 and T=1 capsid assembly. J Mol Biol 342:987–999. https://doi.org/10.1016/j.jmb.2004.07.003

Sangita V, Lokesh GL et al (2005a) Research papers structural studies on recombinant T = 3 capsids of Sesbania mosaic virus coat protein mutants research papers. Acta Crystallogr D Biol Crystallogr D61:1402–1405. https://doi.org/10.1107/S0907444905024029

Sangita V, Satheshkumar PS et al (2005b) Research papers structure of a mutant T = 1 capsid of Sesbania mosaic virus: role of water molecules in capsid architecture and integrity research papers. Acta Crystallogr D Biol Crystallogr D61:1406–1412. https://doi.org/10.1107/S0907444905024030

Savard C et al (2011) Improvement of the trivalent inactivated flu vaccine using PapMV nanoparticles. PLoS One 6:E21522

Schlick T et al (2005) Dual-surface modification of the tobacco mosaic virus. J Am Chem Soc 127(11):3718–3723

Schneemann A et al (1993) Use of recombinant Baculoviruses in synthesis of morphologically distinct Viruslike particles of flock house virus, a Nodavirus. J Virol 67(5):2756–2763

Schoonen L et al (2018) Modular, bioorthogonal strategy for the controlled loading of cargo into a protein Nanocage. Bioconjug Chem 29:1186–1193. https://doi.org/10.1021/acs.bioconjchem.7b00815

Shanks M, Lomonossoff GP (2000) Co-expression of the capsid proteins of cowpea mosaic virus in insect cells leads to the formation of virus-like particles. J Gen Virol 81:3093–3097

Shenton W et al (1999) Inorganic–organic nanotube composites from template mineralization of tobacco mosaic virus. Adv Mater 11(3):253–256

Shukla S, Steinmetz NF (2016) Minireview emerging nanotechnologies for cancer immunotherapy. Exp Biol Med 241:1116–1126. https://doi.org/10.1177/1535370216647123

Shukla S et al (2014a) Increased tumor homing and tissue penetration of the filamentous plant viral nanoparticle potato virus X. Mol Pharmacol 10(1):33–42. https://doi.org/10.1021/mp300240m.Increased.

Shukla S, Dickmeis C et al (2014b) Molecular farming of fluorescent virus-based nanoparticles for optical imaging in plants, human cells and mouse models. Biomater Sci 2:784–797

Shukla S, Wen A et al (2014c) Presentation of HER2 epitopes using a filamentous plant virus-based vaccination platform. J Mater Chem B 2:6249–6258

Shukla S et al (2017) Plant viral nanoparticles-based HER2 vaccine: immune response influenced by differential transport, localization and cellular interactions of particulate carriers. Biomaterials 121:15–27

Singh P et al (2010) Bio-distribution, toxicity and pathology of cowpea mosaic virus nanoparticles in vivo. J Control Release 120(1–2):41–50. https://doi.org/10.1016/j.jconrel.2007.04.003. Bio-distribution

Skwarczynski M, Toth I (2016) Chemical science peptide-based synthetic vaccines. Chem Sci 7:842–854. https://doi.org/10.1039/C5SC03892H

Smith TJ et al (2000) The structure of cucumber mosaic virus and comparison to cowpea Chlorotic mottle virus. Am Soc Microbiol 74(16):7578–7586

Smith MT et al (2013) Enhanced protein stability through minimally invasive, direct, covalent, and site-specific immobilization. Biotechnol Prog 29(1):247–254. https://doi.org/10.1002/btpr.1671

Souli JC et al (1991) Immunogenicity and safety in newborns of a new recombinant hepatitis B vaccine containing the S and pre-S2 antigens. Vaccine 9(8):545–548

Staczek J et al (2000) Immunization with a chimeric tobacco mosaic virus containing an epitope of outer membrane protein F of Pseudomonas aeruginosa provides protection against challenge with P. aeruginosa. Vaccine 18(21):2266–2274

Steinmetz NF, Manchester M (2011) Viral nanoparticles. Pan Stanford Publishing Pte. Ltd., Singapore

Steinmetz NF et al (2010) Buckyballs meet viral nanoparticles – candidates for biomedicine. J Am Chem Soc 131(47):17093–17095. https://doi.org/10.1021/ja902293w.Buckyballs

Storni T et al (2002) Critical role for activation of antigen-presenting cells in priming of cytotoxic T cell responses after vaccination with virus-like particles. J Immunol 168:2880–2886

Storni T et al (2018) Nonmethylated CG motifs packaged into virus-like particles induce protective cytotoxic T cell responses in the absence of systemic side effects. J Immunol 172:1777–1785. https://doi.org/10.4049/jimmunol.172.3.1777

Strable E et al (2009) Unnatural amino acid incorporation into virus-like particles Erica. Bioconjug Chem 19(4):866–875. https://doi.org/10.1021/bc700390r.Unnatural.

Takahashi RU et al (2008) Presentation of functional foreign peptides on the surface of SV40 virus-like particles. J Biotechnol 135:385–392

Tinazzi E et al (2015) Plant-derived chimeric virus particles for the diagnosis of primary Sjögren syndrome. Front Plant Sci 6:1080. https://doi.org/10.3389/fpls.2015.01080

Tumban E et al (2011) A Pan-HPV vaccine based on bacteriophage PP7 VLPs displaying broadly cross-neutralizing epitopes from the HPV minor capsid protein, L2. PLoS One 6(8):e23310

Tyulkina LG et al (2011) New viral vector for Superproduction of epitopes of vaccine proteins in plants. Acta Nat 3(11):4–6

Uchida M et al (2012) Site-directed coordination chemistry with P22 virus-like particles. Langmuir 28(4):1998–2006

Uhde-Holzem K et al (2010) Immunogenic properties of chimeric potato virus X particles displaying the hepatitis C virus hypervariable region I peptide R9. J Virol Methods 166:12–20

Vaculik P et al (2015) New positions for peptide presentation in potato virus X capsid protein. Open Life Sciences 10:133–141. https://doi.org/10.1515/biol-2015-0019

van Eldijk MB et al (2013) Designing two self-assembly mechanisms into one viral capsid protein. J Am Chem Soc 134(16):18506–18509. https://doi.org/10.1021/ja308132z.Designing

Van Hest JCM, Cornelissen JJLM, Koay MST (2014) Using viruses as nanomedicines. Br J Pharmacol 171:4001–4009. https://doi.org/10.1111/bph.12662

Venkataraman S et al (2009) Structure of Seneca Valley Virus-001, an oncolytic picornavirus representing a new genus. Structure 16(10):1555–1561. https://doi.org/10.1016/j.str.2008.07.013.Structure

Venkataraman S, Savithri HS, Murthy MRN (2018) Structural aspects of plant viruses. In: Plant viruses diversity, interaction and management. CRC Press, Boca Raton, pp 3–26

Venter PA et al (2011) Multivalent display of proteins on viral nanoparticles using molecular recognition and chemical ligation strategies. Biomacromolecules 12:2293–2301

Vitiello C, Adhya MC (2005) An amino acid substitution in a capsid protein enhances phage survival in mouse circulatory system more than a 1000-fold. Virus Res 114(1–2):101–103

Wang J, Roden R (2013) Virus-like particles for the prevention of human papillomavirus-associated malignancies. Exp Rev Vaccin 12(2):129–141
Wang X et al (2012) Biomineralization-based virus Shell-engineering: towards neutralization biomineralization-based virus Shell-engineering: towards neutralization escape and tropism expansion. Adv Healthc Mater 1(July):443–449. https://doi.org/10.1002/adhm.201200034.
Wen AM, Steinmetz NF (2016) Design of virus-based nanomaterials for medicine, biotechnology, and energy. Chem Soc Rev 45(15):4074–4126. https://doi.org/10.1039/C5CS00287G
Wen AM et al (2013) Design rules for nanomedical engineering: from physical virology to the applications of virus-based materials in medicine. J Biol Phys 39:301–325. https://doi.org/10.1007/s10867-013-9314-z
Wen AM et al (2015) Tropism of CPMV to professional antigen presenting cells enables a platform to eliminate chronic infections. ACS Biomater Sci Eng 1(11):1050–1054
Wena AM, Steinmetza NF (2016) Design of virus-based nanomaterials for medicine, biotechnology, and energy. Chem Soc Rev:1–140
Whaley SR et al (2000) Selection of peptides with semiconductor binding specificity for directed nanocrystal assembly. Nature 405(6787):665–668
Wu Z et al (2013) Development of viral nanoparticles for efficient intracellular delivery. Nanoscale 4(11):3567–3576. https://doi.org/10.1039/c2nr30366c.
Yaddanapudi K, Mitchell R, Eaton J (2013) Cancer vaccines: looking to the future. Oncoimmunology 2:e23403
Yan D et al (2015) The application of virus-like particles as vaccines and biological vehicles. Appl Microbiol Biotechnol 99:10415–10432. https://doi.org/10.1007/s00253-015-7000-8
Yang CD et al (2007) Induction of protective immunity in swine by recombinant bamboo mosaic virus expressing foot-and-mouth disease virus epitopes. BMC Biotechnol. https://doi.org/10.1186/1472-6750-7-62
Yin Z, Nguyen HG, Chowdhury S, Bentley P, Bruckman MA et al (2012) Tobacco mosaic virus as a new carrier for tumor associated carbohydrate antigens. Bioconjug Chem 23:1694–1703
Yoo SY, Merzlyak A, Lee S (2014) Synthetic phage for tissue regeneration. Mediators of Inflammatio. Hindawi Publishing Corporation, 2014. https://doi.org/10.1155/2014/192790
Yusibov V, Rabindran S (2008) Recent progress in the development of plant – derived vaccines. Expert Rev Vaccin 7:1173–1183
Yusibov V et al (1997) Antigens produced in plants by infection with chimeric plant viruses immunize against rabies virus and HIV-1. Proc Natl Acad Sci 94:5784–5788
Yusibov V et al (2005) Peptide-based candidate vaccine against respiratory syncytial virus. Vaccine 23:2261–2265
Zeng Q et al (2013) Cucumber mosaic virus as drug delivery vehicle for doxorubicin. Biomaterials 34(19):4632–4642
Zhang W et al (2017) Encapsulation of inorganic nanomaterials inside virus-based nanoparticles for bioimaging. Nano 1(4):358–368. https://doi.org/10.7150/ntno.21384
Zilker C et al (2017) Nanoparticle-based B-cell targeting vaccines: tailoring of humoral immune responses by functionalization with different TLR-ligands. Nanomedicine 13(1):173–182

Advances in Phage Inspired Nanoscience Based Therapy

10

Nashwa Saleh El-Gazzar and Gamal Enan

Abstract

The resurgence of bacteria that are resistant to most of all available antibiotics is a major challenge faced by scientists recently. Further research needs to be compelled to decipher other novel therapy to kill such multidrug-resistant bacteria. This nature of bacteriophage and the mechanisms of phage infection of bacteria are discussed herein. Also, we examined an existing body or research indicating the potential for a widespread application of phages as treatment therapy for many bacterial infections. Phages could be exploited as templates in efforts to fabricate nanomaterials for diverse application. Thus, mixed therapy of both phages and nanomaterials are in development and are used recently in vivo. Application of phage therapy-based nanoscience in breast cancer is discussed herein.

Keywords

Phage therapy · Nano cancer · Nano therapy · Virus template nano-materials · Page display and page peptides

10.1 Overview

10.1.1 Therapy

The expression of cancer diseases occurs in various forms; for instance, there are at least various kinds of cancer that target the human breast. Thus, in this scenario, a perfect therapy is needed to get off "bad" cells which must be operative in terms of role and time.

For cure, conventionally a pharmaceutical medicine is given to protect body from the disease, but only when a pharmaceutical application is not effective. Other

N. S. El-Gazzar · G. Enan (✉)
Department of Botany and Microbiology, Faculty of Science, Zagazig University, Zagazig, Egypt

S. K. Saxena, S. M. P. Khurana (eds.), *NanoBioMedicine*,
https://doi.org/10.1007/978-981-32-9898-9_10

methods such as radioactive therapies are applied to remove sickness from the body. In some cases, surgery is needed in which an organ substitute is performed, called implantation. All approaches when used in combination selectively can eliminate disease in the long term. Nanotechnologies have heavy impact on modern drugs. In this regard, some have proven to be clinically effective and now are being introduced in present clinical market (Luisa and Sutherland 2013).

There have been many advances in development of new biologically active drugs and their delivery systems at specific site. DDS (Drug delivery systems) are not new; its research began in the mid-1960s and reached advancement until the present time. DDS can be in the form of pills which are repeatedly injected into the system using implants or other systems (Luisa and Sutherland 2013; Washington et al. 2016).

Structurally, biological macromolecules have a 3D nanoenvironment to mediate specific cell functions. The new drugs layout demands a very itemized conception of nanomedia. Therefore, to gain insight into macromolecule structure on the nanoscale is possible by NMR, X-ray, and electron microscopy for interaction with biological processes and developing new medicines.

10.1.2 Phage Therapy

Phage therapy deals with applications of bacteriophages which hit bacteria in treating bacterial infections. In this aspect, application of phages therapeutically began in the 1930s. It reached an advanced level in Pasteur Institute Paris (PIP) in the early twentieth century and later spread to Europe, Soviet Union, the United States and various parts of the world in short duration. Due to its chemical composition, it was rejected by the West in the 1940s; however, it is still being used in some countries. Nowadays, the resistance of bacteria to most of all obtainable antibiotics is accelerating. This chapter discusses phage therapy from the historical and ecological perspective and covers tools to be improved in future.

Highly specific phages are isolated and introduced specifically against bacterial infections. It is obvious that they play an effective role in maintaining the bacterial equilibrium naturally. In oceans at a moment, some oceanic bacteria are infected by replicating phages. This covers major portion of cycling of nutrients and prevents overgrowth of bacterial species. The phage therapy and therapeutic phage implementation do not have side effects. They also do not interact with other medications. The understanding of phage biology and ecology is well established. Clinical use of phage applications have contaminations recalcitrant for other available treatments in comparison to typically double-blind trials (Alavidze et al. 2007; Sulakvelidze et al. 2001).

There are some encouraging developments. The US FDA in 2006 approved phage preparations which target *Listeria monocytogenes* (Lang 2006). The phage therapy is a standard medical pursuit of Georgia Republic and available in Poland, Russia, and other countries of Eastern Europe. However, it is occasionally extended to Australia, Egypt and France on a compassionate basis. The naturopathic physicians in Washington, the United States, have expressed specifically the use of natural products already approved in others countries. Nestlé Corporation did a detailed study in Bangladesh using phage therapy for control of *Escherichia coli* by targeting diarrhea in children (Brüssow 2012). Simultaneously, two phage formulations such as novel cocktail T4-like phages were obtained from Bangladeshi infant patients along with Russian anti-*E. coli* phage (Microgen) in making oral rehydration solution (ORS).This will be checked only as a placebo. Pirnay et al. (2012) tried introducing phage therapy trials in Western pharmaceutical paradigm.

Presently, many resources are available (Abedon et al. 2011; Kutter et al. 2010, 2013).

10.1.3 Phage Therapy Today

In medical science, pre-clinical trials are done on animals. In case of anti-tumor drugs, experiments are conducted in mice along with other models. The phage therapy may be advantageous in veterinary science, and it has market value. But its efficacy needs to be tested before application on human subjects. A good variety of trials were used on calves, mice, and piglets by mixing coli-phages against pathogenic strains controlling diarrhea in animals. This caused diarrhea in calves severely and even other animals. Smith (1985) states phage treatment proved to be more effective in comparison to a series of antibiotics which were used as controls, as done by d'Herelle, six decades before protecting some calves from *E.coli.*

Smith (1985) conducted experiments along with controls that were reviewed. It stimulated a keen interest in the West. Phage therapy that controls the toxin produced by *E.coli* O157: H15 strain is known to be a result of the two lysogenic phages.

10.2 The History, Ecology, Structure, Functions and Properties of Bacteriophage

10.2.1 Historic Context: Discovery and Early Research

In 1917, Felix d'Herelle and Edward Twort independently published about isolating filterable entities which may infect bacteria and can frame parts or obvious "plaques" in bacterial cultures that were sufficiently diluted, which implies involvement of discrete particles. d'Herelle introduced the name *bacteriophage* for such entities that were collected from soldiers stools, when they had dysentery. In due course of time, he isolated phages which targets avian hyphosis. This infection up to 75% in French chickens. The untreated died. All infected phage-treated chickens survived. In Paris, dysentery was a big challenge at that time. Many assistants swallowed far more phages. The first phage treatment cured a child fully. After this treatment, several children were treated for therapeutic applications.

D'Herelle studied the biology of phage. In *The Bacteriophage: Its Role in Immunity*, d'Herelle (1922) wrote about the fundamentals of phages in the host describing specificity of multiplication and adsorption, and he commented on the role of phages for natural control in microbial infections. He noted that phages were very specific in control of disease organisms. He produced the first ever phage cocktails: Bacté-Intesti-Phage, Bacté-Coli-Phage, Bacté-Pyo-Phage, Bacté-Dysentérie-Phage and Bacté-Rhino-Phage, which were commercially successful. It continued in France till early 1990s when physicians adopted phage cocktails framed by d'Herelle's Lab or through Bacteriophage Service of Lyon and Paris Pasteur Institute branches (Abedon et al. 2011).

Major success of d'Herelle's several progresses was due to focus on phage biology. He wrote books on phage therapy and phages. The vision of phage therapy in practice got more publicity through a translated version of his old book published recently in the name *Bacteriophage 1* (2011).

Over the years, a working Institute of Bacteriophage Research in Tbilisi developed a world center for phage therapy which developed its experimental clinics comprising of scientific and industrial facilities. A campus was developed in 1926 in the woods of River Mtkvari for the project, where d'Herelle deposited large amounts of library materials, supplies, and equipment.

D'Herelle wanted to move to Georgia. The cottage constructed for him stands even now in the institute's campus. But in (1937), Eliava was arrested as "people's enemy" by Beria. d'Herelle was disillusioned and never returned to Georgia. However, their Bacteriophage Institute survived under the leadership of a group of women trained by Eliava and d'Herelle on bacteriophages for therapy and bacterial typing throughout the former Soviet Union. This institute was then shifted to All-Union Ministry of Health in 1951, which took an active role in providing bacteriophages. This was used for therapy and even bacterial typing throughout the former Soviet Union.

10.2.2 Basic Biology and Ecology of Bacteriophages

Bacterial viruses behave like spaceships which are capable of transferring their genomes to a new cell so as to multiply. In bacteriophages, bacteria cells are the target. Some phages are specific for bacterial strains, which target a few strains, while others may infect broader spectrum of strains. It may infect cases more than one species. They cannot multiply in the eukaryotic cells and none are known which can infect both (Gram-negative and Gram-positive) bacteria.

The virus contains genetic material (DNA or RNA) within a protein shell (Fig. 10.1). The bacteriophages have tails. They have tips that bind to one or more of molecules on target bacteria surface. The phage DNA molecule moves rapidly with the help of the tail to enter the host cell and produces the progeny phage. This provides new phage DNA template and transforms host cell to a factory for production of a particular phage. Electron microscopy shows phages have specific shape and sizes. DNA sequencing techniques have revealed several types of the genomes with a wide variety of organisms which provide remarkable strain of sorts and characters having potential therapeutic applications (Alavidze et al. 2007).

According to the literature, 95% of the investigated phages (Bezuidenhout 2017) belong to one of the following three-tailed morph types: podoviridae, stubby-tailed, siphoviridae, which have long tails (flexible) (Fig. 10.2). Myoviridae along with

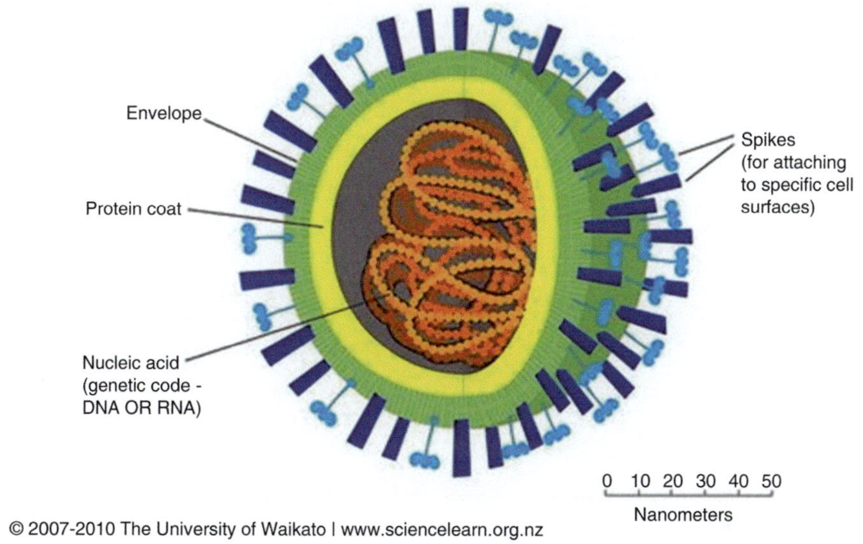

Fig. 10.1 Virus protein shell packed with the genetic material (RNA or DNA)

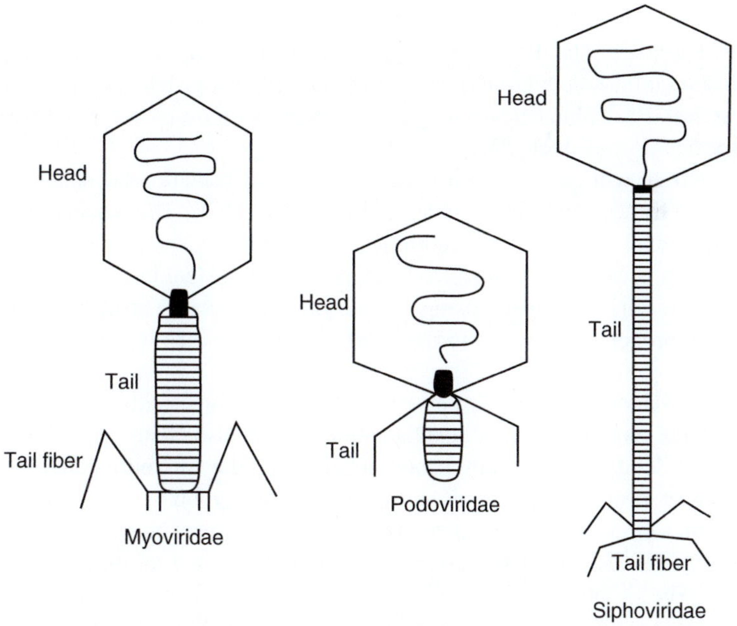

Fig. 10.2 Phages of myoviridae, podoviridae, and siphoviridae

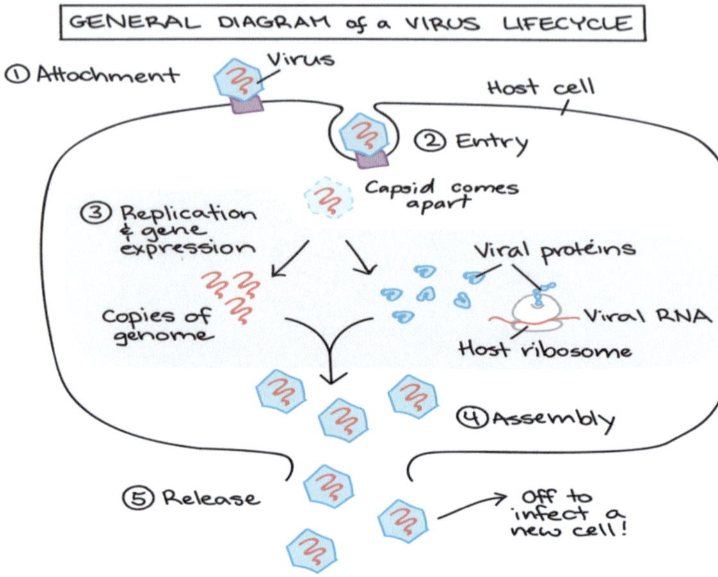

Fig. 10.3 The structure of a bacteriophage

tails consist outer contractile sheath and an inner tube which have connection with complex baseplate. The three morphotypes are obviously more ancient. The groups of each sorts infects both Gram-negative and Gram-positive bacteria. It also infects some Archae. Each phage genus only infects relatively most closely connected bacterial species (Pirnay et al. 2012).

One-step growth curve experiments in relation of bacteria and phages in lytic infections (Grimes et al. 2002) is an addition (Fig. 10.3). This covers an eclipse period in which DNA starts replication where no free phages are present in the cell. It has a latent period and lysis which freed phage to transfere into novel hosts. Life cycle of phage shows active closing or changing host roles. Through T4, approximately 300 progeny phages are formed within 30 min in each bacterial cell (Pirnay et al. 2012).

Bacteriophages (Greek for "eaters of bacteria") are also usually called as phages. A lytic phage can rapidly kill a host bacterial cell and also create copies of itself to attack and kill additional nearby bacterial cells until they are eliminated. The filamentous bacteriophages (Genus *Inovirus*) constitute a group of prokaryotic viruses that contain a circular single-stranded DNA genome encased in a long, flexible, and slender protein coat. Ff-like filamentous phages are known to utilize pili as their receptors. The simplicity of these viruses and the ease to manipulate made them an ideal model to study their structure and interactions. The Ff class of the filamentous bacteriophages (f1, fd, and M13) has been the most extensively studied. Different strains have been characterized, with DNA homology from very little to total, but they have a similar life cycle and structure virion. Filamentous bacteriophages infect mainly Gram-negative bacteria (Deng et al. 1999). Some of these phages are related to bacterial virulence. It is CTX which is best studied. It carries the ctxAB genes which codes for cholera toxin which is the most important virulence for *Vibrio cholerae* (Vafabakhsh et al. 2014). The phage gets attached to *V.cholerae* chromosome which is capable of infecting non-pathogenic strains of Vibrio which results from toxigenic lysogens (Waldor and Mekalanos 1996).

10.2.3 Phages Might Be an Alternative for Antibiotics

Loads of antibiotics are used in food, cattle, chickens, and so on. This results in the antibiotic-resistant isolates of bacteria. At the same time, outbreak of salmonella poisoning (sometimes lethal) occurs due to digestion of such meat. It is true for "rare hamburgers". It may be through vegetables that contamination of *E. coli* and Salmonella occur. The *E. coli* variant 0157:H7 is known to cause "hamburger disease." Children from 1 to 10 years having hemolytic uremic condition were treated with phage therapy in the 1960s, due to infection by *E. coli* which can result in kidney failures. A child gets infected after exhaustion of beverages or polluted food or undercooked ground beef. It is also due to unpasteurized juices, contaminated water, and even due to presence of infected people.

Salmonella serotype enteritis covers maximum outbreaks and related cases causing large number of deaths followed by *Listeria monocytogenes*. The presence of cocktail of phage might be useful in treating cases but it may be also applied against infected food items (Moody 1967). *Klebsiella pneumonia* results in pneumonia when immunity is compromised in individuals. It showed its presence in hospital-acquired infections. It can be stated that *Klebsiella* is a normal human flora which is usually not pathogenic.

Phages have potential to lyse *Klebsiella* infection. Experimental studies indicated that phage is very efficient in managing infections in guinea pigs and mice when inoculated through the nose or by injection. This resulted in no toxicity. A test was conducted in 1992 in humans with *Klebsiella* infected 109 patients treated with phage cocktail.

The phage preparations appeared to be effective and were non-toxic. It can be used in burns treatment. It is an acute problem in terms of severity in times of war and even terrorist attacks. Curing through phage in humans resulted in obvious progress (Nishima et al. 2011). The physicians in Georgia treated battle wounds with bandages dipped in anti-staph phage which resulted in healing of wounds in a short period of time. This outcome from Eliava Institute was marketed and has now spread all around the world (Motterlini and Otterbein 2010).

For replacing treatment with medicines and drug improvement requires a right phage to match the target. It needs prevention of phages which may lysogenize and may introduce toxic bacterial genes thereby increasing pathogenicity than killing bacteria. A recent methodology for selecting phage might be important which contains DNA sequencing for identification of genes precipitated in integration or any toxic genes, which can be done routinely and cheaply with DNA sequencing (Bradley and Wang 2015).

In case of antibiotics, phage resistance was found. This factor discouraged experiments in phage therapy. It is known that mutation resistance rate to phage is 100 times in comparison to resistance to antibiotics. The use of phage cocktail leads to various receptors on bacteria. It might avoid drug resistance development. This may be a new approach for treatment of HIV/ AIDS in which phage cocktail prevents resistance to drugs (O'Flaherty et al. 2009; Pirnay et al. 2012).

The discovery of bacteriophage was done 100 years back when d'Herelle used phage in preventing bacterial infections. Now by using genetic engineering technology one can engineer phages having broad spectrum or narrow spectrum. Three types of hurdles in improvement are as follows: to prevent infection in bacterium; to disguise restriction enzymes interfere of host and for clinical trials to obtain government approval. An alternative method is to use phage as a vaccine (O'Flaherty et al. 2009). This is done through commercial companies in curing viral diseases such as herpes. Also non-scientific reason may be to avoid an improved method (O'Flaherty et al. 2009). For using phage therapy, it is essential to purify phage through advanced technology for removal of bacterial debris. The mixture of phages against different organisms (cocktail) might be introduced in laboratories. Biotech companies are interested in their profit and patents. It is essential to patent the modifications introduced or made through use of antibiotics and their methods of use (Nishima et al. 2011).

10.3 General Approaches for Obtaining Assembled Phage Therapy Particles; Phage Therapy and Nanotechnology

10.3.1 Virus-Template Nanomaterials

Viruses are called supramolecular pathogenic machines which have discrete shape and size which proved effective against infection in host cells (Rossmann and Rao 2012). Because of their stability and symmetry, viral assemblies have been widely used as templates to manufacture nanomaterials for phage display with several applications (Wen and Steinmetz 2016), synthesis from metal nanoparticles (Ueno 2008); intracellular delivery (Ma et al. 2012; Wen and Steinmetz 2016) and energy transfer (Koudelka and Manchester 2010; Wen and Steinmetz 2016). The assembly of virus capsid protein to form virus parts such as bacteriophage M13, TMV, cowpea mosaic virus and cowpea chlorotic mottle virus are mostly applied in this case. In contrast, these viruses and tailed bacteriophages work as complex nanomachines. T4 worked best with bacteriophage (Arisaka et al. 2016). T4 bacteriophage creates more than 40 component proteins that frame variety assembly structures, for example, the tail fiber (Bartual et al. 2010; Granell et al. 2017), capsid head (Rao and Black 2010), the base plate (Taylor et al. 2016; Yap et al. 2016), and the sheath (Aksyuk et al. 2009). There is a needle puncturing cell in the center of bacteriophage T4 baseplate (Kanamaru et al. 2002).

Structural changes in component protein enable efficient genome injection into the host cells (Hu et al. 2015; Arisaka et al. 2016; Taylor et al. 2016). For instance, tail fiber in bacteriophage T4 acted as a binding receptor. Its head helps in storage of DNA and sheath is involved in contraction, while basoplate is associated or causes cell puncture. In bacteriophages, we normally focus on cell-puncturing needle. This puncture of cells may deliver molecular matter to living cells. These results in future designing of nanomachines that depend on alterations in natural supramolecular assemblies.

10.3.2 Nanotechnological Engineering of Bacteriophage T4 Component Proteins

The bacteriophageT4 component proteins can be used for formation of nanomaterials through chemical and genetic engineering. This might be used in different infection process, such as DNA transfer, recognition of target cells, cell puncturing. It can be achieved by which re-assembling three-dimensional structure may be recorded in vitro. This characterizes many examples that used the head, sheath, DNA packaging motor and the tail fiber for diverse implementations (Aksyuk et al. 2009; Sun et al. 2015; Taylor et al. 2016).

10.3.3 Head

The bacteriophage T4 head has a icosahedral (120 × 86 nm) structure which consists of main gene capsid protein and small outer capsid protein which is highly antigenic for gene (Hoc) (Rao and Black 2010). The head is associated with motor for DNA packaging.

It has a highly symmetric structure (Eric et al. 2003). It has DNA (Fig. 10.4) loading function which undergoes several applications, viz., transfer of genes at intracellular level along with proteins (Tao et al. 2013; Liu et al. 2014), cellular

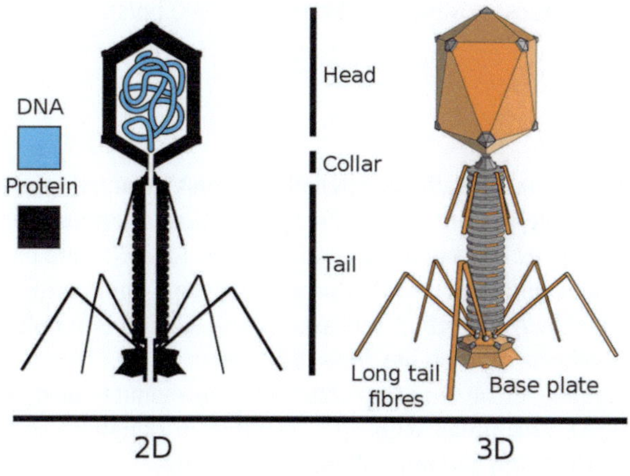

Fig. 10.4 Bacteriophage T4 genome with obvious structures of the head, the sheath, the tail fiber, and the base plate with the needle

imaging (Robertson et al. 2011), metal nanoparticles (Hou et al. 2010), and assembly of sensors (Archer and Liu 2009).

Rao and colleagues used bacteriophageT4 head for delivering DNA and proteins in the cells (Tao et al. 2013). DNA participated in packaging of head motor complex that runs through ATP. The Hoc-fused Soc proteins/peptides get involved on head surface. The changed heads gets fused to cells in a nonspecific way. This may be possible by identification of a receptor and the assembly. The surface associated proteins and participated DNA gets entrapped in cytosol.

The results of investigations recorded that head of T4 bacteriophage can be used in intracellular delivery of peptides, genes, proteins, and inspired nanoparticles.

Rao and colleagues used stable head of T4 bacteriophage highly symmetric to design 3-dimensional arrays of metal nanoparticle. Hou et al. (2010) proposed bacteriophage T4 nanomaterials which were based on the component proteins through genetic and chemical engineering (Hou et al. 2010; Robertson et al. 2011; Tao et al. 2013; Liu et al. 2014).

The metal nanoparticles (diameter 3.0–4.5 nm) take position on T4 bacteriophage head Cys and Met assembly on bacteriophage T4 head surface acts as S-donor. It consists of complex of metal ions. The electrocatalytic action of the Rh, Pt and, Pd assembly was increased through use of bacteriophage T4 head. Fe, Co, and Ni nanoparticles get assembled with bacteriophage T4 head in a similar way. This was because of participation of carboxylate groups on exterior surfaces at 8–9 pH range. The magnetic characters of the resultant metal nanoparticle–Ni/T4 head consists of super paramagnetic properties. Even Fe/T4 and Co/T4 heads showed ferromagnetic behavior.

The bacteriophage T4 DNA packaging motor consists of pentamer of gp17 (Gene porter protein phage) (large terminase protein) which is accumulated on a dodecamer of gp20 which served for translocation of DNA in bacteriophage T4 head through ATP-driven mechanism (Sun et al. 2015). This purified gp17 has DNA packaging action (Pires et al. 2016). The analysis of gp17 molecule revealed packaging of motor to be most powerful and fastest (~2000 bp/s).

10.3.4 Sheath

The bacteriophage T4 tail sheath has gp18 (138 copies) near tail tube (Aksyuk et al. 2009). It plays the role of contraction during infection. It drives tail tube by outer membrane. Both genetic and chemical modifications in gp18 were used to find conformational changes and structure that proceeds during contraction (Efimov et al. 2002). It is well known that gp18 self-assembles and forms tubular structures as poly-sheaths in spite of not having base tail tube or plate (Moody 1967). Daube et al. (2007) revealed self-assembling characters for manufacturing nano architectures and that gp18 assembled with gp15 resulted in doughnut-shaped structures having thickness of 20 nm and outer diameter of 50 nm.

10.3.5 Tail Fibers

The bacteriophage tail fibers T4 are of dimension 3 × 160 nm and are used in making an initial connection to host cells receptor molecules (Leiman et al. 2010). The fibers making tail are formed of gp37, gp36, gp35, gp34 which may develop proximal 70 nm (distal) half-fiber and proximal segments.

There are reports on crystal structures of the gp34 (744–1289 residues) C-terminal region (residues 785–1026) at tip of gp37 (residues 785–1026) (Bartual et al. 2010; Granell et al. 2017). Hyman et al. (2002) developed mesoscale (structures) by use of self-assembly of tail fibers. At initial stage, gp37sequence of T4 bacteriophage was removed partially for formation of mutated phage having short tail fibers which does not affect the structure and even function.

10.3.6 Cell Penetration Occurs Through Protein Needle Motifs

The gp27–gp5 proteins which are placed on base plate of T4 bacteriophage help in cell-puncturing for sneaking to outer membrane of *Escherichia coli* (Keten et al. 2011). The gp27 trimer develops a 5-nm-long cylinder (hollow). The gp5 trimer forms 14-nm-long needle to form the handle structure. The C-terminal end of (gp5)3 develops a robust needle which is made of 3-stranded β-helix. This is framed through VXGXXXXX repeat sequence (Kanamaru et al. 2002). The structure of gp5.4 is now projected. This protein projects at the terminal end of gp5 (Shneider et al. 2013). The gp5.4 and β-helix of C-terminal gp5 during infection connects with outer cell membrane and punctures it.

The value of β-helix gp5 as membrane puncturing has been projected through many methods. Using molecular dynamics and simulations, Buehler and colleagues recorded mechanical stiffness in three stranded β-helix in gp5 which is greater than protein motifs, namely, 1-stranded β-helix and cellular membrane, and gp5 is suitable in cell penetration (Keten et al. 2011). Kitao and colleagues enhanced gp5 penetration in lipid bilayer which comprised of dioleylphosphatidyl ethanolamine (Nishima et al. 2011). Molecular dynamics was obtained through application of force on top of gp5 which will be an attempt to mimic effect of force resulted by bacteriophage T4 sheath. This enhancement projected gp5 β-helix to help develop hole in membrane through screwing motion. This serves in charge interactions strongly with the charged side chains on β-helix surface to raise the lipids upward (Sanghamitra et al. 2014).

10.3.7 Penetration

Caudovirales based on tail morphology also known as tailed phages may be divided into three families, viz., Siphoviridae, Myoviridae, and Podoviridae (Xu and Xiang 2017). Myoviridae phages (T4, P2, ϕ92, Mu) have a common puncturing needle connected to membrane which consists of oligosaccharide binding (OB-fold)

N-terminal oligonucleotide domain having C-terminal tip domain and β-helix (Browning et al. 2012; Harada et al. 2013). The needles are connected to the host cell through fusion of the C-terminal (tip domains). Gp5.4 of T4 bacteriophage gets stabilized through coordination of 3 His residues with a Fe ion. Similarly, in case of gp138 of bacteriophage ϕ92, the tip is stabilized through connection of 6 His with Fe ion (Browning et al. 2012; Shneider et al. 2013). A robust β-PN (β-helical protein needle) was constructed by entering the β-helix domain of gp5 with folding of T4 bacteriophage (Shneider et al. 2013). From C terminal of β-helix of gp5, (Harada et al. 2013) recorded structures of trimer–dimer needle.

Nanomaterials Based on Gp5

For construction of a supramolecular assembly, needle structure of gp5 was formed in different ways (Sugimoto et al. 2006; Ueno et al. 2006).

Through coordination with gold ions in His 6 gp5 triad and reduction through NaBH4, gp5 in tetrapod assembly gold nanoclusters are formed. The tetrapod frame up is constructed selectively which is based upon electrostatic repulsion of negatively (−) charged area in C termini of gp5 (Sugimoto et al. 2006). The cup-shaped look of gp5- gp27 was used as nano environment in catalytic reactions.

It was observed that the cup structure (gp27–gp5) underwent to form an energy transfer system for introducing a cystein residues. FRET (Highly efficient fluorescence resonance energy transfer) appeared through dual modification in donor fluorescein (Fl) molecule and an acceptor TMR (tetra methyl 4 rhodamine) molecule to introduce gp5and gp27cysteine residues. In comparison to this alteration, in sites of TMR and Fl it was observed to stimulate significant fluorescence self-quenching of TMR.

CO Delivery

It has been observed that β-PN intracellularly delivers CO (carbon monoxide) (Inaba et al. 2015a). CO behaves as gas signaling and serves as anti-proliferative, anti-inflammatory having even anti-apoptotic effects (Motterlini and Otterbein 2010). CO releasing molecules (CORMs) act as macromolecular carriers in systems intracellular CO delivery efficiently (Inaba et al. 2015b).

Mostly CORMs are formed through carbonyl metal complexes transition. It has been observed that in CO delivery strategy, Ruthenium carbon monoxide bond Ru (CO)2 moieties get induced on tips of β-PN through reaction of triads of (His)6 at the C-terminals with Ru(CO)3C12]2 (CORM-2) (Inaba et al. 2015a).

Protein Delivery

Gp5 is not an artificial carrier and has been used in protein delivery; since the three lysozyme domains of gp5 got cleaved after inoculation, they spread in periplasm, utilize inner membrane *E. coli* (Kanamaru et al. 2002, 2005). We can expect that β-PN might be useful as carrier in intracellular delivery of exogenous proteins. This is the genetic fusion with β-PN through which sfGFP (Super folder green fluorescent protein sfGFP) is delivered (Inaba et al. 2014).

The PLAL pulsed laser ablation in liquids was broadly used in fabrication of NPs of morphology, various characters and size by not generating any external impurities. Instead, PLAL got improved to be successful in immobilizing native NPs by different fused biomolecules such as peptides, proteins and oligonucleotides (Scibilia et al. 2016). The probes have demerits, viz., folding instability, sensibility to changes, raised synthesis, production cost. The "phageprobes" was used in numerous fields as valid substitute for antibodies or peptides (Lentini et al. 2015). Phageprobes get associated with phage display technology and allow use of random foreign peptide sequences in coat proteins of bacteriophages. They serve as a foreign peptide. Now this is possible for selection of one or more phage clones which would recognize and bind to their target (De-Plano et al. 2017).

The SiNPs (Silicon nanoparticles) are used as expected nanoplatform. This is because of optical properties, high biocompatibility and specificity. Silicon nanoparticles have quantum effects in functioning of biomaterials. This helps bio-functional to M13-engineered bacteriophage and SiNPs which selectively recognize peripheral blood mononuclear cells (PBMC) (Laura et al. 2018).

Basic Principles Correlated to Virus and Nanoparticles in Relation to Auto-assembly; Phage Display with Nanoparticles Paul Ehrlich firstly developed selective targeting of diseased cells for minimizing toxic side effects. If a compound selectively targets a diseased cell, then the toxin gets attached to its targeting moiety. It works as a "magic bullet" which kills only the diseased cells. Antibodies having high affinity have been tested extensively in selective targeting and cancer imaging. Some have been successfully tested, such as 90Y-labeled Zevalin and 131I-labeled Bexxar for CD20-targeted radiolabeled (monoclonal antibodies) in treatment of lymphoma follicular non-Hodgkin's. However, antibodies are slow in tumor penetration, continue for long time, and blood has slow rate of clearance by the hepatobiliary system. Some even show adverse hematological effects and have resistance (Klatersky 2006). Different peptides are being tested for selective targeting because of faster uptake rates and rapid blood clearance. They have high tissue penetration and easy diffusion passes by urinary tract and are not immunogenic. They are easily and cost effectively synthesized.

Most peptides being developed are modifications of natural peptides while some are derived from libraries to find suitable peptides. Recently, the use of phage display that links pheno type to genotype has been tried to develop peptides at any target, holding high affinity targets in patient treatments.

Phage display takes R-DNA technology to form bacteriophages having desired peptide on protein shells. Target peptide agonists and antagonists may be identified experimentally. This enables antibodies engineering and peptides possibility as new drugs. Recent research is on use of phage as nanoparticle imaging systems 142 (Deutscher 2010). Display of phage was discovered by George Smith in which the short peptides may be displayed by fusion of phage coat proteins (Smith 1985). Advantages of phage display are direct link from phenotype to genotype. Through this, library of 109–1011 clones may be formed in a couple of weeks (Pires et al. 2016).

Small foreign peptides are inserted in minor coat protein III (cpIII) N-terminus of fd phage to display five copies of peptide (Smith 1985). These libraries are used to screen unwanted non-specific binders for obtaining peptides. They have high affinity toward the desired target. The advantages of *in vivo* phage selection over *in situ* or *in vitro* methods selection is that it occurs in the same complex milieu. Here selected peptide selectively targets and hits unknown targets. Thus, libraries may be screened for human tumor cells to create peptides for individualized tumor cells.

Galectin-3 is b-galactoside-binding lectins family that binds terminal galactopyranose on carbohydrates, especially the Thomas Friedenreich (TF) antigen. TF is a disaccharide (Ga1b1-3Ga1NAc) antigen and have poor expression on surfaces of most adenocarcinom (Cathy et al. 2013).

Phage Display of Selected Peptides in Prostate and Breast Cancers A peptide that is avid to galectin-3 was extracted to be identified from bacteriophage display (Zou et al. 2005). The peptide has 16 amino acids sequence, ANTPCGPYTHDCPVKR, known as G3-C12. This was linked to DOTA chelator by a tri-peptide linker called Gly-Ser-Gly (GSG). DOTA (GSG)-G3-C12 was labeled successfully in *in vivo* selection of tumor phage. Previous studies had illustrated the development of DTDTPA gold nanoparticles in molecular imaging of 144 C. S with 111In, and the radio-labeled peptide tested in human MDA-MB-435 (breast tumor-bearing model) (Jemal et al. 2011). Previous records demonstrated tumor uptake of 1.2 ± 0.24%, 0.75 ± 0.05%, and 0.6 ± 0.04% ID/g at 0.5, 1.0, and 2 h post injection. The investigations with the non-radio-labeled peptide had 52% reduction at 2 h post injection but have major drawback of 22% high kidney uptake.

The quantitative measurements of small tumors have been hardly detected by SPECT. In a study, the ability to label DOTA(GSG)-G3-C12 with 68Ga was used as an alternative to 111InSPECT imaging. Previous results revealed that 68Ga-DOTA (GSG)-G3-C1 could be successfully framed in 0.1 M ammonium acetate (pH 5) at 100 °C for 30 min. The labeling power was 52%, that is, having an impure mixture of peptide. Additionally, microwave labeling techniques was developed (Cantorias et al. 2009). Yields were similar to the reaction completed in 1 min with additional time for HPLC, normally 15 min. The HPLC-purified peptide was tested further for stability at room temperature. Radio-TLC and HPLC study demonstrated that the 68Ga-labeled peptide remains intact up to 85% up to 4 h.

10.4 Development of Nanomaterials with Bacteriophage Chemical and Genetic Strategies

The components of bacteriophageT4 have been studied for application as scaffolds and molecular templates. The bacteriophage T4 components can synthesize various nanomaterials. It acts as three-dimensional molecular transporters, cell-penetrating materials, and biocatalysts as β-helical protein needle (β-PN) that develops β-helix domain of gp5 in bacteriophage T4. β-PN delivers molecular cargos to living cells and acts as lysozyme. The studies recorded that component proteins may be formed through natural bio nanomachines yet preserving their original actions. The studies on bacterial secretion systems, bacteriocins, cell puncturing needles of bacteriophages and protein injection systems forcibly optimized for cell puncture. It may be useful as cell-penetrating agent (Migliori et al. 2014). The component structures of T4 bacteriophage, viz., head, sheath, tail fiber, baseplate may also be conserved in other bacteriophages (Brackmann et al. 2017; Veesler and Cambillau 2011). The various infection mechanisms in other tailed bacteriophages may also present useful molecular scaffolds (Xu and Xiang 2017). This can predict the work of unknown gene products. Bacteriophage T4 is the natural nanomachines inspiring to improve dynamic functions through extraction of various component structures.

10.4.1 Phage Display Technology and Phage Peptide Library

Smith (1985) pioneered fusing foreign protein genes in the DNA of filamentous phage. The coat protein and foreign protein projected together on outer portion of phage surface and called phage display technology. This united phenotype and genotype into one phage particle. This technology displays the random oligonucleotide fragment sequences and helps in synthesizing them. This put them at tail portion of the flagellar genes of filobacti-virus. Thus, each phage clone may result into a singular peptide for per foreign gene cloned. These phages may comprise library of phage peptides.

10.5 Phage Therapy and Nanomedicine

10.5.1 Phage Display Technology vis-a-vis Breast Cancer

Breast cancer comes in the category of serious diseases. Breast cancer threatens women's health as a top killer. Biological therapy is a powerful tool to cure breast cancer. Phage display technique has many merits in this field. To cure this more and more, peptides have been investigated which projected new antitumor drugs and vaccines (Jemal et al. 2011). Various therapy, viz., radiotherapy, chemotherapy, hormone therapy and targeted cancer therapy are useful in treating breast cancer. They are highly priced. They have also many drawbacks such as human anti-mouse antibody (HAMA) and reaction of regular mAbs which limits its use in clinical trials. Due to this reason, application of phage display technology has potential to separate peptides which bond sensitively to tissues or cells in the patients and result in improved cancer vaccines and newly targeted drugs for breast cancer.

The serum, cancer cells, or endotheliocyte of breast cancer just like other cancer patients may express molecular markers that have poor expression like HER2, bFGF (Li et al. 2012), CD44 (Park et al. 2012). As long as peptides expressed are plenty enough, some more phages may bond to these targets. It has been observed that these markers, which are expressed poorly, may be hard to purify.

10.6 Artificial Bio-nanomachines, a Recent Design Based on Protein Needles from Bacteriophage T4Biophys

T4 Bacteriophage is a natural bio-nanomachine that realizes potent effects on host cells through mutual activity of specific 3D architectures of protein. The relationships between dynamics and protein structures have recently revealed (Kostyuchenko et al. 2003; Sun et al. 2015) with the design in fabrication of nanomachines through bacteriophage T4 component proteins. The gp5 is a needle protein which is present in medium as a baseplate of contractile tail of bacteriophageT4. This punctures host cells. The studies recorded that it has a common motif useful in cell puncture even in injection systems, for example, T6SS. The artificial needle which is based on gp5 β-helical domain is capable of penetrating cells. This may be engineered in delivering various cargos in living cells. The unparalleled bacteriophage T4 components and nanomachines (natural) mainly for employing as molecular scaffolds in efforts to fabricate new bio nanomachines for cell penetration (Sun et al. 2015), the sheath (Aksyuk et al. 2009), the tail fiber (Granell et al. 2017), and the baseplate (Kostyuchenko 2003).

10.6.1 Association, Engineering, and Implementation of Virus-Based Protein Nanoparticles

The *Enterobacteriaceae* covers a group of Gram-negative facultative anaerobic bacteria causing infections, viz., bacteremia, septic arthritis, endocarditis, osteomyelitis, lower respiratory tract, skin and urinary tract, intra-abdominal and ophthalmic infections not only in humans and animals but also in poultry and fish. They cause millions of death every year, leading to extreme economic losses. Medicinal treatment is necessary and it is an effective method to prevent *Enterobacteriaceae* diseases. Due to abuse of antibiotics, drug resistance is reported in *Enterobacteriaceae* infections. Therefore, to find new ways of their control is highly needed. Phage therapy is highly potential, and it can be used instead of antibiotics as an alternate antibacterial mechanism.

Phage therapy is superior to antibiotics on the following lines. The phages multiply specifically in large numbers at host locations during bacterial killing and participate in establishing the phage dose. Besides cost, phage formation is relatively low (Abedon and Thomas-Abedon 2010). Phages have specific host that directs them to split targeted pathogenic bacteria with minimal effect of the normal flora. In comparison, chemical antibiotics show a broad spectrum which may induce super infections (Haellman and Fussenegger 2016).

As against antibiotics, no phages offer toxicity to flora and medium (Haellman and Fussenegger 2016). However, phage therapy also has certain limitations, majorly it is not safe. All phages are characterized for fitness before they are used for clinical trial. The rapid improvements in sequencing technologies can be proved for safety of phages use in therapy. The other problem is that it is a rigid host but this may be annulled through improved phage cocktails. Another limitation is the instability of the phage as a therapeutic agent. Wide range of studies are essential to improve upon this factor, besides developing phase-resistance bacteria during their growth with phages. It can also be mentioned that result may not be same in future. The bacteria will develop multi-phage resistance. Thus, there is an urgent need for improving this facet of phage therapy with the help of nanotechnology.

10.7 Phage Therapy and the Future

Apart from international websites that promote phage therapy, there are at least a dozen companies that develop phage therapy, including two using phage collection at Eliava Institute (Georgia). In the United States, Intralytex (a company in Maryland) got grants from USDA and U.S. Army for development of phage therapy. They are using phage as a probiotic for shigella infection among soldiers. They also used phages for preventing food-borne diseases caused by salmonella and pseudomonas. Another company participated is Gangagen, introducing phages and phage products (lysins) against *Staphylococcus auries*. Nestlé Food Corporation is involved in testing phages against strains of *E. coli* and is currently curing childhood diarrhea. There are plenty of data that phage therapy would be an alternative for antibiotics in curing various diseases and also antibiotic-resistant bacteria (Haellman and Fussenegger 2016).

Viruses and their protein capsids act as biologically produced nanomachines with multiple, complex biological functions through their mechano-chemical actions during the infection cycle. The merits of nanoscience and nanotechnology have opened up large number of new possibilities to exploit engineered viral capsids as protein-based nanoparticles for various nanotechnological or biomedical and biotechnological applications.

Various sections provide outlines on: (1) The historic, the ecology the structure, properties and functions of bacteriophage (2) general approaches for getting assembled phages for therapeutic uses; (3) basic principles and biochemical events in assembly of virus nanoparticles; (4) chemical and genetic strategies for engineering the particles; (5) Improved phages for therapy and nanomedicine in practice (6) Genetic improvement by engineering of phages to alter their properties and suitability for various application.

References

Abedon ST, Thomas-Abedon C (2010) Phage therapy pharmacology. Curr Pharm Biotechnol 11:28–47

Abedon S, Kuhl S, Blasdel B, Kutter E (2011) Phage treatment of human infections. Bacteriophage 1:66–85

Aksyuk AA, Leiman PG, Kurochkina LP, Shneider MM, Kostyuchenko VA, Mesyanzhinov VV, Rossmann MG (2009) The tail sheath structure of bacteriophage T4: amolecular machine for infecting bacteria. EMBO J 28:821–829. https://doi.org/10.1038/emboj.2009.36

Alavidze Z, Meiphariani A, Dzidzishvili L, Chkonia I, Goderdzishvili M, Kvatadze N, Jgenti D, Makhatadze N, Gudumidze N, Gvasalia G (2007) Treatment of the complicated forms of infl amatory wounds by bacteriophages. Med J Ga 2:123–127

Archer MJ, Liu JL (2009) Bacteriophage T4 nanoparticles as materials in sensor applications: variables that influence their organization and assembly on surfaces. Sensors 9:6298–6311. https://doi.org/10.3390/s90806298

Arisaka F, Yap ML, Kanamaru S, Rossmann MG (2016) Molecular assembly and structure of the bacteriophage T4 tail. Biophys Rev 8:385–396. https://doi.org/10.1007/s12551-016-0230-x

Bartual SG, Otero JM, Garcia-Doval C, Llamas-Saiz AL, Kahn R, Fox GC, van Raaij MJ (2010) Structure of the bacteriophage T4 long tail fiber receptor-binding tip. Proc Natl Acad Sci USA 107:20287–20292. https://doi.org/10.1073/pnas.1011218107

Bezuidenhout CC (2017) Stubby- tailed podoviridae, the siphoviridae with long, often flexible tails and the myoviridae, ZAGPPHC, 332; 808/8/15.

Brackmann M, Nazarov S, Wang J, Basler M (2017) Using force to punch holes: mechanics of contractile nanomachines. Trends Cell Biol 27:623–632. https://doi.org/10.1016/j.tcb.2017.05.003

Bradley RW, Wang B (2015) Designer cell signal processing circuits for biotechnology. New Biotechnol 32:635–643

Browning C, Shneider MM, Bowman VD, Schwarzer D, Leiman PG (2012) Phage pierces the host cell membrane with the iron-loaded spike. Structure 20:326–339. https://doi.org/10.1016/j.str.2011.12.009

Brüssow H (2012) What is needed for phage therapy to become a reality in Western medicine? Virology 434(2):138–142

Cantorias M, Figueroa SD, Quinn TP, Lever JR, Hoffman TJ, Watkinson LD, Carmack TL, Cutler CS (2009) Development of a high specific activity 68Ga-labeled DOTA-Rheniumcyclized a-MSH peptide peptide analog to target MC1 receptors overexpressed by melanoma tumors. Nucl Med Biol 36(5):505–513

Cathy SC, Chanda N, Shukla R, Sisay N, Cantorias M, Zambre A, McLaughlin M, Kelsey J, Upenandran A, Robertson D, Deutscher S, Kannan R, Katti K (2013) Nanoparticles and phage display selected peptides for imaging and therapy of cancer. Recent Results Cancer Res 194. https://doi.org/10.1007/978-3-642-27994-2_8. Springer, Berlin Heidelberg

Daube SS, Arad T, Bar-Ziv R (2007) Cell-free co-synthesis of protein nanoassemblies: tubes, rings, and doughnuts. Nano Lett 7:638–641. https://doi.org/10.1021/nl062560n

De Laura MP, Scibilia S, Rizzo MG, Crea S, Franco D (2018) One-step production of phage–silicon nanoparticles by PLAL as fluorescent nanoprobes for cell identification. Applied Physics A 124:222

De Plano LM, Carnazza S, Messina GML, Rizzo MG, Marletta G, Guglielmino SPP (2017) Specific and selective probes for *Staphylococcus aureus* from phage-displayed random peptide libraries. Colloids Surf B Biointerfaces 157:473–480

Deng W, Leaper K, Bownes M (1999) A targeted gene silencing technique shows that Drosophila myosin VI is required for egg chamber and is maginalsic morphogenesis. J Cell Sci 112(21):3677–3690

Deutscher SL (2010) Phage display in molecular imaging and diagnosis of cancer. Chem Rev 110:3196–3211. https://doi.org/10.1021/cr900317

Efimov AV, Kurochkina LP, Mesyanzhinov VV (2002) Engineering of bacteriophage T4 tail sheath protein. Biochemistry (Mosc) 67:1366–1370. https://doi.org/10.1023/A:1021857926152

Eric SM, Kutter E, Mosig G, Arisaka F, Kunisawa T, Rüger W (2003) Microbiology and Molecular Biology Reviews – American Society for microbiology. https://doi.org/10.1128/MMBR.67.1.86-156

Granell M, Namura M, Alvira S, Kanamaru S, van Raaij MJ (2017) Crystal structure of the carboxy-terminal region of the bacteriophage T4 proximal long tail fiber protein gp34. Viruses. https://doi.org/10.3390/v9070168

Grimes S et al (2002) Bacterial infections by bacteriophage with a schematic structure of a bacteriophage. Adv Virus Res 58:255

Haellman V, Fussenegger M (2016) Synthetic biology—toward therapeutic solutions. J Mol Biol 428:945–962

Harada K, Yamashita E, Nakagawa A, Miyafusa T, Tsumoto K, Ueno T, Toyama Y, Takeda S (2013) Crystal structure of the C-terminal domain of mu phage central spike and functions of bound calcium ion. Biochim Biophys Acta 1834:284–291. https://doi.org/10.1016/j.bbapap.2012.08.015

Herelle F (1917) An invisible microbe that is antagonistic to the dysentery bacillus. CR Acad Sci Paris 165:373–375.3

Herelle F (1922) The bacteriophage: its role in immunity. Williams and Wilkins, Baltimore

Hou L, Gao F, Li N (2010) T4 virus-based toolkit for the direct synthesis and 3D organization of metal quantum particles. Chem Eur J 16:14397–14403. https://doi.org/10.1002/chem.201000393

Hu B, Margolin W, Molineux IJ, Liu J (2015) Structural remodeling of bacteriophage T4 and host membranes during infection initiation. Proc Natl Acad Sci USA 112:E4919–E4928. https://doi.org/10.1073/pnas.1501064112

Hyman P, Valluzzi R, Goldberg E (2002) Design of protein struts for self assembling nanoconstructs. Proc Natl Acad Sci USA 99:8488–8493. https://doi.org/10.1073/pnas.132544299

Inaba H, Sanghamitra NJM, Fukai T, Matsumoto T, Nishijo K, Kanamaru S, Arisaka F, Kitagawa S, Ueno T (2014) Intracellular protein delivery system with protein needle–GFP construct. Chem Lett 43:1505–1507. https://doi.org/10.1246/cl.140481

Inaba H, Sanghamitra NJM, Fujita K, Sho T, Kuchimaru T, Kitagawa S, Kizaka-Kondoh S, Ueno T (2015a) A metal carbonyl-protein needle composite designed for intracellular CO delivery to modulate NFkB activity. Mol Bio Syst 11:3111–3118. https://doi.org/10.1039/c5mb00327j

Inaba H, Fujita K, Ueno T (2015b) Design of biomaterials for intracellular delivery of carbon monoxide. Biomater Sci 3:1423–1438. https://doi.org/10.1039/C5BM00210A

Jemal A, Bray F, Center MM (2011) Global cancer statistics. CA Cancer J Clin 61:69–90

Kanamaru S, Leiman PG, Kostyuchenko VA, Chipman PR, Mesyanzhinov VV, Arisaka F, Rossmann MG (2002) Structure of the cell-puncturing device of bacteriophage T4. Nature 415:553–557. https://doi.org/10.1038/415553a

Kanamaru S, Ishiwata Y, Suzuki T, Rossmann MG, Arisaka F (2005) Control of bacteriophage T4 tail lysozyme activity during the infection process. J Mol Biol 346:1013–1020. https://doi.org/10.1016/j.jmb.2004.12.042

Keten S, Xu Z, Buehler MJ (2011) Triangular core as a universal strategy for stiff nanostructures in biology and biologically inspired materials. Mater Sci Eng C 31:775–780. https://doi.org/10.1016/j.msec.2011.01.004

Klatersky J (2006) Adverse effects of the humanized antibodies used as cancer therapeutics. J Curr Opin Oncol 18:316

Kostyuchenko VA, Leiman PG, Chipman PR, Kanamaru S, van Raaij MJ, Arisaka F, Mesyanzhinov VV, Rossmann MG (2003) Threedimensional structure of bacteriophage T4 baseplate. Nat Struct Biol 10:688–693. https://doi.org/10.1038/nsb970

Koudelka KJ, Manchester M (2010) Chemically modified viruses: principles and applications. Curr Opin Chem Biol 14:810–817. https://doi.org/10.1016/j.cbpa.2010.10.005

Kutter E, De Vos D, Gvasalia G, Alavidze Z, Gogokhia L, Kuhl S, Abedon ST (2010) Phage therapy in clinical practice: treatment of human infections. Curr Pharm Biotechnol 11:69–86

Kutter E, Borysowski J, Międzybrodzki R, Górski A, Kutateladze M, Alavidze Z, Goderdzishvili M, Adamia R (2013) Clinical phage therapy. In: Borysowski J et al (eds) Phage therapy: current research and applications. Caister Academic Press (in press)

Lang LH (2006) FDA approves use of bacteriophages to be added to meat and poultry products. Gastroenterology 131(5):1370

Leiman PG, Basler M, Ramagopal UA, Bonanno JB, Sauder JM, Pukatzki S, Burley SK, Almo SC, Mekalanos JJ (2010) Type VI secretion apparatus and phage tail-associated protein complexes share a common evolutionary origin. Proc Natl Acad Sci USA 106:4154–4159. https://doi.org/10.1073/pnas.0813360106

Lentini G, Fazio E, Calabrese F, De Plano LM, Puliafico M, Franco D, Nicolff MS, Carnazza S, Trusso S, Allegra A, Neri F, Musolino C, Guglielmino SP (2015) Phage-AgNPs complex as SERS probe for U937 cell identification. Biosens Bioelectron 74:398–405

Liu JL, Dixit AB, Robertson KL, Qiao E, Black LW (2014) Viral nanoparticle-encapsidated enzyme and restructured DNA for cell delivery and gene expression. Proc Natl Acad Sci USA 111:13319–13324. https://doi.org/10.1073/pnas.1321940111

Luisa F, Sutherland D (2013) Principles, Applications, Implications and Hands-on Activities EUROPEAN COMMISSION Directorate-General for Research and Innovation Industrial technologies (NMP) Unit G.4 — Nano Sciences and Nano Technologies *European Commission B-1049 Brussels*

Ma Y, Nolte RJM, Cornelissen JJLM (2012) Virus-based nanocarriers for drug delivery. Adv Drug Deliv Rev 64:811–825. https://doi.org/10.1016/j.addr.2012.01.005

Migliori AD, Keller N, Alam TI, Mahalingam M, Rao VB, Arya G, Smith DE (2014) Evidence for an electrostatic mechanism of force generation by the bacteriophage T4 DNA packaging motor. Nat Commun 5:4173. https://doi.org/10.1038/ncomms5173

Moody MF (1967) Structure of the sheath of bacteriophage T4: I. Structure of the contracted sheath and poly sheath. J Mol Biol 25:167–200. https://doi.org/10.1016/0022-2836(67)90136-2

Motterlini R, Otterbein LE (2010) The therapeutic potential of carbon monoxide. Nat Rev Drug Discov 9:728–743. https://doi.org/10.1038/nrd3228

Nishima W, Kanamaru S, Arisaka F, Kitao A (2011) Screw motion regulates multiple functions of T4 phage protein gene product 5 during cell puncturing. J Am Chem Soc 133:13571–13576. https://doi.org/10.1021/ja204451g

O'Flaherty S, Ross RP, Coffey A (2009) Bacteriophage and their lysins for elimination of infectious bacteria. FEMS Microbiol Rev 33(4):801–819

Park HY, Lee KJ, Lee SJ (2012) Screening of peptides bound to breast cancer stem cell specific surface marker CD44 by phage display. Mol Biotechnol 51:212–220

Pires DP, Cleto S, Sillankorva S, Azeredo J, Lu TK (2016) Genetically engineered phages: a review of advances over the last decade. Microbiol Mol Biol Rev 80:523–543

Pirnay JP, Verbecken G, Rose R, Jennes S, Zizl M, Huys I, Lavigne R, Merabishvili M, Vaneechoutte M, Buckling A, De Vos D (2012) Introducing yesterday's phage therapy in today's medicine. Future Virol 7:379–390

Rao VB, Black LW (2010) Structure and assembly of bacteriophage T4 head. Virol J 7:356. https://doi.org/10.1186/1743-422X-7-356

Robertson KL, Soto CM, Archer MJ, Odoemene O, Liu JL (2011) Engineered T4 viral nanoparticles for cellular imaging and flow cytometry. Bioconjug Chem 22:595–604. https://doi.org/10.1021/bc100365

Rossmann MG, Rao VB (eds) (2012) Viral molecular machines. Springer, New York

Sanghamitra NJM, Inaba H, Arisaka F, Wang DO, Kanamaru S, Kitagawa S, Ueno T (2014) Plasma membrane translocation of a protein needle based on a triple-stranded b-helix motif. Mol Bio Syst 10:2677–2683. https://doi.org/10.1039/C4MB00293H

Scibilia S, Lentini G, Fazio E, Franco D, Neri F, Mezzasalma AM, Guglielmino SPP (2016) Self-assembly of silver nanoparticles and bacteriophage. Sens Bio-Sens Res 7:146–152

Shneider MM, Buth SA, Ho BT, Basler M, Mekalanos JJ, Leiman PG (2013) PAAR-repeat proteins sharpen and diversify the type VI secretion system spike. Nature 500:350–353. https://doi.org/10.1038/nature12453

Smith GP (1985) Filamentous fusion phage: novel expression vectors that display cloned antigens on the virion surface. Science 228:1315–1317

Sugimoto K, Kanamaru S, Iwasaki K, Arisaka F, Yamashita I (2006) Construction of a ball-and-spike protein supramolecule. Angew Chem Int Ed 45:2725–2728. https://doi.org/10.1002/anie.200504018

Sulakvelidze A, Alavidze A, Morris J (2001) Bacteriophage therapy. Antimicrob Agents Chemother 45:649–659

Sun L, Zhang X, Gao S, Rao PA, Padilla-Sanchez V, Chen Z, Sun S, Xiang Y, Subramaniam S, Rao VB, Rossmann MG (2015) Cryo- EM structure of the bacteriophage T4 portal protein assembly at near-atomic resolution. Nat Commun 6:7548. https://doi.org/10.1038/ncomms8548

Tao P, Mahalingam M, Marasa BS, Zhang Z, Chopra AK, Rao VB (2013) In vitro and in vivo delivery of genes and proteins using the bacteriophage T4 DNA packaging machine. Proc Natl Acad Sci USA 110:5846–5851. https://doi.org/10.1073/pnas.1300867110

Taylor NMI, Prokhorov NS, Guerrero-Ferreira RC, Shneider MM, Browning C, Goldie KN, Stahlberg H, Leiman PG (2016) Structure of the T4 baseplate and its function in triggering sheath contraction. Nature 533:346–352. https://doi.org/10.1038/nature17971

Ueno T (2008) Functionalization of viral protein assemblies by selfassembly reactions. J Mater Chem 18:3741–3745. https://doi.org/10.1039/b806296

Ueno T, Koshiyama T, Tsuruga T, Goto T, Kanamaru S, Arisaka F, Watanabe Y (2006) Bionanotube tetrapod assembly by in situ synthesis of a gold nanocluster with (gp5–His6)3 from bacteriophage T4. Angew Chem Int Ed 45:4508–4512. https://doi.org/10.1002/anie.200504588

Vafabakhsh R, Kondabagil K, Earnest T, Lee KS, Zhang Z, Dai L, Dahmen KA, Rao VB, Ha T (2014) Single-molecule packaging initiation in real time by a viral DNA packaging machine from bacteriophage T4. Proc Natl Acad Sci USA 111:15096–15101. https://doi.org/10.1073/pnas.1407235111

Veesler D, Cambillau C (2011) A common evolutionary origin for tailedbacteriophage functional modules and bacterial machineries. Microbiol Mol Biol Rev 75:423–433. https://doi.org/10.1128/MMBR.00014-11

Waldor MK, Mekalanos JJ (1996) Lysogenic conversion by a filamentous phage encoding cholera toxin. Science 272(5270):1910–1914

Washington DC, Karimi M, Mirshekari H, Basri SMM, Bahrami S, Moghoofei M, Hamblin MR (2016) Bacteriophages and phage-inspired nanocarriers for targeted delivery of therapeutic cargos. Adv Drug Deliv Rev 106:45–62. https://doi.org/10.1016/j.addr.2016.03.003

Wen AM, Steinmetz NF (2016) Design of virus-based nanomaterials for medicine, biotechnology, and energy. Chem Soc Rev 45:4074–4126. https://doi.org/10.1039/C5CS00287G

Xu J, Xiang Y (2017) Membrane penetration by bacterial viruses. J Virol 91:e00162–e00117. https://doi.org/10.1128/JVI.00162-17

Yap ML, Klose T, Arisaka F, Speir JA, Veesler D, Fokine A, Rossman MG (2016) Role of bacteriophage T4 baseplate in regulatingassembly and infection. Proc Natl Acad Sci USA 113:2654–2659. https://doi.org/10.1073/pnas.1601654113

Zou J, Glinsky VV, Landon LA, Matthews L, Deutscher SL (2005) Peptides specific to the galatin-3carbohydrate recognition domain inhibit metastasis-associated cancer cell adhesion. Carcinogenesis 26:309–318

Part IV

NanoBioMedicine in Diagnosis and Therapy

Diagnosis of Tuberculosis: Nanodiagnostics Approaches

11

Anil Kumar Gupta, Amit Singh, and Sarman Singh

Abstract

Tuberculosis (TB) remains one of the most devastating infectious diseases worldwide. The burden of TB is alarmingly high in developing countries, where diagnosis latent TB infection (LTBI), Extra-pulmonary tuberculosis (EPTB), drug-resistant tuberculosis (DR-TB), HIV-associated TB, and paediatric TB is still a challenge. This is mainly due to delayed or misdiagnosis of TB, which continues to fuel its worldwide epidemic. The ideal diagnostic test is still unavailable, and conventional methods remain a necessity for TB diagnosis, though with poor diagnostic ability. The nanoparticles have shown potential for the improvement of drug delivery, reducing treatment frequency and diagnosis of various diseases. The engineering of antigens/antibody nanocarriers represents an exciting front in the field of diagnostics, potentially flagging the way toward development of better diagnostics for TB. This chapter discusses the presently available tests for TB diagnostics and also highlights the recent advancement in the nanotechnology-based detection tests for *M. tuberculosis*.

Keywords

M. tuberculosis · Nano diagnostics · Nanoparticles

Anil Kumar Gupta and Amit Singh have contributed equally to this chapter.

A. K. Gupta · A. Singh · S. Singh (✉)
Division of Clinical Microbiology and Molecular Medicine, All India Institute of Medical Sciences, New Delhi, India

Department of Microbiology, All India Institute of Medical Sciences, Bhopal, India

S. K. Saxena, S. M. P. Khurana (eds.), *NanoBioMedicine*,
https://doi.org/10.1007/978-981-32-9898-9_11

11.1 Introduction

Tuberculosis (TB), caused by an aerobic, acid-fast bacillus, that is, *Mycobacterium tuberculosis (M. tuberculosis)*, to humans, and it is still a major public health problem worldwide (Kerantzas and Jacobs 2017). For many decades, it has continued to pose a significant threat to human health (WHO 2017). The situation becomes critical by the increasing incidence of multidrug-resistant (MDR) forms of *M. tuberculosis,* that is, resistance to both isoniazid (INH) and rifampicin (RIF) and now, extensively Drug-Resistant Tuberculosis (XDR-TB) strains, that is, MDR TB strains plus resistance to any fluoroquinolone and at least one of three injectable second-line drugs (amikacin, kanamycin, or capreomycin) that is virtually untreatable (CDC 2006). It has been estimated that almost billions of peoples will be newly affected with TB between 2000 and 2020. WHO estimated 10.4 million new cases and more than 1.3 million deaths in 2016 (WHO 2017). Among the estimated 10.4 million incident cases, 10% were children and 35% were female. In 2016, 153,119 cases of multidrug-resistant TB and rifampicin-resistant TB (MDR/RR-TB) were reported to the World Health Organization (WHO) globally (WHO 2017). Most of the TB cases reported in 2016 occurred in Asia (56%) and the African Region (29%). The smaller proportions of cases were reported in the Eastern Mediterranean Region (8%), European Region (4%), and the Region of the Americas (3%).

A fast and reliable laboratory diagnosis would help in the control of TB especially in the high burden countries. Current TB diagnostics mostly depend on the identification of *M. tuberculosis* by acid-fast bacilli (AFB) staining directly from clinical specimens, culture, and molecular tests. Although, smear microscopy permits the rapid detection of mycobacteria in clinical samples, it has comparatively low sensitivity and requires at least 5×10^3 bacilli/ml in the specimen (Desikan 2013). Also, it has a higher failure rate in children and immuno-compromised groups such as Acquired Immuno Deficiency Syndrome (AIDS) (Desikan 2013; Tuberculosis Division 2005).

The solid culture method is based on the visible appearance of growth on the medium, but it takes very long time for detection (up to 60 days), while automated liquid culture system, that is, Mycobacterial Growth Indicator Tube (MGIT) 960 system (Recommended by WHO for liquid culture and drug susceptibility test) is slightly better than solid culture due to reduced time duration required (42 days) for bacilli detection (Chihota et al. 2010; Lawson et al. 2013). However, it requires longer duration to obtain the results and also specific laboratory facilities, which may be unreachable by resource-limited or poor countries. Immunological approaches, such as the Tuberculin Skin Test (TST) and IFN-Gamma Release Assay (IGRA) have been developed for the detection of TB/latent TB infection (LTBI) (Lagrange et al. 2013, 2014; Pai et al. 2004). However, both TST and IGRA failed to distinguish between latent TB and active TB infection in the high burden countries (Chegou et al. 2009; Ra et al. 2011).

Current nucleic acid amplification-based tests (NAAT), that is, Polymerase Chain Reaction (PCR) (Gopinath and Singh 2009), Xpert MTB/Rif assay (Rufai et al. 2017), Loop-mediated isothermal amplification (LAMP) (Kumar et al. 2014) and Line Probe Assay (LPA) (Rufai et al. 2014) are able to detect *M. tuberculosis* within few hours to days in suspected TB patients than culture methods and play an important role in the patient care and TB control programmes.

Despite all advances in TB diagnosis landscape, there is no accurate, rapid, inexpensive, point-of-care assay available for *M. tuberculosis* detection, well-matched for children, extrapulmonary TB (EPTB) and HIV associated TB (HIV-TB)(Kozel and Burnham-Marusich 2017). Furthermore, in developing countries, like India and Pakistan, where resources are very limited and the requirement of sophisticated, costly instruments becomes an extraburden due to the requirement of trained technicians to perform the tests, which directly or indirectly increases the diagnostic cost. Therefore, an improvised version and/or new diagnostic test/techniques are urgently required for the prevention and treatment of *M. tuberculosis* infection to fulfil the unmet demands (Singh et al. 2015). From these viewpoints, diagnostic test based on nanotechnology can offer fast and efficient alternative methods for TB detection (Caliendo et al. 2013).

Nanotechnology, known as general purpose technology, utilizes nanoscale molecules ranging from 1 nm to100 nm. It plays a key role in the development of many fields such as automotive, textile, electronics, food, healthcare, and due to its unique characteristics, it is useful in optical, mechanical, magnetic, catalytic, and electrical perspectives (Chaturvedi et al. 2012). For the past several decades, biomedical applications such as tissue engineering, drug delivery, bioimaging, and nanodiagnostics have been developed by utilizing the concept of nanotechnology. Among these applications, nanodiagnostics-based rapid test has drawn more and more attention for infectious diseases due to its unique characteristics in the early detection with high sensitivity and specificity (Wang et al. 2017).

These potential of nanodiagnostics opened the door for development of portable, robust, and affordable POCs, which can detect infectious diseases very efficiently (Sharma and Bhargava 2013; Wang et al. 2017; Singh et al. 2017). In this direction, various innovative and efficient nanodiagnostics have been developed by researchers for infectious diseases including TB. Laksanasopin et al. (2015) developed a smartphone-based POC to diagnose infectious diseases by connecting traditional immunoassay into a smartphone via accessories such as dongle (Laksanasopin et al. 2015). Hence, nanodiagnostics-based POCs are promising tools for rapid detection of infectious diseases and could be exploited in the near future for different clinical requirements.

In this chapter, we highlight prospects of the advances in the nanotechnology-based diagnostic methods that can offer better solutions for diagnosis of *M. tuberculosis* infections.

11.2 Diagnosis of Tuberculosis

The TB diagnostic method can be divided into three categories: conventional methods, immunological methods, and new diagnostic methods.

11.2.1 Conventional Methods

Microscopy is possibly the earliest and most rapid procedure that can be performed in the laboratory to detect the presence of AFB, Adenosine Deaminase Activity (ADA), culture (egg-based solid media like Lowenstein–Jensen medium, agar-based medium like Middlebrook 7H10), which is shown in Table 11.1.

11.2.2 Immunological Methods

Enzyme-linked immunosorbent assay (ELISA), tuberculin skin test (TST), interferon-gamma determination, and tuberculin test are discussed in Table 11.1 above.

11.2.3 New Diagnostic Methods

Automated culture methods (BECTEC 460 TB (Aggarwal et al. 2008), BECTEC MGIT™ 960 (Rodrigues et al. 2009), Versa TREK and BacT/ALERT 3D) (Mirrett et al. 2007), Nucleic acid amplification methods (amplified MTD, amplified *M. tuberculosis* direct test (AMTD) (Goessens et al. 2005; Reischl et al. 1998), transcriptase-mediated amplification system and amplicor MTB test (Wang and Tay 1999), Multiplex Polymerase Chain Reaction (PCR) (Gopinath and Singh 2009), LAMP (Yadav et al. 2017), Real-time PCR (Watanabe Pinhata et al. 2015), LPA (Desikan et al. 2017), Xpert MTB/RIF assay (Osman et al. 2014). Genetic identification methods: PCR restriction-enzyme analysis, RFLP (Gómez Marín et al. 1995), Spoligo typing (Mistry et al. 2002), DNA probes (Badak et al. 1999) and DNA sequencing (Brown et al. 2015), etc. are shown in Table 11.1. Other molecular tests that are under development or under evaluation have been mentioned in Table 11.2.

11.3 Diagnostic Gaps Between Existing Technologies and Its Unmet Clinical Need

M. tuberculosis was identified more than a century ago, and its diagnosis in the developing world still remains a major healthcare issue owing to a number of challenges, listed below. First, *M. tuberculosis* is a slow-growing bacterium, and therefore it cannot provide direction for on-site patient care. Second, the PTB patient do not develop symptoms at the early stage of infection, which lead to delays in seeking patient care (Parsons et al. 2011; Kritski et al. 2013). Third, even the active PTB cases often exhibit low bacteria count of sputum thus making it difficult to detect with smear microscopy and other commonly used POC diagnostic tests in the developing world. Fourth, use of sputum and other invasive body fluids in the diagnosis

Table 11.1 Summary of the available diagnostic tests/methods for TB

Technology, test	Stage of development	Developer(s)/ supplier(s)	Level of the health system	DST utility
A: Conventional method				
Direct visualization (Microscopy)				
Conventional microscopy with acid-fast staining	In routine use	Multiple	Microscopy	No
Fluorescent microscopy with nonspecific cell-wall staining	In routine use	Multiple	Microscopy	No
Fluorescent microscopy with LED light source	In routine use	Various	Microscopy	No
Growth-based detection (Culture)				
Conventional solid media LJ, Middlebrook 7H10/7H11 agar, 7H9/7H12/Dubos medium	Commercialized reagents and prepared media	Multiple	Referral	Yes
Automated liquid culture systems MGIT 960, BacT/ ALERT 3D, VersaTREK Myco, etc.	Commercialized, under study for feasibility and impact of use in resource-limited settings	BD, BioMerieux, Thermofisher	Referral	Yes
MODS assay, thin-layer culture	Academic evaluations published	Non-commercial testing methods	Referral	Yes
Phage-based detection	Commercialized, improved test in development	Biotec	Referral	Yes
B: Immunological method				
Latent Tuberculosis Infection detection				
Tuberculin skin test with PPD	Commercialized	Multiple	Microscopy	No
Whole-blood IFN-γ release assay	Commercialized; in evaluation for disease-endemic countries	Cellestis	Referral	No
ELISPOT IFN-γ release assay	Commercialized; in evaluation for disease-endemic countries	Oxford Immunotech	Referral	No
Antigen detection (Immunodiagnosis)				
TB-derived antigen detection in urine or other clinical material	In development	Various	Research centre	No

(continued)

Table 11.1 (continued)

Technology, test	Stage of development	Developer(s)/ supplier(s)	Level of the health system	DST utility
TB-derived antigen detection in exhaled air/breath	In evaluation	Rapid Biosensor Systems	Health centre	No
Antibody detection (Immunodiagnostic)				
Detection of diagnostic antibody responses to TB	WHO banned all existing commercial test	Various	Health centre	No
C: Molecular detection				
Automated, non-integrated NAAT	Commercialized	GenProbe,Roche,	Referral	No
Automated, integrated NAAT	Commercialized	Cepheid	Referral	Yes
Simplified manual NAAT (LAMP)	In evaluation	Eiken	Referral	No
Non-amplified probe detection	In development	Investigen,	Microscopy	No
GeneXpert	Commercialized	Cepheid, USA	Referral	Yes

Modified from Pai and O'Brien (2008)

Table 11.2 Tuberculosis diagnostics pipeline (2016): Products in later-stage development or on track for evaluation by WHO

New molecular diagnostics					
S. No.	Test	Type	Developer(s)/ supplier(s)	Status	Comments
1.	BD MAX MTB assay	qPCR for MTB in automated BD MAX	Becton, Dickinson	100% sensitivity and 97.1% specificity with smear-positive samples (Rocchetti et al. 2016)	Under the evaluation stage
2.	EasyNAT	Isothermal DNA amplification / lateral flow to detect MTB	Ustar	Poor sensitivity, especially for smear-negative specimens, in Tanzanian field study (Bholla et al. 2016; Mhimbira et al. 2015)	
3.	FluoroType MTB	Semi-automated direct MTB detection; PCR in a closed system; results in 3 h	Hain Lifescience	Sensitivity 88%, specificity- 98% (Bwanga et al. 2015; Obasanya et al. 2017)	Marketed

(continued)

Table 11.2 (continued)

New molecular diagnostics					
S. No.	Test	Type	Developer(s)/ supplier(s)	Status	Comments
4.	GeneChip	RT-PCR for RIF + INH DR	CapitalBio	CCDCP and University of Georgia published a paper on 1400 samples from SW China (Sensitivity 83–94.6%, specificity 91.3–98%) (Zhang et al. 2018; Zhu et al. 2015)	Marketed
5.	Genedrive MTB/RIF	Portable RT-PCR for MTB + RIF resistance	Epistem	Lower sensitivity (45.4%) (Shenai et al. 2016)	Marketed in India
6.	GenoType MTBDR*plus*	Line probe assay for RIF + INH resistance	Hain Life science	(Sensitivity 90.3% and specificity 98.5%) (Nathavitharana et al. 2016)	WHO recommended
7.	LiPA pyrazinamide	Line probe assay for PZA resistance	Nipro	High sensitivity (65.9–100%) and specificity (98.2–100%) (Rienthong et al. 2015)	Marketed
8.	LiPA MDR-TB	Line probe assay for RIF + INH resistance	Nipro	Sensitivity 89% and 99.4% Specificity(Havumaki et al. 2017)	Marketed
9.	REBA MTB-MDR	Line probe assay for RIF + INH resistance	YD Diagnostics	(Havumaki et al. 2017)	Marketed
10.	REBA MTB-XDR	Line probe assay for FQ + SLID DR	YD Diagnostics	Initial study 2015 (Jaksuwan et al. 2018; Lee et al. 2015)	Marketed
11.	MeltPro TB/INH	Closed-tube RT-PCR for INH DR	Zeesan Biotech	3-site evaluation of 1096 clinical Isolates (Liang et al. 2018; Pang et al. 2016)	Chinese FDA-approved
12.	MeltPro TB/STR	Closed-tube RT-PCR for streptomycin DR	Zeesan Biotech	3-site evaluation of 1056 clinical Isolates (Zhang et al. 2015)	WHO guidance pending
13.	PURE-LAMP	Manual NAAT by LAMP for MTB detection	Eiken	Eddabra and AitBenhassou (2018)	WHO review

(continued)

Table 11.2 (continued)

New molecular diagnostics					
S. No.	Test	Type	Developer(s)/ supplier(s)	Status	Comments
14.	Xpert MTB/ RIF Ultra	Next-generation cartridge-based detection of MTB + RIF resistance	Cepheid	Sensitivity (77–90%) and specificity (96–98%) (Arend and Soolingen 2018; Dorman et al. 2018)	Recommended by WHO (2017)
15.	XpertOmini	Single cartridge mobile platform	Cepheid	FIND's study results awaited	
16.	Truenat MTB	Chip-based NAAT with RT-PCR on a handheld device for MTB	Molbio Diagnostics, Bigtec Labs	Sensitivity (99%) (Nikam et al. 2014)	FIND & ICMR studies underway
17.	TB-LAMP	Single tube detection system	Eiken-Chemical Company, Japan	The sensitivity of TB-LAMP is lower than Xpert MTB/RIF assay but greater than smear microscopy (Gray et al. 2016; Pham et al. 2018)	Recommended by WHO

of TB with existing techniques is more complex compared to blood and urine samples (Sharma et al. 2015).

The unavailability of accurate and validated biomarkers (for Active TB and LTBI infection) either derived from host or pathogen are due to inadequate knowledge of the host–pathogen interaction, pathogenesis, and protected immune response generated by *M. tuberculosis* during infection, which limited utility of rapid diagnostic test of TB (Goletti et al. 2016).

Despite exiting technologies, development of simple POCs test in the near future is still challenging in the current TB diagnostics pipeline (Pai and Nathavitharana 2014). Although Xpert MTB provides same-day detection, its use is limited by its cost and poor detection rate in extra-pulmonary tuberculosis (EPTB) (Rufai et al. 2017). Hence, there is an urgent need of inexpensive TB diagnostic test for resource-limited settings to miniaturize TB diagnosis, which can be done by using a novel nanotechnology approach.

11.4 Nanotechnology

11.4.1 Nanoparticles

Nanoparticle (NP) is a small particle less than 100 nm in diameter. The unique property depends on the size and composition of the particles compared to atoms and other materials. These properties includes: (1) large to volume ratio (metal NP, in

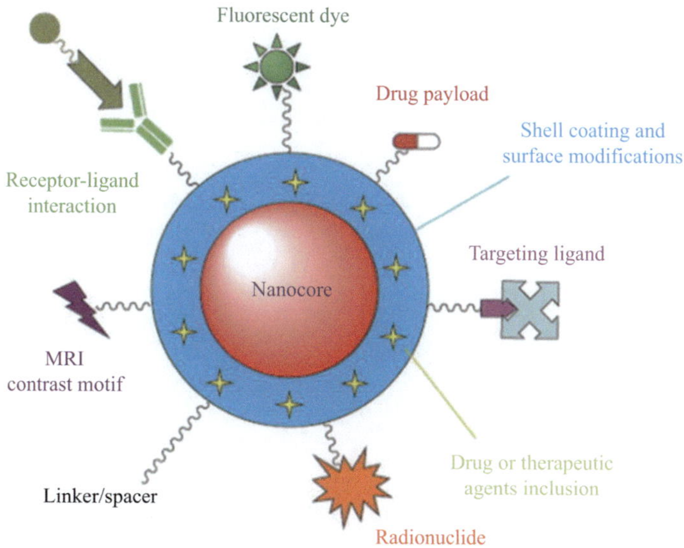

Fig. 11.1 Outline of multifunctional nanoparticle in human health including molecular imaging, disease diagnostics, drug delivery and therapy (Adopted from Chatterjee et al. 2014)

particular gold NP), (2) surface plasmon resonance, (3) Surface-Enhanced Raman Scattering (SERS), (4) super-magnetization or ferromagnetic nanoparticles (e.g. iron oxide), (5) enhanced photoluminescence (semiconductor quantum dots), (6) high electric and heat conductivity, (7) potent surface catalytic activity (Gatoo et al. 2014; Khan et al. 2019). The combination of nanoparticles with biology has led to the development of various diagnostic test/devices, contrast agents, analytical tools, physical therapy, and drug delivery systems. Since biomolecules and cellular organelles lie in the nanosized range, NPs can be altered with various biomolecules, such as antibodies, nucleic acid, peptides (Jacob and Deigner 2018; Wang and Wang 2014). Such manipulations enable NPs to be extremely useful in both *in vivo* and *in vitro* biomedical research and applications (Curtis and Wilkinson 2001). A schematic presentation of a core/shell nanoparticle for multipurpose biomedical applications is shown in Fig. 11.1.

11.4.2 Types of Nanoparticle-Based Platforms

The designing of nanodiagnostics are based on the binding of a labelled nanoparticle or probe to the target biomolecule, generates a quantifiable electric signal characteristic of the target biomolecules (Alharbi and Al-sheikh 2014). The most promising approaches include nanoparticles (carbon and gold nanoparticles), nanotubes, nanoshells, nanopores, quantum dots (QDs), and nanocantilever technologies, which display promising activity in the diagnostic applications (Capek 2016;

Table 11.3 Types of nanodevices used in clinical applications

S. No	Nonodevices		Applications
1.	Cantilevers	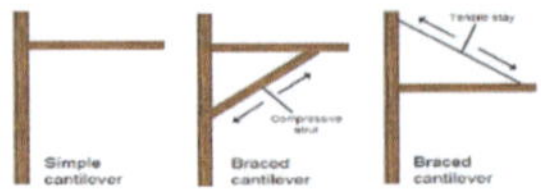	High thoughtful screening Protein biomarkers detection SNPs Gene expression detection
2.	Carbon Nanotubes	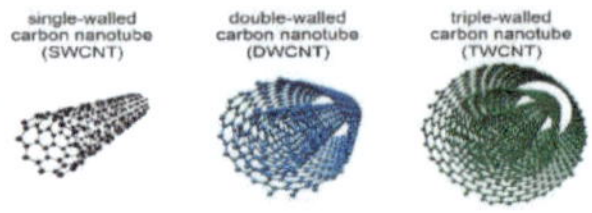	SNPs Protein biomarkers detection
3.	Dendrimers	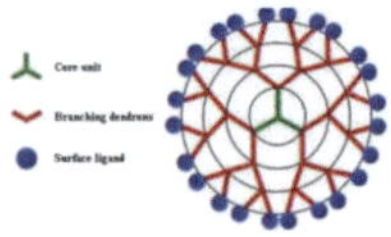	Image contrast agents
4.	Nanocrystals	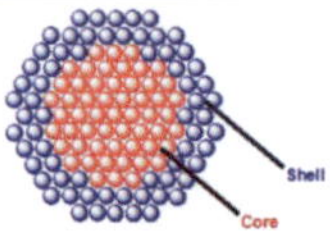	Improved formulation for poorly soluble drugs
5.	Nanoparticles	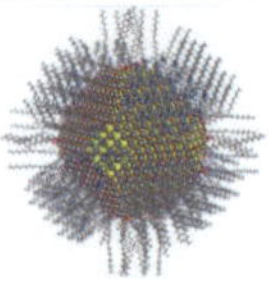	Target drug delivery MRI, USG image contrast agents Reporters of apoptosis
6.	Nanoshell	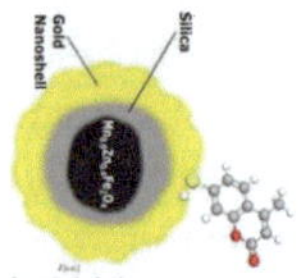	Tumour-specific imaging
7.	Nanowires	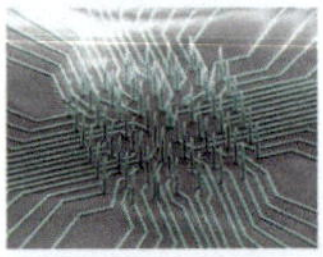	High thoughtful screening Disease protein biomarkers detection SNPs Gene expression analysis
8.	Quantum dots		Optical detection of genes and proteins in animal model Cell assays Visualization of tumour and lymph node in human

Mancebo 2009). The QDs are semiconductor nanocrystals which are characterized by strong light absorbance, and they can be used as fluorescent labels/tag for the detection of biomolecules. The cantilevers and QDs are the most promising nanostructures, which are mainly characterized by high photostability, single-wavelength excitation, and size-tunable emission (Azzazy et al. 2006; Rizvi et al. 2010). Different types of nanostructure or nanodevices that are used for specific purposes are listed in Table 11.3.

11.4.3 Nanoparticle-Based Diagnostics

Nano diagnostics, referred to as the use of nanotechnology in diagnostic applications, has been widely explored for the development of diagnostic test with high sensitivity and prior detection of infection. The nanoscale size and high surface-to-volume ratio of nanoparticles makes this field superior and indispensable in multi-field of human action. The unique properties of nanomaterials or nanostructures deliberate the nanodiagnostic platforms and ability of rapid detection by utilizing very small volumes of clinical samples (Jackson et al. 2017). The technology itself is variegated, and several options are available, for instance, nanosuspensions, nanoemulsions, niosomes (nonionic surfactant-based vesicles). Therefore, nanodiagnostic approaches have strong potential to be cost-effective, user-friendly, and robust (Azzazy et al. 2006; Kumar et al. 2011; Wang et al. 2017).

The significant progress has been made in the field of nanotechnology in the last two decade, which showed its wide potential and advantageous applications in the field of biomedicine, biotechnology, human and animal health including nanodiagnostics and nanomedicines. Majority of the nanodiagnostic work has been carried out in the field of cancer diagnostics, but this technique has also been contributed significantly to the diagnosis of various infectious diseases presently (Kumar et al. 2011; Yukuyama et al. 2017). Most of the infectious disease-causing agents such as bacteria (*M. tuberculosis*), virus (SARS), and fungi may sometimes cause an epidemic outbreak, resulting in higher morbidity and mortality (Mathuria 2009; Nasiruddin et al. 2017; Xu et al. 2018). Thus, initiation of nano-based diagnostic platforms in a clinical setting is gaining importance these days. This is because of the ability of nanodiagnostics to achieve consistency, quick conclusions with simple and movable devices by using various body fluids, such as blood, sputum, or urine samples from patients (Banyal et al. 2013; Wang et al. 2017).

In addition, the highly sensitive nanodiagnostics platforms, with strong potential must be robust, cost effective, and reproducible and could be extremely applicable for the diagnosis of infectious diseases, especially in resource-limited areas in the developing countries.

11.4.4 Gold Nanoparticle (AuNPs)-Based Diagnostics for TB

The gold nanoparticles (AuNPs) pose unique physiochemical (inert and nontoxic) and optical characteristics making them most appropriate nanomaterial for clinical diagnosis, treatments, and other multidisciplinary research. The optical property of AuNPs with antibody or antigen and other biomolecules enable their utility in the diagnosis of various pathogens. Moreover, AuNPs do not disturb the functional activity even after antigen immobilization (Choi and Frangioni 2010; Sonawane and Nimse 2016). The antibody–antigen reaction is enhanced by the surface functionalization of gold nanoparticles, thereby increasing immunoassay signals, which ultimately increase the test sensitivity (Kim et al. 2018). It offers an easy, low-cost assay, which allows simultaneously numerous sample testing. The assay has been found to be very specific and produce reliable results even with tiny amount of mycobacterial DNA. The colorimetric detection of target gene/sequence from test DNA samples via AuNP probes (thiol-linked single-stranded DNA, or ssDNA, modified gold nanoparticles) offer a low-cost alternative method for detection (Chandra et al. 2010; Cordeiro et al. 2016).

The utilization of AuNPs was firstly reported in TB diagnosis by (Baptista et al. 2008), which utilized DNA probes (oligonucleotide derived from the gene sequence of the *M. tuberculosis* RNA polymerase subunit) coupled with AuNPs for the colorimetric detection of *M. tuberculosis*. Principally, at wavelength 526 nm, if the complementary DNA is present, the nanoprobe solution remains pink in colour (no DNA probe aggregation), while the solution turns purple (due to nanoprobe aggregation at a high NaCl concentration) in the absence of complementary DNA in the samples. The method is more accurate when compared to other diagnostic methods, that is, InnoLiPA-Rif-TB, which gave 100% concordance (Baptista et al. 2008). The test was proved to be more sensitive than smear microscopy and can be simply visualized for detection. The major advantage of this method is that the chances of contamination is very less (carried out in a single tube reducing contamination), rapid (takes approximately 15 min per sample).

Subsequently, activity of this method was also compared with automated liquid culture system (BACTEC™ MGIT™) and semi-nested PCR, which shows greater sensitivity and specificity of the test in the detection of *M. tuberculosis* complex (Baptista et al. 2008; Cordeiro et al. 2016). Insertion sequence (IS6110) of *M. tuberculosis* was also used to increase the sensitivity of this test along with microfluidics technology, which utilized calorimetric detection of AuNPs coupled with IS6110 sequence (Tsai et al. 2017).

Surface Plasmon Resonance (SPR) has attracted much attention for novel metal (Au), which gives a red colour to the AuNPs colloid. The method is based on the real-time monitoring of changes happening in the surface refractive index, formed by association or dissociation of the molecules from the sensor (Khan et al. 2019). The major advantage of the SPR-based test is its optical sensor sensitivity making it capable of detecting even tiny amount of disease-specific analyte from the complex fluid without any specific procedure (Masson 2017; Nguyen et al. 2015; Wang and Fan 2016). Due to these advantages, SPR has emerged as a powerful optical tool, which can provide valuable data in the analysis of biomedical and chemical analyses. The SPR-based CFP-10 antigen detection system was developed in clinical samples by Yang et al. (2014), which showed reputable usefulness in TB diagnostics (Hsieh et al. 2012; Yang et al. 2014).

Zhu et al. (2017) developed AuNPs modified indium tin oxide (ITO) electrode for the direct detection of *M. tuberculosis* using genomic DNA (gDNA) isolated directly from clinical samples. The method utilized two probes: capture probe and gold nanoprobe coupled with alkaline phosphatase (ALP) enzyme as detection probe. First, ITO probe is activated via capture probe, then activated probe is immersed in the gDNA containing hybridization buffer to form double strand DNA (dsDNA) via hybridization of probe and target nucleotide sequence. Finally, ITO is placed as electrode in the buffer containing detection probe to generate hybridization sandwich. The electric signals are then recorded using voltametry (Zhu et al. 2017).

11.4.5 AuNP-Mediated Dipstick Assay

The colloidal AuNPs were coated with the *M. tuberculosis* antigen using alkanethiols derivatives and anti-MTB rabbit antibodies. These antigen-coated AuNPs act as a counter or detector reagent in this assay. The serum samples or antibody immobilized on the nitrocellulose (NC) membrane binds to the *M. tuberculosis* antigen coated on AuNPs. Resultant binding could be visually detected by naked eye, due to the development of the red colour formed by the gold nanoparticles on the nitrocellulose membrane (NC) (Stephen et al. 2015).

11.4.6 Silica Nanoparticles-Based Detection

The application of mesoporous silica nanoparticle (SiO_2NPs) has been reported in the various fields, that is, imaging, drug delivery, and biosensors (Sun et al. 2015). The indirect immunofluorescence microscopy has been developed by utilizing nanoparticle coupled with fluorescent dye for the detection of *M. tuberculosis*. The technology consists of SYBR Green I mediated assay, which stained only bio-conjugated fluorescent silica nanoparticles. The intensity of fluorescent signals is five-fold higher than conventional fluorescence isothiocyanate (FITC)-based detection method. This assay gives promising results within 2 h and therefore is considered to be a promising method for the rapid detection of *M. tuberculosis* (Qin et al. 2007).

11.4.7 Magnetic Nanoparticles-Based Detection

The magnetic nanoparticles (MNPs), nanoscale-sized molecules are present in nature. They harbour favourable features for their usage in the nano-biomedicine, that is, imaging therapy (Akbarzadeh et al. 2012). The surface of MNPs can be easily modified with recognition moieties, that is, antibodies, antibiotics, and carbohydrate, which enable their use for bacterial detection. Super paramagnetic iron oxide nanoparticle (Iron oxide nanoparticles [IONPs], composed of magnetite [Fe_3O_4] or maghemite [γ-Fe_2O_3] nanoparticles) is commonly used in the field of drug therapy, cell tracking, drug delivery by magnetic resonance imaging (MRI) (Cristea et al. 2017; Sabale et al. 2017). Using IONPs coupled with IgG has allowed enhanced detection limit (10^4 CFU/mL) of bacterial cells significantly by using nano-MALDI platforms (Chiu, 2014). Various studies have reported the use of diagnostic magnetic resonance (DMR) along with iron oxide nanoparticles for the detection of *M. tuberculosis* DNA (Kaittanis et al. 2010; Vallabani and Singh 2018).

Engstrom and his co-workers developed a novel platform using streptavidin tagged magnetic nanobeads labelled with biotin, for the detection of rifampicin mutation in the *rpoB* gene of *M. tuberculosis*. The assay comprised of 11 padlock probes (PLPs) targeting 23S ITS region of *M. tuberculosis* (Engstrom et al. 2013).Of which, one probe was for MTBC detection and another PLP for wild-type and a remaining mixture of nine PLPs are designed for identification of a common mutation in the RRDR-*rpoB* gene. The detection system is based on the Brownian relaxation principal, and signal is detected *via* AC susceptometry (Engström et al. 2013).

Efficacy of super-paramagnetic iron oxide (SPIO) nanoparticles has also been tried for improvement of the sensitivity and specificity of MRI systems in TB detection (Sabale et al. 2017). This method is more effective for the diagnosis of TB at the molecular level and also provides a valuable tool for the analysis of antibody–antigen and parasite–host interactions. The procedure includes activation of SPIO nanoparticles using an anti-MTB surface antibody to form conjugates. Then, conjugate was incubated with mycobacterium followed by MRI imaging, which reveals specific target recognition by reducing signal intensity. This method is more specific for the detection of EPTB (Musculoskeletal TB, Central nervous system TB, abdominal TB) (Skoura et al. 2015).

11.4.8 Quantum Dots-Based Detection System

Quantum dots, also known as semiconductor nanocrystals, possess unique optical and physical properties making them suitable for diagnostics developments (Kairdolf et al. 2013; Smith and Nie 2010). The broad absorption spectra, narrow emission spectra, slow excited-state decay rates, and broad absorption cross-sections are major advantages of quantum dots over other fluorescence-based methods (Rizvi et al. 2010). Also, it can identify multiple targets at the same time, which makes it a much sought after application in the identification of various pathogens in single clinical samples (Rizvi et al. 2010).

The hybrid detection system (Quantum dots and magnetic beads) uses *M. tuberculosis*-specific molecular probes for TB detection. One probe binds to the *23S rRNA* gene of the mycobacterium very precisely and the second probe precisely recognizes IS900 conserved sequence in mycobacterium, which was treated on sulphurous acid chromium quantum dots. Subsequently, sandwich is formed after hybridization with target gene sequences of mycobacterium DNA, isolated from suspected samples of TB patients. Then, quantum dot-magnetic bead conjugates are exposed to ultraviolet (UV) light, which emits red fluorescence (visible to the naked eye).

These conjugate detection systems are also highly versatile molecular probes, which can be easily modified according to their diagnostic utility/purposes. The method can identify the unamplified DNA of *M. tuberculosis* complex directly from clinical samples (Gazouli et al. 2012). A similar study was reported by Liandris and his co-workers who utilized quantum dots of CdSeO3 coupled with streptavidin and species-specific probes, which detect surface antigen of mycobacterium species.

The gDNA of mycobacterial was targeted by a sandwich hybridization, which consisted of two biotinylated probes that would recognize and detect the target DNA specifically. The detection limit is approximately 10^4cell/mL of the sample (Liandris et al. 2011).

11.4.9 Magnetic Barcode Assays

The principal of magnetic barcode (MB) assays is more or less similar to QDs. The assay used specific complementary DNA sequences of *M. tuberculosis* as probes for TB detection (Liong et al. 2013; Wang et al. 2017).

The major differences are the necessity for DNA extraction and PCR amplification, which are not required in quantum dots assay. After DNA is captured by probes, the resultant conjugate is then marked by complementary magnetic nanoparticle probes, which is then detected by nuclear magnetic resonance (NMR) techniques (Chen et al. 2017) (Fig. 11.2).

11.4.10 Biosensors-Based Detection System

A biosensor is an analytical system developed for the detection of presences/absences of a specific biological analyte via integrating a bio-recognition element (transduction system, amplifiers, and display unit). The biosensor consists of an analytical device coupled with biological analytes, which report physio-chemical changes in the sensing area (Bhalla et al. 2016; van den Hurk and Evoy 2015; Mehrotra 2016). The biosensor is based on the detection of short nucleotide sequences of *M. tuberculosis* DNA. TB biosensor can be divided into one of the following categories: mass/piezoelectric, biochemical, electrical, and optical sensors (Table 11.4). These sensing platforms are based on the detecting antibody–antigen interaction, nucleic acid hybridization, and whole mycobacterium bacilli (Lim et al. 2015; Prabhakar et al. 2008; Zhou et al. 2011).

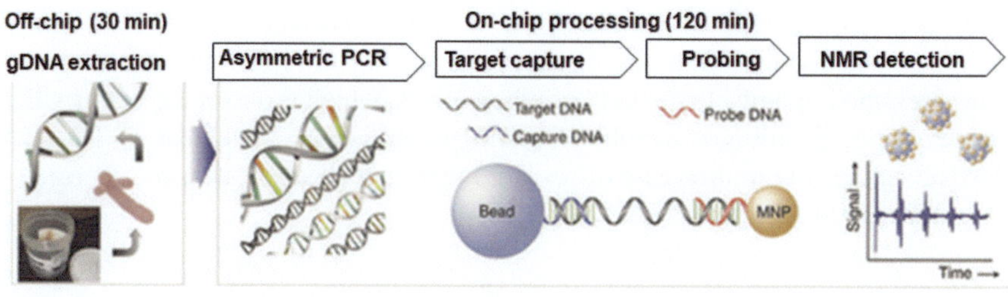

Fig. 11.2 Framework of the procedure of magnetic barcode (MB) assays

Table 11.4 Comparison of different biosensors developed for *M. tuberculosis* detection

S. No	Technology	Biomarkers	Detection Limit	References
1.	QCM	Whole MTB bacilli	10^5 CFU/mL	Kaewphinit et al. (2010)
2.	MSPQC	NH3 & CO2 absorption	10^2CFU/mL	Mi et al. (2012)
3.	RBS breath analyser	Ag85 B antigen	94% sensitivity 79% specificity	Camilleri (2015) and McNerney et al. (2010)
4.	Interferometric biosensor	38 kDa antigen	_	Wang et al. (2013)
5.	SPR	ssDNA	115 ng/mL (28fM ssDNA)	Hsu et al. (2013) and Prabowo et al. (2016)
6.	SPCE	Ag360 & Ag231	1 ng/mL	Wang et al. (2013)
7.	Enzymatic sensor	Mycolic acid antibody	_	Wang et al. (2013)
8.	Electro-osmosis microchip sensor	Whole *M. tuberculosis* Bacilli	100 CFU/mL	Hiraiwa et al. (2015) and Khairulina et al. (2017)

Abbreviations: *QCM* quartz crystal microbalance, *MSPQC* multi-channel series piezoelectric quartz crystal, *RBS* rapid biosensor system, *SPR* surface plasmon resonance, *SPCE* screen-printed carbon electrode, *BES* bioelectric sensor, *CFU* colony forming unit

11.5 Conclusion and Future Perspectives

Presently, approximately 40–50% of TB cases still remain undetected either due to non-availability of diangstic services, poor awareness in the masses or due to scanty or absence of tubercle bacilli in the clinical samples. In such conditions, tests taregetting the whole bacilli in clinical samples are missing many cases due to their poor detection sensitivity. This suggested that detection of bacillary by-products or detection of triggered changes in the host-immune response might be an alternative diagnostic approach for TB detection. Although several attempts have been made in this direction, none of these attempts has displayed clear clinical utility.

Nanotechnology is a fast progressing field, which attracts multi-disciplinary teams to target various healthcare challenges in the diagnosis and treatment of infectious diseases, cancer, and cardiovascular diseases. These technologies have contributed significantly in the diagnosis of various bacterial and viral diseases. Most significantly the nano-based technologies help miniaturizing the diagnostic devices and implants. In the field of tuberculosis, which is one of the major killer disease, the application of nanobiotechnology can help management of TB with added advantage of rapidity, ease of performing test, at cheaper rates, and especially useful for resource-limited countries.

References

Aggarwal P, Singal A, Bhattacharya SN, Mishra K (2008) Comparison of the radiometric BACTEC 460 TB culture system and Löwenstein-Jensen medium for the isolation of mycobacteria in cutaneous tuberculosis and their drug susceptibility pattern. Int J Dermatol 47:681–687

Akbarzadeh A, Samiei M, Davaran S (2012) Magnetic nanoparticles: preparation, physical properties, and applications in biomedicine. Nanoscale Res Lett 7:144

Alharbi KK, Al-sheikh YA (2014) Role and implications of nanodiagnostics in the changing trends of clinical diagnosis. Saudi J Biol Sci 21:109–117

Arend SM, van Soolingen D (2018) Performance of Xpert MTB/RIF Ultra: a matter of dead or alive. Lancet Infect Dis 18:8–10

Azzazy HME, Mansour MMH, Kazmierczak SC (2006) Nanodiagnostics: a new frontier for clinical laboratory medicine. Clin Chem 52:1238–1246

Badak FZ, Goksel S, Sertoz R, Nafile B, Ermertcan S, Cavusoglu C, Bilgic A (1999) Use of nucleic acid probes for identification of Mycobacterium tuberculosis directly from MB/BacT bottles. J Clin Microbiol 37:1602–1605

Banyal S, Malik P, Tuli HS, Mukherjee TK (2013) Advances in nanotechnology for diagnosis and treatment of tuberculosis. Curr Opin Pulm Med 19:289–297

Baptista P, Pereira E, Eaton P, Doria G, Miranda A, Gomes I, Quaresma P, Franco R (2008) Gold nanoparticles for the development of clinical diagnosis methods. Anal Bioanal Chem 391:943–950

Bhalla N, Jolly P, Formisano N, Estrela P (2016) Introduction to biosensors. Essays Biochem 60:1–8

Bholla M, Kapalata N, Masika E, Chande H, Jugheli L, Sasamalo M, Glass TR, Beck H-P, Reither K (2016) Evaluation of Xpert® MTB/RIF and Ustar EasyNAT™ TB IAD for diagnosis of tuberculous lymphadenitis of children in Tanzania: a prospective descriptive study. BMC Infect Dis 16:246

Brown AC, Bryant JM, Einer-Jensen K, Holdstock J, Houniet DT, Chan JZM, Depledge DP, Nikolayevskyy V, Broda A, Stone MJ et al (2015) Rapid whole-genome sequencing of Mycobacterium tuberculosis isolates directly from clinical samples. J Clin Microbiol 53:2230–2237

Bwanga F, Disqué C, Lorenz MG, Allerheiligen V, Worodria W, Luyombya A, Najjingo I, Weizenegger M (2015) Higher blood volumes improve the sensitivity of direct PCR diagnosis of blood stream tuberculosis among HIV-positive patients: an observation study. BMC Infect Dis 15:1–5

Caliendo AM, Gilbert DN, Ginocchio CC, Hanson KE, May L, Quinn TC, Tenover FC, Alland D, Blaschke AJ, Bonomo RA et al (2013) Better tests, better care: improved diagnostics for infectious diseases. Clin Infect Dis Off Publ Infect Dis Soc Am 57:S139–S170

Camilleri D (2015) Point-of-care disease diagnosis: a breath screen for active tuberculosis at the point-of-care. In: BioOptics World. https://www.bioopticsworld.com/biomedicine/article/16429686/pointofcare-disease-diagnosis-a-breath-screen-for-active-tuberculosis-at-the-pointofcare

Capek I (2016) Nanosensors based on metal and composite nanoparticles and nanomaterials. In: Parameswaranpillai J, Hameed N, Kurian T, Yu Y (eds) Nanocomposite materials: synthesis, properties and applications. CRC Press, Boca Raton, pp 1–40

CDC (2006) Notice to readers: revised definition of extensively drug-resistant tuberculosis. MMWR 55(43):1176

Chandra P, Das D, Abdelwahab AA (2010) Gold nanoparticles in molecular diagnostics and therapeutics. Digest J Nanomater Biostruct 5:363–367

Chatterjee K, Sarkar S, Jagajjanani Rao K, Paria S (2014) Core/shell nanoparticles in biomedical applications. Adv Colloid Interf Sci 209:8–39

Chaturvedi S, Dave PN, Shah NK (2012) Applications of nano-catalyst in new era. J Saudi Chem Soc 16:307–325

Chegou NN, Black GF, Kidd M, van Helden PD, Walzl G (2009) Host markers in QuantiFERON supernatants differentiate active TB from latent TB infection: preliminary report. BMC Pulm Med 9:21

Chen Y-T, Kolhatkar AG, Zenasni O, Xu S, Lee TR (2017) Biosensing using magnetic particle detection techniques. Sensors 17

Chihota VN, Grant AD, Fielding K, Ndibongo B, van Zyl A, Muirhead D, Churchyard GJ (2010) Liquid vs. solid culture for tuberculosis: performance and cost in a resource-constrained setting. Int J Tuberc Lung Dis Off J Int Union Tuberc Lung Dis 14:1024–1031

Chiu T-C (2014) Recent advances in bacteria identification by matrix-assisted laser desorption/ionization mass spectrometry using nanomaterials as affinity probes. Int J Mol Sci 15:7266–7280

Choi HS, Frangioni JV (2010) Nanoparticles for biomedical imaging: fundamentals of clinical translation. Mol Imaging 9:291–310

Cordeiro M, Ferreira Carlos F, Pedrosa P, Lopez A, Baptista PV (2016) Gold nanoparticles for diagnostics: advances towards points of care. Diagnostics 6:1–20

Cristea C, Tertis M, Galatus R (2017) Magnetic nanoparticles for antibiotics detection. Nanomaterials 7:1–25

Curtis A, Wilkinson C (2001) Nantotechniques and approaches in biotechnology. Trends Biotechnol 19:97–101

Desikan P (2013) Sputum smear microscopy in tuberculosis: Is it still relevant? Indian J Med Res 137:442–444

Desikan P, Panwalkar N, Mirza SB, Chaturvedi A, Ansari K, Varathe R, Chourey M, Kumar P, Pandey M (2017) Line probe assay for detection of Mycobacterium tuberculosis complex: an experience from Central India. Indian J Med Res 145:70–73

Dorman SE, Schumacher SG, Alland D, Nabeta P, Armstrong DT, King B, Hall SL, Chakravorty S, Cirillo DM, Tukvadze N et al (2018) Xpert MTB/RIF Ultra for detection of Mycobacterium tuberculosis and rifampicin resistance: a prospective multicentre diagnostic accuracy study. Lancet Infect Dis 18:76–84

Eddabra R, Ait Benhassou H (2018) Rapid molecular assays for detection of tuberculosis. Pneumonia 10:4

Engström A, de la Torre TZG, Strømme M, Nilsson M, Herthnek D (2013) Detection of rifampicin resistance in Mycobacterium tuberculosis by padlock probes and magnetic nanobead-based readout. PLoS One 8:e62015

Gatoo MA, Naseem S, Arfat MY, Mahmood Dar A, Qasim K, Zubair S (2014) Physicochemical properties of nanomaterials: implication in associated toxic manifestations. Biomed Res Int:1–8

Gazouli M, Lyberopoulou A, Pericleous P, Rizos S, Aravantinos G, Nikiteas N, Anagnou NP, Efstathopoulos EP (2012) Development of a quantum-dot-labelled magnetic immunoassay method for circulating colorectal cancer cell detection. World J Gastroenterol WJG 18:4419–4426

Goessens WHF, de Man P, Koeleman JGM, Luijendijk A, te Witt R, Endtz HP, van Belkum A (2005) Comparison of the COBAS AMPLICOR MTB and BDProbeTec ET assays for detection of Mycobacterium tuberculosis in respiratory specimens. J Clin Microbiol 43:2563–2566

Goletti D, Petrucciani E, Joosten SA, Ottenhoff TH (2016) Tuberculosis biomarkers: from diagnosis to protection. Infect Dis Rep 8(2):6568

Gómez Marín JE, Rigouts L, Villegas Londoño LE, Portaels F (1995) Restriction fragment length polymorphism (RFLP) analysis and tuberculosis epidemiology. Bull Pan Am Health Organ 29:226–236

Gopinath K, Singh S (2009) Multiplex PCR assay for simultaneous detection and differentiation of Mycobacterium tuberculosis, Mycobacterium avium complexes and other Mycobacterial species directly from clinical specimens. J Appl Microbiol 107:425–435

Gray CM, Katamba A, Narang P, Giraldo J, Zamudio C, Joloba M, Narang R, Paramasivan CN, Hillemann D, Nabeta P et al (2016) Feasibility and operational performance of TB LAMP in decentralized settings — results from a multi-center study. J Clin Microbiol 54(8):1984–1991

Havumaki J, Hillemann D, Ismail N, Omar SV, Georghiou SB, Schumacher SG, Boehme C, Denkinger CM (2017) Comparative accuracy of the REBA MTB MDR and Hain MTBDRplus line probe assays for the detection of multidrug-resistant tuberculosis: a multicenter, non-inferiority study. PLoS One 12:e0173804

Hiraiwa M, Kim J-H, Lee H-B, Inoue S, Becker AL, Weigel KM, Cangelosi GA, Lee K-H, Chung J-H (2015) Amperometric immunosensor for rapid detection of Mycobacterium tuberculosis. J Micromech Microeng Struct Device Syst 25:055013

Hsieh S-C, Chang C-C, Lu C-C, Wei C-F, Lin C-S, Lai H-C, Lin C-W (2012) Rapid identification of Mycobacterium tuberculosis infection by a new array format-based surface plasmon resonance method. Nanoscale Res Lett 7:180

Hsu S-H, Lin Y-Y, Lu S-H, Tsai I-F, Lu Y-T, Ho H-T (2013) Mycobacterium tuberculosis DNA detection using surface plasmon resonance modulated by telecommunication wavelength. Sensors 14:458–467

Jackson TC, Patani BO, Ekpa DE (2017) Nanotechnology in diagnosis: a review. Adv Nanopart 06:93

Jacob LJ, Deigner H-P (2018) Chapter 10: Nanoparticles and nanosized structures in diagnostics and therapy. In: Deigner H-P, Kohl M (eds) Precision medicine. Academic Press, pp 229–252

Jaksuwan R, Patumanond J, Tharavichikul P, Chuchottaworn C, Pokeaw P, Settakorn J (2018) The prediction factors of Pre-XDR and XDR-TB among MDR-TB patients in Northern Thailand. J Tuberc Res 06:36–48

Kaewphinit T, Santiwatanakul S, Promptmas C, Chansiri K (2010) Detection of non-amplified Mycobacterium tuberculosis genomic DNA using piezoelectric DNA-based biosensors. Sensors 10:1846–1858

Kairdolf BA, Smith AM, Stokes TH, Wang MD, Young AN, Nie S (2013) Semiconductor quantum dots for bioimaging and biodiagnostic applications. Annu Rev Anal Chem Palo Alto Calif 6:143–162

Kaittanis C, Santra S, Perez JM (2010) Emerging nanotechnology-based strategies for the identification of microbial pathogenesis. Adv Drug Deliv Rev 62:408–423

Kerantzas CA, Jacobs WR (2017) Origins of combination therapy for tuberculosis: lessons for future antimicrobial development and application. MBio 8:e01586–e01516

Khairulina K, Chung U, Sakai T (2017) New design of hydrogels with tuned electro-osmosis: a potential model system to understand electro-kinetic transport in biological tissues. J Mater Chem B 5:4526–4534

Khan I, Saeed K, Khan I (2019) Nanoparticles: properties, applications and toxicities. Arab J Chem 12:908–931

Kim J, Oh SY, Shukla S, Hong SB, Heo NS, Bajpai VK, Chun HS, Jo C-H, Choi BG, Huh YS et al (2018) Heteroassembled gold nanoparticles with sandwich-immunoassay LSPR chip format for rapid and sensitive detection of hepatitis B virus surface antigen (HBsAg). Biosens Bioelectron 107:118–122

Kozel TR, Burnham-Marusich AR (2017) Point-of-care testing for infectious diseases: past, present, and future. J Clin Microbiol 55:2313–2320

Kritski A, Fujiwara PI, Vieira MA, Netto AR, Oliveira MM, Huf G, Squire SB (2013) Assessing new strategies for TB diagnosis in low- and middle-income countries. Braz J Infect Dis 17:211–217

Kumar A, Mazinder Boruah B, Liang X-J (2011) Gold nanoparticles: promising nanomaterials for the diagnosis of cancer and HIV/AIDS. J Nanomater:1–17

Kumar P, Pandya D, Singh N, Behera D, Aggarwal P, Singh S (2014) Loop-mediated isothermal amplification assay for rapid and sensitive diagnosis of tuberculosis. J Infect 69:607–615

Lagrange PH, Thangaraj SK, Dayal R, Deshpande A, Ganguly NK, Girardi E, Joshi B, Katoch K, Katoch VM, Kumar M et al (2013) A toolbox for tuberculosis (TB) diagnosis: an Indian multicentric study (2006–2008). Evaluation of QuantiFERON-TB gold in tube for TB diagnosis. PLoS One 8:e73579

Lagrange PH, Thangaraj SK, Dayal R, Deshpande A, Ganguly NK, Girardi E, Joshi B, Katoch K, Katoch VM, Kumar M et al (2014) A toolbox for tuberculosis (TB) diagnosis: an Indian multi-centric study (2006–2008); evaluation of serological assays based on PGL-Tb1 and ESAT-6/CFP10 antigens for TB diagnosis. PLoS One 9:e96367

Laksanasopin T, Guo TW, Nayak S, Sridhara AA, Xie S, Olowookere OO, Cadinu P, Meng F, Chee NH, Kim J et al (2015) A smartphone dongle for diagnosis of infectious diseases at the point of care. Sci Transl Med 7:273re1

Lawson L, Emenyonu N, Abdurrahman ST, Lawson JO, Uzoewulu GN, Sogaolu OM, Ebisike JN, Parry CM, Yassin MA, Cuevas LE (2013) Comparison of Mycobacterium tuberculosis drug susceptibility using solid and liquid culture in Nigeria. BMC Res Notes 6:215

Lee YS, Kang MR, Jung H, Choi SB, Jo K-W, Shim TS (2015) Performance of REBA MTB-XDR to detect extensively drug-resistant tuberculosis in an intermediate-burden country. J Infect Chemother Off J Jpn Soc Chemother 21:346–351

Liandris E, Gazouli M, Andreadou M, Sechi LA, Rosu V, Ikonomopoulos J (2011) Detection of pathogenic mycobacteria based on functionalized quantum dots coupled with immunomagnetic separation. PLoS One 6:e20026

Liang B, Tan Y, Li Z, Tian X, Du C, Li H, Li G, Yao X, Wang Z, Xu Y et al (2018) Highly Sensitive Detection of Isoniazid Heteroresistance in Mycobacterium tuberculosis by DeepMelt Assay. J Clin Microbiol 56:e01239-17

Lim JW, Ha D, Lee J, Lee SK, Kim T (2015) Review of micro/nanotechnologies for microbial biosensors. Front Bioeng Biotechnol:1–3

Liong M, Hoang AN, Chung J, Gural N, Ford CB, Min C, Shah RR, Ahmad R, Fernandez-Suarez M, Fortune SM et al (2013) Magnetic barcode assay for genetic detection of pathogens. Nat Commun 4:1752

Mancebo SM (2009) Synthesis and applications of nanoparticles in biosensing systems. Nanobioelectronics and biosensors Group, Catalan Institute of NanotechnologyBellaterra, Barcelona, Spain (PhD thesis), 1–108. https://www.tdx.cat/bitstream/handle/10803/3300/smm1de3.pdf;sequence=1

Masson J-F (2017) Surface plasmon resonance clinical biosensors for medical diagnostics. ACS Sens 2:16–30

Mathuria JP (2009) Nanoparticles in tuberculosis diagnosis, treatment and prevention -a hope for future. Dig J Nanomater Biostruct 4(2):309–312

McNerney R, Wondafrash BA, Amena K, Tesfaye A, McCash EM, Murray NJ (2010) Field test of a novel detection device for Mycobacterium tuberculosis antigen in cough. BMC Infect Dis 10:161

Mehrotra P (2016) Biosensors and their applications – a review. J Oral Biol Craniofac Res 6:153–159

Mhimbira FA, Bholla M, Sasamalo M, Mukurasi W, Hella JJ, Jugheli L, Reither K (2015) Detection of Mycobacterium tuberculosis by EasyNAT diagnostic kit in sputum samples from Tanzania. J Clin Microbiol 53:1342–1344

Mi X, He F, Xiang M, Lian Y, Yi S (2012) Novel phage amplified multichannel series piezoelectric quartz crystal sensor for rapid and sensitive detection of Mycobacterium tuberculosis. Anal Chem 84:939–946

Mirrett S, Hanson KE, Reller LB (2007) Controlled clinical comparison of VersaTREK and BacT/ALERT blood culture systems. J Clin Microbiol 45:299–302

Mistry NF, Iyer AM, D'souza DTB, Taylor GM, Young DB, Antia NH (2002) Spoligotyping of Mycobacterium tuberculosis isolates from multiple-drug-resistant tuberculosis patients from Bombay, India. J Clin Microbiol 40:2677–2680

Nasiruddin M, Neyaz MK, Das S (2017) Nanotechnology-based approach in tuberculosis treatment. Tuberc Res Treat 4920209:1–12

Nathavitharana RR, Hillemann D, Schumacher SG, Schlueter B, Ismail N, Omar SV, Sikhondze W, Havumaki J, Valli E, Boehme C et al (2016) Multicenter noninferiority evaluation of hain genoType MTBDRplus version 2 and Nipro NTM+MDRTB line probe assays for detection of rifampin and isoniazid resistance. J Clin Microbiol 54:1624–1630

Nguyen HH, Park J, Kang S, Kim M (2015) Surface plasmon resonance: a versatile technique for biosensor applications. Sensors 15:10481–10510

Nikam C, Kazi M, Nair C, Jaggannath M, M M, R V, Shetty A, Rodrigues C (2014) Evaluation of the Indian TrueNAT micro RT-PCR device with GeneXpert for case detection of pulmonary tuberculosis. Int J Mycobacteriol 3:205–210

Obasanya J, Lawson L, Edwards T et al (2017) FluoroType MTB system for the detection of pulmonary tuberculosis. ERJ Open Res:1–3

Osman M, Simpson JA, Caldwell J, Bosman M, Nicol MP (2014) GeneXpert MTB/RIF version G4 for identification of rifampin-resistant tuberculosis in a programmatic setting. J Clin Microbiol 52:635–637

Pai M, O'Brien R (2008) New diagnostics for latent and active tuberculosis: state of the art and future prospects. Semin Respir Crit Care Med 29:560–568

Pai M, Riley LW, Colford JM (2004) Interferon-gamma assays in the immunodiagnosis of tuberculosis: a systematic review. Lancet Infect Dis 4:761–776

Pai M, Nathavitharana R (2014) Extrapulmonary tuberculosis: new diagnostics and new policies. Indian J Chest Dis Allied Sci 56:71–73

Pang Y, Dong H, Tan Y et al (2016) Rapid diagnosis of MDR and XDR tuberculosis with the MeltPro TB assay in China. Sci Rep 6:25330

Parsons LM, Somoskövi Á, Gutierrez C, Lee E, Paramasivan CN, Abimiku A, Spector S, Roscigno G, Nkengasong J (2011) Laboratory diagnosis of tuberculosis in resource-poor countries: challenges and opportunities. Clin Microbiol Rev 24:314–350

Pham TH, Peter J, Mello FCQ, Parraga T, Lan NTN, Nabeta P, Valli E, Caceres T, Dheda K, Dorman SE et al (2018) Performance of the TB-LAMP diagnostic assay in reference laboratories: Results from a multicentre study. Int J Infect Dis 68:44–49

Prabhakar N, Arora K, Arya SK, Solanki PR, Iwamoto M, Singh H, Malhotra BD (2008) Nucleic acid sensor for M. tuberculosis detection based on surface plasmon resonance. Analyst 133:1587–1592

Prabowo BA, Alom A, Secario MK, Masim FCP, Lai H-C, Hatanaka K, Liu K-C (2016) Graphene-based portable SPR sensor for the detection of Mycobacterium tuberculosis DNA strain. Procedia Eng 168:541–545

Qin D, He X, Wang K, Zhao XJ, Tan W, Chen J (2007) Fluorescent nanoparticle-based indirect immunofluorescence microscopy for detection of Mycobacterium tuberculosis. J Biomed Biotechnol 2007:89364

Ra SW, Lyu J, Choi C-M, Oh Y-M, Lee S-D, Kim WS, Kim DS, Shim TS (2011) Distinguishing tuberculosis from Mycobacterium avium complex disease using an interferon-gamma release assay. Int J Tuberc Lung Dis Off J Int Union Tuberc Lung Dis 15:635–640

Reischl U, Lehn N, Wolf H, Naumann L (1998) Clinical evaluation of the automated COBAS AMPLICOR MTB assay for testing respiratory and nonrespiratory specimens. J Clin Microbiol 36:2853–2860

Rienthong S, Boonin C, Chaiyasirinrote B, Satproedprai N, Mahasirimongkol S, Yoshida H, Kondo Y, Namwat C, Rienthong D (2015) Evaluation of a novel line-probe assay for genotyping-based diagnosis of Mycobacterium tuberculosis in Thailand. Int J Tuberc Lung Dis Off J Int Union Tuberc Lung Dis 19:817–822

Rizvi SB, Ghaderi S, Keshtgar M, Seifalian AM (2010) Semiconductor quantum dots as fluorescent probes for *in vitro* and *in vivo* bio-molecular and cellular imaging. Nano Rev 1:1–10

Rocchetti TT, Silbert S, Gostnell A, Kubasek C, Widen R (2016) Validation of a multiplex realrime PCR assay for detection of Mycobacterium spp., Mycobacterium tuberculosis complex, and Mycobacterium avium complex directly from clinical samples by use of the BD Max open system. J Clin Microbiol 54:1644–1647

Rodrigues C, Shenai S, Sadani M, Sukhadia N, Jani M, Ajbani K, Sodha A, Mehta A (2009) Evaluation of the bactec MGIT 960 TB system for recovery and identification of Mycobacterium tuberculosis complex in a high volume tertiary care centre. Indian J Med Microbiol 27:217

Rufai SB, Kumar P, Singh A, Prajapati S, Balooni V, Singh S (2014) Comparison of Xpert MTB/RIF with line probe assay for detection of rifampin-monoresistant Mycobacterium tuberculosis. J Clin Microbiol 52:1846–1852

Rufai SB, Singh S, Singh A, Kumar P, Singh J, Vishal A (2017) Performance of Xpert MTB/RIF on ascitic fluid samples for detection of abdominal tuberculosis. J Lab Physician 9:47–52

Sabale S, Kandesar P, Jadhav V, Komorek R, Motkuri RK, Yu X-Y (2017) Recent developments in the synthesis, properties, and biomedical applications of core/shell superparamagnetic iron oxide nanoparticles with gold. Biomater Sci 5:2212–2225

Sharma P, Bhargava M (2013) Applications and characteristics of nanomaterials in industrial environment. Int J Civ, Struct, Environ Infrastructure Eng Res Dev (IJCSEIERD) 3:63–72. ISSN 2249-6866

Sharma S, Zapatero-Rodríguez J, Estrela P, O'Kennedy R (2015) Point-of-care diagnostics in low resource settings: present status and future role of microfluidics. Biosensors 5:577–601

Shenai S, Armstrong DT, Valli E, Dolinger DL, Nakiyingi L, Dietze R, Dalcolmo MP, Nicol MP, Zemanay W, Manabe Y et al (2016) Analytical and clinical evaluation of the epistem genedrive assay for detection of Mycobacterium tuberculosis. J Clin Microbiol 54:1051–1057

Singh A, Gopinath K, Sharma P, Bisht D, Sharma P, Singh N, Singh S (2015) Comparative proteomic analysis of sequential isolates of Mycobacterium tuberculosis from a patient pulmonary tuberculosis turning from drug sensitive to multidrug resistant. Indian J Med Res 141:27–45

Singh A, Gupta AK, Gopinath K, Sharma P, Singh S (2017) Evaluation of 5 Novel protein biomarkers for the rapid diagnosis of pulmonary and extra-pulmonary tuberculosis: preliminary results. Sci Rep 7:44121. https://doi.org/10.1038/srep44121(2017)

Skoura E, Zumla A, Bomanji J (2015) Imaging in tuberculosis. Int J Infect Dis 32:87–93

Smith AM, Nie S (2010) Semiconductor nanocrystals: structure, properties, and band gap engineering. Acc Chem Res 43:190–200

Sonawane MD, Nimse SB (2016) Surface modification chemistries of materials used in diagnostic platforms with biomolecules. J Chem:1–19

Stephen BJ, Singh SV, Datta M, Jain N, Jayaraman S, Chaubey KK, Gupta S, Singh M, Aseri GK, Khare N et al (2015) Nanotechnological approaches for the detection of mycobacteria with special references to Mycobacterium avium subspecies paratuberculosis (MAP). Asian J Anim Vet Adv 10:518–526

Sun R, Wang W, Wen Y, Zhang X (2015) Recent advance on mesoporous silica nanoparticles-based controlled release system: intelligent switches open up new horizon. Nano 5:2019–2053

Tsai T-T, Huang C-Y, Chen C-A, Shen S-W, Wang M-C, Cheng C-M, Chen C-F (2017) Diagnosis of tuberculosis using colorimetric gold nanoparticles on a paper-based analytical device. ACS Sens 2:1345–1354

Tuberculosis Division IUTALD (2005) Tuberculosis bacteriology–priorities and indications in high prevalence countries: position of the technical staff of the Tuberculosis Division of the International Union Against Tuberculosis and Lung Disease. Int J Tuberc Lung Dis Off J Int Union Tuberc Lung Dis 9:355–361

Vallabani NVS, Singh S (2018) Recent advances and future prospects of iron oxide nanoparticles in biomedicine and diagnostics. 3 Biotech 8:279

van den Hurk R, Evoy S (2015) A review of membrane-based biosensors for pathogen detection. Sensors 15:14045–14078

Wang D-S, Fan S-K (2016) Microfluidic surface plasmon resonance sensors: from principles to point-of-care applications. Sensors (Basel) 16(8)

Wang EC, Wang AZ (2014) Nanoparticles and their applications in cell and molecular biology. Integr Biol Quant Biosci Nano Macro 6:9–26

Wang S, Inci F, De Libero G, Singhal A, Demirci U (2013) Point-of-care assays for tuberculosis: role of nanotechnology/microfluidics. Biotechnol Adv 31:438–449

Wang SX, Tay L (1999) Evaluation of three nucleic acid amplification methods for direct detection of Mycobacterium tuberculosis complex in respiratory specimens. J Clin Microbiol 37:1932–1934

Wang Y, Yu L, Kong X, Sun L (2017) Application of nanodiagnostics in point-of-care tests for infectious diseases. Int J Nanomedicine 12:4789–4803

Watanabe Pinhata JM, Cergole-Novella MC, Moreira dos Santos Carmo A, Ruivo Ferro e Silva R, Ferrazoli L, Tavares Sacchi C, Siqueira de Oliveira R (2015) Rapid detection of Mycobacterium tuberculosis complex by real-time PCR in sputum samples and its use in the routine diagnosis in a reference laboratory. J Med Microbiol 64:1040–1045

WHO (2017) WHO | Global tuberculosis report 2017

Xu K, Liang ZC, Ding X, Hu H, Liu S, Nurmik M, Bi S, Hu F, Ji Z, Ren J et al (2018) Nanomaterials in the prevention, diagnosis, and treatment of Mycobacterium Tuberculosis infections. Adv Healthc Mater 7:1700509

Yadav R, Sharma N, Khaneja R, Agarwal P, Kanga A, Behera D, Sethi S (2017) Evaluation of the TB-LAMP assay for the rapid diagnosis of pulmonary tuberculosis in Northern India. Int J Tuberc Lung Dis Off J Int Union Tuberc Lung Dis 21:1150–1153

Yang H, Qin L, Wang Y, Zhang B, Liu Z, Ma H, Lu J, Huang X, Shi D, Hu Z (2014) Detection of Mycobacterium tuberculosis based on H R binding peptides using surface functionalized magnetic microspheres coupled with quantum dots – a nano detection method for Mycobacterium tuberculosis. Int J Nanomedicine 17:77–88

Yukuyama MN, Kato ETM, Lobenberg R, Bou-Chacra NA (2017) Challenges and future prospects of nanoemulsion as a drug delivery system. Curr Pharm Des 23:495–508

Zhang M-J, Ren W-Z, Sun X-J, Liu Y, Liu K-W, Ji Z-H, Gao W, Yuan B (2018) GeneChip analysis of resistant Mycobacterium tuberculosis with previously treated tuberculosis in Changchun. BMC Infect Dis 18:234

Zhang T, Hu S, Li G, Li H, Liu X, Niu J, Wang F, Wen H, Xu Y, Li Q (2015) Evaluation of the MeltPro TB/STR assay for rapid detection of streptomycin resistance in Mycobacterium tuberculosis. Tuberc Edinb Scotl 95:162–169

Zhou L, He X, He D, Wang K, Qin D (2011) Biosensing technologies for Mycobacterium tuberculosis detection: status and new developments. Clin Dev Immunol 193963:1–8

Zhu L, Liu Q, Martinez L, Shi J, Chen C, Shao Y, Zhong C, Song H, Li G, Ding X et al (2015) Diagnostic value of GeneChip for detection of resistant Mycobacterium tuberculosis in patients with differing treatment histories. J Clin Microbiol 53:131–135

Zhu L, Song L, Liu W, Yan X, Xiao J, Chen C (2017) Gold nanoparticles-modified indium tin oxide microelectrode for in-channel amperometric detection in dual-channel microchip capillary electrophoresis. Anal Methods 29:4319–4326

Molecular Mechanisms of Drug Resistance in *Mycobacterium tuberculosis*: Role of Nanoparticles Against Multi-drug-Resistant Tuberculosis (MDR-TB)

12

Amit Singh, Anil Kumar Gupta, and Sarman Singh

Abstract

Tuberculosis (TB) is a leading chronic bacterial infection caused by *Mycobacterium tuberculosis* (*M. tuberculosis*) and an increasing public health threat. The current therapeutic management of *M. tuberculosis* is insufficient due to the prolonged course of treatment, side effects of drugs, and unorganized therapy, and these aspects can lead to therapeutic failure and development of drug-resistant tuberculosis. The multi-drug-resistant (MDR), extensively drug-resistant (XDR), and total drug-resistant (TDR) tuberculosis pose significant challenges to the diagnosis, lengthy course of treatment, higher side effect, cost, and control of tuberculosis worldwide. Drug resistance to the anti-TB drugs has existed since the commencement of the antibiotic era. The understanding of the entire mechanisms of drug resistance helps in the development of newer rapid diagnostic tools and newer drug with novel targets for drug-resistant TB. The newer diagnostics and drug target tools help to improve the existing treatment, management, and prevent emergence of TB. The recent advances in the new-generation sequencing (NGS) help to unravel the novel gene mutations to understand the mechanism of drug resistance. The physiognomies of the nanoparticle in the treatment of MDR-TB and XDR-TB are discussed. The targeted nanoparticle-based treatment may further increase the efficacy with less dosage and reduced toxic side effects of drugs. This chapter summarises the molecular mechanism of drug resistance and novel drug delivery systems for treatment of the drug-resistant and susceptible TB.

Amit Singh, Anil Kumar Gupta and Sarman Singh have contributed equally to this chapter.

A. Singh · A. K. Gupta · S. Singh (✉)
Division of Clinical Microbiology and Molecular Medicine,
All India Institute of Medical Sciences, New Delhi, India

Department of Microbiology, All India Institute of Medical Sciences, Bhopal, India

S. K. Saxena, S. M. P. Khurana (eds.), *NanoBioMedicine*,
https://doi.org/10.1007/978-981-32-9898-9_12

Keywords
Tuberculosis · Drug resistant · Nanotechnology · MDR-TB · Nanoparticle

12.1 Introduction

Tuberculosis (TB) is an infectious disease caused by *Mycobacterium tuberculosis* (*M. tuberculosis*), killing maximum number of humans. According to the latest estimates of 2017, there are almost 10.4 million new TB cases and 1.3 million deaths per annum (WHO 2017). Worldwide, the latent TB infections has exaggerated over one-third of the population. *M. tuberculosis* causes ill health among millions of people every year in developing and developed countries and ranks as the second leading cause of death in the infectious disease category worldwide, after the emergence of Human Immunodeficiency Virus (HIV) and the development of multi- and extensively drug-resistant (MDR and XDR) tuberculosis strains, which cause a major hurdle in treatment across the globe (Banerjee et al. 1994).

This is despite the allocation of massive resources to support the WHO recommended DOTS programs which are meant to prevent drug resistance in TB. MDR-TB refers to simultaneous resistance to rifampicin (RIF) and isoniazid (INH), whereas XDR-TB refers to MDR-TB plus the additional resistance to at least one injectable drug and a quinolone. Both MDR and XDR forms of TB are difficult to treat and expensive, lasting for a long duration, more likely to fail and result in death of patient. The need for extended therapy using various combinations of drugs remains a practical hindrance to effective tuberculosis control (Baulard et al. 2000). Factors such as aging populations, poverty, malnutrition, and the current epidemic of HIV all together have created an immense pool of individuals vulnerable to drug-resistant TB.

The management of MDR-TB with traditional drug delivery system has seen an increase in drug resistance. This resistance has resulted from noncompliance of lower socioeconomic patient due to the high dose of drugs-related side effects and prolonged treatment. The development of drug resistance, MDR, and XDR TB indicates the urgent need for understanding of drug-resistance mechanism and new drug delivery systems such as nanoparticles or nanoparticles containing drugs for anti-tubercular treatment (ATT). These nanoparticles will help to overcome the hitches associated with the drug delivery system of conventional drugs in the tuberculosis treatment. Drug resistance against all classes of natural and synthetic existing antibiotics leads to an urgent requirement for newer drugs and alternatives. In the last 40 years, only a few antibiotics have been discovered (Projan 2003).

12.2 Mechanisms of Drug Resistance

The occurrence of drug resistance represents one of the most serious problems in the control of infectious diseases. The resistance is defined as temporary or permanent decline in the efficacy of a drug against an organism's population previously susceptible to that compound. All drugs used in the treatment of TB tend to select drug-resistant strain. Following are the types of drug resistance: (a) Primary resistance: It is defined as the occurrence of drug resistance in a TB patient who has never received any prior treatment with anti-TB drug. (b) Secondary/acquired drug resistance: resistance that arises through or after a course of chemotherapy, as a result of non-adherence to the suggested drug regimen or faulty prescription. (c) Initial or transitional drug resistance: This type originated during treatment where sometimes a few colonies of resistant culture are obtained just before sputum conversion. These bacilli do not multiply nor does their presence influences treatment outcome.

Drug resistance is the decrease in effectiveness of an antimicrobial drug. Microorganisms have established various ways to resist the toxic effects of drug. (A) Drug modifying and inactivating enzymes: The mycobacterial genome codes for several enzymes that inactivate antibiotics by hydrolysis, phosphorylation, acetylation or adenylation of the drug compounds, and by formation of inactive derivatives (Davies 1994), for example, β-lactamases and aminoglycoside modifying enzymes. (B) Prevention of drug influx: The *M. tuberculosis* cell wall consists of complex lipids, and it acts as a permeability barrier for entry of drugs into the cell. (C) Overactive drug efflux system: Once the drug has entered the cell, these can be actively effluxed out with the help of active proteins (Louw et al. 2009; Da Silva et al. 2011). Efflux pumps are energy-dependent and may utilize the transmembrane electrochemical gradient of protons or ATP to drive the extrusion of drug from the cell. (D) Mutations (Target alteration): Spontaneous mutations in the *M. tuberculosis* genome can give rise to enzymes/proteins modification that alter the affinity of the antibiotic for the target and make the bacterium resistant to drug, depending on the action of drug (Fig. 12.1).

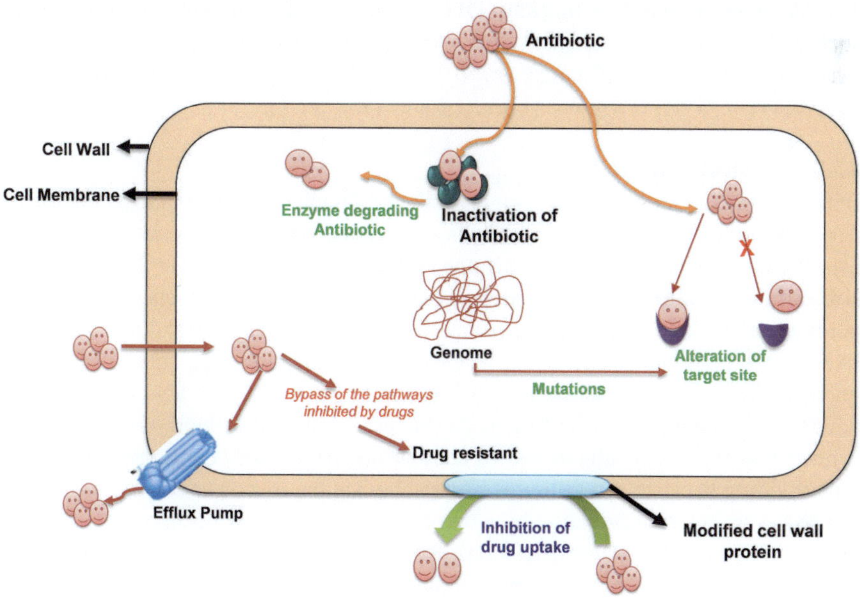

Fig. 12.1 Mechanism of drug resistance in *Mycobacterium tuberculosis*

MDR-TB mainly occurs where sequential drug resistance follows constant treatment failure. The treatment of tuberculosis can be divided into first-line and second-line drugs. In order to control epidemic of the drug resistance, it is necessary to gain perception of how *M. tuberculosis* develops drug resistance? This information will help us in understanding better how to prevent the incidence of drug resistance as well as identifying proteins/genes associated with drug resistance of anti-tuberculosis drugs.

M. tuberculosis is somewhat uncommon among bacterial pathogens where drug resistance is not associated with specific transacting "resistance genes" acquired horizontally from other bacteria. Instead, drug resistance arises wholly through mutation of mycobacterial housekeeping genes coding for drug targets, for proteins involved in drug uptake, efflux pump, or for the activation of pro-drugs. Multi-drug resistance seems to be the outcome of sequential accumulation of such mutations during failed chemotherapy. Mutations in the genes of *M. tuberculosis* that can cause resistance to anti-TB drugs occur spontaneously with an estimated rate of 3.5×10^{-6} for INH, 3.1×10^{-8} for RIF, 2.29×10^{-8} for STR, and 1.10×10^{-7} for ETH (Dooley and Simone 1994).

12.2.1 Molecular Mechanism of Resistance to First-Line Drugs

The drugs used in anti-TB treatment are supposed to have effective sterilizing action that is capable of shortening the treatment period. According to WHO/RNTCP guidelines, the four major drugs prescribed for Category I (Cat I) patients are Isoniazid (INH), Rifampicin (RIF), pyrazinamide (PZA), and ethambutol (EMB). The drug resistance to first-line anti-TB drugs has been linked to mutations in genes, such as *katG, ahpC, inhA, kasA,* and *ndh* for INH resistance*; rpoB* for RIF resistance, *pncA* for PZA resistance, *embB* for EMB resistance, and *rpsL* and *rrs* for streptomycin (STR) resistance (Fig. 12.2 and Table 12.1).

12.2.1.1 Isoniazid (INH)

INH is one of the most effective drugs for the treatment of *M. tuberculosis* infection. The Minimum Inhibitory Concentration (MIC) of INH against the *M. tuberculosis* is 0.05 μg/ml. INH is a pro-drug that enters into the mycobacterial cell through passive diffusion (Bardou et al. 1998). It is activated by a *katG* gene encoded catalase peroxidase enzyme. The peroxidase activity of the enzyme is important to activate INH into a toxic substance in the bacterial cell, and this toxic substance subsequently affects intracellular targets such as inhibition of mycolic acid biosynthesis, which are an important part of the mycobacterial cell wall. A lack of mycolic acid synthesis ultimately results in loss of cellular integrity (Barry et al. 1998) and kills only dividing bacteria. No killing is observed when the mycobacteria are in stationary phase or growing in anaerobic conditions (Mitchison and Selkon 1956). INH drug is activated by catalase-peroxidase hemoprotein (katG). The action of INH is bacteriostatic for the first 24 hours of treatment, after which the action turns bactericidal (Vilchèze and Jacobs Jr 2007; Singh et al. 2014a, b). Despite its simple

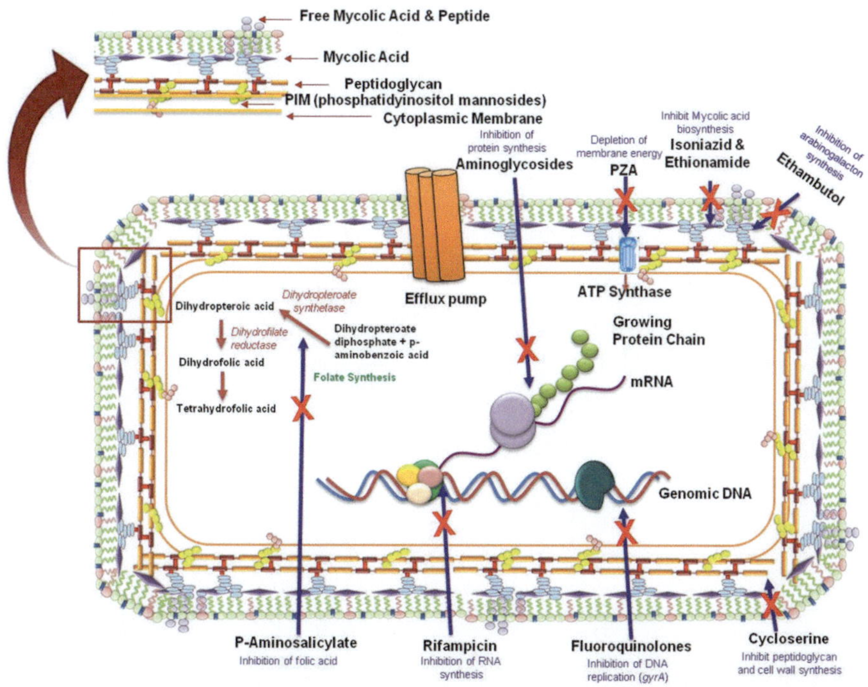

Fig. 12.2 Molecular mechanism of drug resistance in tuberculosis (figure adopted from doctoral thesis of Dr. Amit Singh, AIIMS, New Delhi)

structure, the mechanism by which INH kills *M. tuberculosis* is complex. It induces morphological alterations in mycobacteria, such as surface bulging and wrinkles or loss of internal structure (Takayama et al. 1973). While activated, INH binds with InhA-nicotinamide adenine dinucleotide (NADH) and forms INH-NADH complex to form a ternary complex that results in inhibition of mycolic acid biosynthesis. The role of the β-ketoacyl ACP synthase (*kasA* gene product) and the specific nature of the INH metabolites responsible for activity is still to be predicted. The crystal structures of InhA and the related enzyme MabA of *M. tuberculosis* are solved (Dessen et al. 1995; Cohen-Gonsaud et al. 2002).

The drug resistance to INH is due to an alteration of target genes (such as *inhA, katG, kasA, ahpC, ndh*). The mutations occur in the activating enzyme (*katG*) and in the target gene (*inhA*). The expression of proteins which are not targeted resulted in drug resistance. Conspicuous among these non-target resistance determinants are the enzymes that can no longer activate the pro-drug, covalently modify the target, or modify the antibiotic or active efflux of drug from the cell. Sometimes the molecular target itself can generate high-level resistance to INH as in the case of *InhA* over expression. Specific mutations or overexpression of the target *inhA* gene may generate organisms having increased MICs of INH and ethionamide (ETA), with MICs at least five times higher than wild-type (WT) (Vilchèze et al. 2006). The transfer of *inhA* mutant gene S94A into WT *M. tuberculosis* was sufficient to confer

Table 12.1 Molecular mechanism and properties of drug resistance to first-line anti-TB drugs

S. no.	Drug class	Drugs	Year of drug discovery	Drug structure	Mechanism of action	Gene(s) involved in resistance	Gene(s) function	Mutation frequency
1.	Isonicotinic acid hydrazide	Isoniazid	1952	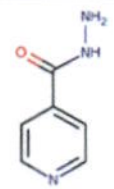	Inhibition of mycolic acid biosynthesis and other effects	*KatG*	Catalase-peroxidase	50–95%
						InhA	Enoyl ACP reductase	
						acpH	Alkyl hydroperoxide reductase	8–43%
						KasA & B		10–15%
2.	Rifamycin	Rifampicin	1966	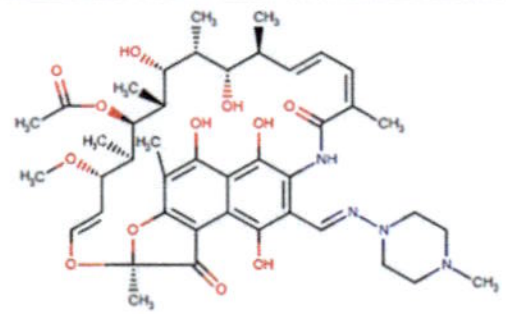	Inhibition of DNA-dependent RNA polymerase; inhibition of RNA synthesis	*rpoB*	β-subunit of RNA polymerase	8%
3.	Unspecified	Ethambutol	1961	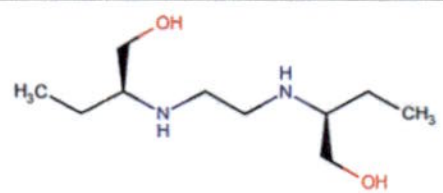	Inhibition of arabinogalacton synthesis; inhibition of the transfer of mycolic acids into the cell wall and also the synthesis of spermidine	*embCAB*	Arabinosyl transferase	47–65%

4.	Aminoglycosides	Streptomycin	1944	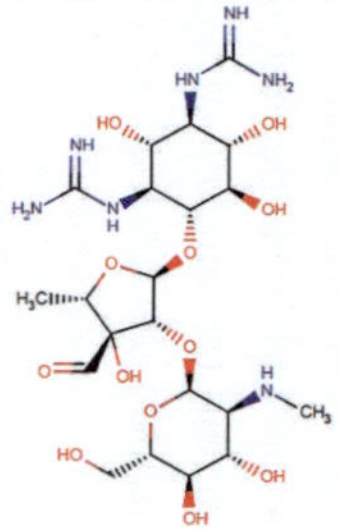	Inhibition of protein synthesis	*rpsL*	S12 ribosomal protein	52–59%
						rrs	16s rRNA	8–21%
						gidB	rRNA methyltransferase	–
5.	Synthetic derivative of nicotinamide	Pyrazinamide	1952	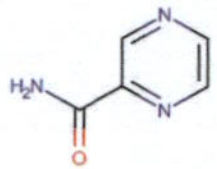	Depletion of membrane energy and also inhibition of the activity of FAS I	*pncA*	Nicotinamidase/ Pyrazinamidase	72-97%

Table adopted from doctoral thesis of Dr. Amit Singh, AIIMS, New Delhi

resistance to INH and ETA, demonstrating that this is the target for INH (Vilchèze et al. 2006). Six-point mutations at codons 16 (I-T), 21 (I-T and I-V), 47 (I-T), 78 (V-A), and at 95 (I-P) are also associated with INH resistance. Substitution of mutation at codon 94 (S-T) results in decreased binding affinity of *inhA* for NADH, resulting in the inhibition of mycolic acid synthesis. The mutations in *inhA* promoter regions are frequently observed at positions -15(C-T), -16(A-G), -8(T-G/A), and -24(G-T). These mutations at the promoter region result in the over expression of *inhA* leading to low level INH resistance (Ramaswamy and Musser 1998). The mutations in *katG* gene lead to high-level resistance in INH-resistant clinical isolates which carry mutations at codon 138, 328, 315 (S-T), and 463 (R-L) (Cohen-Gonsaud et al. 2002). INH-resistant mutants with alteration in the genes *ahpC* (alkyl hydroperoxide reductase), *ndh* (NADH dehydrogenase), and *kasA* have also been observed. The mutation in codon 315 (S-T) causes loss of *katG* activity and is accompanied by increased expression of *ahp*C protein, which protects the bacteria against oxidative damage and is capable of detoxifying organic peroxides within the cell, but it does not provide protection against INH (Sherman et al. 1996).

The *kasA* gene encodes a ß-ketoacyl-ACP synthase involved in the synthesis of mycolic acid. Mutations have been described in *kasA* gene, which confer low levels of INH resistance. Genotypic study of the *kas*A gene discloses four different amino acid changes involving codons 66 (G-A), 269 (G-A), 312 (G-A), and 413 (C-A) (Mdluli et al. 1998; Ramaswamy and Musser 1998) reported similar mutation in INH sensitive isolates. NADH dehydrogenase binds with the active site of *inhA* to form the ternary complex with activated INH encoded by *ndh* gene. Structural studies have shown that a reactive form of INH binds with NAD(H) co-factor and generates a covalent adduct of INH-NAD. The mutations in the *ndh* gene cause defects in the enzymatic activity, which results in NADH accumulation and NAD depletion (Lee et al. 2001). The accumulation of NADH can then inhibit the binding of the adduction of INH-NAD to the active site of the InhA enzyme. Earlier *ndh* gene mutation analysis showed changes in codons at position 110 (T to A) and 268 (R to H) in 9.5% of INH-resistant isolates (Lee et al. 2001).

12.2.1.2 Rifampicin (RIF)

RIF is a semi-synthetic antibiotic produced from *Streptomyces mediterranei*. It is a broad spectrum antibacterial having excellent sterilizing activity against *M. tuberculosis* (Rattan et al. 1998). RIF is an important first-line drug used along with INH and PZA. It is the backbone of short-course chemotherapy (DOTS). RIF inhibits the essential *rpoB* gene, a product of ß-subunit of DNA-dependent RNA polymerase activity, early acting in transcription (Wehrli et al. 1968). RNA polymerase is composed of four different subunits (α, ß, ß', and σ) encoded by *rpoA, rpoB, rpoC,* and *rpoD* genes, respectively. The binding of RIF with ß-subunit closes the RNA/DNA channel and blocks the transit of growing RNA chain.

Drug resistance to RIF is mainly associated with the mutations in the *rpoB* gene, which confer conformational variations leading to defective binding of drugs and consequently resistance. It is estimated that more than 95% of *M. tuberculosis* clinical isolates resistant to RIF have a mutation in an 81-bp region of *rpoB* known as the rifampicin resistance–determining region (RRDR) between codons 507 and 533, and these mutations are associated with a high level of resistance to RIF (Ramaswamy and Musser 1998). Mutations are dominated by single nucleotide changes, resulting in substitutions of single amino acid (especially mutations at codons 531, 526, and 516). However, in-frame deletion and insertion mutations also occur at lower frequencies at codons 511, 516, 518, and 522. In RRDR region of *rpoB* gene, 16 single base changes, two insertions, one deletion, and five multiple mutations (with two or three codons implicated) were reported. It was also observed that eight strains had no mutations, suggesting the existence of alternate mechanisms to RIF resistance in *M. tuberculosis* (Singh et al. 2015, 2016a; b). Future studies/experiments on these strains on the differential mechanism of RIF's resistance will present great scope in understanding the resistance mechanism leading to identifying a new drug target and intern-effective treatment of MDR-TB (Rattan et al. 1998; Aragón et al. 2006; Comas et al. 2012).

12.2.1.3 Pyrazinamide (PZA)

PZA is a pro-drug, bactericidal in nature and active at acidic pH. The mechanism of PZA action is poorly understood. It targets an enzyme involved in the fatty acid synthesis and is responsible for killing persistent bacilli of tuberculosis in the initial intensive phase of chemotherapy (Somoskovi et al. 2001). PZA diffuses into bacteria in a passive manner and gets activated into pyrazinoic acid (POA) by pyrazinamidase (PZAase) encoded by *pncA*. The accumulation of POA drops the pH to a suboptimal level, which leads to inactivation of a vital target enzyme, such as fatty acid synthase FAS I (Boshoff et al. 2002) and also may help to disrupt membrane transport function. This interferes with the *M. tuberculosis* ability to synthesize new fatty acids, required for replication and growth. The activity of PZA is highly specific for TB, as it has no effect on other mycobacteria. *M. bovis* is naturally resistant to PZA due to a unique point mutation in codon 169 (C to G) of the *pncA* gene.

The mutations in *pncA* gene in *M. tuberculosis* conferred drug resistance of PZA (Scorpio and Zhang 1996). The mutations are unusually located and spread throughout the gene, but there are three hot spots of clustered mutations around amino acids 3–71, 61–85, and 132–142. A small number of PZA mutations occur outside the *pncA* gene (coding for PZAase) but these have not been characterized (Scorpio and Zhang 1996). PZA-resistant isolates without *pncA* mutations have also been observed, which suggest the existence of another mechanism conferring PZA resistance. The mutations in *pncA* at codon 114 (T to M) are not associated with PZA resistance. The accurate data of PZA susceptibility testing is not available, because it is not done routinely in many countries due to technical problems. Hence, the exact data of global incidence to PZA resistance is unknown.

12.2.1.4 Ethambutol (EMB)

EMB is a bacteriostatic antibiotic that is active against growing bacilli, but do not have any effect on non-replicating or dormant bacilli (Takayama and Kilburn 1989). EMB inhibits arabinosyl-transferases (*embB*) involved in cell-wall biosynthesis (Telenti et al. 1997). It inhibits the polymerization of cell wall components *i.e.* arabinan of arabino-galactan (AG) and of lipo-arabinomannan (LAM), induces the accumulation of D-arabinofuranosyl-P-decaprenol (intermediate product of arabinan biosynthesis). AG forms part of the mucolyl-AG-peptidoglycan layer which anchors the peptidoglycan layer to the lipid-mycolic acid outer layer encoded by *embA* and *embB* genes. The LAM that appears to be attached to the cell membrane via phosphatidyl-inositol was encoded by *embC* (Deng et al. 1995; Sreevatsan et al. 1997). During EMB treatment, more than 50% of the cell arabinan were released from the bacteria, whereas no galactan was released (Deng et al. 1995). There are multiple arabinosyl-transferase enzymes existing, but the *embB* gene product seems to be the main target in *M. avium.* However, knockouts study in *M. smegmatis* of *embA*, *embB,* and *embC* were all viable; the knock-out mutant of *embB* is the slowest growing (Escuyer et al. 2001).

Mutations of *embB* result in the development of drug resistance to EMB. The most common mutations in *embB* gene are at codon 306 (A to G, A to C, G to A, and G to C), which result in three different amino acid substitutions (V, L, and I) in resistant isolates. Mutants in multiple *emb* genes may have even higher MICs. EMB mutants (~25% in some studies) with no changes in the *emb* genes have also been identified (Sreevatsan et al. 1997; Ramaswamy et al. 2000). These five mutations are associated with 70–90% of all EMB resistant isolates (Ramaswamy and Musser 1998). Missense mutations were also identified in three different codons at 285 (F to L), 330 (F to V), and 630 position (T to I) in EMB-resistant isolates. MICs were commonly higher for strains with 306 position (M to L; M to V; P to V; and T to I) mutations than those organisms with codon 306 (M to I). Mutations other than codon 306 are also present, but are quite rare. There are approximately 30% clinical isolates which lack the mutation in *embB* but have phenotypic resistance to EMB (Johnson et al. 2006). Therefore, there is a need for further study to understand the drug-resistance mechanism of the EMB in clinical isolates. There is no cross-resistance of EMB with other drugs, which has come to light.

12.2.1.5 Streptomycin (STR)

Streptomycin was the first aminoglycoside antibiotic identified produced by the soil actinomycetes *Streptomyces griseus.* It is an alternative first-line ATT recommended by the WHO (Cooksey et al. 1996). It is bactericidal due to effects that are not fully understood. It is in the same class as kanamycin (KAN) and amikacin (AMI), which are used for second-line ATT. It acts by binding to 30S ribosomal subunit of susceptible organisms and disrupts the initiation and elongation steps in protein synthesis. Specifically, STR binds to the 16S rRNA four nucleotides of a single amino acid of S12 protein (*rrs* and *rpsL).* This interferes with decoding spot in the vicinity of nucleotide 1400 in 16S rRNA of 30S subunit (Sreevatsan et al. 1996; Chan et al. 2000). This region interacts with the wobble base in the anticodon of tRNA. This leads to intrusion with the initiation complex, misreading of mRNA and, in turn, leading to insertion of incorrect amino acids into the polypeptides leading to non-functional or toxic peptides, and the breakup of polysomes into non-functional monosomes.

Drug resistance mechanisms were associated with ribosomal changes in the 16S rRNA and ribosomal protein S12 (*rrs* and *rpsL*) inducing ribosomal changes, which lead to misreading of the mRNA and inhibition of protein synthesis in *M. tuberculosis.* A point mutation in *rrs* and *rpsL* gene in 65–67% of STR-resistant isolates were reported (Ramaswamy and Musser 1998) in the *rrs* gene of a C-T transition at positions 491, 512, and 516, and an A-C/T transversion at position 513 was observed in the highly conserved 530 loop. This loop region is part of the aminoacyl tRNA binding site and is involved in the decoding process. The C-T alteration at codon 491 is not responsible for resistance to STR as it occurs in both STR sensitive as well as resistant isolates (Victor et al. 2001). Other mutation in the 915 loop at 903 (C to A/G) and 904 (A to G) was also reported to have an association with STR resistance (Carter et al. 2000). The most commonly detected mutations in the *rpsL* gene are at codon 43 (AAG to AGG/ACG) and codon 88 (AAG to AGG/CAG) reported in 50% of STR-resistant strains (Nair et al. 1993). Recently, it was demonstrated that mutations in *gidB* (Rv3919c), which encode a conserved 7-methylguanosone methyltransferase specific for the 16s rRNA can invoke a low level of STR resistance (Okamoto et al. 2007). In addition, low levels of STR resistance are also being associated with altered cell permeability, efflux pump (Spies et al. 2008), or rare mutations which lie outside of the *rrs* and *rpsL* genes. Cross-resistance with other class members (AMI and KAN) and the macrocycle polypeptide capreomycin (CAP) also exists (Law et al. 2014).

12.2.2 Mechanism of Resistance to Second-Line Drugs

According to the WHO and RNTCP guidelines, the following drugs can be classified as second-line drugs: fluoroquinolones (ofloxacin, ciprofloxacin, moxifloxacin, levofloxacin, and gatifloxacin), aminoglycosides (kanamycin and amikacin), thionamides (ethionamide and prothionamide), D-cycloserine, and polypeptides (capreomycin, viomycin, and enviomycin). Unfortunately, second-line drugs are more toxic and less effective than first-line drugs and are mostly used in the treatment of MDR-TB and resulted in prolong duration of treatment time from 6 to 9 months. The existing understanding of the molecular mechanisms associated with resistance to second-line drugs are summarized in Table 12.2 and Fig. 12.2. The phenotypic methods to detect drug resistance to second line are not well established, and the molecular mechanisms of resistance are also less defined.

12.2.2.1 Fluoroquinolones (FQs)

This class of drug is bactericidal and consists of ciprofloxacin (CIP), ofloxacin (OFL), levofloxacin (LEV), gatifloxacin (GATI), moxifloxacin (MOXI), and so on. Fluoroquinolones (FQs) are used as second-line drugs for treatment of MDR-TB. The quinolones targets and inactivates ATP-dependent enzyme topoisomerase II (DNA gyrase) and topoisomerase IV.

In most bacteria, gyrase facilitates DNA untwisting required to replicate one DNA double helix into two, and topoisomerase IV activates decatenation (active in

Table 12.2 Molecular mechanism and properties of drug resistance to second-line anti-TB drugs

Second-line ATT drug								
6.	Fluoroquinolones (aminoquinolines and derivatives)	Ciprofloxacin	1963	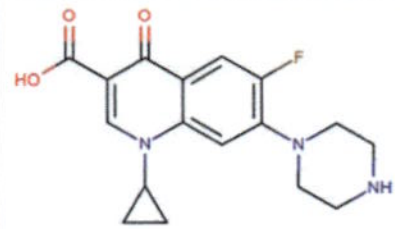	Halting DNA replication by inhibition of gyrA	*GyrA/B*	DNA gyrase subunit A/B	75–94%
		Ofloxacin	1963	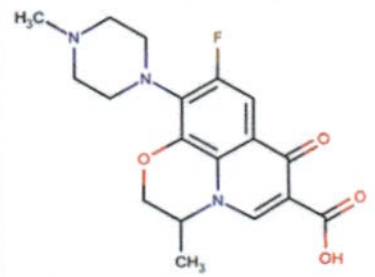	Halting DNA replication by inhibition of gyrA	*GyrA/B*	DNA gyrase subunit A/B	75–94%
		Moxifloxacin	1963	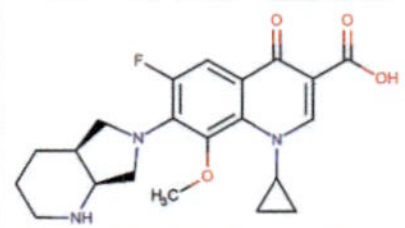	Halting DNA replication by inhibition of gyrA	*GyrA/B*	DNA gyrase subunit A/B	75–94%
7.	Derivative of isonicotinic acid	Ethionamide	1956	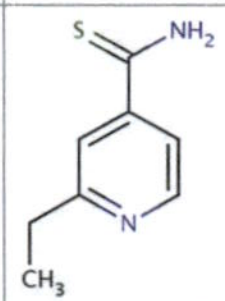	Inhibition of mycolic acid synthesis	*inhA* *ethA* *ethR*	Enoyl-ACP reductase Flavin monooxygenase Transcriptional repressor	37%

8.		D-Cycloserine	1952	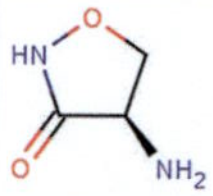	Inhibition of peptidoglycan and cell wall synthesis	*Alr*	Alanine racemose	Unknown
						*ddl**		
9.	Cyclic polypeptide	Capreomycin	1960	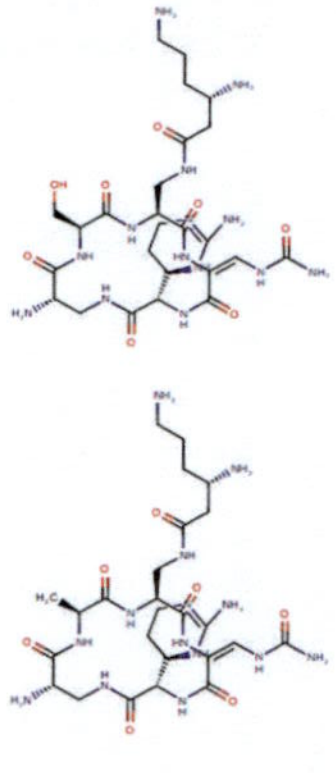 Capreomycin 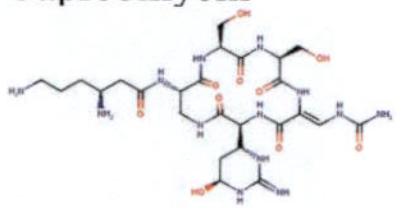 Viomycin		*rrs*	rRNA methyl transferase	75–94%
		Viomycin				*tlyA*	2’-O-methyltransferase	

(continued)

Table 12.2 (continued)

Second-line ATT drug								
10.	Aminoglycosides	Amikacin	1957	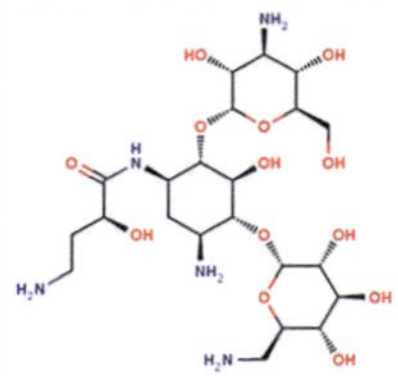	Inhibition of protein synthesis	*rrs* (16sRNA)	16s rRNA	76%
		Kanamycin	1957	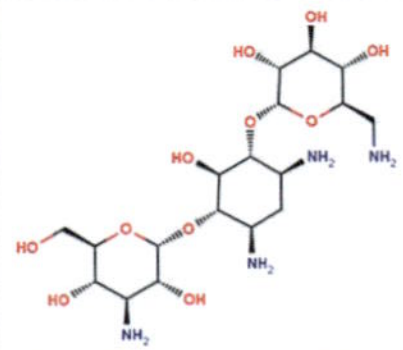	Inhibition of protein synthesis	*rrs* (16sRNA)	16s rRNA	76%
			–			*eis* promoter	Acetylation; involved in intracellular survival	–
11.	Salicylic acid – anti-folate	PAS	1946	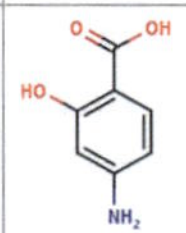	Inhibition of folic acid and iron metabolism	*thyA*	Thymidylate synthase A	36%

Table adopted from doctoral thesis of Dr. Amit Singh, AIIMS, New Delhi

the partitioning of the chromosomal DNA during cell division). DNA gyrase is a tetrameric A2B2 protein (Champoux 2001), an essential enzyme involved in replication, transcription, and repair of bacterial DNA. It consists of two components arranged in a GyrA2/GyrB2 complex encoded by the *gyrA* and *gyrB* genes and introduces negative supercoils in closed circular DNA molecules (Ramaswamy and Musser 1998; Rattan et al. 1998), while topoisomerase IV is coded by *parC* and *pare* (Gordon et al. 2001).

Drug resistance to FQs is due to mutations in quinolone-resistant determining region of *garA* and very rarely in *garB*. There is no known cross-resistance between FQs and other classes of antimicrobials; however, cross-resistance has been observed between FQs. Resistance (3-5 fold higher than wild-type [WT]) can arise in *M. tuberculosis* from changes in either *gyrA*(320bp) or *gyrB*(375bp). Furthermore, specific mutations in *gyrA* (Missense mutations in codon 90, 91, and 94) resulted in hypersensitivity to quinolones. Many MDR clinical strains are sensitive to MOXI even though they are also resistant to OFL; however, it was also observed that *M. tuberculosis* strains were resistant to both MOXI and OFL in which the mutation was often found in *gyrA* (A94G) (Kam et al. 2006). While polymorphism at *gyrA* codon 95 (Ser95>Thr95) occurred in both resistant and sensitive isolates, it is not associated with FQ resistance (Rattan et al. 1998). Another mechanism for FQs resistance may be a reduction in cell-wall permeability to drugs, drug efflux pump, drug sequestration, or even drug activation (Zhang and Yew 2009). It is also reported that a FQs-resistant protein from *M. tuberculosis* mimics DNA. MfpA is a founding member of the pentapeptide repeat protein family that confers resistance to FQs by binding to DNA gyrase and inhibits its activity (Hegde et al. 2005). However, the significance of the MfpA protein to FQs resistance in clinical isolates has not been evaluated.

12.2.2.2 Aminoglycosides

This class of drugs is bactericidal in nature and consists of Kanamycin (KAN) and Amikacin (AMI) of second-line drugs. Aminoglycosides is an inhibitor of protein synthesis. Therefore, it cannot be used against dormant *M. tuberculosis*. Aminoglycosides firmly binds to the conserved A site of 16s rRNA in the 30S ribosomal subunit and disrupts the elongation of the peptide chain in the bacteria. This binding interferes with mRNA binding and tRNA acceptor sites leading to the production of non-functional or toxic peptides (Suzuki et al. 1998). Drug resistances to aminoglycosides are due to modification of ribosomal structures at the 16S rRNA, and mutations in the *rrs* gene at positions 1400, 1401, and 1483 encoding for 16s rRNA are associated with resistance to KAN and AMI. The *eis* promoter, which has been verified to enhance the intracellular survival of *M. smegmatis*, has also been considered for its utility as a marker for KAN resistance, when mutated (Suzuki et al. 1998; Zaunbrecher et al. 2009). An A to G change at codon 1400 in the *rrs* gene showed resistance to KAN (Suzuki et al. 1998).

12.2.2.3 Ethionamide (ETA)

ETA is bacteriostatic and an important drug in the treatment of MDR-TB, and this drug is mechanistically and structurally analogous to INH. Like INH, ETA is also thought to be a pro-drug activated by *katG*-independent mechanism (bacterial mechanism) leading to the formation of sulfoxide metabolite (Baulard et al. 2000). Activated drug inhibits the *inhA* gene product enoyl-ACP reductase and then disrupts cell-wall biosynthesis by inhibiting mycolic acid biosynthesis (Johnsson et al. 1995). Various experiments performed on *M. tuberculosis*, *M. bovis,* and *M. smegmatis* revealed that *etaA* (also called *ethA*) codes for a flavin mono-oxygenase and is responsible for activation of ETA, and its expression is controlled by *ethR* (also called *etaR*), a transcriptional repressor gene (Baulard et al. 2000; DeBarber et al. 2000). Drug resistance for ETA and INH are due to mutations in the promoter of the *inhA* gene (Morlock et al. 2003). *M. tuberculosis* mutants overexpressing *etaA* were hypersensitive to the drug, whereas *etaR* overexpressing were drug resistant (DeBarber et al. 2000). Naturally occurring ETA mutants harbour changes in the enzymes responsible for drug activation (Baulard et al. 2000). Laboratory-derived mutants of *inhA* show cross-resistance between ETA and INH. There is complete cross-resistance between ETA and prothinonamide. Thiocarlide and thiacetazone are possibly activated by e*thA*, but have a different mode of action (Wang et al. 2007).

12.2.2.4 D-Cycloserine (CYS)

CYS is a broad spectrum antibiotic, which is either bactericidal or bacteriostatic. It is the cyclic analog of D-alanine amino acid. It interferes with an early step in bacterial cell-wall synthesis in the cytoplasm by competitive inhibition of two enzymes: L- alanine racemase (*Alr* converts L-alanine to D-alanine) and D-alanine (D-alanine ligase (*Ddl*) synthesizes the pentapeptide core using D-alanine). Both the enzymes are vital in the synthesis of peptidoglycan, and subsequently in cell-wall biosynthesis and maintenance (Ramaswamy and Musser 1998; Feng and Barletta 2003). The drug resistance mechanism has not been identified in *M. tuberculosis.* Overexpression of the alanine racemase (*alr*) gene in *M. smegmatis* is necessary and sufficient to confer CYS resistance. A transversion of G to T in *alr* promoter may lead to the overexpression of alanine recemase (Ramaswamy and Musser 1998; Feng and Barletta 2003).

12.2.2.5 Peptides (Viomycin and Capreomycin)

Viomycin (VIO) and capreomycin (CAP) are basic polypeptide antibiotics. They have the capacity to kill bacteria, but the mode of action is not fully understood. However, it is known that capreomycin interacts with the ribosome and inhibits protein synthesis. The microarray experiment in *M. tuberculosis* demonstrated the up-regulation of various ribosomal proteins (e.g. *RplI* binds to 23S rRNA; *RpsR* helps in attachment of the 5S RNA into the large ribosomal subunit; *RplJ* is involved in translation mechanism; and *RplY* binds to the 50S rRNA), *Rv1988* probable 23S rRNA methyl transferase encoded by *erm* and *Rv2907c* (16s rRNA processing protein, *RimM*). It supports the interaction of drugs with ribosomal proteins (Fu and Shinnick 2007). CAP is potently active against the persistent forms of *M. tuberculosis*; the drug may have a primary or secondary target outside the ribosome. The drug resistance mechanism is associated with mutation in *rrs* gene that encodes the 16s rRNA, specifically G to A or G to T nucleotide change at codon 1473 and *tlyA* (Slayden and Barry 2002). It is shown that in *M. smegmatis* exhibits resistance due to variations in the 70S ribosomal unit (Taniguchi et al. 1997). Mutations in the *rrs* gene (encodes the 16S rRNA) is associated with resistance to CAP and VIO, specifically a G-A or G-T nucleotide change at codon 1473 (Taniguchi et al. 1997). CAP also binds to component in the bacterial cell which results in the production of abnormal proteins. These proteins are necessary for the survival of bacteria. Therefore, the production of these abnormal proteins is finally fatal for bacterial cell. There is a possible cross-resistance to streptomycin, and no cross-resistance has been observed between CAP and INH, PAS, CYS, ETA, or ETH (Law et al. 2014).

12.2.2.6 Para Aminosalicylic Acid (PAS)

PAS is an anti-mycobacterial, bacteriostatic agent used with other anti-tuberculosis drugs (most often INH) for the treatment of all forms of active tuberculosis due to sensitive strains of TB. It also inhibits the onset of bacterial resistance to STR and INH. There are two mechanisms responsible for PAS bacteriostatic action against the *M. tuberculosis*. First, aminosalicylic acid inhibits folic acid synthesis. The binding of p-aminobenzoic acid (PABA) to pteridinesynthetase acts as the first step in folic acid synthesis. PAS binds pteridinesynthetase with greater affinity than PABA, effectively inhibiting the synthesis of folic acid. As bacteria are incapable of using external sources of folic acid, their cell growth and multiplication reduces. Second, PAS may inhibit the synthesis of the cell wall component, mycobactin; hence decreasing iron uptake by the *M. tuberculosis*. PAS was previously thought to target dihydropteroatevsynthase (DHPS), the target of sulfonamide drugs, but Nopponpunth et al. (Nopponpunth et al. 1999) confirmed that PAS was a poor *in vitro* inhibitor of the enzyme. Another study demonstrated that transposon directed disruption of the *M. bovis* thymidylate synthase gene, *thyA*, which results in resistance to PAS (Rengarajan et al. 2004). Enzyme activity of thymidylate synthase in the PAS-resistant transposon mutants was reduced. PAS could also interfere with iron acquisition by the bacilli. Recent study on ABC transporters and virulence in *M. tuberculosis* and carboxymycobactin inhibitors reveals the targets of iron uptake mechanism as a drug target (Rodriguez and Smith 2006). Mutations in *thyA* have been also associated with clinical *M. tuberculosis* PAS-resistant isolates (Rengarajan et al. 2004).

12.3 Nanoparticles

Nanotechnology has become one of the most promising technologies globally and is being applied in all areas of science. Nanoparticles are small particles ranging between 1 and 100 nm in diameter. The conversion of any material to nanoparticle resulted in alteration of its biological, mechanical, physiological, optical, and electronic properties. Metal nanoparticles have received special attention globally due to their widespread application in the physiochemical and biomedical fields. The physiochemical approaches for synthesis of metal nanoparticles are restricted due to heavy metal pollution in the environment. Therefore, synthesizing nanoparticles by biological sources (*i.e.* microorganisms and plants) is advantageous in easy scaling up, nontoxicity, and reproducibility. Several plants as well as microorganisms have been explored for synthesis of metal nanoparticles (Singh et al. 2016a, b). Its uniqueness lies in its size and composition of the particles when compared to atoms and other material. These characteristics find use in different useful applications. The combination of nanoparticles with biology has led to the development of diagnostic devices, contrast agents, analytical tools, physical therapy applications, and drug delivery vehicles. Nanoparticles are used in both *in vivo* and *in vitro* biomedical research and applications (Singh et al. 2016a, b).

12.3.1 Natural Nanoparticles

Nature has an amazing ability to produce nanoscale elements by self-assembly. They are found in larger amount due to various natural processes, including photochemical reactions, volcanic eruptions, forest fires, etc. Plants, microbes, and animals also produce huge amount of nanoparticles *via* natural process, such as shedding of skin and hair. However, air pollution, which is generated by human activities, industry charcoal burning and natural disasters such as sand storms, volcanic eruptions, and forest fires, can produces massive quantities of nanoparticulate material that severely affect air quality worldwide.

12.3.2 Microbial Nanoparticles

Microorganisms have the strong potential to produce nanoparticle and are displayed as eco-friendly and cost-effective tools naturally. They may synthesize these nanoscale materials to fulfill high-energy demands required for the physiochemical process. The microorganisms have the ability to accumulate and detoxify hazardous heavy metals by the sequential action of reductase enzymes, resulting in production of narrow-sized metal nanoparticles by reducing metal salts (Singh et al. 2016a, b). The detailed mechanism of synthesis of nanoparticles in microorganisms is depicted in Fig. 12.3.

A number of bacteria, such as *Pseudomonas deceptionensis*, *B. methylotrophicus*, and *Brevibacterium frigoritolerans,* have been reported for the synthesis of silver and gold nanoparticles. Similarly, numerous Bacillus species, like *Bacillus licheniformis, Bacillus amyloliquefaciens, Rhodobactersphaeroides, Bacillus subtilis,* and *Streptomyces anulatus* were also reported, which are more stable and show excellent antimicrobial activity against human pathogens (Elbeshehy et al. 2015; Soni and Prakash 2015). Moreover, some other bacterial genera, such as *Klebsiella, Escherichia coli, Enterobacter, Aeromonas, Corynebacterium, Lactobacillus Pseudomonas, Rhodobacter, Rhodococcus, Brevibacterium,* and *Pyrobaculum* have

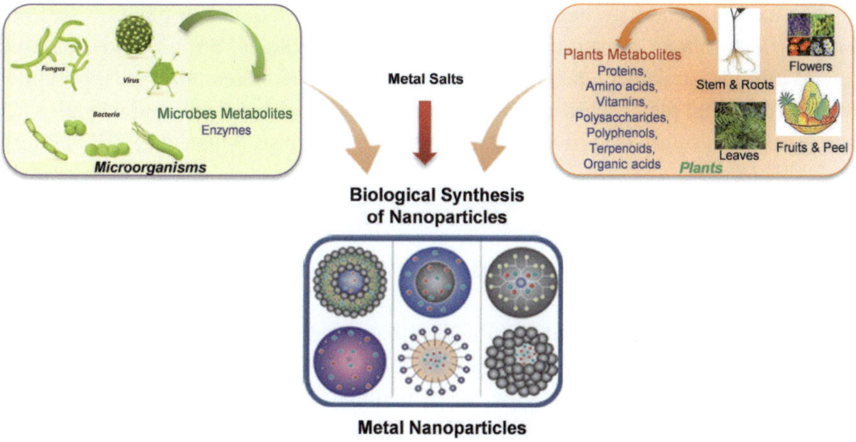

Fig. 12.3 Natural sources of nanoparticles

been shown to synthesize metal nanoparticle (Singh et al. 2016a, b). The synthesis of nanoparticles has been reported by both intracellular and extracellular processes, that is bacterial biomass, supernatant, and derived components. The synthesis of nanoparticles by extracellular process received much attention due to elimination of the downstream processing step, which is mandatory for the recovery of nanoparticles in intracellular procedures, including cell-wall disruption by sonication, centrifugations with additional washing steps required for the purification of nanoparticles. Additionally, presence of metal-resistant genes, proteins, peptides, enzymes, cofactors, and organic/inorganic materials plays major roles in the long-term stability by acting as reducing agents that prevent aggregation of the nanoparticles (Das et al. 2017).

12.3.3 Plants Nanoparticles

Phyto nanotechnology, that is synthesis of nanoparticles using plant parts such as leaves, fruits, stems, roots, offers a simple, rapid, cost-effective, and eco-friendly method. Presently, it has various applications such as biocompatibility, scalability, especially utility in the medical field (Noruzi 2015). Accordingly, nanoparticles derived from plants are nontoxic in nature, and their easy availability is extremely useful for fulfilling the high demand of nanoparticles in the biomedical and environmental fields.

Recently, gold and silver nanoparticles have been synthesized successfully by the leaf and root extracts of various plants such *as Panax ginseng*, *B. prionitis, P. zeylanica, and S. cumini*, suggesting the utility of medicinal plants as alternate resources for treatment of various deadly diseases, such as TB, HIV (Chang et al. 1979; Singh et al. 2016a, b; Baker et al. 2013). Additionally, various plant parts – including fruits, stems, roots, leaves and their extracts – have been used for the synthesis of metal nanoparticles (Singh et al. 2016a, b).

The exact mechanism behind the synthesis of nanoparticles in plants is still not known. It has been presumed that biomolecules (proteins, amino acids, organic acid, vitamins) and secondary metabolites (such as flavonoids, alkaloids, polyphenols, terpenoids) play important roles in the reduction of metal salt and may also act as capping and stabilizing agents in nanoparticles synthesis (Duan et al. 2015; Philip et al. 2011).

12.4 Nanotechnological Approach as Antimicrobial Activity

The increase in the number of MDR-TB threatens global TB control and is a foremost public health concern in numerous countries. The occurrence of MDR-TB alone or with HIV has imposed the use of nanoparticles for the treatment of tuberculosis. Several existing anti-tubercular drugs can be combined into nanocarrier systems by using nanotechnology (Hemeg 2017). The azole group of anti-fungal drugs like clotrimazole and econazole has shown effective anti-tubercular activity against drug resistance and latent TB (Ahmad et al. 2006). The method of nanoencapsulation for formulation of various drugs intended for oral administration increases the bioavailability, therapeutic efficacy, and toxicity of drugs as opposed to the free form of drug (Pandey and Khuller 2007).

Several concepts and clarifications have been projected for numerous nanoparticles for their microbicidal activity (Fig. 12.4) and have been deliberate on the basis of phenotypic and structural changes in the mycobacterial cells. Nanoparticles have the ability to get attached with the cell of bacteria and subsequently enter in it causing structural changes in the permeability of cell membrane, leading to cell death (Sondi and Salopek-Sondi 2004). Several bimetals, metal oxide, metals, and metal halide in nanoparticles form have been known for their antimicrobial nature (Hemeg 2017). The bacteria are less likely to acquire drug resistance against nanoparticles. The nanoparticles may contain Au, Ag, Cu, Zn, Ti, Mg, Ni, Ce, Al, Cd, Pd, and super paramagnetic Fe and Y. Amongst the metallic nanoparticles, Au nanoparticles have good antibacterial activity unless their surface is modified. Among all metal nanoparticles, silver nanoparticles are the most effective against microbial infection due to their large surface area (Rai et al. 2012; Dakal et al. 2016; Hemeg 2017). Furthermore, resistance to the silver nanoparticles has been reported due to genetic mutation or alteration in microbes (Graves et al. 2015; Hemeg 2017).

Silver nanoparticles also get deposited in lungs, liver, spleen, and other organs causing damage and dysfunction of organs. The metal oxide nanoparticles may promote horizontal conjugate transfer of resistant genes, increasing resistance to antibiotics and also causes oxidative lesions and DNA damage (Hemeg 2017). Nanoparticles also emerge to fight as a biofilm of anti-microbial for treatment of various bacterial infections. Nanoparticles may be microbicidal or microbiostatic in action against bacteria (Fig. 12.5). Nanomaterials also have affinity towards the phosphorus and sulphur containing biomolecules present in microbial cells, e.g., DNA bases. They can act on the bases and damage DNA leading to cell death.

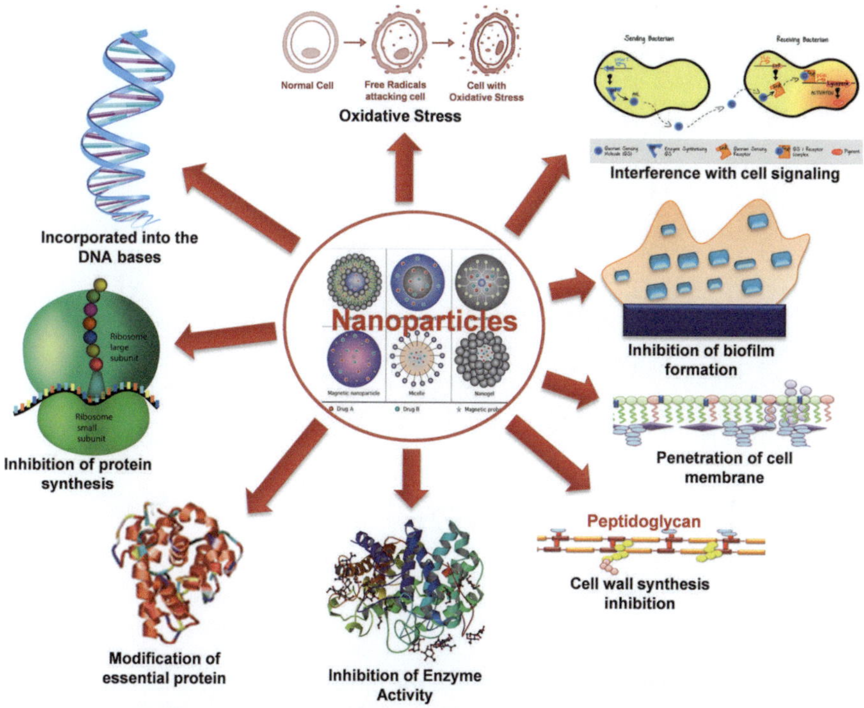

Fig. 12.4 Mechanism of antimicrobial activity of nanoparticles

Nanoparticles also lead to disruption of bacterial signal transduction, disruption of cellular function, collapsing the membrane potential, and inhibiting the ATPase activity to decrease the level of ATP also inhibiting the binding of tRNA to ribosome (Figs. 12.4 and 12.5) (Hemeg 2017; Singh et al. 2014a, b).

12.5 Mode of Action of Metal-Based Nanoparticles

Metallic nanoparticles use complex mechanism simultaneously to battle microbes (Fig. 12.5), depreciating the possibility of resistance development, as it would involve multiple gene mutations simultaneously in the same microbe for evolvement of drug resistance. Various molecular mechanisms contributing to killing of drug resistant bacteria by metal-based nanoparticles have been extensively reviewed in Table 12.3 (Dakal et al. 2016; Durán et al. 2016; Rai et al. 2012). The toxicity of nanoparticles are also attributed to ROS production such as superoxide anion, hydrogen peroxide, and hydroxyl radicles, which inhibit the replication of DNA, amino acid synthesis, and damage microbial cell membrane.

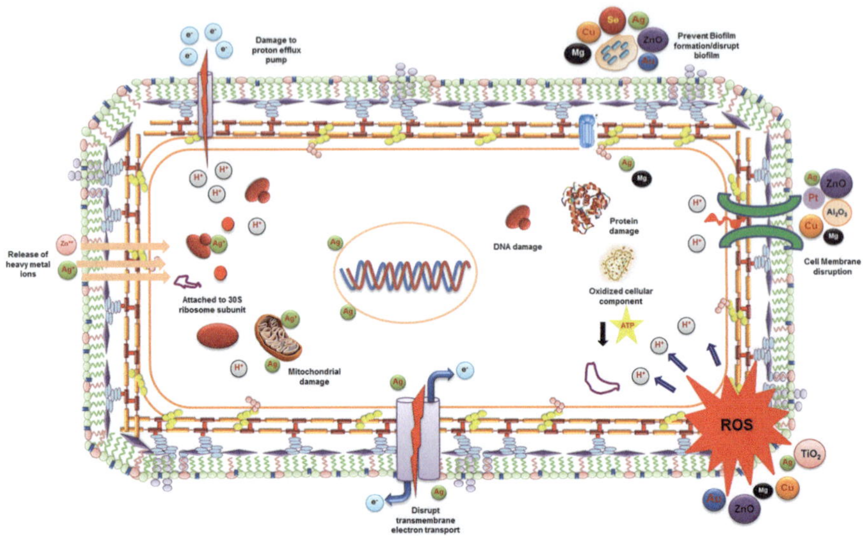

Fig. 12.5 The bactericidal effects of nanomaterials

The release of metal ions disrupts the microbial membrane. The interaction of nanomaterials with bacterial DNA leads to inhibition of DNA replication and cell division. It may also adsorb on the bacterial cells, resulting in death due to oxidative stress. Nanoparticles may also alter the permeability of cell membrane, ribosome destabilization, and disruption of biofilm (Fig. 12.5) (Hemeg 2017). The inorganic nanoparticle used for the photo-thermal therapy is an excellent method used for killing of microbial cells. A nanoparticle absorbs the electromagnetic radiation and converts them into heat energy and transfers it to the microbial cells, leading to death. Gold nanoparticles were broadly studied in the photo-thermal therapy in cancer treatment and selective killing of pathogenic bacteria (Singh et al. 2014a, b). However, inorganic nanoparticles may be costly and may have serious side effects; to overcome this, ecofriendly nanoparticles synthesized using green nanotechnology were invented, which are cost effective with lesser side effects. The green nanoparticles avoid the toxic by-products generation and are eco-friendly. Metallic nanoparticles synthesize biologically are potentially safe and biocompatible for use in human therapeutic. Silver nanoparticle from *Helicteres isora* fruit extract has been reported to have potential activity against XDR *P. aeruginosa* clinical isolates (Singh et al. 2014a, b). Various other green nanoparticles were used for the treatment of microbial disease mentioned in Table 12.3. The antimicrobial activity of green nanoparticles that are biologically synthesized is because of metal ions release from the nanoparticle coupled with bio-organic compound (Hemeg 2017).

Table 12.3 Antibacterial activity of nanoparticles

Nanoparticles	Particle size (nm)	Antimicrobial effects
Silver	5–100	Lipid peroxidation, reactive oxygen species (ROS) generation, inhibition of cytochromes, inhibition of cell-wall synthesis, bacterial membrane disintegration, increase in membrane permeability, dissipation of proton gradient resulting in lysis, ribosome destabilization, adhesion to cell surface causing lipid and protein damage, intercalation between DNA bases, disruption of biofilms
Gold	5–400	Disruption of respiratory chain, loss of membrane potential, reduced ATPase activity, bacterial membrane disruption, decline in subunit of ribosome for tRNA binding
Zinc oxide	12–60	Inhibition of biofilm, ROS generation, Zn^{2+} release, enhanced membrane permeability, disruption of membrane, adsorption to cell surface, lipids, and protein damage, inhibition of microbial biofilm formation
Copper	1–100	Inhibition of DNA replication, ROS generation, disorganization of membrane, lipid peroxidation, DNA degradation, protein oxidation
Copper oxide	20–95	The Cu^{2+} ions released by the CuO nanoparticles are able to disrupt cell membrane and enzyme function
Selenium	30	Biofilm inhibition
Titanium oxide	20	Adsorption to cell surface, ROS generation, inhibition of biofilm
Nickel oxide	4–80	Increase in bacterial cell-wall permeability
Nitric oxide	20–100	RNOS irreversibly binds to heme protein resulting in heme removal from protein
Fe_3O_2@TiO_2 magnetic nanoparticles	6–90	ROS damages cell component
Iron oxide	4–9	ROS-generated oxidative stress leads to protein and DNA damage
Cadmium sulphide	2–60	Antibiofilm activity
Magnesium fluoride	5 nm	ROS generation, penetration of cell envelope, lipid peroxidation, biofilm inhibition
Aluminium oxide	30–60	Cell-wall damage, enters cytoplasm
Copper-zinc bimetal		Antioxidant activity

(continued)

Table 12.3 (continued)

Nanoparticles	Extracted from	Antimicrobial effects
Silver bio-nanoparticles	*Phyllanthusamarus*	Inactivation of enzymes by interaction with thiol groups, membrane damage, release of free ions
	*Helicteresisora*fruit	Turbulence of membrane permeability, lipid peroxidation, leakage of reducing sugars and proteins, respiratory chain dehydrogenases inactivation
	Artemisia cappilaris	Membrane damage, release of free ions
	Aloe vera	ROS production, DNA damage, release of free ions, increase in membrane permeability,
	Acalyphaindica leaf	Alteration in membrane permeability and respiratory chain
	Rhizopusoryzae	Alteration in membrane permeability, ROS production, membrane damage
	Cocusnucifera influorescence	Interference with the molecular build-up of bacterial cell wall
Silver-zinc	*Caltropisprocera* fruits or leaves	Restraining biofilm formation, inhibition of adenyl cyclase
Gold	*Citrulluslanatus*rind	Antioxidant activities
Silver, Gold, Silver-gold bimetallic	*Plumbagozeylanica*	Biofilm inhibition
Aluminium oxide	leaf extract of lemongrass	Intracellular oxidative stress contributing to loss of cell membrane integrity

Nanoparticles	Antibiotics	Antimicrobial effects
Silver	Chloramphenicol, rifampicin, Ampicillin, ciprofloxacin, vancomycin	Active against MRSA, *Salmonella, Staphylococcus aureus, Pseudomonos aeruginosa,* etc.
Gold	Ampicillin, Kanamycin, Streptomycin, Gentamycin, Neomycin	Active against *Streptococcus bovis, Staphylococcus epidermidis, Enterobacter aerogenes, Staphylococcus aureus, Micrococcus luteus, Escherichia coli, Pseudomonos aeruginosa*
Zinc oxide	Ciprofloxacin	Active against MDR *Acinetobacter baumannii*

12.6 Nanoparticles Conjugated with Antibiotics

The mutations of genetic alteration in bacterial genome may result in rapid evolution of drug resistance and also to silver nanoparticles. Therefore, the nanoparticle functionalization with existing antibiotics to combat mycobacterial resistance reduced the dose of drugs; hence, toxicity and side effects may also decrease. The silver nanoparticle coupled with rifampicin or ciprofloxacin are used for effectively killing the bacterial cells. The inconsistency of drug-resistance with respect to the individual treatment is responsible for the limited data on nanotechnological approach in drug-resistant TB (Hemeg 2017; Iram et al. 2016; Wan et al. 2016). The effect of gold nanoparticle and kanamycin has been observed against *Streptococcus bovis* and shows very good antimicrobial activity. Various combinations of antibiotics and nanoparticles are mentioned in Table 12.3.

The development of MDR and XDR-TB has put enormous pressure on scientist and pharma industries to search novel antimicrobial agents or repurposing of existing drugs. In the area of drug resistance, nanotechnology has vast potential to prevent the spread of drug-resistant strains and for disease management. The chemical and biological metallic nanoparticles have shown to be potential agent in the treatment of bacterial disease. The nanoparticles are now considered as an alternative of antibiotics due to their bactericidal and immunopotentiating properties. The metallic nanoparticles when used with the existing antibiotics for the treatment of microbial infection reduced the antibiotic doses to be taken and minimized the toxicity and also reduced the probability of multi-drug resistance development. They also help in revival of the antibacterial activity of old-generation antibiotics against which most of the pathogens develop resistance. Furthermore, studies are required to delineate the molecular mechanism of nanoparticles bactericidal and toxicity activity. Nanoparticles are better option to combat the MDR and XDR-TB.

12.7 Conclusion and Prospects

Tuberculosis is an increasing public health threat as they have evolved to resist against various available drugs. Still antibiotics remain the backbone of the fight against TB, but the emergence of MDR/XDR and total drug resistance tuberculosis is creating an alarm globally. The over-prescribing of antibiotics, over use in livestock and farming, and incomplete treatment are mainly responsible for the growing rise. This also creates pressure for the selection of the fittest strain to thrive, causing the emergence of MDR/XDR TB and emphasizing the urgent need of surrogate therapeutic preferences. Nanoparticles are now being considered as better option to antibiotics due to their microbicidal activity. The nanoparticles used with current antibiotics lower the dosage of these drugs to be administered. Moreover, it minimizes the cost and toxicity and lowers the probability of resistant strains development. The nanoparticles synthesized naturally are eco-friendly and cost effective. The nanoparticle alone or synergistically with antibiotics play an important role for future therapeutics in the field of nano medicine. Repurposing the existing antibiotics with nanoparticles may serve as adjunct to the current therapies for tuberculosis. However, in-depth pharmacokinetic and pharmacodynamics study of nanoparticles is required for its translation to the patients.

References

Ahmad Z, Sharma S, Khuller GK, Singh P, Faujda J, Katoch VM (2006) Antimycobacterial activity of econazole against multidrug-resistant strains of *Mycobacterium tuberculosis*. Int J Antimicrob Agents 28:543–544

Aragón LM, Navarro F, Heiser V, Garrigó M, Español M, Coll P (2006) Rapid detection of specific gene mutations associated with isoniazid or rifampicin resistance in *Mycobacterium tuberculosis* clinical isolates using non-fluorescent low-density DNA microarrays. J Antimicrob Chemother 57:825–831

Baker S, Rakshith D, Kavitha KS, Santosh P, Kavitha HU, Rao Y, Satish S (2013) Plants: emerging as nanofactories towards facile route in synthesis of nanoparticles. Bioimpacts 3:111–117

Banerjee A, Dubnau E, Quemard A, Balasubramanian V, Um KS, Wilson T, Collins D, de Lisle G, Jacobs WR (1994) *inhA*, a gene encoding a target for isoniazid and ethionamide in *Mycobacterium tuberculosis*. Science 263:227–230

Bardou F, Raynaud C, Ramos C, Laneelle MA, Laneelle G (1998) Mechanism of isoniazid uptake in *Mycobacterium tuberculosis*. Microbiology 144(Pt 9):2539–2544

Barry CE, Lee RE, Mdluli K, Sampson AE, Schroeder BG, Slayden RA, Yuan Y (1998) Mycolic acids: structure, biosynthesis and physiological functions. Prog Lipid Res 37:143–179

Baulard AR, Betts JC, Engohang-Ndong J, Quan S, McAdam RA, Brennan PJ, Locht C, Besra GS (2000) Activation of the pro-drug ethionamide is regulated in mycobacteria. J Biol Chem 275:28326–28331

Boshoff HI, Mizrahi V, Barry CE (2002) Effects of pyrazinamide on fatty acid synthesis by whole mycobacterial cells and purified fatty acid synthase I. J Bacteriol 184:2167–2172

Carter AP, Clemons WM, Brodersen DE, Morgan-Warren RJ, Wimberly BT, Ramakrishnan V (2000) Functional insights from the structure of the 30S ribosomal subunit and its interactions with antibiotics. Nature 407:340–348

Champoux JJ (2001) DNA topoisomerases: structure, function, and mechanism. Annu Rev Biochem 70:369–413

Chan ED, Chatergee D, Iseman D, Heifets B (2000) Pyrazinamide, ethambutol, ethionamide, and aminoglycosides. In: Tuberculosis: a comprehensive international approach, 2nd edn. CRC Press, Boca Raton, pp 780–781

Chang MW, Tasaka H, Kuwabara M, Watanabe T, Matsuo Y (1979) Effects of panax ginseng extracts on the growth of *Mycobacterium tuberculosis* H37Rv. Hiroshima J Med Sci 28:115–118

Cohen-Gonsaud M, Ducasse S, Hoh F, Zerbib D, Labesse G, Quemard A (2002) Crystal structure of MabA from *Mycobacterium tuberculosis*, a reductase involved in long-chain fatty acid biosynthesis. J Mol Biol 320:249–261

Comas I, Borrell S, Roetzer A, Rose G, Malla B, Kato-Maeda M, Galagan J, Niemann S, Gagneux S (2012) Whole-genome sequencing of rifampicin-resistant *Mycobacterium tuberculosis* strains identifies compensatory mutations in RNA polymerase genes. Nat Genet 44:106–110

Cooksey RC, Morlock GP, McQueen A, Glickman SE, Crawford JT (1996) Characterization of streptomycin resistance mechanisms among *Mycobacterium tuberculosis* isolates from patients in New York City. Antimicrob Agents Chemother 40:1186–1188

Da Silva PEA, Von Groll A, Martin A, Palomino JC (2011) Efflux as a mechanism for drug resistance in *Mycobacterium tuberculosis*. FEMS Immunol Med Microbiol 63:1–9

Dakal TC, Kumar A, Majumdar RS, Yadav V (2016) Mechanistic basis of antimicrobial actions of silver nanoparticles. Front Microbiol 7:1831

Das RK, Pachapur VL, Lonappan L, Naghdi M, Pulicharla R, Maiti S (2017) Biological synthesis of metallic nanoparticles: plants, animals and microbial aspects. Nanotechnol Environ Eng 2:18. https://doi.org/10.1007/s41204-017-0029-4

Davies J (1994) Inactivation of antibiotics and the dissemination of resistance genes. Science 264:375–382

DeBarber AE, Mdluli K, Bosman M, Bekker LG, Barry CE (2000) Ethionamide activation and sensitivity in multidrug-resistant *Mycobacterium tuberculosis*. Proc Natl Acad Sci U S A 97:9677–9682

Deng L, Mikusova K, Robuck KG, Scherman M, Brennan PJ, McNeil MR (1995) Recognition of multiple effects of ethambutol on metabolism of mycobacterial cell envelope. Antimicrob Agents Chemother 39:694–701

Dessen A, Quemard A, Blanchard JS, Jacobs WR, Sacchettini JC (1995) Crystal structure and function of the isoniazid target of *Mycobacterium tuberculosis*. Science 267:1638–1641

Dooley SW, Simone M (1994) The extent and management of drug-resistant tuberculosis: the American experience. In: Clinical tuberculosis. Chapman & Hall, London, pp 171–189

Duan H, Wang D, Li Y (2015) Green chemistry for nanoparticle synthesis. Chem Soc Rev 44:5778–5792

Durán N, Durán M, de Jesus MB, Seabra AB, Fávaro WJ, Nakazato G (2016) Silver nanoparticles: a new view on mechanistic aspects on antimicrobial activity. Nanomed Nanotechnol Biol Med 12:789–799

Elbeshehy EK, Elazzazy AM, Aggelis G (2015) Silver nanoparticles synthesis mediated by new isolates of Bacillus spp., nanoparticle characterization and their activity against bean yellow mosaic virus and human pathogens. Front Microbiol 6:453

Escuyer VE, Lety MA, Torrelles JB, Khoo KH, Tang JB, Rithner CD, Frehel C, McNeil MR, Brennan PJ, Chatterjee D (2001) The role of the *embA* and *embB* gene products in the biosynthesis of the terminal hexaarabinofuranosyl motif of *Mycobacterium smegmatis* arabinogalactan. J Biol Chem 276:48854–48862

Feng Z, Barletta RG (2003) Roles of *Mycobacterium smegmatis* D-alanine: D-alanine ligase and D-alanine racemase in the mechanisms of action of and resistance to the peptidoglycan inhibitor D-cycloserine. Antimicrob Agents Chemother 47:283–291

Fu LM, Shinnick TM (2007) Genome-wide exploration of the drug action of capreomycin on *Mycobacterium tuberculosis* using Affymetrix oligonucleotide Gene Chips. J Infect 54:277–284

Gordon SV, Eiglmeier K, Garnier T, Brosch R, Parkhill J, Barrell B, Cole ST, Hewinson RG (2001) Genomics of *Mycobacterium bovis*. Tuberc Edinb 81:157–163

Graves JL, Tajkarimi M, Cunningham Q, Campbell A, Nonga H, Harrison SH, Barrick JE (2015) Rapid evolution of silver nanoparticle resistance in *Escherichia coli*. Front Genet 6:42

Hegde SS, Vetting MW, Roderick SL, Mitchenall LA, Maxwell A, Takiff HE, Blanchard JS (2005) A fluoroquinolone resistance protein from *Mycobacterium tuberculosis* that mimics DNA. Science 308:1480–1483

Hemeg HA (2017) Nanomaterials for alternative antibacterial therapy. Int J Nanomedicine 12:8211–8225

Iram S, Akbar Khan J, Aman N, Nadhman A, Zulfiqar Z, Arfat Yameen M (2016) Enhancing the anti-enterococci activity of different antibiotics by combining with metal oxide nanoparticles. Jundishapur J Microbiol 9:e31302

Johnson R, Jordaan AM, Pretorius L, Engelke E, van der Spuy G, Kewley C, Bosman M, van Helden PD, Warren R, Victor TC (2006) Ethambutol resistance testing by mutation detection. Int J Tuberc Lung Dis 10:68–73

Johnsson K, King DS, Schultz PG (1995) Studies on the mechanism of action of isoniazid and ethionamide in the chemotherapy of tuberculosis. J Am Chem Soc 117:5009–5010

Kam KM, Yip CW, Cheung TL, Tang HS, Leung OC, Chan MY (2006) Stepwise decrease in moxifloxacin susceptibility amongst clinical isolates of multidrug-resistant *Mycobacterium tuberculosis*: correlation with ofloxacin susceptibility. Microb Drug Resist 12:7–11

Law V, Knox C, Djoumbou Y, Jewison T, Guo AC, Liu Y, Maciejewski A, Arndt D, Wilson M, Neveu V et al (2014) Drug Bank 4.0: shedding new light on drug metabolism. Nucleic Acids Res 42:D1091–D1097

Lee AS, Teo AS, Wong SY (2001) Novel mutations in ndh in isoniazid-resistant *Mycobacterium tuberculosis* isolates. Antimicrob Agents Chemother 45:2157–2159

Louw GE, Warren RM, Gey van Pittius NC, McEvoy CRE, Van Helden PD, Victor TC (2009) A balancing act: efflux/influx in mycobacterial drug resistance. Antimicrob Agents Chemother 53:3181–3189

Mdluli K, Slayden RA, Zhu Y, Ramaswamy S, Pan X, Mead D, Crane DD, Musser JM, Barry CE (1998) Inhibition of a *Mycobacterium tuberculosis* beta-ketoacyl ACP synthase by isoniazid. Science 280:1607–1610

Mitchison DA, Selkon JB (1956) The bactericidal activities of antituberculous drugs. Am Rev Tuberc 74:109–116; discussion, 116–123

Morlock GP, Metchock B, Sikes D, Crawford JT, Cooksey RC (2003) *ethA, inhA*, and *katG* loci of ethionamide-resistant clinical *Mycobacterium tuberculosis* isolates. Antimicrob Agents Chemother 47:3799–3805

Nair J, Rouse DA, Bai GH, Morris SL (1993) The rpsL gene and streptomycin resistance in single and multiple drug-resistant strains of Mycobacterium tuberculosis. Mol Microbiol 10:521–527

Nopponpunth V, Sirawaraporn W, Greene PJ, Santi DV (1999) Cloning and expression of *Mycobacterium tuberculosis* and *Mycobacterium leprae* dihydropteroate synthase in *Escherichia coli*. J Bacteriol 181:6814–6821

Noruzi M (2015) Biosynthesis of gold nanoparticles using plant extracts. Bioprocess Biosyst Eng 38:1–14

Okamoto S, Tamaru A, Nakajima C, Nishimura K, Tanaka Y, Tokuyama S, Suzuki Y, Ochi K (2007) Loss of a conserved 7-methylguanosine modification in 16S rRNA confers low-level streptomycin resistance in bacteria. Mol Microbiol 63:1096–1106

Pandey R, Khuller GK (2007) Nanoparticle-based oral drug delivery system for an injectable antibiotic-streptomycin. Evaluation in a murine tuberculosis model. Chemotherapy 53:437–441

Philip D, Unni C, Aromal SA, Vidhu VK (2011) Murraya Koenigii leaf-assisted rapid green synthesis of silver and gold nanoparticles. Spectrochim Acta A Mol Biomol Spectrosc 78:899–904

Projan SJ (2003) Why is big Pharma getting out of antibacterial drug discovery? Curr Opin Microbiol 6:427–430

Rai MK, Deshmukh SD, Ingle AP, Gade AK (2012) Silver nanoparticles: the powerful nanoweapon against multidrug-resistant bacteria. J Appl Microbiol 112:841–852

Ramaswamy S, Musser JM (1998) Molecular genetic basis of antimicrobial agent resistance in *Mycobacterium tuberculosis*: 1998 update. Int Union Tuberc Lung Dis 79:3–29

Ramaswamy SV, Amin AG, Goksel S, Stager CE, Dou SJ, El Sahly H, Moghazeh SL, Kreiswirth BN, Musser JM (2000) Molecular genetic analysis of nucleotide polymorphisms associated with ethambutol resistance in human isolates of *Mycobacterium tuberculosis*. Antimicrob Agents Chemother 44:326–336

Rattan A, Kalia A, Ahmad N (1998) Multidrug-resistant *Mycobacterium tuberculosis*: molecular perspectives. Emerg Infect Dis 4:195–209

Rengarajan J, Sassetti CM, Naroditskaya V, Sloutsky A, Bloom BR, Rubin EJ (2004) The folate pathway is a target for resistance to the drug para-aminosalicylic acid (PAS) in mycobacteria. Mol Microbiol 53:275–282

Rodriguez GM, Smith I (2006) Identification of an ABC transporter required for iron acquisition and virulence in *Mycobacterium tuberculosis*. J Bacteriol 188:424–430

Scorpio A, Zhang Y (1996) Mutations in *pncA*, a gene encoding pyrazinamidase/nicotinamidase, cause resistance to the antituberculous drug pyrazinamide in tubercle bacillus. Nat Med 2:662–667

Sherman DR, Mdluli K, Hickey MJ, Arain TM, Morris SL, Barry CE, Stover CK (1996) Compensatory ahpC gene expression in isoniazid-resistant *Mycobacterium tuberculosis*. Science 272:1641–1643

Singh A, Gopinath K, Singh N, Singh S (2014a) Deciphering the sequential events during in vivo acquisition of drug resistance in *Mycobacterium tuberculosis*. Int J Mycobacteriol 3:36–40

Singh R, Smitha MS, Singh SP (2014b) The role of nanotechnology in combating multi-drug resistant bacteria. J Nanosci Nanotechnol 14:4745–4756

Singh A, Gopinath K, Sharma P, Bisht D, Sharma P, Singh N, Singh S (2015) Comparative proteomic analysis of sequential isolates of *Mycobacterium tuberculosis* from a patient pulmonary tuberculosis turning from drug sensitive to multidrug resistant. Indian J Med Res 141:27–45

Singh P, Kim YJ, Zhang D, Yang DC (2016a) Biological synthesis of nanoparticles from plants and microorganisms. Trends Biotechnol 34(7):588–599. https://doi.org/10.1016/j.tibtech.2016.02.006

Singh R, Nawale L, Arkile M, Wadhwani S, Shedbalkar U, Chopade S, Sarkar S, Chopade BA (2016b) Phytogenic silver, gold, and bimetallic nanoparticles as novel antitubercular agents. Int J Nanomedicine 11:1889–1897

Slayden RA, Barry CE (2002) The role of *Kas*A and *Kas*B in the biosynthesis of meromycolic acids and isoniazid resistance in *Mycobacterium tuberculosis*. Tuberc Edinb 82:149–160

Somoskovi A, Parsons LM, Salfinger M (2001) The molecular basis of resistance to isoniazid, rifampin and pyrazinamide in *Mycobacterium tuberculosis*. Respir Res 2:164–168

Sondi I, Salopek-Sondi B (2004) Silver nanoparticles as antimicrobial agent: a case study on *E. coli* as a model for Gram-negative bacteria. J Colloid Interface Sci 175:177–182

Soni N, Prakash S (2015) Antimicrobial and mosquitocidal activity of microbial synthesised silver nanoparticles. Parasitol Res 114:1023–1030

Spies FS, da Silva PEA, Ribeiro MO, Rossetti ML, Zaha A (2008) Identification of mutations related to streptomycin resistance in clinical isolates of *Mycobacterium tuberculosis* and possible involvement of efflux mechanism. Antimicrob Agents Chemother 52:2947–2949

Sreevatsan S, Escalante P, Pan X, Gillies DA, Siddiqui S, Khalaf CN, Kreiswirth BN, Bifani P, Adams LG, Ficht T et al (1996) Identification of a polymorphic nucleotide in oxyR specific for *Mycobacterium bovis*. J Clin Microbiol 34:2007–2010

Sreevatsan S, Stockbauer KE, Pan X, Kreiswirth BN, Moghazeh SL, Jacobs WR, Telenti A, Musser JM (1997) Ethambutol resistance in *Mycobacterium tuberculosis*: critical role of embB mutations. Antimicrob Agents Chemother 41:1677–1681

Suzuki Y, Katsukawa C, Tamaru A, Abe C, Makino M, Mizuguchi Y, Taniguchi H (1998) Detection of kanamycin-resistant *Mycobacterium tuberculosis* by identifying mutations in the 16S rRNA gene. J Clin Microbiol 36:1220–1225

Takayama K, Kilburn JO (1989) Inhibition of synthesis of arabinogalactan by ethambutol in *Mycobacterium smegmatis*. Antimicrob Agents Chemother 33:1493–1499

Takayama K, Wang L, Merkal RS (1973) Scanning electron microscopy of the H37Ra strain of *Mycobacterium tuberculosis* exposed to isoniazid. Antimicrob Agents Chemother 4:62–65

Taniguchi H, Chang B, Abe C, Nikaido Y, Mizuguchi Y, Yoshida SI (1997) Molecular analysis of kanamycin and viomycin resistance in *Mycobacterium smegmatis* by use of the conjugation system. J Bacteriol 179:4795–4801

Telenti A, Philipp WJ, Sreevatsan S, Bernasconi C, Stockbauer KE, Wieles B, Musser JM, Jacobs WR (1997) The *emb* operon, a gene cluster of *Mycobacterium tuberculosis* involved in resistance to ethambutol. Nat Med 3:567–570

Victor TC, van Rie A, Jordaan AM, Richardson M, van Der Spuy GD, Beyers N, van Helden PD, Warren R (2001) Sequence polymorphism in the *rrs* gene of *Mycobacterium tuberculosis* is deeply rooted within an evolutionary clade and is not associated with streptomycin resistance. J Clin Microbiol 39:4184–4186

Vilchèze C, Jacobs WR Jr (2007) The mechanism of isoniazid killing: clarity through the scope of genetics. Annu Rev Microbiol 61:35–50

Vilchèze C, Wang F, Arai M, Hazbón MH, Colangeli R, Kremer L, Weisbrod TR, Alland D, Sacchettini JC, Jacobs WR (2006) Transfer of a point mutation in *Mycobacterium tuberculosis inhA* resolves the target of isoniazid. Nat Med 12:1027–1029

Wan G, Ruan L, Yin Y, Yang T, Ge M, Cheng X (2016) Effects of silver nanoparticles in combination with antibiotics on the resistant bacteria *Acinetobacterbaumannii*. Int J Nanomedicine 11:3789–3800

Wang F, Langley R, Gulten G, Dover LG, Besra GS, Jacobs WR Jr, Sacchettini JC (2007) Mechanism of thioamide drug action against tuberculosis and leprosy. J Exp Med 204:73–78

Wehrli W, Knusel F, Schmid K, Staehelin M (1968) Interaction of rifamycin with bacterial RNA polymerase. Proc Natl Acad Sci U S A 61:667–673

WHO (2017) WHO|Global tuberculosis report 2017. WHO, Geneva

Zaunbrecher MA, Sikes RDJ, Metchock B, Shinnick TM, Posey JE (2009) Overexpression of the chromosomally encoded aminoglycoside acetyltransferaseeis confers kanamycin resistance in *Mycobacterium tuberculosis*. Proc Natl Acad Sci U S A 106:20004–20009

Zhang Y, Yew WW (2009) Mechanisms of drug resistance in *Mycobacterium tuberculosis*. Int J Tuberc Lung Dis 13:1320–1330

Nanophytotherapeutic Potential of Essential Oils Against *Candida* Infections

13

Zeeshan Fatima, Shweta Singh, and Saif Hameed

Abstract

Candidal infections are increasing at an alarming pace resulting in increase in mortality and morbidity rates. They can cause life-threatening infections which can range from oral to systemic infections in immunocompromised patients. With concomitant emergence of multidrug resistance (MDR), there is an intense need to search for novel compounds as the current therapeutic regimes are limited by their toxicity concerns and cost-effectiveness. Natural sources represent an attractive reservoir of compounds which have the potential to overcome the MDR problem. Essential oils (EOs) can be used in their edible form or as medicinal oils with diverse benefits and lesser side effects. This book chapter presents the summarized information on the health benefits and antifungal mode of action of natural compounds isolated from EOs. Additionally, the use of nanosystems as a carrier and delivery system of various EOs on the target pathogen is also discussed.

Keywords

MDR · Essential oils · Antifungal · Cell wall · Cell membrane · Biofilms

Zeeshan Fatima and Shweta Singh have contributed equally to this chapter.

Z. Fatima · S. Singh · S. Hameed (✉)
Amity Institute of Biotechnology, Amity University Haryana, Gurugram (Manesar), India

S. K. Saxena, S. M. P. Khurana (eds.), *NanoBioMedicine*,
https://doi.org/10.1007/978-981-32-9898-9_13

13.1 Introduction

13.1.1 Human Fungal Infections

Fungal infections are the most prevalent infections among the microbial infections nowadays. Among the many fungal species, *Candida* species are causing many infections ranging from superficial to invasive infections as they reside in a commensal manner in the oral, gastrointestinal tract, etc. (Gow and Yadav 2017). It has been reported that vulvovaginal candidiasis affects nearly 75% females once in a lifetime (Sobel 2007). *Candida albicans* is the most frequent in causing infections, but now even non- *albicans* species like *C. tropicalis*, *C. glabrata*, *C. parapsilosis*, and *C. krusei* are becoming prevalent in causing life-threatening infections (Claudia and Dario 2013). Candidiasis most frequently occurs in newborns and patients having immunocompromised conditions like AIDS, chemotherapy, gastrointestinal surgery, and extensive burns and also due to prolonged usage of antibiotics and intravascular administration of indwelling devices (Singh et al. 2015). The most common infection is oral candidiasis which is commonly called thrush which affects human oral cavity (Lalla et al. 2013). There are many antifungal drugs available for treating candidal infections which mainly belong to three classes, namely, azoles, polyenes, and echinocandins (Kathiravan et al. 2012). But due to the development of resistance against antifungal drugs by the phenomenon of multidrug resistance (MDR) (Tanwar et al. 2014), there is now a pressing need to search some alternatives.

13.1.2 Essential Oils and Its Significance

Essential oils (EOs) are the complex volatile compounds synthesized as secondary metabolites in plants with multitude of functions including antimicrobial activities. They function by protecting plants from insects and microbes and also against herbivorous animals by reducing their appetite for such plants. But on the other hand, these compounds attract the insects which carry pollen grains or seeds for dispersion. They can be synthesized in the plant organs like flowers, twigs, buds, leaves, seeds, roots, wood, and fruits and stored in secretory cells. They are composed of complex, variable compounds which belong to many classes like phenolic compounds, terpenes, alkaloids, alcohols, etc. and derived from intermediates of many metabolic pathways. Although they are hydrocarbons with oxygenated derivatives, some may also contain nitrogen or sulfur. They can be monoterpenes, sesquiterpenes, and even diterpenes representing the main constituents of many EOs (Bakkali et al. 2008). In addition, phenylpropanoids, fatty acids, and their esters or their decomposition products are also volatile components. These oils are extracted through the various distillation methods, solvent extraction. EOs are a rich source of compounds showing pharmaceutical properties like antimicrobial, antiparasitic, antiseptic, antioxidant, spasmolytic, and anti-inflammatory activity (Jassbi et al. 2012). Since Vedic times, these are widely used in foods and medicines.

At present, 3000 essential oils are known, but only 300 are commercially employed to be used in pharmaceutical, food, cosmetic, agronomic, perfume, and sanitary industries. EOs are potential alternatives to be used as a substitute for synthetic medicines as they have minimal side effects. They have also been used in aromatherapy since ages along with being used in cooking and as food additives (Bakkali et al. 2008). The nanoparticles are also used as an effective tool for delivery of EOs to the pathogen. The combination of nanoparticles and EOs is also used as phytonanocoating materials to inhibit the candidal biofilms (Anghel et al. 2013).

13.2 Essential Oil Classifications

Essential oils may be broadly classified into edible oils and non-edible oils.

Edible Essential Oils Essential oils can be consumed directly by the user in cooking in many countries. Some of the oils are described here:

1. *Mustard oil*: It is extracted from black mustard (*Brassica nigra*), brown Indian mustard (*B. juncea*), and white mustard (*B. hirta*) by grinding and distillation methods. The oil makes up to the 30% of mustard seeds. It is traditionally widely used as a cooking oil for food preparation and in making pickles in many parts of the world including North India, Eastern India, Nepal, Bangladesh, and Pakistan. It is also used in salad dressing along with lemon and honey. According to studies, it has strong anticancerous properties because it contains a good amount of linoleic acid (Pan et al. 2012). It also contains an ample amount of monounsaturated and polyunsaturated acids like MUFA and PUFA along with the omega-3 and omega-6 fatty acids which can reduce the risk of ischemic heart disease (Sengupta and Ghosh 2011). The selenium present in the oil contributes to its anti-inflammatory property which is good for treating arthritis. It is also used in the formulation of known diclofenac which cures joint and muscular pains. It has known antimicrobial properties due to the presence of glucosinolate which inhibits the bacterial growth. It has antifungal properties and is used in treating skin infections, rashes, allergies, and asthma (Ghosh et al. 2012). It has anti-aging properties due to the presence of vitamin E and is used for hair and skin problems. When used with turmeric powder, it is very efficient in treating denture and teeth problems and provides relief from gingivitis and periodontitis (Nagpal and Sood 2013).
2. *Olive oil*: It is extracted by pressing the fruit from the olive plant (*Olea europaea*) which is a traditional tree crop native to the Mediterranean Basin. It is commonly used for cooking purposes and salad dressing in European countries. It consists mainly of oleic acid (83%), with smaller amounts of other fatty acids including linoleic acid (21%) and palmitic acid (20%). It has a good amount of vitamin E, an antioxidant which is used in making non-sticky moisturizers, protects skin from psoriasis and skin cancer and sunrays, and treats skin inflam-

mation. The presence of oleuropein provides anticancerous properties and helps in treating breast cancer (Bendini et al. 2007). It is beneficial in treating diabetes as it contains mono- and polyunsaturated fats which were reported in *The American Journal of Clinical Nutrition*. According to the scientific American report, it is beneficial in helping to prevent Alzheimer's disease due to the presence of oleocanthal. It has anti-inflammatory properties and is helpful in strengthening the bones and treating depression by increasing the serotonin levels (Abuznait et al. 2013). It is also helpful in controlling cholesterol levels as olive oil has the highest level of monounsaturated fat (75–80%), which helps in maintaining good cholesterol and HDL in the body. It has also pain relief properties, treats constipation and stomachaches, maintains good digestive system, prevents stroke, and aids weight loss (Somova et al. 2003).

3. *Coconut oil*: It is extracted from the kernel of mature coconuts from coconut palm tree (*Cocos nucifera*) by dry or wet processing which are grown in the coastal areas. It is commonly used in food preparation, skin and body care, household and medicinal uses. It contains mainly lauric acid, capric acid, and caprylic acid, along with myristic acid, linoleic acid, oleic acid, phenolic acid, vitamin E, vitamin K, and iron. A 12-carbon lauric acid which when digested can break down into monolaurin can inhibit the growth of pathogens like *Candida albicans* and *Staphylococcus aureus*. It reduces the risk of heart diseases by increasing HDL cholesterol and converting the LDL into a less harmful form. It contains fatty acids which can increase ketone level in the blood which can supply energy for brain cells in Alzheimer's patients. Also, medium-chain triglycerides (MCTs), fatty acids present in coconut oil, can increase the concentration of ketone bodies in the blood which can help in reducing the seizures in epileptic children (Henderson et al. 2009). It is a rich source of lauric acid, capric acid, and caprylic acid which have strong antiviral, antifungal, antimicrobial, and antibacterial properties which boost up immunity. It naturally promotes the growth of good bacteria in the gut which aids in good digestion, reduces the risk of heart disease and diabetes, and relieves constipation (Ogbolu et al. 2007).
4. *Groundnut oil*: It is also known as peanut oil which is derived from seeds of legume crop (*Arachis hypogaea*), and oil is extracted by pressing the peanut kernels. It contains fatty acids which are oleic acid, linoleic acid, and palmitic acid along with stearic acid, arachidic acid, behenic acid, lignoceric acid, and other fatty acids. It has high smoke point which is well suited for high heat cooking and for stir fries. It is used to make soaps through mixing with vegetable oil through saponification. It contains vitamin E which is good for skin health (Anyasor et al. 2009). It is rich in monounsaturated fatty acid (MUFA) which lowers "bad cholesterol" and increases "good cholesterol" in the blood. It reduces the risk of coronary artery disease and heart attacks and maintains a healthy blood lipid profile. The presence of linoleic acid helps in maintaining the blood flow as it is a precursor of prostaglandins which maintain contraction and dilation of blood vessels (Ozcan 2010). It is helpful in reducing diarrhea,

constipation, and digestive problems and helps in maintaining the blood glucose level. It is also helpful in treating dandruff and keeping up the healthy hair.

5. *Sunflower oil*: It is extracted from the seeds of sunflower plant (*Helianthus annuus*) by expeller pressing. It contains palmitic acid, stearic acid, oleic acid, linoleic acid, lecithin, tocopherols, carotenoids, and waxes (Skorić et al. 2008). It is considered as one of the healthiest oils due to presence of a variety of health-enhancing nutrients as it is rich in vitamins E, B1, B5, B6, and C and minerals like copper, phosphorus, zinc, and magnesium. It also contains folate, potassium, calcium, riboflavin, iron, and niacin which provide several health benefits. It has a high vitamin E content which is beneficial for preventing asthma and colon cancer (Quiles et al. 2002). It contains choline and phenolic acid which reduces the risk of cardiovascular diseases and heart attacks, atherosclerosis, and arthritis. The presence of vitamin E in sunflower oil is helpful in improving skin health and regenerating cells; thus, sunflower oil is commonly used in cosmetic applications. The presence of gamma alpha linolenic acid (GLA) which helps in preventing thinning of hair reducing the chances of baldness, and used in treatment of hair loss. It is also rich in vitamin A and carotenoids which help in the prevention of uterine, lung, and skin cancers and aids in cataracts. It also helps in curing fungal infections like athlete's foot which starts between the toes when applied in a topical manner (Guo et al. 2017).
6. *Almond oil*: It is extracted (by cold pressing methods) from dried seeds from almond fruit (*Prunus dulcis*) of the tree native to the Middle East, the Indian subcontinent, and North Africa areas. It contains monounsaturated oleic acid (an omega-9 fatty acid), linoleic acid (a polyunsaturated omega-6 essential fatty acid), and saturated fatty acid. Almond oil is a rich source of vitamin E providing 261% of the daily value per 100 ml. Sweet almond oil is rich in folic acids, unsaturated fats, protein, and potassium and monounsaturated fatty acids (Ahmad 2010) These fats boost up the heart health and prevent cardiovascular diseases. It regulates blood pressure and maintains the cholesterol levels, type 2 diabetes, and liver health. It helps to fight common infections, works as an effective laxative, aids in digestion, and reduces the risk of cancer. It contains omega-3 fatty acids and potassium which nourishes the nervous system. It has an ample amount of vitamin D which helps to absorb calcium and strengthens bones. It also helps in easing the pain and stress from strained muscle due to its analgesic properties. The presence of linoleic and linolenic acids in almond oil helps in reducing inflammation (Kamil and Chen 2012).
7. *Canola oil*: It is derived from seeds of rapeseed (*Brassica rapa*) by slightly heating and then crushing the seed. It has the least amount of saturated fat and no trans-fat or cholesterol. It has a high erucic acid content. Alpha-linolenic acid (ALA) is an omega-6 fatty acid which protects against strokes and heart problems. It has the highest level of the plant sterols, beta-sterol and campesterol, which reduces the risk of heart diseases. It has a good amount of monounsaturated fats (MUFA), which helps in reducing LDL (bad cholesterol) and increases good cholesterol (HDL), thus lowering cholesterol. Linoleic acid

(LA) is an essential omega-6 fatty acid which is important for the brain and for the growth and development of infants. It has also anti-inflammatory properties and helps in reducing the inflammation caused during asthma and bowel disorders. It is also used to treat arthritis because it reduces tenderness and stiffness of joints. Being rich in vitamins E and K, it helps to cure skin and hair problems (Lin et al. 2013).

8. *Soyabean oil*: It is derived from seeds of soyabean (*Glycine max*). It is rich in polyunsaturated fats, omega-6 proteins, and amino acid and low in saturated fats. It contains β-sitosterol, which can cause a reduction in cholesterol storage. It contains omega-3 fatty acids which protect the body from various cardiovascular diseases. It is high in unsaturated fat with zero cholesterol, and presence of monounsaturated and polyunsaturated fatty acids in soybeans prevents transportation of cholesterol in the bloodstream. It has a good amount of vitamin K which aids in improving Alzheimer's disease. It has phytic acid which acts as an antioxidant and helps in combating many diseases like cancer, diabetes, inflammation, and tumor. It also contains vitamin D which helps in increasing the strength of bones. It also contains vitamin E which helps in maintaining the skin and hair health (Ivanov et al. 2010).
9. *Rice bran oil*: Rice bran oil is the oil extracted from the hard outer brown layer of rice after chaff (rice husk). It is popularly used as a cooking oil and in salad dressings in the several countries like Asian countries, including Bangladesh, Japan, India, and China. It helps in reducing the cholesterol levels in our body as it lowers the LDL levels. It helps in reducing the risk of cancer due to the presence of tocopherols and tocotrienols, which are substances that are antimutagenic in nature. It is rich in antioxidants which naturally boost up the immune system. It contains vitamin E which helps in maintaining the hormone balance and thus improves endocrine system. It helps in reducing the risk of heart diseases as it contains a good amount of omega-3 and omega-6 fatty acids (Sohail et al. 2017).
10. *Corn oil*: It is extracted from the germ of corn (maize) extracted by expeller pressing method and mainly used in cooking. It is composed mainly of polyunsaturated fatty acids (PUFAs) and low in saturated fats. It helps in reducing LDL levels, thus maintaining the cholesterol levels. It contains linoleic acid (omega-6) which is required for some immune system functions. It is rich in vitamin E and antioxidants. It has omega-3 and omega-6 which are known to work against inflammation and also helps in relieving the symptoms of arthritis, as well as headaches, gastrointestinal problems, and even inflammatory conditions of the skin. It can help in reducing allergic reactions when used topically. It is rich in flavonoids and antioxidants, such as lutein, which can reduce free radical activity.

Non-edible Oil This category includes the oils which are used for medicinal purposes. This class includes the following essential oils. Some of the oils are described here:

1. *Basil oil*: It is extracted from basil leaves (*Ocimum basilicum*) commonly called Tulsi through steam distillation. Basil oil consists of various chemical compounds like linalool α-pinene, citronellol, camphene, β-pinene, limonene, cis-ocimene, camphor, methyl chavicol, γ-terpineol, myrcene, geraniol, methyl cinnamate, and eugenol. It also contains vitamin A, magnesium, potassium, iron, and calcium. It has analgesic, antidepressant, anti-venomous, digestive, insecticide, tonic, and stimulant properties. It is effective in treating cold and fever due to its antispasmodic property. It is ophthalmic in nature and can relieve from bloodshot eyes. It has antibacterial and antiviral properties and works to detoxify the urinary and digestive tracts and can speed up the healing process for painful infections. It has antioxidant effects which can help in fighting chronic infections like bronchitis and respiratory illnesses. It is beneficial in getting relief from healing fatigued or aching muscles (Sakkas and Papadopoulou 2017). The emulsion-based delivery systems are very well used for the delivery of limonene to the pathogen.
2. *Eucalyptus oil*: This oil is extracted and purified from eucalyptus tree (*Eucalyptus globulus*) (Tasmanian blue gum) which is an evergreen tree native to Australia. It is used in cold and cough medications, influenza, other respiratory infections, rhinitis, and sinusitis. The oil contains 1,8-cineole promoting anti-inflammation which can be used to treat bronchial asthma. It has antimicrobial, antifungal, and antiviral properties which are effective in treating wounds, acne, boils, minor cuts, insect bites, and skin infections. Being antifungal in nature, it is beneficial for treating fungal-infected toenails and superficial onychomycosis when used topically. It is used in cosmetics, insecticides, and disinfectants. It is found to be very effective against cavities, dental plaque, gingivitis, and other dental infections owing to its germicidal properties (Dhakad et al. 2017).
3. *Tea tree oil*: It is extracted from tea tree *Melaleuca alternifolia* which is native to New South Wales, Australia. It is used for treating lice since traditional times. It is a terpinen-4-ol-type oil which consists of many components. It has known antibacterial, antimicrobial, antiseptic, antiviral, balsamic, expectorant, fungicide, insecticide, and stimulant properties. A study on the formation of novel PEG-stabilized lipid nanoparticles which are loaded with terpinen-4-ol is found to be effective against *C. albicans* by inhibiting biofilm formation, blocking the respiration, and ascribing to the mitochondrial membrane (Sun et al. 2012).

   It is useful in treating urinary tract infections. It is found to be very efficient to treat root canal pain, ease pneumonia, accelerate healing in cellulitis, and help in reducing gingivitis inflammation. It has antimicrobial and anti-inflammatory property and might help to heal blepharitis condition (eyelid inflammation) (Carson et al. 2006).
4. *Clove oil*: It is extracted from the clove plant (*Syzygium aromaticum*) which is native to Asian, African, and the Near and Middle East countries. The oil can be extracted from leaf, bud, and twig. It is used for flavoring the food and used as medicine for toothache, dental problems, etc. since ancient times. It mainly consists of the eugenol (70–90%), beta-caryophyllene, gallotannic acid, methyl

salicylate, eugenin, kaempferol, rhamnetin, oleanolic acid, and stigmasterol. It is popularly used as an antiseptic for treating infections like athlete foot, wounds, and cuts. It can be used to treat cold, cough, asthma problems, sinusitis, and headache. It helps in maintaining blood glucose level, thus in curing diabetes. It has antiviral and antioxidant properties which help in the boosting of the immune system. It can also be used as an insect repellent (Chaieb et al. 2007). It has been reported that *Eugenia caryophyllata* EO in solid lipid nanoparticle is found to be effective against human fungal pathogens (Fazly Bazzaz et al. 2018). In order to stabilize the antibiofilm activity of these EOs, there was a development of hybrid nanomaterial on catheter surface pellicles. This suggests the usage of nanosystems for the antibiofilm coatings for biomedical applications (Grumezescu et al. 2012).

5. *Lemongrass oil*: It is extracted from lemongrass (*Cymbopogon*) which is native to Asian and African countries and is used as an insecticide since ancient times. Lemongrass essential oil is extracted from partially dried leaves by a process of steam distillation. It mainly consists of geraniol, citronellal, citral, myrcene, geranyl acetate, nerol, neral, and limonene. It has medicinal properties which are utilized to treat muscular pain, headaches, etc. It has antimicrobial properties, which can be used in the treatment of typhoid, food poisoning, skin diseases, body odor, and malaria. It can work as a tonic for nerves and thus can help to cure nervous disorders, Parkinson's disease, vertigo, Alzheimer's disease, etc. It has antipyretic, carminative, anti-inflammatory, and anticarcinogenic properties (Avoseh et al. 2015).
6. *Cajuput oil*: It is extracted from leaves of the cajuput tree (*Melaleuca cajuputi*) through *steam* distillation of its twigs and leaves. The main components are caryophyllene, alpha-pinene, limonene, alpha-terpineol, linalool, etc. It is beneficial in fighting infections, making it a bactericide, antiviral, and fungicide. It is effective against skin infections like psoriasis and scabies. It has anti-inflammatory, antiseptic, antipyretic, carminative, antispasmodic, and antioxidant properties. It is also used as an insect repellent. It also helps in providing relief from intestinal worms (Carson et al. 2006).
7. *Lavender oil*: It is extracted from the lavender flower by distillation methods. It mainly consists of linalool and linalyl acetate along with lavandulyl acetate, terpinen-4-ol, lavandulol, and 1,8-cineole. It has antibacterial, anti-diabetic, antioxidant, and anticancer properties. It can be used to get relief from migraines, headaches, depression, and anxiety issues. It works as a bug repellent, useful to repel mosquitoes, midges, and moths. It can work as an analgesic which can provide relief from pain in sore and tense muscles, muscular aches, rheumatism, sprains, backache, and lumbago. It is beneficial in treating indigestion, skin, and hair problems (Cavanagh and Wilkinson 2002).
8. *Thyme oil*: It is extracted from the herb called thyme (*Thymus vulgaris*). It mainly contains thymol, p-cymene, myrcene, borneol, and linalool. It has well known antispasmodic, antirheumatic, bactericidal, bechic, cardiac, carminative, cicatrisant, diuretic, emmenagogue, expectorant, hypertensive, and insecticidal properties. It has antiseptic properties due to caryophyllene and camphene

which help in healing cuts and wounds. It is also used to treat eczema and other skin infections. It is helpful in maintaining dental health and curing skin and hair problems. It also helps in removing toxins from the body and normalizes the heart. It is also beneficial in improving vision, reducing cellulite, and preventing hair loss. It is very beneficial in treating cough and chest infections and used in cough syrups (Sakkas and Papadopoulou 2017).

9. *Peppermint oil*: It is extracted from the peppermint plant which is a hybrid mint, between watermint and spearmint. The main constituents are menthol and menthone along with menthyl acetate, 1,8-cineole, limonene, beta-pinene, beta-caryophyllene, eriocitrin, hesperidin, and kaempferol. It is used as a flavoring agent in juices and beverages. It is beneficial in treating nausea and sinus and retarding the cavity and aids in digestion problems, bowel issues, and urinary tract infections. It works as a decongestant which clears the airway in the throat and nose and helps in reducing bronchial asthma. It contains menthol which helps in reducing the risk of prostate cancer (Marwa et al. 2017). The menthol-loaded nanostructured lipid carriers are found to be effective against candidal growth in food spoilage which in turn helps in food preservation (Trinetta et al. 2017; Piran et al. 2017).
10. *Chamomile oil*: It is extracted from flowers of the chamomile plant. It contains alpha-pinene, caryophyllene, myrcene, pinocarvone, farnesol, and propyl angelate. It helps in reducing anxiety by promoting relaxation in muscles. It has antimicrobial, antiseptic, antibiotic, and antioxidant properties which help to get relief from allergies and retard the growth of intestinal worms. It is very effective in treating wounds, ulcers, gout, skin irritations, bruises, burns, and canker sores. It contains anodyne compounds which have antispasmodic properties which can be used to treat or relieve digestive issues, such as gas, leaky gut, acid reflux, indigestion, diarrhea, and vomiting. It has anticancer property which kills the cancer cells by inducing apoptosis (Stanojevic et al. 2016).

13.3 Mechanism of Action of Compounds in Essential Oils

Every drug has its own mechanism of action to inhibit the growth of microbes. There are several mechanisms through which the drug hampers the growth. It can inhibit the cell wall, ergosterol biosynthesis, and mitochondrial dysfunction. It can be genotoxic, or have oxidative effect, as well as effect on cell membrane, cell signaling, or can target the virulence traits like, yeast to hyphal transition, biofilm formation and cell adherence. The following sections will deal with the description of EOs and their component with diverse mode of actions based on targets (Fig. 13.1).

13.3.1 Essential Oils Targeting Cell Wall Integrity

Cell wall and membrane are among the most important target of fungal cell. Candidal cell wall contains virulence factors, antigens, etc. and is composed mostly

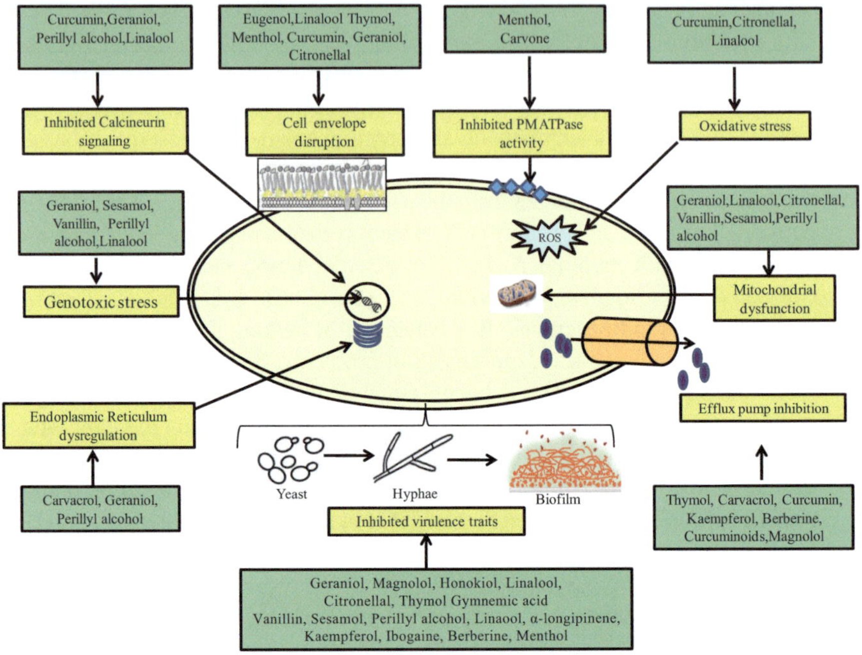

Fig. 13.1 Summary of mode of antifungal actions for EOs and their components. Yellow and green colors denote the affected targets and compounds, respectively

of carbohydrates which could be mannose, glucans, mannoproteins, and chitin (Ruiz-Herrera et al. 2006). Many antifungal drugs are known to target the cell wall and cell membrane.

Eugenol is a phenylpropene which is isolated from clove essential oil. It has known pharmacological properties and has been proven beneficial in combating disease caused by pathogen *Salmonella typhi* and *Proteus mirabilis*. It has been reported that it affects *Candida* by damaging cell wall. It disrupts the cell wall integrity and increases the cell permeability and fluidity. It denatures protein in the wall and reacts with the phospholipid bilayer of cytoplasmic membrane, thereby causing irreplaceable damage to the cell (Latifah-Munirah et al. 2015). Linalool is a monoterpene alcohol which is isolated from the lavender. It has been reported that it targets the cell wall, interferes in the ergosterol biosynthesis, and disrupts the cell membrane integrity. It has been also reported that it even inhibits the growth of candidal strains isolated from oral candidiasis patients (Dias et al. 2017). Thymol is a monoterpene phenol which is isolated from *Thymus vulgaris*, and it has known antiseptic and antimicrobial properties. It is lipophilic in nature which makes it capable of protruding inside the cell membrane of fungus. Through atomic force studies, it has been concluded that the action of thymol on candidal cell membrane results in the formation of folds and holes. It induces asymmetric and membrane tensions in the cell membrane. It intercalates with the fatty acyl chains and disturbs the lipid packing resulting in disrupted cell wall integrity and changes in membrane fluidity (Braga and Ricci 2011). Menthol is isolated from the essential oil from *Mentha piperita* EO and is known for its analgesic property. It has been studied for its anticandidal activity. It was well known to disrupt the membrane integrity which results in loss of structure and function of the membrane. This disruption results in the release of intracellular components like radicals, cytochrome C, calcium, potassium, and magnesium ions, proteins, and nucleic acids, leading to cell death. It is also known to reduce the ergosterol levels and increase the cell acidification by decrease in PM-ATPase activity, thus resulting in cell death (Samber et al. 2015). Carvone is isolated from mint oil of *Mentha piperita* which has known antimicrobial properties. It disrupts the cell membrane integrity, reduces ergosterol levels, and induces reduction in PM-ATPase activity. It also synergizes with FLC and inhibits the growth of FLC-resistant strains (Samber et al. 2015). Similarly, beta-caryophyllene is one of the main constituents of clove oil. It also disrupts the cell wall integrity by reducing the ergosterol levels and has considerable activity against the FLC-resistant *Candida* strains. Monoterpenoids named as geraniol and citronellal which are isolated from lemongrass oil disrupts the cell membrane integrity, deplete ergosterol levels, and alter pH homeostasis (Singh et al. 2016a, b). Capric and lauric acids are saturated fatty acids which have been isolated from groundnut oil. They disintegrate the cell membrane and induce the disorganization and shrinking of cytoplasm, due to development of hydrostatic pressure within the cell (Bergsson et al. 2001). Another compound called as curcumin which is isolated from the rhizome of *Curcuma longa* also alters the cell wall, cell membrane integrity, and cell viability (Lee and Lee 2014).

Overexpression of efflux pumps in *C. albicans* is one of the major causes for the development of MDR. Efflux pumps are also one of the major targets for combating candidal infections. Many compounds isolated from essential oil are reported to target efflux pumps by either modulating the efflux pump activity or inhibiting the activity of efflux pump transporters cdr1p/cdr2p and downregulation of efflux pump genes CDR1. Thymol, carvacrol, curcumin, kaempferol, berberine, curcuminoids, and magnolol are reported to inhibit the efflux pump activity of CDR1 (Singh et al. 2017).

13.3.2 Essential Oils Targeting Cell Signaling

Regulation of signaling pathways plays an important role in the pathogenesis of candidal infections. Many pathways are involved in maintaining the decorum in the cell. Ion signaling networks are crucial for the development and virulence for the regulation of major functions in *Candida* like gene expression, host association, filamentation, invasion, pathogen stress response, and survival. Calcium homeostasis pathways play a vital role in the regulation of calcineurin signaling pathways. Similarly, H^+ ion homeostasis is also important for pathogenesis. Carvacrol is a monoterpene phenol which is also found to synergize with the azoles. ER is the main source for trafficking system in a cell. It disrupts the morphology and integrity of endoplasmic reticulum and induces the unfolded protein response in *C. albicans*. Its disruption might result in a collapsing vesicle trafficking network. It also disturbs calcium homeostasis which can result in the disturbance of signaling pathways related to calcium like calcineurin signaling and TOR signaling pathway that regulates the cell growth (Rao et al. 2010). Linalool is also involved in inducting oxidative stress in *C. albicans*. It causes disturbance in mitochondrial complexes which results in decrease in ATP level and cell viability. Through mitochondrial-dependent O_2 and membrane-dependent lipid peroxidation, there is an accumulation of peroxides. There is a formation of reactive oxygen species including superoxide radicals, peroxides, lipid peroxides, and hydroxyl radicals which interferes with proteins and nucleic acid and disrupts the structures. ROS-induced stress can result in arresting cell cycle at G1 phase, which results in DNA damage, apoptosis, and necrosis. It also modulates the activity of membrane-bound signaling proteins and thus signaling pathways (Mátè et al. 2017). Geraniol and citronellal despite being monoterpenes affect calcineurin and oxidative stress signaling pathways, respectively (Singh et al. 2016a, b; Saibabu et al. 2017). Perillyl alcohol is another monoterpene albeit a monocyclic which is isolated from the essential oil of peppermint and affects calcineurin signaling pathway (Ansari et al. 2016). Curcumin is known to induce oxidative stress which in turn leads to apoptosis (Sharma et al. 2010).

13.3.3 Essential Oils Targeting Virulence Traits

Virulence traits are involved in host recognition, binding to the host, and clearance from the host after infection. It also includes the expression of genes involved in traits like the yeast to hyphal transition, cell adherence, and biofilm formation. Morphological switching or yeast to hyphal transition executed is by external signals. This dimorphic switching is one of the major contributors for candidal infections. Yeast to hyphal switching plays a role in escaping from phagocytosis. The hyphal form is favorable for the invasion in the host cell. Hyphal cell expresses the wall protein and favors for deep-seated invasions, and it is critical for systemic infections. Cell adherence is the primary step for the biofilm formation. Thus the inhibition of adherence to either biotic (expression of adhesins/cell surface proteins) or abiotic (like surface hydrophobicity) surfaces can be a promising factor. *Candida* possesses a key feature to adapt to the different habitats by developing microbial communities which attaches to the biotic or abiotic surfaces known as biofilms embedded in extracellular matrix. Biofilms are a major concern for hospitals because there is a rise in mortality rates due to nosocomial infections in immunocompromised patients. These biofilms can inhabit in the indwelling devices like urinary catheters, dental materials, artificial heart valves, joint prostheses, contact lenses, penile implants, and intrauterine devices. They are very hard to eradicate as this strong structure becomes resistant to the action of antifungal agents by blocking the entry of drugs. Mature biofilms are very hard to treat as they have a complex network of microbial community. There are many EOs that have been tested on the virulence traits of *Candida*. Two neolignan compounds magnolol and honokiol isolated from bark and stem of *Magnolia officinalis* are known to be used in the treatment of asthma, liver disease, diarrhea, etc. It has been known to suppress the adhesion, yeast to hyphal transition, and biofilm formation. It has been found that there is a downregulation of the genes *RAS1*, *EFG1*, *TEC1*, and *CDC35* involved in Ras1-cAMP-Efg1 pathway (Sun et al. 2015). This pathway plays a crucial role in yeast to hyphal switching. The treatment of these lignan compounds also downregulates the genes (*HWP1*, *ALS3*, and *ECE1*) encoding for adhesin proteins. The expression of these genes is also regulated by this pathway. Triterpenoid saponin compound gymnemic acid isolated from leaves of *Gymnema sylvestre* is used for treating diabetes. It is found to affect the yeast to hyphal morphogenesis through downregulation of hyphal induction genes and related pathway (Vediyappan et al. 2013). The in vivo studies on *C. elegans* have shown that it is effective for treating invasive hyphal growth. Thymol is a major component on thyme oil (*Thymus vulgaris*) isolated from the *Satureja hortensis* plant that effects the candidal biofilm growth. Linalool and α-longipinene are isolated from cascarilla bark oil and helichrysum oil, respectively. Linalool is a major component of lavender oil and has been used as a food additive. It is found to downregulate the adhesin genes *HWP1* and *ALS3*. Linalool also downregulates the expression level of *CYR1* and *CPH1*, which encode for components of the cAMP-PKA and MAPK hyphal regulatory pathway (Hsu et al. 2013). It inhibits the cell adherence, germ tube formation, and biofilm formation. It has been also reported that citronella isolated from the *Cymbopogon winterianus* plant and *Cinnamon cassia* (cinnamon) has been found to inhibit the biofilm formation in the denture cleansing. Kaempferol, ibogaine, and berberine are reported to suppress the extracellular enzyme activity which is necessary for establishing the hyphal form from yeast form for causing infections (Yordanov et al. 2008). 1,8-cineole is the main component of eucalyptus oil extracted from eucalyptus tree that inhibits both planktonic growth and biofilm formation. Menthol isolated from essential oil from *Mentha piperita* EO was reported to inhibit the cell adherence and the metabolic activity in the biofilm formation of *C. albicans*.

There are certain natural compounds isolated from essential oils which have diverse and varied modes of action to inhibit the growth of *C. albicans*. For instance, geraniol and citronellal extracted from lemongrass oil have been studied and have effect on cell membrane and all virulence traits (Singh et al. 2016a, b). Similarly, sesamol isolated from sesame seeds also disrupts cell wall integrity and virulence traits (Ansari et al. 2014), but perillyl alcohol not only affects virulence traits but affects the glyoxylate cycle in *C. albicans* (Ansari et al. 2016).

13.4 Essential Oils in Comparison to Known Antifungals

Nature has bestowed us with the diverse classes of natural compounds which have broad antimicrobial properties. Conventional drugs have known side effects in comparison with the natural compounds. Natural compounds from EOs are known to be much more beneficial owing to their natural origin. Compounds from essential oils can also be investigated more in future for better understanding of their antimicrobial properties. Although EOs have high MIC values in comparison with the known antifungal drugs, there are wide ranges of studies depicting that they work synergistically with the known antifungal drugs belonging to azoles, polyenes, and echinocandins. Even at some instances where they are not synergistic, they showed potentiation of antifungal activities of known drugs. EOs can be used in combination with drug therapy along with known drugs.

13.5 Conclusions

EOs are a huge treasure of natural compounds which are proven to be potent in combating fungal infections. EOs can serve the mankind in maintaining good health as they have edible and medicinal properties. The mode of action of many natural compounds isolated from EOs is needed to be further elucidated through more research. Thus there is a need to search more on the structural, functional, and mode of action of EOs to exploit their phytotherapeutic potential. Parallel to it, there is a need to search the novel nanosystems for more potent delivery and application of EOs in biomedical therapeutics.

References

Abuznait AH, Qosa H, Busnena BA, El Sayed KA, Kaddoumi A (2013) Olive-oil-derived oleocanthal enhances β-amyloid clearance as a potential neuroprotective mechanism against Alzheimer's disease: in vitro and in vivo studies. ACS Chem Neurosci 4:973–982

Ahmad Z (2010) The uses and properties of almond oil. Complement Ther Clin Pract 16:10–12

Anghel I, Grumezescu AM, Holban AM, Ficai A, Anghel AG, Chifiriuc MC (2013) Biohybrid nanostructured iron oxide nanoparticles and Satureja hortensis to prevent fungal biofilm development. Int J Mol Sci 14:18110–18123. https://doi.org/10.3390/ijms140918110

Ansari MA, Fatima Z, Hameed S (2014) Sesamol: a natural phenolic compound with promising anticandidal potential. J Pathog 2014:895193. https://doi.org/10.1155/2014/895193

Ansari MA, Fatima Z, Hameed S (2016) Anticandidal effect and mechanisms of monoterpenoid, perillyl alcohol against Candida albicans. PLoS One 11(9):e0162465. https://doi.org/10.1371/journal.pone.0162465

Anyasor GN, Ogunwenmo KO, Oyelana OA, Ajayi D, Dangana J (2009) Chemical analyses of Groundnut (Arachis hypogaea) oil. Pak J Nutr 8:269–272

Avoseh O, Oyedeji O, Rungqu P, Nkeh-Chungag B, Oyedeji A (2015) Cymbopogon species; ethnopharmacology, phytochemistry and the pharmacological importance. Molecules 20:7438–7453

Bakkali F, Averbeck S, Averbeck D, Idaomar M (2008) Biological effects of essential oils – a review. Food Chem Toxicol 46:446–475

Bendini A, Cerretani L, Carrasco-Pancorbo A, Gómez-Caravaca AM, Segura-Carretero A, Fernández-Gutiérrez A, Lercker G (2007) Phenolic molecules in virgin olive oils: a survey of their sensory properties, health effects, antioxidant activity and analytical methods. An overview of the last decade. Molecules 12:1679–1719

Bergsson G, Arnfinnsson J, Steingrímsson O, Thormar H (2001) In vitro killing of Candida albicans by fatty acids and monoglycerides. Antimicrob Agents Chemother 45:3209–3212

Braga PC, Ricci D (2011) Thymol-induced alterations in Candida albicans imaged by atomic force microscopy. Methods Mol Biol 736:401–410

Carson CF, Hammer KA, Riley TV (2006) Melaleuca alternifolia (Tea Tree) oil: a review of antimicrobial and other medicinal properties. Clin Microbiol Rev 19:50–62

Cavanagh HM, Wilkinson JM (2002) Biological activities of lavender essential oil. Phytother Res 16:301–308

Chaieb K, Hajlaoui H, Zmantar T, Kahla-Nakbi AB, Rouabhia M, Mahdouani K, Bakhrouf A (2007) The chemical composition and biological activity of clove essential oil, Eugenia caryophyllata (Syzigium aromaticum L. Myrtaceae): a short review. Phytother Res 21:501–506

Claudia S, Dario L (2013) Candida infections, causes, targets, and resistance mechanisms: traditional and alternative antifungal agents. Biomed Res Int 2013:204237. https://doi.org/10.1155/2013/204237

Dhakad AK, Pandey VV, Beg S, Rawat JM, Singh A (2017) Biological, medicinal and toxicological significance of Eucalyptus leaf essential oil: a review. J Sci Food Agric. https://doi.org/10.1002/jsfa.8600

Dias IJ, Trajano ERIS, Castro RD, Ferreira GLS, Medeiros HCM, Gomes DQC (2017) Antifungal activity of linalool in cases of Candida spp. isolated from individuals with oral candidiasis. Braz J Biol. https://doi.org/10.1590/1519-6984.171054

Fazly Bazzaz BS, Khameneh B, Namazi N, Iranshahi M, Davoodi D, Golmohammadzadeh S (2018) Solid lipid nanoparticles carrying Eugenia caryophyllata essential oil: the novel nanoparticulate systems with broad-spectrum antimicrobial activity. Lett Appl Microbiol 66:506–513. https://doi.org/10.1111/lam.12886

Ghosh V, Mukherjee A, Chandrasekaran N (2012) Mustard oil microemulsion formulation and evaluation of bactericidal activity. Int J Pharm Pharm Sci 4:497–500

Gow NAR, Yadav B (2017) Microbe Profile: Candida albicans: a shape-changing, opportunistic pathogenic fungus of humans. Microbiology 163:1145–1147

Grumezescu AM, Chifiriuc MC, Saviuc C, Grumezescu V, Hristu R, Mihaiescu DE, Stanciu GA, Andronescu E (2012) Hybrid nanomaterial for stabilizing the antibiofilm activity of Eugenia carryophyllata essential oil. IEEE Trans Nanobiosci 11:360–365. https://doi.org/10.1109/TNB.2012.2208474

Guo S, Ge Y, Jom KN (2017) A review of phytochemistry, metabolite changes, and medicinal uses of the common sunflower seed and sprouts (Helianthus annuus L.). Chem Cent J 11:95. https://doi.org/10.1186/s13065-017-0328-7

Henderson ST, Vogel JL, Barr LJ, Garvin F, Jones JJ, Costantini LC (2009) Study of the ketogenic agent AC-1202 in mild to moderate Alzheimer's disease: a randomized, double-blind, placebo-controlled, multicenter trial. Nutr Metab (Lond) 6:31. https://doi.org/10.1186/1743-7075-6-31

Hsu CC, Lai WL, Chuang KC, Lee MH, Tsai YC (2013) The inhibitory activity of linalool against the filamentous growth and biofilm formation in Candida albicans. Med Mycol 51:473–482

Ivanov DS, Lević JD, Sredanović SA (2010) Fatty acid composition of various soybean products. Food Feed Res 37:65–70

Jassbi AR, Asadollahi M, Masroor M, Schuman MC, Mehdizadeh Z, Soleimani M, Miri R (2012) Chemical classification of the essential oils of the Iranian Salvia species in comparison with their botanical taxonomy. Chem Biodivers 9:1254–1271. https://doi.org/10.1002/cbdv.201100209

Kamil A, Chen CY (2012) Health benefits of almonds beyond cholesterol reduction. J Agric Food Chem 60:6694–6702

Kathiravan MK, Salake AB, Chothe AS et al (2012) The biology and chemistry of antifungal agents: a review. Bioorg Med Chem 20:5678–5698

Lalla RV, Patton LL, Dongari-Bagtzoglou A (2013) Oral candidiasis: pathogenesis, clinical presentation, , diagnosis and treatment strategies. J Calif Dent Assoc 41:263–268

Latifah-Munirah B, Himratul-Aznita WH, Zain NM (2015) Eugenol, an essential oil of clove, causes disruption to the cell wall of Candida albicans (ATCC 14053). PLoS One 10(2):e0117695. https://doi.org/10.1371/journal.pone.0117695.

Lee W, Lee DG (2014) An antifungal mechanism of curcumin lies in membrane-targeted action within Candida albicans. IUBMB Life 66:780–785

Lin L, Allemekinders H, Dansby A, Campbell L, Durance-Tod S, Berger A, Jones PJ (2013) Evidence of health benefits of canola oil. Nutr Rev 71:370–385

Marwa C, Fikri-Benbrahim K, Ou-Yahia D, Farah A (2017) African peppermint (Mentha piperita) from Morocco: Chemical composition and antimicrobial properties of essential oil. J Adv Pharm Technol Res 8:86–90

Máté G, Kovács D, Gazdag Z, Pesti M, Szántó Á (2017) Linalool-induced oxidative stress processes in the human pathogen Candida albicans. Acta Biol Hung 68:220–231

Nagpal M, Sood S (2013) Role of curcumin in systemic and oral health: an overview. J Nat Sci Biol Med 4:3–7

Ogbolu DO, Oni AA, Daini OA, Oloko AP (2007) In vitro antimicrobial properties of coconut oil on Candida species in Ibadan, Nigeria. J Med Food 10:384–387

Ozcan MM (2010) Some nutritional characteristics of kernel and oil of peanut (Arachis hypogaea L.). J Oleo Sci 59:1–5

Pan A, Chen M, Chowdhury R, Wu JH, Sun Q, Campos H, Mozaffarian D, Hu FB (2012) α-Linolenic acid and risk of cardiovascular disease: a systematic review and meta-analysis. Am J Clin Nutr 96:1262–1273

Piran P, Kafil HS, Ghanbarzadeh S, Safdari R, Hamishehkar H (2017) Formulation of menthol-loaded nanostructured lipid carriers to enhance its antimicrobial activity for food preservation. Adv Pharm Bull 7:261–268. https://doi.org/10.15171/apb.2017.031

Quiles JL, Ramírez-Tortosa MC, Gómez JA, Huertas JR, Mataix J (2002) Role of vitamin E and phenolic compounds in the antioxidant capacity, measured by ESR, of virgin olive, olive and sunflower oils after frying. Food Chem 76:461–468

Rao A, Zhang Y, Muend S, Rao R (2010) Mechanism of antifungal activity of terpenoid phenols resembles calcium stress and inhibition of the TOR pathway. Antimicrob Agents Chemother 54:5062–5069

Ruiz-Herrera J, Elorza MV, Valentín E, Sentandreu R (2006) Molecular organization of the cell wall of Candida albicans and its relation to pathogenicity. FEMS Yeast Res 6:14–29

Saibabu V, Singh S, Ansari MA, Fatima Z, Hameed S (2017) Insights into the intracellular mechanisms of citronellal in Candida albicans: implications for reactive oxygen species-mediated necrosis, mitochondrial dysfunction, and DNA damage. Rev Soc Bras Med Trop 50(4):524–529

Sakkas H, Papadopoulou C (2017) Antimicrobial activity of basil, oregano, and Thyme essential oils. J Microbiol Biotechnol 27:429–438

Samber N, Khan A, Varma A, Manzoor N (2015) Synergistic anti-candidal activity and mode of action of Mentha piperita essential oil and its major components. Pharm Biol 53:1496–1504

Sengupta A, Ghosh M (2011) Hypolipidemic effect of mustard oil enriched with medium chain fatty acid and polyunsaturated fatty acid. Nutrition 27:1183–1193

Sharma M, Manoharlal R, Puri N, Prasad R (2010) Antifungal curcumin induces reactive oxygen species and triggers an early apoptosis but prevents hyphae development by targeting the global repressor TUP1 in Candida albicans. Biosci Rep 30:391–404

Singh S, Fatima Z, Hameed S (2016a) Insights into the mode of action of anticandidal herbal monoterpenoid geraniol reveal disruption of multiple MDR mechanisms and virulence attributes in Candida albicans. Arch Microbiol 198:459–472

Singh S, Fatima Z, Hameed S (2016b) Citronellal induced disruption of membrane homeostasis in Candida albicans and attenuation of its virulence attributes. Rev Soc Bras Med Trop 49:465

Singh S, Fatima Z, Hameed S (2015) Predisposing factors endorsing Candida infections. Infrez Med 23:211–223

Singh S, Hans S, Fatima Z, Hameed S (2017) Overcoming fungal multidrug resistance by natural compounds targeting efflux pumps. Front Anti-Infect Drug Discov 7:1–12

Skorić D, Jocić S, Sakac Z, Lecić N (2008) Genetic possibilities for altering sunflower oil quality to obtain novel oils. Can J Physiol Pharmacol 86:215–221

Sobel JD (2007) Vulvovaginal candidiasis. Lancet 369:1961–1971

Sohail M, Rakha A, Butt MS, Iqbal MJ, Rashid S (2017) Rice bran nutraceutics: a comprehensive review. Crit Rev Food Sci Nutr 57:3771–3780

Somova LI, Shode FO, Ramnanan P, Nadar A (2003) Antihypertensive, antiatherosclerotic and antioxidant activity of triterpenoids isolated from Olea europaea, subspecies africana leaves. J Ethnopharmacol 84:299–305

Stanojevic LP, Marjanovic-Balaban ZR, Kalaba VD, Stanojevic JS, Cvetkovic DJ (2016) Chemical Composition, Antioxidant and Antimicrobial Activity of Chamomile Flowers Essential Oil (Matricaria chamomilla L.). J Essent Oil Bear Plants 19:2017–2028

Sun L, Liao K, Wang D (2015) Effects of magnolol and honokiol on adhesion, yeast-hyphal transition, and formation of biofilm by Candida albicans. PLoS One 10(2):e0117695. https://doi.org/10.1371/journal.pone.0117695

Sun LM, Zhang CL, Li P (2012) Characterization, antibiofilm, and mechanism of action of novel PEG-stabilized lipid nanoparticles loaded with terpinen-4-ol. J Agric Food Chem 60:6150–6156. https://doi.org/10.1021/jf3010405

Tanwar J, Das S, Fatima Z, Hameed S (2014) Multidrug resistance: an emerging crisis. Interdiscip Perspect Infect Dis 2014, 541340. https://doi.org/10.1155/2014/541340

Trinetta V, Morgan MT, Coupland JN, Yucel U (2017) Essential oils against pathogen and spoilage microorganisms of fruit juices: use of versatile antimicrobial delivery systems. J Food Sci 82(2):471–476

Vediyappan G, Dumontet V, Pelissier F, d'Enfert C (2013) Gymnemic acids inhibit hyphal growth and virulence in Candida albicans. PLoS One 8(9):e74189. https://doi.org/10.1371/journal.pone.0074189

Yordanov M, Dimitrova P, Patkar S, Saso L, Ivanovska N (2008) Inhibition of Candida albicans extracellular enzyme activity by selected natural substances and their application in Candida infection. Can J Microbiol 54:435–440

Part V

Intersection of NanoBioMedicine with Therapeutics and Diagnostics

Emerging Trends in Nanotheranostics

14

Deepa Suhag, Mehak Chauhan, Adeeba Shakeel,
and Souvik Das

Abstract

In recent times, the population is more prone to getting sick than the times before. This widespread occurrence of diseases warrants a diagnostic system in place which can help us acknowledge its presence even during the early stages, especially in the case of life-threatening ones. Nanotheranostics is a field which helps us with the same; it not only helps us to diagnose a disease in its earlier stages but also helps treat it at the point of care itself. Nanotheranostics include diagnosis at a nanoscale level, using tools such as MRI, PET-CT scan, etc., as well as providing therapy at the nanoscale level using tools like chemotherapy, radiotherapy, photodynamic therapy, etc. Both systems are conjugated on a nanocarrier which can be a polymer, lipid, or inorganic material depending upon the properties required. This new age medical system has also opened avenues for something called personalized medicine.

Keywords

Nanomedicine · Theranostics · Therapy · Imaging and precision medicine

D. Suhag (✉)
Amity Institute of Biotechnology, Amity University Haryana,
Manesar, Panchgaon, Haryana, India

M. Chauhan · A. Shakeel
Amity Institute of Biotechnology, Amity University Uttar Pradesh,
Noida, Uttar Pradesh, India

S. Das
French National Institute of Health Unit, University Center for Health Research,
Amiens, France

S. K. Saxena, S. M. P. Khurana (eds.), *NanoBioMedicine*,
https://doi.org/10.1007/978-981-32-9898-9_14

14.1 Introduction

Nanomedicine is a new field wherein nanotechnology has joined hands with medicine. It involves tools in nanotechnology that are used for the treatment of any disease. Any material in a nanoscale range having biocompatibility can be used as a platform for nanomedicine.

Nanomedicine involves drug delivery by nanoparticles which is much more efficient and faster than conventional drug delivery.

However, the efficacy depends upon certain factors that include:

1. Drug encapsulation
2. Drug carrier delivery at the targeted site
3. Successful drug release

All of these abovementioned factors are often interdependent but are also conditional to the choice of nanoparticle carrier, the matrix the drug is embedded into, and the physiological conditions of the target site. Hence, while developing a nanomedicine, all these features must be taken into consideration, as even a single flaw would result in failure of treatment (Kunjachan et al. 2013).

Although nanomedicine alone has been employed in treatment of various diseases, recently its application in a new field has been gaining importance.

We understand that the framework of every person is different, may it be their genetics, their general physiology, their immune system, or even the environment they live in. Owing to these variations, every individual responds differently to a treatment. A drug which might be a miracle drug for one person may prove to be detrimental or ineffectual for another. This is exactly the case in diseases like cancer, which themselves are heterogenous in nature.

Therefore, an approach where the medicines are designed, keeping just an individual in mind is necessary, especially in the case of life-threatening diseases, wherein even a single-failed course of treatment could result in fatality due to the shortage of time. It is at this critical point that theranostic approaches step in.

Theranostics is a term that is coined by the amalgamation of two words, therapeutics and diagnostics. The root words are self-explanatory. Theranostics integrate the science of diagnosis with therapy. A combined technique helps to make the process of disease diagnosis and treatment faster and cost-effective with minimal side effects.

This is a new class of treatment that has been rapidly garnering a lot of attention due to its ability to provide a personalized and specific treatment for several diseases as we all know that one drug cannot cure everyone.

A theranostic system can be defined as nanomedicine model which combines diagnosis with targeted drug delivery. This system is also referred to as personalized medicine. One of the key features of this type of medicinal approach is that they diagnose the disease at the site of action and deliver the drug on the target site itself, basis of the diagnosis. This approach of personalize medicine has moved the diagnosis from a pathology laboratory to the site of ailment itself (Burguma et al. 2018).

The diagnosis is done through MRI, ultrasound, and all the other techniques that are used in a laboratory, while the therapy involves targeted drug release, radiotherapy, multimodal therapy, etc.

Since the advent of nanotechnology, developing theranostic tools has become more efficient, as now a nanoparticle can carry out both the therapy and imaging, thereby integrating nanomedicine within itself.

This new class of theranostics is better than the traditional one as it allows the imaging not only before and after the delivery of drug but also during the process of treatment. This entire process is termed as co-delivery. The nanomaterials were earlier used for diagnosis, now, however by just adding a therapeutic device to it, making it a theranostic tool. Apart from providing a suitable platform for personalized medicine, other areas where combining diagnosis and therapy have proven to be helpful are as follows (Sharma et al. 2017):

I. The heterogeneity of a disease cannot be understood by conventional means.
II. The biopsy results may sometimes be prone to error during sampling.
III. Studying only a part of tumor rather than the entire area by the conventional biopsies or other diagnostic echniques may lead to misdiagnosis.

Hence, a diagnosis done at the molecular level ensures that sampling error does not occur and the disease is studied as a whole, rather than just examining a part or product of it.

14.2 Components of a Theranostic System

The assembly should not just include a diagnostic probe and a drug carrier, but it should also bear an imagining agent or a signal emitter and a chemosensitizer to nullify any resistance. Together all these components make a multifunctional theranostic system (Sharma et al. 2017).

Signal emitter has optical or magnetic properties, the drug can be chemotherapeutic, and the carrier would have polymer matrix modified with functional groups. All of this should be encapsulated in a synthetic or natural biocompatible biopolymer (Figs. 14.1 and 14.2).

Development of diagnostic probes which can help us in visualizing the real-time status of tumors is an integral part of the development of a theranostic model. It is this real-time analysis that enables us to study the disease based on its molecular profile and progression rate for everyone, separately.

An ideal diagnostic probe should have the following features:

1. Can identify the target location
2. Is nontoxic
3. Has high signal-to-noise ratio
4. High specificity for the target
5. Can perform early detection

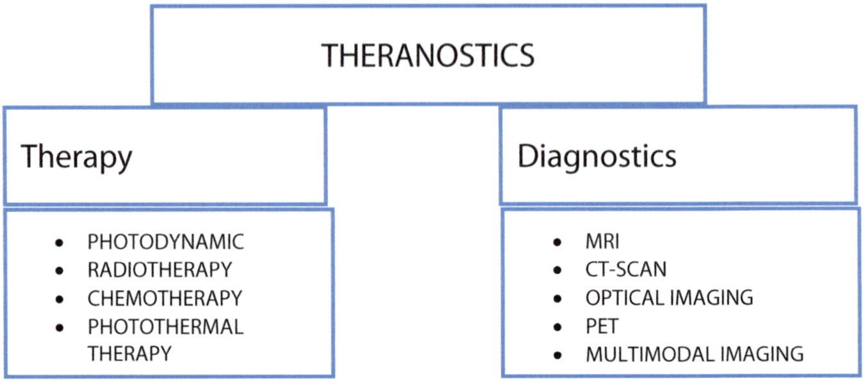

Fig. 14.1 Theranostics as a combination of diagnosis and therapy

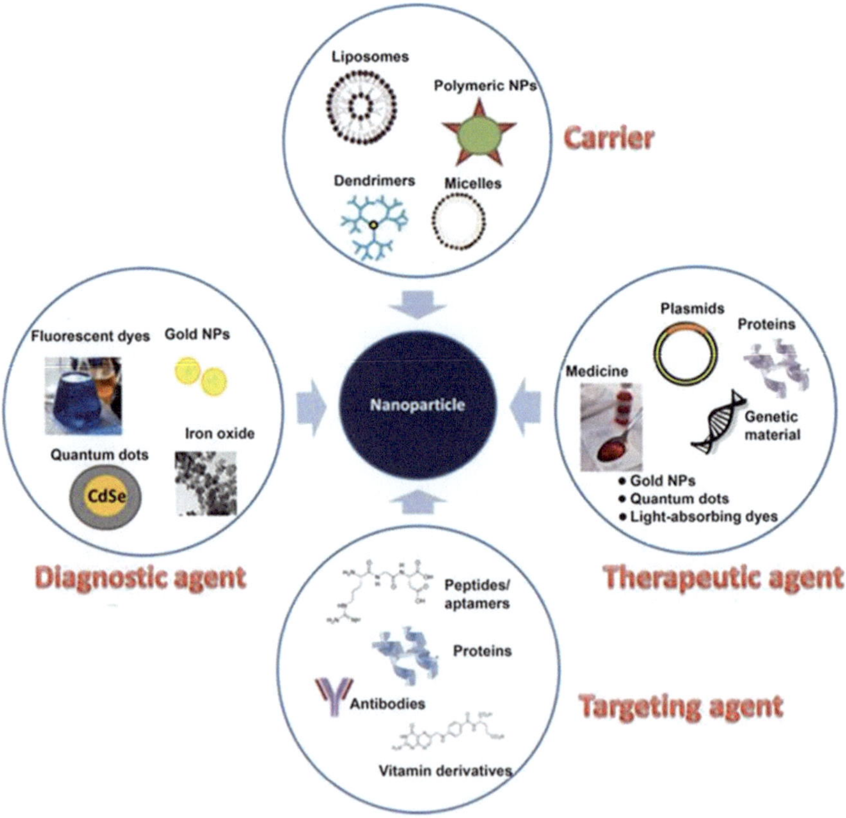

Fig. 14.2 Components of theranostic system (Menon 2013)

Drug carriers always have specific targeting agents attached to their surface. These target agents have high affinity for the tumor cells and hence help in the attachment of the theranostic system to the tumor. This precise attachment of the theranostic system to the tumor in focus then results in the drug release at the specific tumor site. Additionally, this phenomenon of site-specific drug release also reduces the occurrence of side effects.

Moreover, the drug release is heavily governed by the microenvironment. For instance, when the pH around the target site is toward acidic, it could lead to the cleavage of drug carriers in an enzymatic reaction and cause the release of drug. In such instances, nonenzymatic drug carriers show sustained drug delivery (Lammers et al. 2010).

We will read about the carriers and diagnostic agents in detail in the following section of the chapter.

14.3 Applications of Theranostic Systems

1. Detection of heterogenous diseases (e.g., cancer)
2. Individualistic treatment (i.e., personalized medicine)
3. Real-time monitoring of drug release
4. Therapy and post-therapy imaging
5. Nanobiomaterials providing a larger surface area for loading of probes and large surface area also corresponding to a better uptake

14.4 Features/Characteristics of an Ideal Theranostics System:

I. Materials used should be ideally biodegradable and biocompatible.
II. Pharmacokinetics and pharmacodynamics should be well known and understood.
III. According to the route of administration, the formulations should be made.
IV. Chemicals should be intra- and intercompatible.

14.5 Platforms Employed for Nanomedicine Development

Platforms as the name suggests are support systems that can host and carry the therapeutic agent along with the imaging modalities and contrast agents to the target site. Depending upon various factors like the physiological conditions of the area to be targeted, the chemical nature of drug to be delivered, and the imaging modality being used, various platforms can be chosen. Some of the most commonly used platforms are discussed as follows:

14.6 Inorganic Platforms

14.6.1 Gold Nanoparticles (AuNPs)

AuNPs are made up of gold and are novel in their properties. They are prepared after hydrogen tetrachloroaurate is chemically altered. These are prepared in the form of spheres, rods, wires, or cubes. The ease with which they can be modified into different shapes leads to their use in nanotheranostics. Different shapes show different properties. For example, the spheroids which are around 10 nm in size absorb UV at 520 nm, while the nano-rods absorb near-infrared radiation (NIR) in the range of 690–900 nm.

Features that make it a suitable nanotheranostic platform are (Sonali et al. 2018):

I. Inherent optical properties
II. Diagnostic properties
III. Biocompatibility and low toxicity
IV. Allows binding to biomaterials via Au-S bonds
V. Tunable core size
VI. Monodispersivity
VII. Light-scattering properties

14.6.2 Magnetic Nanoparticles (MNPs)

Their use in nanotheranostics is ascribed to their ability of being traceable by magnetic resonance imaging (MRI). Owing to this property, the magnetic fields can be exploited to direct the nanoparticles toward the targeted sites.

One of the most widely employed nanotheranostic MNPs is iron oxide nanoparticles (IONPs) (Blau et al. 2016). The wide acceptance of IONPs is attributed to their biocompatibility, superparamagnetic behavior, and cost-effectiveness. These nanoparticles have a magnetic core of hematite or magnetite with a polymer encapsulation. The surface of IONPs is modified by inorganic molecules. The formation of these nanoparticles is carried out by thermal decomposition and co-precipitation techniques (Zhang and Kievit 2011).

14.6.3 Quantum Dots (QDs)

QDs were first prepared by Bell et al. at Bell Laboratories in 1983. These are inorganic semiconductors which are nanoscale (<10 nm) nanocrystals. They have emerged as a popular tool in nanotherapy and diagnosis. They are made up of cadmium selenide (CdSe) with a coating of zinc sulfide layer. Their action is found to be better than organic fluorophores (Sonali et al. 2018). Their properties can be modified by tuning in their size and shape, e.g., depending upon their shape and size, their emission wavelength can be in any range from 450 to 1800 nm.

QD's surface can be modified by cross-linking target moieties. The method used for the conjugation of targeting ligands on QD surface is avidin-biotin cross-linking. QDs can be used for fluorescence imaging both in vivo and in vitro. However, they tend to get metabolized and hence pose a certain toxicity to human beings so their use is limited.

14.7 Carbon-Based Platforms

14.7.1 Carbon Nanotubes (CNTs)

They are carbon nanoparticles wherein each carbon is covalently bonded to three other carbon atoms to make a hexagon. They are cylindrical in shape and are flexible in nature. You can bend them and they will not break, but when released, they would come back to their original form (Sonali et al. 2018).

They are used for electrical detection and are highly sensitive. They are label-free platforms.

Their functionalized form is preferred to the non-functionalized form as the latter is more toxic in nature.

Research by Xu et al. shows that they can compete with tumor in the recruitment of macrophage from systemic sources, and hence, it decreases their level in the tumor cells, in turn inhibiting the metastasis.

14.7.2 Carbon Dots (CDs)

These are carbon nanoparticles whose surface has been modified by adding functional groups. Due to these modifications, the CDs are enabled to exhibit fluorescence properties, which can be further exploited for imaging practices. In an experiment, Sun et al. used these CDs with and without ZnS for performing optical imaging experiments in mice. They are thought to be at par with QDs which have a CdSe/ZnS moiety and are being considered as an alternative to CDs since they do not have any heavy metals conjugated to them.

14.8 Polymeric Platforms

14.8.1 Dendrimers

Dendrimers are hyperbranched polymers. To begin with, a dendrimer mixture of amines and amides was made, and it came to be known as poly(amidoamine)

(PAMAM) in 1985 and gained popularity under the commercial name of Starburst. Later, in 2006, another new dendrimer was synthesized. They were known as polypropylene imine (PPI) dendrimers and were usually small (Urvashi et al. 2014).

The dendrimers have a 3D structure which enables them to carry the drug and act as a drug delivery system. Their surface can also be modified by adding various functional groups that enable loading of various therapeutic agents.

They are preferred over other delivery systems as:

I. They display size uniformity.
II. They have low macrophage uptake.
III. They enhanced target ability.
IV. They can enter the cell rapidly.
V. Even after surface modifications, they are less cytotoxic.
VI. They are exceptionally biocompatible.

14.8.2 Polymeric Nanoparticles

These are polymeric colloidal particles which are in the range of submicron. These are used as vehicles for the delivery of drugs because of their excellent biocompatibility and less toxicity. These can be easily modified by adding different functional groups on their surface.

They are very stable and can easily entrap drug molecules to avoid their degradation.

There are mainly two types of polymeric nanoparticles (Sonali et al. 2018):

I. Nanospheres
II. Nanocapsules

Nanospheres are composed of a matrix that is polymeric in nature and have drug dispersed homogeneously, while nanocapsules are vesicle-like systems which have drugs present in the center, surrounded by a protective layer of polymer.

14.9 Lipid-Based Platforms

14.9.1 Liposomes

Lipids can be used in the form of vesicles or micelles as delivery systems. Lipids can be used as delivery systems due to their excellent biocompatibility as most of our cells are composed of phospholipids so they can easily cross the membrane barriers and traverse unhindered in between the cells.

Liposomes are lipid-based nanoparticles which are made up of phospholipids and are of 90–150 nanoscale range. They are made from the rehydration of any lipid film, followed by the application of physical stress. They possess an aqueous core which is well surrounded by a lipid layer infused with cholesterol. They can accommodate both the hydrophilic and hydrophobic molecules. The hydrophilic molecules can be kept inside facing the hydrophilic head of the bilayer, while the hydrophobic ones can be kept in contact with the hydrophobic tails of lipid bilayers (Sharma et al. 2017).

They are used to carry many chemotherapeutic compounds such as paclitaxel, annamycin, and lipopeptide. Additionally, they are also preferred because they allow a controlled and sustained release of the drug and then aid in their accumulation at the target site.

14.9.2 Micelles

Micelles are self-assembled nanocarriers and have attracted a lot of attention due to their uniform size. They lack an aqueous core, and hence, the contrast agents and drugs have to be conjugated to the polymers during manufacturing and further attached to an anchor molecule (Sonali et al. 2018). They can be synthesized from an array of amphiphilic molecules, but an aqueous environment is imperative (Sharma et al. 2017).

A micelle is composed of hydrophobic tails having hydrophilic heads that are arranged together in a cluster; hence it can easily transport a hydrophilic molecule through the lipid bilayer that is hydrophobic in nature. Vice versa is also true when we need to deliver a hydrophobic molecule through an aqueous environment; it guarantees that the hydrophobic molecule is not acted upon by enzyme and in turn altered or degraded.

The hydrophobic molecules usually employed are poly(d-l lactic acid), poly(l-aspartic acid), poly(propylene oxide), and PCL, and hydrophilic molecules include PEG (Sharma et al. 2017).

The properties of various nanoplatforms are described in Fig. 14.3.

14.10 Therapy

14.10.1 Chemotherapy

Chemotherapy has been the standard of care for cancer treatment and has contributed greatly in reducing mortality associated with cancer. In a field which was largely dominated by surgery and radiotherapy until the 1960s, we have thereafter progressed by leaps and bounds in the way we target this complex constellation of a disease (DeVita and Chu 2008). However, chemotherapeutic agents come with its share of adverse side effects which range from mild nausea to severe organ damage, which can be fairly attributed in part to drug resistance and lack of drug specificity.

	Particle type	Composition/Structure	Properties	Applications
	Polymer	e.g., PLGA, glycerol, chitosan, DNA; monomers, copolymers, hydrogels	Some biodegradable	Drug delivery; passive release (diffusion), controlled release (triggered)
	Dendrimer	PAMAM, etc.	Low polydispersity, cargo, biocompatible	Drug delivery
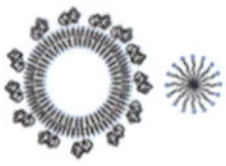	Lipid	Liposomes, micelles	Can carry hydrophobic cargo, biocompatible, typically 50–500 nm	Drug delivery
	Quantum dots	CdSe, CuInSe, CdTe, etc.	Broad excitation, no photobleaching, tunable emission, typically 5–100 nm	Optical imaging
	Gold	Spheres, rods, or shells	Biocompatibility, typically 5–100 nm	Hyperthermia therapy, drug delivery
	Silica	Spheres, shells, mesoporous	Biocompatibility	Contrast agents, drug delivery (encapsulation)
	Magnetic	Iron oxide or cobalt-based; spheres, aggregates in dextran or silica	Superparamagnetic, ferromagnetic (small remanence to minimize aggregation), superferromagnetic (~10 nm), paramagnetic	Contrast agents (MRI), hyperthermia therapy
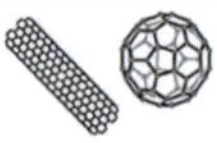	Carbon-based	Carbon nanotubes, buckyballs, graphene	Biocompatible	Drug delivery

Fig. 14.3 Composition and properties of various nanoparticles (Dawidczyk et al. 2014)

To mitigate the pitfalls, the use of nanostructures in cancer therapy has been around since the last decade and has given rise to a plethora of treatment options which we shall delve into in the following sections.

Traditionally, the rationale of using nanochemotherapy (NCT) has been to improve pharmacokinetics of the drugs and reduce systemic toxicities, thereby increasing their therapeutic index. The process starts with encapsulation of the drugs, followed by delivery of the conjugate to the region of interest, and, finally, maximal drug release. At the heart of this therapy is the nanoparticulate system, and they can be broadly classified as, but not limited to, liposomes, polymeric micelles and polymer-drug conjugates, dendrimers, oil nanoemulsions, inorganic nanocarriers, and, their combinations thereof (Glasgow and Chougule 2015). We will now have an overview of the major categories.

Liposomes consist of amphiphilic phospholipid bilayers, i.e., it is a structure having a lipid bilayer with an inner aqueous core which can load lipophilic or hydrophilic drugs (Bangham et al. 1965). As they are generally made of natural/synthetic lipids and cholesterol, they are quite biocompatible and biodegradable (Al-Jamal and Kostarelos 2011). They can also be modified in various ways, for example, with polyethylene glycol (PEG) stabilizers and coated ligands to improve their targeting potential (Wu and McMillan 2009). Doxil®, a formulation containing doxorubicin, is an example of such PEGylated nanoliposomal system with FDA approval, and it was the first of its kind to enter the market (Working and Dayan 1996). Thus, nanoliposomal carriers remain the most clinically approved nanosystems for chemotherapy. Some products already in the market for various types of cancer are Doxil®, DaunoXome®, DepoCyt®, and Onco-TCS®. They are not limited to chemotherapeutics for cancer, but also some formulations have been in the market for severe fungal infections, hepatitis A, and influenza like Abelcet®, Ambisome®, Epaxal®, and Inflexal® V to name a few (Bulbake et al. 2017). Research in recent years is geared toward delivering multiple drugs inside a common liposomal structure. This resonates well with the ongoing clinical trials, where drugs encapsulated in liposomes are combined with free drugs for combination chemotherapy. A recent phase III trial using non-PEGylated liposome-encapsulating doxorubicin (a mainstay of treating breast cancer) plus free cyclophosphamide had a better safety profile and good efficacy in treating metastatic breast cancer (Lorusso et al. 2014). Research in the preclinical stages is witness to an interesting concept of combining nanoparticles with liposomes, which can be used to engineer nanodelivery systems to improve theranostics; the liposomes can make these nanoparticles more biocompatible and diagnostically efficacious (Al-Jamal and Kostarelos 2011).

Micelles are colloidal nanoparticles made up of amphiphilic molecules which can self-aggregate in aqueous medium (Cho et al. 2015). Thermodynamic processes drive the formation of these aggregates with the increase in their concentration; critical micelle concentration (CMC) is the concentration at which this sequestration occurs forming hydrophobic cores with hydrophilic coronas (Torchilin 2006). Pharmaceutical formulations have been using these as excipients to increase aqueous solubility of drugs. PEG is an example of one of the most commonly used hydrophilic polymers, and the hydrophobic inner block uses polymers such as L-lysine, caprolactone, L-lactic acid, and spermine among others (Blanco et al. 2009). However, majority of the polymeric micelles cannot be technically classified as conjugates as the drug and the carrier form no covalent bonds. Mikhail et al. have been able to chemically conjugate docetaxel to poly(ethylene glycol)-b-poly(epsilon-caprolactone) polymeric micelles which is one of the many examples of aforementioned conjugates (Mikhail and Allen 2010). The trend in this field has been toward developing "smart" micelles leveraging the advances in the field of nanotechnology. "Smart" has been the go-to word in the field of contemporary theranostics, referring to the ability of the system to target specific tissues and be responsive to biological stimuli. An example of such a system can be found in literature where a pH-sensitive polymeric micelle was made by conjugating doxorubicin with self-assembled poly(styrene-co-maleic anhydride) derivative and adipic dihydrazide, encapsulating disulfiram, to target drug-resistant breast cancer cells (Duan et al. 2013). Another use of the micellar system was made to deliver siRNA, having poly(ethylene glycol)-block-poly(aspartic acid) and calcium phosphate as the building blocks (Kakizawa et al. 2004). To put it into perspective, most of the micellar formulations are in their preclinical phase, but some have made it to the trials. Some of the formulations and their most current trial reports are summarized in Table 14.1:

Although most formulations using nanocarriers to deliver chemotherapeutic agents making it into clinical trials fall in the above categories, there are other types of nanocarriers harboring potential in making the cut in the foreseeable future.

Table 14.1 List of first-generation and next-generation polymeric micelles in clinical trials

Name	Drug in combination	Formulation	Condition(s)	Phase
First generation				
SP1049C	Doxorubicin	Pluronic® L61 and L127	Various gastrointestinal cancers	I, II, and III
NK911	Doxorubicin	PEG-b-poly(α,β aspartic acid)	Various	II (results pending)
NK105	Paclitaxel	PEG and modified pAsp	Metastatic or recurrent breast cancer	I and II completed; III failed (to provide substantial benefits)
Next generation				
NC-6300	Doxorubicin	PEG-b-p(b-aspartic acid)	Advanced solid tumors	I (no results available)
NC-6004	Cisplatin	PEG-b-p(L-glutamic acid)	Various	I, II and III underway for various indications
NC-4016	DACHPt	PEG-poly(γ-benzyl-L-glutamate)	Advanced solid tumors and lymphoma	I
NK012	SN-38 (7-ethyl10-hydroxycamptothecin)	PEG-b-poly(L-glutamic acid)	Various	I and II for different indications
CriPec®	Docetaxel	CriPec® (patented)	Solid tumors, metastatic cancer	I

Summarized from a review by Varela-Moreira et al. (2017)

Dendrimer-based nanoparticle is one such system. They are basically hyper-branched polymeric macromolecules with repeating units forming a star-like structure. The drugs/imaging agents/targeting moiety are usually conjugated to the surface or the inner space between the branched units.

Thus, their unique structure allows maximal biological interaction. Majority of the work on dendrimers have focused on drug encapsulation and formulation. However, clinical translation has been scanty thus far, giving concerns over their cytotoxicity and biocompatibility. Research in this field is now mostly focused toward improving these aspects and, in addition, addressing the difficulties and expenses associated with their synthesis.

Inorganic nanocarriers are comprised of inorganic metals (like gold and silver) or metal oxides (like aluminum oxide, titanium oxide, zinc oxide, etc.). They find their use mainly in diagnostics, imaging, photothermal therapy, and photodynamic therapy (discussed separately in other sections). Mostly, gold and iron oxide dominate this spectrum for chemotherapeutic drug delivery. Gold nanocarriers are advantageous as they are more chemically stable and can be cast into various shapes like nano-shells, nano-rods, nano-cages, and nanospheres (Dai 2016). They were prepared with the common antitumor drugs like doxorubicin, camptothecin, and 5-fluorouracil among others, for targeted delivery (Dai 2016 and Pedrosa et al. 2015). Iron nanocarriers have been used as "smart" nanocarriers as they can be magnetically guided, tracked through vasculature, and designed with various coatings – offering the possibility of diagnosis and therapy at the same time (Kudr et al. 2017). Another category within inorganic nanocarriers is mesoporous silica nanoparticles (MSN), which set itself apart in their high loading capacity, tunable pore/particle size, and characteristically low in vivo toxicity (Watermann and Brieger 2017). They are generally synthesized using sol-gel chemistry or the evaporation-induced self-assembly (EISA) process (Hao et al. 2014). The only clinical study underway is a phase I study (NCT02106598) titled "Targeted Silica Nanoparticles for Real-Time Image-Guided Intra-operative Mapping of Nodal Metastases." Their multifunctional and modular design points to the next generation of NCTs – carriers having multicomponent cargoes, thereby combining molecular machinery with biomimetic support for an unprecedented engineering of nanotheranostic platforms.

Considerable research in combining chemotherapy with nanotechnology has led us to a stage where it is helping us overcome the limitations of mono- or one-dimensional therapy. Biocompatibility, targeting ability, low toxicity, pH/light/temperature sensitivity, and improved pharmacokinetic profile are testaments to the carefully engineered nanoparticles and a plethora of surface modifications. However, translation from bench to bedside is an aspect that needs to be focused on, with an emphasis on advanced characterization and long-term toxicity studies. The regulatory hurdles encountered are becoming more complex as multifunctional properties of nanocarriers have multiple additional components which claim to treat multiple indications using a single nanoparticle. Nonetheless, there is an unmet need for a panacea of sorts, which is only possible by integrating knowledge from various fields and relevant sectors, subsequently aiding the ascent of nanotheranostic platforms.

14.10.2 Photothermal Therapy

The eponymous photothermal therapy (PTT) is an emerging therapeutic strategy combining PTT agents and near-infrared radiation (NIR) (Shanmugam et al. 2014 and Hussein et al. 2018). PTT works by thermal excitation of PTT agents, which, on absorbing the light energy from NIR, gains kinetic energy and results in heating up the area in its vicinity. It is noninvasive in nature, and the induced thermal energy ablates the surrounding tumor cells by either denaturing proteins or disrupting membranes. Advances in the field of nanotechnology have only bolstered the application of PTT, especially noble metal nanoparticles such as gold (Au) and silver (Ag). These particles exhibit unique light to heat conversion using a phenomenon called localized surface plasmon resonance (LSPR). Briefly, the free electrons in conducting band around the surface of the particles can oscillate, and at a certain electromagnetic frequency, they reach maximum oscillation (surface plasmon resonance effect), from which they cool off by dissipating heat into their surroundings (Hussein et al. 2018 and Xie et al. 2017). Concerns may arise over the exposure of healthy tissue to electromagnetic radiation, but NIR laser has a poor absorbance and can penetrate the skin without causing damage to local tissues.

Au nanostructures are one of the most versatile nanomaterials (as discussed previously) with tunable optical properties. Recently, Ali et al. used Au nano-rods on human oral squamous cell carcinoma and reported that cell apoptosis was caused by high levels of phenylalanine (Ali et al. 2016). A long-term in vivo study to find out the efficiency of Au nano-rods as PTT agents found that Au nano-rods when conjugated to rifampicin were safer and more effective than when conjugated to PEG (Ali et al. 2017). Excessive reactive oxygen species (ROS) could be produced during PTT and can be a cause of concern as they cause cellular damage. A study using Pt-coated Au nano-rods demonstrated comparable efficacy to traditional Au nano-rods while being able to scavenge the heat stress-induced ROS (Aioub et al. 2017). Other recent studies involved testing Au-attapulgite composite on human adenocarcinoma cells and bioconjugating 2-naphthalenethiol and folic acid with Au nano-pyramids to target mammary adenocarcinoma (Wu et al. 2016 and Feng et al. 2017). The clinical challenges of using Au nanoparticle are agglomeration and instability inside the body. The animal models for testing such nanoparticles were highly specific, thereby hard to achieve in a clinical setting, but more research in multiple xenograft models and detailed readouts of their action in the body should pave the way for developing more clinically relevant applications (Dai 2016).

Iron oxide nanoparticles have been useful in cancer treatment using alternating magnetic field (AFM) to induce high temperatures, but NIR laser is still preferred because of its safety profile (Sohail et al. 2017 and Chu et al. 2013). Zhou et al. reported a triple functional (targeting, PTT, and imaging) iron/iron oxide core/shell nanoparticles on a xenograft HeLa model with improved MRI-guided targeting and efficient ablation of cancer cells (Zhou 2014). An interesting study elucidated the multifunctionality of iron-platinum nanoparticles which could be heated up to ~100 °C in picoseconds and caused intracellular explosions in breast cancer cells (Chen et al. 2013). Concerns regarding this mode of therapy is mostly around the use of highly intense laser irradiation but has been mitigated to some extent using iron oxide nanocubes (Espinosa et al. 2016).

There has been a race to harness the capabilities of graphene, and it hasn't shied away from entering the PTT domain. It brings us to a new class of compounds – organic nanomaterials. Organic nanomaterials have permeated into biomedical field, and they are attractive as PTT agents because of their efficient light-to-heat conversion. Single-walled carbon nanotubes were PEGylated and shown to be effective in destroying the primary tumor as well as disseminated cells in the lymph node using PTT (Liang et al. 2014). Graphene oxide was coated with PEG attached to $BaGdF_5$ nanoparticles by Zhang et al. and used for both MRI and PTT in a HeLa tumor xenograft; histological examination of tumors revealed thermonecrosis and cell shrinkage (Zhang et al. 2015). Chen et al. were also able to show a high synergistic effect of PTT and radiotherapy of radiolabeled reduced graphene oxide coated with PEG in a murine breast cancer model with no apparent toxicity (Chen et al.

2015a). Many organic dyes like cyanine and porphyrin derivatives have been reported as PTT agents but have restricted clinical applications as they are poorly photostable under strong irradiation (Hussein et al. 2018).

Before concluding, we will take a brief look at the most recent class of novel 2D materials which have forayed into this domain, and although it takes a long way from finding its way to clinical settings, it gives a glimpse into where this field is heading. MXenes are an example of such 2D materials, where M denotes a transition metal and X denotes carbon or nitrogen, with a general formula $M_{n+1}X_n$. Ti_3C_2 (MXene) which is one of such powerful PTT agents, showing similar LSPR effect and high light-to-heat conversion rate (Hussein et al. 2018 and Karlsson et al. 2015). Lin et al. applied this technique both in vitro and in vivo. In 4 T1 tumor-bearing mice, Ti_3C_2-soybean phospolipid nanoflakes could decrease the tumor size after a few minutes of exposure to NIR laser (Hussein et al. 2018 and Lin et al. 2017). Dai et al. have also prepared MXene-based hybrids like MnO_x-Ti_3C_2 to destroy tumors and for MRI (Hussein et al. 2018 and Dai et al. 2017).

These examples of current research in the field of PTT using various nanomaterials are not meant to be exhaustive but are an overview of the state of the art and meant to indicate rising trends. PTT is enough to destroy cancerous cells in tumors or disseminated ones in the lymphatic system, but due to inhomogeneous heat distribution arising from aberrant tissue structure, PTT agents should be designed for better penetration and target ability. The affinity of PTT agents for abnormal cells cannot be ignored and can be used for advanced imaging techniques like visualizing distant metastatic lesions deeply embedded into tissues. PTT and nanotechnology have been synergistic in driving this field forward, and the benefits have only started becoming apparent. More sophisticated designing and engineering of PTT agents are the need of the hour. PTT can be used with preexisting treatment options and can really be leveraged for a multimodal approach when dealing with difficult diseases such as cancer (Table 14.2).

14.10.3 Photodynamic Therapy

The essence of photodynamic therapy (PDT), i.e., the use of light and chemical agents for treatment of diseases, can be traced back to ancient Egypt where diseases like psoriasis and vitiligo were ameliorated with vegetable-derived substances (like psoralens) exposed to sunlight (Spikes 1985). The importance of light in disease treatment took off when Niels Finsen was awarded a Nobel Prize in Physiology or Medicine (1903) for his work on phototherapy. However, the advent of PDT was heralded by scientists like H. von Tappeiner and A. Jesionek, who also coined the term "photodynamic action" in their report of tumoricidal effects of eosin exposed to white light (Von Tappeiner 1903). Modern-day PDT can be attributed to Dougherty et al. who they successfully treated skin cancer in 1975 and in 1978 performed the first controlled clinical trials in humans (Dougherty et al. 1975; Celli et al. 2010).

Table 14.2 A list of interventional studies available on clinicaltrials.gov explicitly mentioning PTT for various indications

Study name	Condition(s)	Techniques	Trial status
MacTel laser study	Macular telangiectasia	577-nm PASCAL laser	Underway (recruiting)
Non-damaging Photothermal therapy of non-exudative age related macular degeneration	Macular degeneration, retinal drusen	Macular laser treatment	Underway (recruiting)
Plasmonic nano-photothermal therapy of atherosclerosis	Stable angina, heart failure, atherosclerosis, multivessel coronary, artery disease	Transplantation of iron-bearing nanoparticles	Completed with results
RCT of laser therapy for GSM	Genitourinary syndrome of menopause	Photothermal non-ablative erbium:YAG-laser; fractional microablative CO2-laser	Underway (recruiting)
Comparison between treatment with yellow micropulse laser and green conventional laser in diabetic macular edema	Diabetic macular edema	Micropulse laser treatment	Unknown
Pilot study of AuroLase (tm) therapy in refractory and/or recurrent tumors of the head and neck	Head and neck cancer	AuroLase therapy	Completed with results
Efficacy study of AuroLase therapy in subjects with primary and/or metastatic lung tumors	Primary or metastatic lung tumors	AuroLase therapy	Terminated
2940 nm Er:YAG laser vs. benzoyl peroxide gel for the treatment of inflammatory acne	Acne vulgaris	2940 nm Er:YAG laser; benzoyl peroxide gel	Completed

PDT can be broken down into three components, viz., the photosensitizer (PS), oxygen, and light source; by itself, they are nontoxic components unlike chemotherapeutic drugs. The PS interacts with the light and can act in 2 ways: (i) it can be excited to its triplet state, which reacts with the biomolecules, transfers H-atoms, and generates free radicals which react with molecular oxygen to produce ROS, and (ii) the excited PS in its triplet state directly interacts with oxygen to generate singlet oxygen species. As these reactions have a short half-life, it becomes imperative for the PS to be directly in the region for the ROS/singlet oxygen to interact with the biological substrates, and thus, localization is of utmost importance in drug release studies (Allison and Moghissi 2013 and Calixto et al. 2016). PDT can attack the tumor by directly killing the tumor cells, causing vascular damage to the tumor resulting in various cascades leading to tumor hypoxia and immunostimulation (Lucky et al. 2015).

PSs form the heart of this therapy, and before delving into the nanomaterial aspect of it, we shall have a brief look at the categories. They can be divided into first-generation PS (mainly includes porphyrin-based PSs), second-generation PS (porphyrinoid compounds and non-porphyrinoid compounds like 5-aminolaevulinic acid, chlorin, lutetium texaphyrin, and zinc phthalocyanine to name a few), and third-generation PS (like metallo-porphyrins, metallochlorins/bacteriochlorins, metallo-phthalocyanines, metallo-naphthalocyanines, etc. which are mainly aimed at improving the second-generation technology). Despite their advantages, PSs suffer from challenges like low-water solubility and strong aggregation in aqueous media which reduce the photodynamic effectiveness. Also, the lack of selectivity of PSs can lead to various side effects, and slow in vivo clearance rates can have harmful effects on eyes, skin, and normal tissues (Li et al. 2018). Nanoparticles may, in part, help overcome inherent limitations of such classical PSs. Some properties like high surface-to-volume ratios, amphiphilicity, surface functionalization, and providing a multifunctional nanoplatform make them a logical candidate for advancing this field.

Nanoparticles are being used for PDT either passively (by surface modification with targeting moieties) or actively (by direct involvement in the excitation of PSs). An excellent example of how nanoparticles and PSs can be combined for bioimaging and PDT was demonstrated very recently by Sun et al. wherein they modified nanographene oxide encapsulating aggregation-induced emission (AIE) nanoparticles with PEG (Sun et al. 2018). Conjugated polymers have also been a recent highlight in this field because they are highly photostable, and encapsulating PSs inside these structures amplifies the singlet oxygen generation leading to a higher quantum yield. Clinical safety is still yet to be addressed as organic/inorganic nanoparticles in the body are a potential health risk (Meng et al. 2018). Another latest development in this field deals with supramolecular chemistry, which deals with species resulting from the interaction between two or more chemical species held together by noncovalent forces. These assemblies can rapidly respond to a diverse range of external stimuli which may be biological, physical, or chemical in nature (Li et al. 2018). As an example, recently Li et al. displayed "one-for-all" nanostructured phthalocyanine assemblies (NanoPcTBs) which had protein-based switchable structures. Systemic administration of these NanoPcTBs led to their accumulation in xenograft tumors in mice and responded well to laser irradiation (Li et al. 2017). Like these, literature is awash with a multitude of reports on innovative and novel nanoconjugate PDT systems which are well beyond the scope of this discussion.

It is often prudent to look at the market studies to gauge how well a field is faring. BCC Research LLC revealed in its latest report that photomedicine is rapidly becoming one of the most exciting fields in biomedical research. Their analyses indicate that the global PDT market will grow from $346 million in 2016 to $436 million by 2021, with the highest growth rate in North America. From the perspective of clinical trials, clinicaltrials.gov lists around 400 studies on PDT, of which ~50% have a "completed" status, ~17% are either recruiting and/or ongoing, and ~13% have been terminated or withdrawn. ~14% have managed to make it to Phase III or Phase IV and have a "completed" status. While our knowledge about material

Table 14.3 Advantages and disadvantages of PDT

Advantages	Disadvantages
Less chances of adverse effects	Photosensitivity after treatment
Less invasive	Reduced efficacy in hypoxic microenvironment of tumors
Short treatment time	Treatment efficacy depends on accurate light delivery to the tumor
Usable in outpatient settings	Treating metastatic cancers with current technology is a long shot
Tunable selectivity	
No or only slight trauma	
Lower costs than other treatments	
Multimodal	

Adapted from Calixto et al. (2016)

chemistry, nanotechnology, and nanosynthesis has evolved in leaps and bounds, we are yet to thoroughly understand the complex interplay between all the collective entities in a biological system. The advantages and disadvantages of PDT have been summarized in Table 14.3. However, the sheer volume of research and the market analyses only indicate an upward trajectory for this field. The need of the hour is frequent facilitation between academia, industry, and clinical settings to bring PDT to the forefront in theranostics.

14.10.4 Radiotherapy

Radiation therapy is the most widely used therapy in case of cancer treatment. In radiation therapy, a high-energy beam is targeted on the cancer cells due to which the DNA of these tumor cells is damaged: not only is the chromosomal DNA effected but the mitochondrial DNA is damaged along with the various parts of the cells due to the production of radioactive species.

External radiotherapy includes X-rays, gamma rays, or any other high-energy wave which is used to bombard the cancer cells. External radiotherapy can be done by using an external radioactive source, a beam from which can be directed toward the tumor using an accelerator.

The disadvantages of radiotherapy include the side effects that occur due to the damage caused to the surrounding cells, which is ascribed to the nonspecific radiation exposure of all the cells present in that area. Additionally, repeated exposure to these radiations may also lead to the cells developing a resistance to the radiation therapy altogether.

To overcome these obstacles of the radiation therapy, radiations are now being clubbed with diagnostic approaches and drug delivery systems as a part of theranostics.

In theranostic systems, radiation is produced internally rather than being produced by an external source. In this technique, a radionucleotide nanometal is introduced in the tumor cells which produces radioactive waves and hence internally destroys the tumor.

The efficiency of internal radiotherapy depends upon the accumulation of radionucleotides inside the tumor. It can be achieved when we combine our drug delivery system to it such as it attaches itself to the antigens. Due to this, the cytotoxicity for nontumor cells is also reduced, and side effects are not caused.

Platforms used for RT:

1. Gold is a metal that is widely used in the form of nanoparticles for performing radiation therapy; its sensitivity can be increased by employing a metal that has high absorbance capacity. They are able to radiosensitize after absorbing and hence releasing energy in the surroundings in form of X-rays and photoelectrons. Certain modifications to gold nanoparticles have been done; P-GNPs or polyethylene glycosylation-modified gold particles are a modified version which remains in the systemic circulation for a prolonged time (Ma et al. 2013).
2. Other than gold nanoparticles, the superparamagnetic iron oxide nanoparticle-loaded polymeric micelles are also being used for radiotherapy. They have shown excellent radiotherapeutic effects.
3. Bi-based nanoparticles have also shown promising results.

14.10.5 Image-Guided Therapy

Targeted molecular imaging is one of the most important parts of nanomedicine bestowing us with the combined benefits of noninvasive imaging, specific targeting, and tracking biomarkers involved in the disease progression by providing real-time whole-body scan.

In nanotheranostics, we integrate therapeutic agents and imaging probes into a single nanoparticle for simultaneous delivery and diagnosis. A typical nanotheranostic system comprises of a drug carrier made from a biocompatible and biodegradable material, an imaging label, and a therapeutic or a bioactive molecule. Sophisticated nanotheranostic systems like quantum dots, gold and magnetic nanoparticles, polymeric nanoparticles, carbon nanotubes, polymeric miscelles, etc. have been developed for the successful detection and treatment of many deadly diseases specially cancers by employing imaging techniques like magnetic resonance imaging (MRI), computed tomography (CT), positron emission tomography (PET), and optical imaging (OI).

This chapter focuses on the imaging modalities used to monitor drug delivery, release kinetics, biodistribution, bioavailability, localization in the healthy organs or tissues, accumulation at the target site, and efficacy of various drug delivery systems used in nanotheranostics (Figs. 14.4 and 14.5).

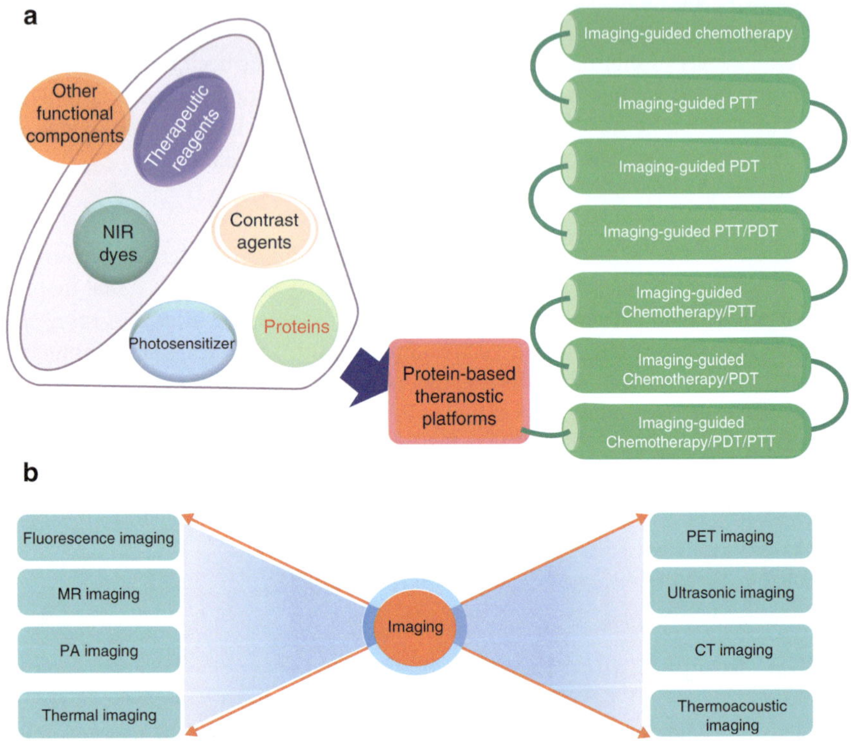

Fig. 14.4 Image guided therapy (Gou et al. 2018)

14.10.6 Positron Emission Tomography and Computed Tomography

Positron emission tomography (PET) is a class of nuclear medicine imaging where images are acquired by imaging the decay of radioisotopes bound to molecules with known biological properties; PET utilizes positron-emitting radionuclides like ^{11}C, ^{13}N, ^{15}O, ^{44}Sc, ^{18}F, ^{62}Cu, ^{64}Cu, ^{18}Ga, ^{72}As, ^{76}Br, ^{86}Y, ^{82}Rb, ^{89}Zr, and ^{124}I which are visualized and quantified (Bar-Shalom et al. 2000; Chopra 2004; Pressly et al. 2007; Devaraj et al. 2009; Herth et al. 2010; Roesch 2012; Müller et al. 2014; Chakravarty et al. 2014). This radiopharmaceutical is either inhaled or injected, and the scan is taken after a delay of either few seconds or minutes allowing enough time for the uptake by the organ of interest. Upon decaying, these isotopes emit two single photons of 511 keV energy which are detected with a detector fixed in PET scanner. Computer analysis then creates a 3D image of the tracer concentration. Being a very highly sensitive imaging modality, PET requires only trace amounts of radioisotopes and helps to identify the disease at a metabolic level. Owing to high-tissue penetration, outstanding sensitivity, useful images, and quantitative information, PET is routinely used in nanotheranostics for monitoring the biodistribution,

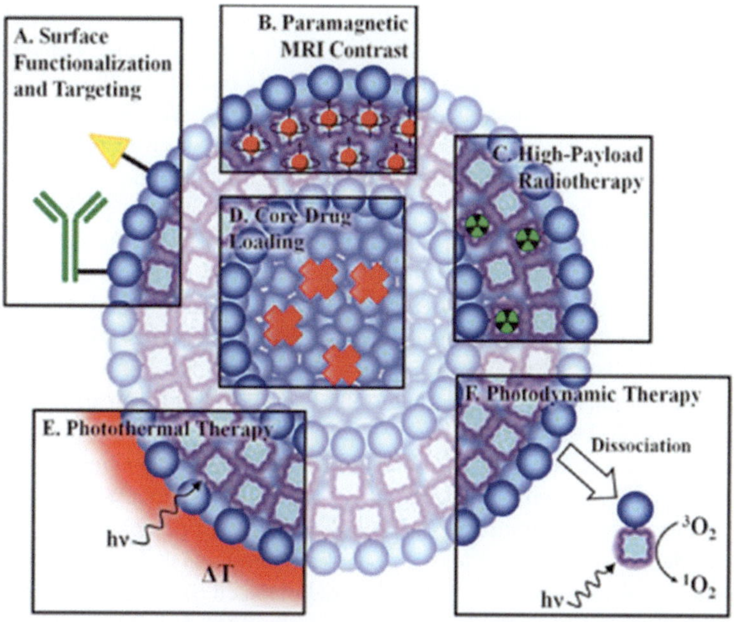

Fig. 14.5 Multimodal imaging (Chen et al. 2015b)

bioavailability, bioactivity, and pharmacokinetics of the nanomedicines. In order to select the radionuclide for a particular study, its physical half-life ($T_{1/2}$) plays an important role for the measurements in the desired time frame. For studies involving intravenous administration for short time frame measurements, radionuclides having short half-life like ^{68}Ga ($T_{1/2}$ = 67.7 min), nitrogen ($T_{1/2}$ = 9.97 min), etc. can be used. The role of positron emission tomography (PET) during the past decade has evolved rapidly from a pure research tool to a methodology of enormous clinical potential. Perhaps, the most striking development is the use of PET in oncology. PET imaging is approved in the United States for lung, lymphoma, colon, and melanoma cancer imaging. Data are accumulating rapidly to attest the efficacy of fluorine-18 fluorodeoxyglucose (FDG) imaging in a wide variety of malignant tumors with sensitivities and specificities often as high as 90. FDG uptake has been shown in tumors of the head and neck, ovary, breast, musculoskeletal system, and neuroendocrine system as well. The major role of PET has emerged as a reliable method for evaluating and staging recurrent disease. But it also has an important role in differentiating benign and malignant primary tumors. This has been shown particularly well in the differential diagnosis of solitary lung nodules. Although FDG has emerged as the dominant radiopharmaceutical for PET imaging in oncology, numerous other compounds are being evaluated. It is likely that more specific and efficacious compounds will be introduced during the next decade. F-18, because of its highly favorable physical characteristics, is likely to become the technetium of PET imaging. The next decade will witness an explosive growth of PET technology in oncologic imaging (Herth et al. 2009; Pérez-Campaña et al. 2013; Locatelli et al. 2012). On the other hand, studies involving prolonged time frame, radionuclides with long half-life like ^{64}Cu ($T_{1/2}$ = 12.7 h), ^{76}Br ($T_{1/2}$ = 16.2 h), ^{124}I ($T_{1/2}$ = 4.1 h), etc. are used. PET finds application in tumor staging, tumor proliferation, neurology, GI imaging, cardiology, etc.

^{18}F is the most commonly used positron-emitting radionuclide used in the form of 2-[^{18}F]fluoro-2-deoxy-D-glucose ([^{18}F] FDG). Having high positron energy (635 keV), half-life of 109.8 min, and abundance availability, ^{18}F is the best suited for PET imaging. ^{18}F-labeling of core/shell quantum dots (QDs) was successfully described by the Duong et al. using these labeled QDs for whole body imaging. QDs were encapsulated in functionalized phospholipid miscelles which were then labeled with ^{18}F. The authors then used these new nanoprobes for whole-body biodistribution and pharmacokinetic monitoring using PET and their cellular uptake (Ducongé et al. 2008).

Similarly, a scenario utilizing "^{64}Cu-radiolabeled solid lipid nanoparticles (SLN) containing bovine serum albumin (BSA) via 6-[p-(bromoacetamido)benzyl]-1,4,8,11-tetra azacyclotetradecane-N, N′, N″, N‴-tetraacetic acid (BAT) chelator was presented by Andreozzi et al. for whole-body biodistribution and pharmacokinetic analysis (Andreozzi et al. 2011). The radiolabeled SLNs were intravenously injected to mice, and static PET images were acquired at 0.5, 3, 20, and 48 h postinjection. Organ distribution was done by gamma counting by measuring the amount of radioactivity in different organs after excision. The study successfully showed that the in vivo PET imaging can be qualitatively evaluated accurately as validated by ex vivo gamma counting.

Liposomes or lipid nanoparticles have also earned a good name in the area of diagnostic imaging agents for their fast and accurate visualization and monitoring of cancer tumors and metastasis. Liposomes loaded with the radionucleotide ^{64}Cu were prepared and injected into human colon adenocarcinoma xenograft mice and monitored for the in vivo performance of these liposome-based nanomedicines (Petersen et al. 2011).

Perez-Medina et al. investigated the therapeutic outcome of a nanoliposomal doxorubicin (Doxil), a nanoparticle drug formulation, in breast cancer mouse model using PET nanoreporter technology (Pérez-Medina et al. 2016). Doxil and radioisotope zirconium-89 (^{89}Zr) were co-injected, and tumor uptake of Doxil in mouse breast cancer model was monitored via Doxil quantification using PET liposomal nanoreporter. ^{89}Zr having a half-life of 78 h allowed prolonged PET-CT imaging.

Computed tomography (CT) is an X-ray-based imaging procedure that enables the cross-sectional 2D and 3D visualization of organs, bones, and tissues of interest. CT generates cross-sectional, anatomical images by using highly electron-dense contrast agents like iodine, which can be reformatted in multiple planes. CT finds vast applications in monitoring disease differentiation, perfusion analysis, and angiography. CT serves as a very important tool for the study of cancer and tumors in the human body in addition to diagnosis of circulatory disorders, kidney and bladder stones, abscesses, inflammatory diseases, bone injuries, and internal organs.

Recently, emphasis has been given on the development of nanoparticles like micelles, lipoproteins, polymeric nanoparticles, solid metal nanoparticles, etc. (van Schooneveld et al. 2010; Cormode et al. 2010; Cormode et al. 2008; Torchilin et al. 1999; Trubetskoy et al. 1997; Kinsella et al. 2011; and Cai et al. 2007). Nanoparticles hold many advantages, such as prolonged blood circulation half-life (Cai et al. 2007), biocompatibility, better in vivo cell tracking (Arifin et al. 2011), and targeted imaging application, to be used as contrast agents when compared to other small molecules.

A. de Vries and colleagues prepared and evaluated the retention time in systemic circulation and organ accumulation of iodine containing polymeric nanoemulsions using high-resolution micro-CT imaging to monitor angiography and perfusion (de Vries et al. 2010). Poly(butadiene)-b-poly-(ethylene glycol) (PBD-PEG) block copolymer self-assemblies loaded with iodine were intravenously injected into healthy mice, and the signal changes produced by the contrast agents were monitored in blood, urine, heart, liver, spleen, and kidney. Retention within systemic circulation was validated by the transversal and coronal CT scans acquired of the contrast agents present in the heart. Moreover, organ accumulation was studied by the organs of the mononuclear phagocytic systems (MPS) specially spleen.

In addition to this, CT imaging has been successfully employed to study the biodistribution of the nanomedicines. Zheng et al. prepared liposomes containing iohexol and injected into rabbit tumor model for assessing their distribution using quantitative CT imaging (Zheng et al. 2009).

In a similar study, Dunne and colleagues monitored the tumor-targeting potential of iohexol-loaded PEGylated liposomes functionalized with NGR peptides, targeting the tumor vasculature (Dunne et al. 2011).

These efforts, therefore, demonstrates that even though the contrast agents have low sensitivity, CT imaging still successfully provides longitudinal assessment of nanomedicine biodistribution, cell-tracking and tumor accumulation.

14.10.7 Magnetic Resonance Imaging (MRI)

A medical application of Nuclear Magnetic Resonance (NMR), MRI is an imaging technique where the spins of specific atomic nuclei are visualized within the body. An MRI scan uses a magnetic field, radio waves, and a computer to create a detailed, cross-sectional image of internal organs and structures. Along with its use in disease diagnosis, differentiation and therapy, it is nowadays being routinely used in the field of nanotheranostics for monitoring drug release, biodistribution, cell tracking studies and pharmacokinetics. Two types of MR contrast agents are used in MRI; T_1 contrast agents who shorten the pin lattice relaxation time of nearby protons, and T_2 contrast agents that enhance the spin-spin relaxation to reduce the signal of media containing structures. In contrast to T_1, T_2 contrast agents are based on superparamagnetic iron oxide (SPIO) nanoparticles that remain intravascular for longer time giving a delayed image-acquisition time window.

Due to higher contrast agent sensitivity and easier quantification procedures, PET and CT scan are preferred for biodistribution and target site accumulation; however, MRI is considered more suitable for monitoring drug release and drug efficacy. T_1 MR contrast agents, like radionuclides, generate contrast images by using freely diffusing water molecules leading to the generation of different signals when present inside a nanocarrier and different for outside, giving better drug release assessment (Terreno et al. 2010). This is in contrast to radionuclides which generate the same signals of similar intensity whether inside or outside the nanocarrier. This characteristic of MR contrast agents was successfully exploited by de Smet et al. for visualizing and quantifying release of local doxorubicin (DOX) from temperature-sensitive liposomes (TLS) and coencapsulating DOX and gadolinium-based T_1 contrast agent (Gd-HPDO3) under MR image guidance mediated by high-intensity-focused ultrasound (HIFU)-mediated hypothermia in tumor-bearing Fisher rats (de Smet et al. 2011). The local temperature-triggered release of [Gd(HPDO3A)(H_2O)] was studied with interleaved T_1 mapping of the tumor tissue and was correlated with corelease of DOX. This implied the use of MRI for probing the in vivo release of DOX from TSLs by temporally and spatially analyzing the relaxation time of the coreleased MRI contrast agent. MRI also allows better monitoring of drug efficacy or therapeutic responses due to its high soft tissue contrast providing highly accurate detection of tissues, organs, and tumors.

The clinical diagnosis of cancer, specially brain tumors via MRI, is dependent on the efficacy of the contrast agents. Conventionally used low-molecular-weight contrast agents like gadolinium ion (Gd (III)) possess many drawbacks such as non-specificity, rapid renal clearance, and low contrast efficiency (Luo et al. 2009). Therefore, development of better contrast agents is a big challenge for tumor diagnosis. The potential of nanocarriers such as dendrimer-based macromolecules is being explored to overcome the shortcomings of the existing contrast agents. Among these, polyamidoamine (PAMAM) (Nwe et al. 2009; Swanson et al. 2008; Boswell et al. 2008; and Xu et al. 2007), hydroxyl-terminated polyester dendrimer (Chen et al. 2010b), and peptide dendritic macromolecules have gathered much attention. $_L$-lysine dendritic macromolecules conjugated with Gd chelates and galactosyl moieties are endowed with highly controlled structures and a twofold increase in T_1 relaxivity when compared to Gd-DTPA and thus have been successfully employed as probes in liver imaging (Luo et al. 2009).

A macromolecular MRI contrast agent based on a dendrigraft poly- $_L$-lysine (DGLs) coupled with a tumor-specific ligand chlorotoxin (CTX) was evaluated to monitor the target site accumulation of nanomedicines in a rat model of glioblastoma. These CTX-conjugated macromolecular contrast agents exhibited longer accumulation, prolonged retention in tumors, and higher signal enhancement within tumors, allowing more accurate tumor diagnosis by discriminating between formulations with different tumor localization kinetics (Huang et al. 2011).

Magnetic nanoparticles (MNPs) have been used for various biomedical applications more specifically as contrast agents for MRI (Wang et al. 2001). Multifunctional abilities of these MNPs are now being explored that range from imaging enhancement property of MNPs to the simultaneous drug delivery giving real-time monitoring of drug distribution to the target tissue as well as provide therapeutics (Huh et al. 2005 and Medarova et al. 2006). One such application was proposed by Mikhaylov and colleagues where they developed universal lapidated magnetic nanocarrier (ferri-liposome) that not only gave enhanced MRI contrast properties but were effectively taken up by tumor cells and the surrounding stromal cells. These ferri-liposomes, comprising of MNPs clustered inside a liposome, were used to deliver cathepsin protease inhibitor to mammary tumor and the surrounding stromal cells in a mouse model that resulted into appreciable decrease in the size of the tumor when compared to systemic delivery of the same drug (Mikhaylov et al. 2011).

de Vries and colleagues explored yet another interesting feature of MRI of imaging cell-based vaccination therapies (de Vries et al. 2005). MRI was used for tracking therapeutic dendritic cell (DC) vaccines labeled with superparamagnetic iron oxide (SPIO) nanoparticles and ^{111}In-oxine injected in melanoma patients. Localization and retention within the primary lymph node and migration to several neighboring lymph nodes were monitored and studied. Similar results were published by Lory et al. where they monitored the migration of inactivated B16 melanoma cancer cell-based vaccines to cytotoxic T cells via afferent lymphatics for successful antitumor immunotherapy (Long et al. 2009).

Moreover, SPIO-containing NPs are being extensively employed for magnetic drug targeting, wherein the magnetic fields are used to propel iron oxide-containing nanomaterials to the target site and/or to retain them at the site of accumulation for a longer time and more efficiently. An elegant example of such study was put forward by Fortin-Ripoche and colleagues where PEGylated magnetoliposomes (ML) were synthesized and were assessed for their accumulation in mice-bearing tumor on their left and right flanks using an extracorporeal magnet (Fortin-Ripoche et al. 2006). Magnet was placed beside the tumor on the right flank, while the other side was left without any magnetic influence. A significant improvement in the accumulation and retention of the ML was seen in the right tumor when exposed to magnet than seen in the contralateral (left) controlled tumor. Therefore, such findings elaborate the usefulness of MRI for noninvasive and quantitative monitoring of target site accumulation and drug delivery.

14.10.8 Optical Imaging (OI)

Recently, biomedical OI has been extensively used for a variety of applications ranging from clinical diagnosis to molecular biology and extending up to the evaluation of the biodistribution of nanomedicine formulations. It is very sensitive (pico molar range) and is flexible in utilizing a variety of target nanoplatforms like peptide, antibody, protein, nanoparticles, etc. In vivo OI uses contrast that interacts with the visible and near-infrared wavelength of light in living tissues. OI is divided into diffusive and ballistic imaging systems. Diffusive OI uses near-infrared spectroscopy (NIRS) or fluorescence-based methods. Ballistic optical imaging uses ballistic photons which are light photons that travel in a straight line through a scattering medium.

OI provides several advantages over other imaging modalities like probing both functional and structural changes with a high spatial, noninvasive, non-ionizing, and more economical resolution, provides information in both macroscopic and microscopic scale, and provides qualitative information with follow-ups. OI can efficiently monitor the target site accumulation of near-infrared fluorophore (NIRF)-labeled nanomedicines especially in the case of subcutaneous tumors and rheumatoid arthritis (Wunder et al. 2003; Licha and Olbrich 2005; and Mountz et al. 2012).

Fluorescence reflectance imaging (FRI) and fluorescence molecular tomography (FMT) are the most extensively used OI techniques to monitor drug delivery process by providing information on the NIRF-labeled nanomedicine giving a semi-quantitative comparison of the accumulation of free vs. nanomedicine-conjugated fluorophores at the target site and their biodistribution (Ntziachristos and Weissleder 2001; Kunjachan et al. 2014; and Ntziachristos et al. 2003). An example of simultaneous drug release and imaging using OI was given by Ferber et al. by synthesizing a dual polymeric system (Ferber et al. 2014). The diagnostic part is made up of high loading self-quenched turn-on system with NIR fluorescent dye Cy5 (SQ-Cy5), while the chemotherapeutic agent paclitaxel (PTX) comprises the therapeutic part, both conjugated to N-(2-hydroxypropyl)methacrylamide (HPMA) copolymer via GFLG linkage. Cathepsin B, overexpressed by breast cancer cells, cleaves the HPMA copolymer-PTX/SQ-Cy5 enabling site-specific release of PTX giving a simultaneous fluorescent signal which helps in the detection of tumor and monitoring of the therapeutic efficacy.

OI was employed to visualize the (hydro)cyanine-containing chitosan-based nanocarrier accumulation in the subcutaneous SCC7 xenografts. Increased number of ROS produced by the tumor-associated immune responses oxidizes hydrocyanine present in the nanomedicine to yield cyanine, and this transition can be sensitively detected by the OI. Therefore, both hydrocyanine- and cyanine-containing nanoparticles elaborated the efficiency of OI to sensitively monitor the differences in tumor physiology-dependent target site accumulation (Kim et al. 2013).

J. Chen et al. used quantum dots (QD) based on OI for exploring the membrane protein structures with a diffraction-unlimited spatial resolution (Chen et al. 2010a). The QDs having excellent photostability and high luminescence improve the signal-to-noise ratio and reproducibility of the optical data. Herein, QD labeling based on near-field OI was used to deduce the nanoscale organization of hyaluronan receptor CD44 molecules of fixed mesenchymal stem cells (MSCs) in air, thereby elaborating their underlying mechanism of functions.

14.10.9 Multimodal Imaging

Multimodal imaging is a combination of imaging modalities aimed at providing a solution to cover up the limitations of the independent techniques. With a rapid advancement of nanotechnology, a more synchronization is needed in the diagnostic and therapeutic functions into a single nanotheranostic platform. Multimodal imaging integrates the functional and/or structural information from several imaging techniques giving more accurate diagnosis. On the other hand, the combined therapeutic merits of different techniques give a better synergistic therapeutic effects and therapeutic efficiency (Lin et al. 2016; Huang et al. 2014; Xiao et al. 2013; Yan et al. 2015; and Tang et al. 2015). The most common multimodal imaging techniques being used for both clinical and research purposes are SPECT-CT, PET-CT, and PET-MR.

An analogue of PET, single photon emission-computed tomography (SPECT) is based on non-coincident gamma-rays generated by radionuclides like ^{99m}Tc, ^{111}In, ^{123}I, and ^{201}Tl with higher sensitivity, highly quantitative results, and high penetration depth. The need for using radioactive probes, low spatial resolution, and lack of anatomical information are few drawbacks associated with SPECT which are now overcome by combining SPECT with CT. An interesting example of this was illustrated by Chrastina et al. where they used SPECT-CT for monitoring the nanomedicine-based drug targeting the lungs (Chrastina et al. 2011). G5-PAMAM dendrimers were first functionalized with antibiotic targeted to aminopeptidase P2 to target the lungs and then radiolabeled with ^{125}I and intravenously administered to healthy mice that were subjected to whole-body SPECT-CT imaging. This combination of molecular SPECT and anatomical CT provided efficient monitoring of in vivo distribution of passively vs. actively targeted nanomedicines or nanocarriers. In spite of high progress made in OI, a major drawback underlying the technique is that the fluorescence signals detected often cannot be accurately converted to specific anatomical information. This limitation leads to the development of FMI-OI hybrids, where the anatomical information produced by the micro-CT is used to construct better fluorescence data obtained from FMT to monitor and quantify probe accumulation more accurately (Nahrendorf et al. 2009; Hyde et al. 2009; Ntziachristos et al. 2002; Ale et al. 2012).

Not every time a single imaging modality can provide all the necessary information for a situation. For instance, even endowed with high sensitivity, PET is still limited with relatively poor spatial resolution. This can be overcome if PET is combined with MRI that offers not only excellent spatial resolution but also better soft tissue contrast. Therefore, PET-MRI finds clinical applications like neurological studies, cancer studies, and stem cell therapy by incorporating both PET isotopes and MRI contrast agents (Pichler et al. 2008; Glaus et al. 2010; and Wehrl et al. 2009). Yang and colleagues developed cRGD-functionalized ^{64}Cu-labeled SPIO nanocarriers loaded with DOX for both tumor-targeted drug delivery and PET-MRI imaging (Yang 2011).

14.11 Perspective

14.11.1 Nanotheranostics for Precision Medicine

Precision medicine is a new domain that is garnering huge attention recently. It is also referred to as personalized medicine. As the name suggests, it is based on the idea that no one drug can be suitable for the entire population. For instance, some people might show side effects to a certain drug, while others may respond very well to it.

As discussed previously in this chapter, theranostics can help us develop medicines that are exclusive to a single patient (Kim et al. 2013). A personalized medicine is developed after studying the condition of the patients at a molecular level while being mindful of their genetic framework and overall physiology. After taking all the abovementioned characteristics of a patient into account, a customized drug is developed, which would then cater to the needs of that particular suffering individual. Therefore, instead of a general prescription, the personalized medicine makes the treatment individual centric. In other words, this approach of precision medicine has the acumen to identify the subgroups of patients.

However, precision medicine was not gaining much momentum until their role in nanotheranostics was delineated. Nanotheranostics, which offers a combined system of imaging, diagnosis, and therapy in single nanoparticle unit, imparts precision medicine a unique multidirectional approach. Combining these fields of clinical treatments promises to pave a path for pioneering breakthroughs in the domain of precision medicine and biomedical research.

By bringing both together, many aspects like early diagnosis, staging of a disease, selection of a personalized treatment, treatment follow-ups, recognizing the side effects if any early itself, and changing the course of treatment can be done at a nanoscale level integrating nanosensors and nanomedicine.

Using these nanoscale components can help us study and treat the disease at the molecular level as required in the precision medicine. The nanosensors can study a biosample that is very little in volume (Jo et al. 2016). The nanomedicine delivers the high dosage of drug and still causes lesser or no side effects because the drug is being delivered directly to the target site.

Precision medicine has an application in the treatment of cancer and diseases that have a genetic predisposition. Genomics has been used in oncology to determine the suitable drug as oncologists look at genetic markers to select the right drug that would be effective against the tumor (Vats et al. 2017). Personalized medicine is just an extension of this very rationale.

Precision medicine offers the following:
(Wishart 2016)

I. A more specific diagnosis.
II. Lesser side effects.
III. Stopping diseases like Parkinson's, Alzheimer's, etc. in their early stages itself.
IV. Better chances at cure.

Unfortunately, even after all the new technological advancements, we are still scratching the surface when it comes to precision medicine. Leading researchers world over are working toward a quick, large-scale, and economic implementation of the systems that would be capable of performing an individualistic diagnosis.

References

Aioub M, Panikkanvalappil SR, El-Sayed MA (2017) Platinum-coated gold nanorods: efficient reactive oxygen scavengers that prevent oxidative damage toward healthy, untreated cells during plasmonic photothermal therapy. ACS Nano 11(1):579–586

Ale A, Ermolayev V, Herzog E, Cohrs C, de Angelis MH, Ntziachristos V (2012) FMT-XCT: in vivo animal studies with hybrid fluorescence molecular tomography-X-ray computed tomography. Nat Methods 9(6):615–620

Ali MRK et al (2016) Simultaneous time-dependent surface-enhanced raman spectroscopy, metabolomics, and proteomics reveal cancer cell death mechanisms associated with gold nanorod photothermal therapy. J Am Chem Soc 138(47):15434–15442

Ali MRK et al (2017) Efficacy, long-term toxicity, and mechanistic studies of gold nanorods photothermal therapy of cancer in xenograft mice. Proc Natl Acad Sci U S A 114(15):E3110–E3118

Al-Jamal T, Kostarelos K (2011) Liposomes: from a clinically established drug delivery system to a nanoparticle platform for theranostic nanomedicine. Acc Chem Res 44(10):1094–1104

Allison RR, Moghissi K (2013) Photodynamic therapy (PDT): PDT mechanisms. Clin Endosc 46(1):24–29

Andreozzi E, Seo JW, Ferrara K, Louie A (2011) Novel method to label solid lipid nanoparticles with ^{64}Cu for positron emission tomography imaging. Bioconjug Chem 22(4):808–818

Arifin DR et al (2011) Trimodal gadolinium-gold microcapsules containing pancreatic islet cells restore normoglycemia in diabetic mice and can be tracked by using US, CT, and positive-contrast MR imaging. Radiology 260(3):790–798

Bangham D, Standish MM, Watkins JC (1965) Diffusion of univalent ions across the lamellae of swollen phospholipids. J Mol Biol 13(1):238–252

Bar-Shalom R, Valdivia AY, Blaufox MD (2000) PET imaging in oncology. Semin Nucl Med 30(3):150–185

Blanco E, Kessinger CW, Sumer BD, Gao J (2009) Multifunctional micellar nanomedicine for cancer therapy. Exp Biol Med (Maywood) 234(2):123–131

Blau R, Krivitsky A, Epshtein Y, Satchi-Fainaro R (2016) Are nanotheranostics and nanodiagnostics-guided drug delivery. Drug Resist Updat Elsevier 27:39–58

Boswell A et al (2008) Synthesis, characterization, and biological evaluation of integrin alphavbeta3-targeted PAMAM dendrimers. Mol Pharm 5(4):527–539

Bulbake U et al (2017) Liposomal formulations in clinical use: an updated review. Pharmaceutics 9(2):12

Burguma MJ, Evansa SJ, Jenkins GJ, Doak SH, Clift MJD (2018) Considerations for the human health implications of nanotheranostics. Handbook of nanomaterials for cancer theranostics. Elsevier

Cai Q-Y et al (2007) Colloidal gold nanoparticles as a blood-pool contrast agent for X-ray computed tomography in mice. Investig Radiol 42(12):797–806

Calixto GMF et al (2016) Nanotechnology-based drug delivery systems for photodynamic therapy of cancer: a review. Molecules 21(3):342

Celli JP et al (2010) Imaging and photodynamic therapy: mechanisms, monitoring and optimization. Chem Rev 110(5):2795–2838

Chakravarty R et al (2014) Matching the decay half-life with the biological half-life: ImmunoPET imaging with (44)Sc-labeled cetuximab Fab fragment. Bioconjug Chem 25(12):2197–2204

Chen J, Pei Y, Chen Z, Cai J (2010a) Quantum dot labeling based on near-field optical imaging of CD44 molecules. Micron 41(3):198–202

Chen W-T et al (2010b) Dynamic contrast-enhanced folate-receptor-targeted MR imaging using a Gd-loaded PEG-dendrimer-folate conjugate in a mouse xenograft tumor model. Mol Imaging Biol MIB Off Publ Acad Mol Imaging 12(2):145–154

Chen C-L et al (2013) Photothermal cancer therapy via femtosecond-laser-excited FePt nanoparticles. Biomaterials 34(4):1128–1134

Chen L et al (2015a) Radionuclide (131)I labeled reduced graphene oxide for nuclear imaging guided combined radio- and photothermal therapy of cancer. Biomaterials 66:21–28

Chen Q, Ke H, Dai Z, Liu Z (2015b) Nanoscale theranostics for physical stimulus-responsive cancer therapies. Biomaterials Elsevier 73:214–230

Cho H, Lai TC, Tomoda K, Kwon GS (2015) Polymeric micelles for multi-drug delivery in cancer. AAPS PharmSciTech 16(1):10–20

Chopra (2004) [74As]-labeled monoclonal antibody against anionic phospholipids. In: Molecular imaging and contrast agent database (MICAD). National Center for Biotechnology Information (US), Bethesda

Chrastina A, Massey KA, Schnitzer JE (2011) Overcoming *in vivo* barriers to targeted nanodelivery: overcoming barriers to targeted nanodelivery. Wiley Interdiscip Rev Nanomed Nanobiotechnol 3(4):421–437

Chu M et al (2013) Near-infrared laser light mediated cancer therapy by photothermal effect of Fe3O4 magnetic nanoparticles. Biomaterials 34(16):4078–4088

Cormode P et al (2008) Nanocrystal core high-density lipoproteins: a multimodality contrast agent platform. Nano Lett 8(11):3715–3723

Cormode P et al (2010) Atherosclerotic plaque composition: analysis with multicolor CT and targeted gold nanoparticles. Radiology 256(3):774–782

Dai Z (ed) (2016) Advances in nanotheranostics I: design and fabrication of theranosic nanoparticles. Springer-Verlag, Berlin/Heidelberg

Dai H, Lin GX, Liu Z, Wu R, Chen Y (2017) Biocompatible 2D titanium carbide (MXenes) composite nanosheets for pH-responsive MRI-guided tumor hyperthermia. Chem Mater 29(20):8637–8652

Dawidczyk CM, Russell LM, Searson PC (2014) Nanomedicines for cancer therapy: state-of-the-art and limitations to pre-clinical studies that hinder future developments. Front Chem 2:69

de Smet M, Heijman E, Langereis S, Hijnen NM, Grüll H (2011) Magnetic resonance imaging of high intensity focused ultrasound mediated drug delivery from temperature-sensitive liposomes: an in vivo proof-of-concept study. J Control Release 150(1):102–110

de Vries JM et al (2005) Magnetic resonance tracking of dendritic cells in melanoma patients for monitoring of cellular therapy. Nat Biotechnol 23(11):1407–1413

de Vries A, Custers E, Lub J, van den Bosch S, Nicolay K, Grüll H (2010) Block-copolymer-stabilized iodinated emulsions for use as CT contrast agents. Biomaterials 31(25):6537–6544

Devaraj K, Keliher EJ, Thurber GM, Nahrendorf M, Weissleder R (2009) 18F labeled nanoparticles for in vivo PET-CT imaging. Bioconjug Chem 20(2):397–401

DeVita VT, Chu E (2008) A history of cancer chemotherapy. Cancer Res 68(21):8643–8653

Dougherty J, Grindey GB, Fiel R, Weishaupt KR, Boyle DG (1975) Photoradiation therapy. II. Cure of animal tumors with hematoporphyrin and light. J Natl Cancer Inst 55(1):115–121

Duan X et al (2013) Smart pH-sensitive and temporal-controlled polymeric micelles for effective combination therapy of doxorubicin and disulfiram. ACS Nano 7(7):5858–5869

Ducongé F et al (2008) Fluorine-18-labeled phospholipid quantum dot micelles for *in vivo* multimodal imaging from whole body to cellular scales. Bioconjug Chem 19(9):1921–1926

Dunne M, Zheng J, Rosenblat J, Jaffray DA, Allen C (2011) APN/CD13-targeting as a strategy to alter the tumor accumulation of liposomes. J Contr Release 154(3):298–305

Espinosa R, Di Corato J, Kolosnjaj-Tabi P, Flaud TP, Wilhelm C (2016) Duality of iron oxide nanoparticles in cancer therapy: amplification of heating efficiency by magnetic hyperthermia and photothermal bimodal treatment. ACS Nano 10(2):2436–2446

Feng J et al (2017) Bioconjugation of gold nanobipyramids for SERS detection and targeted photothermal therapy in breast cancer. ACS Biomater Sci Eng 3(4):608–618

Ferber S et al (2014) Polymeric nanotheranostics for real-time non-invasive optical imaging of breast cancer progression and drug release. Cancer Lett 352(1):81–89

Fortin-Ripoche J-P et al (2006) Magnetic targeting of magnetoliposomes to solid tumors with MR imaging monitoring in mice: feasibility. Radiology 239(2):415–424

Glasgow DK, Chougule MB (2015) Recent developments in active tumor targeted multifunctional nanoparticles for combination chemotherapy in cancer treatment and imaging. J Biomed Nanotechnol 11(11):1859–1898

Glaus C, Rossin R, Welch MJ, Bao G (2010) In vivo evaluation of 64Cu-labeled magnetic nanoparticles as a dual-modality PET/MR imaging agent. Bioconjug Chem 21(4):715–722

Gou Y, Zhou M, Miao D, Su G (2018) Bio-inspired protein-based nanoformulations for cancer theranostics. Front Pharmacol 9(421)

Hao N, Li L, Tang F (2014) Shape-mediated biological effects of mesoporous silica nanoparticles. J Biomed Nanotechnol 10(10):2508–2538

Herth MM et al (2009) Radioactive labeling of defined HPMA-based polymeric structures using [18F]FETos for in vivo imaging by positron emission tomography. Biomacromolecules 10(7):1697–1703

Herth MM, Barz M, Jahn M, Zentel R, Rösch F (2010) 72/74As-labeling of HPMA based polymers for long-term in vivo PET imaging. Bioorg Med Chem Lett 20(18):5454–5458

Huang R et al (2011) Chlorotoxin-modified macromolecular contrast agent for MRI tumor diagnosis. Biomaterials 32(22):5177–5186

Huang P et al (2014) Dye-loaded ferritin nanocages for multimodal imaging and photothermal therapy. Adv Mater Deerfield Beach Fla 26(37):6401–6408

Huh Y-M et al (2005) In vivo magnetic resonance detection of cancer by using multifunctional magnetic nanocrystals. J Am Chem Soc 127(35):12387–12391

Hussein A, Zagho MM, Nasrallah GK, Elzatahry AA (2018) Recent advances in functional nanostructures as cancer photothermal therapy. Int J Nanomedicine 13:2897–2906

Hyde D et al (2009) Hybrid FMT–CT imaging of amyloid-β plaques in a murine Alzheimer's disease model. NeuroImage 44(4):1304–1311

Jo SD, Ku SH, Won Y-Y, Kim SH, Kwon IC (2016) Targeted nanotheranostics for future personalized medicine: recent progress in cancer therapy. Theranostics Ivyspring 6(9):1362–1377

Kakizawa Y, Furukawa S, Kataoka K (2004) Block copolymer-coated calcium phosphate nanoparticles sensing intracellular environment for oligodeoxynucleotide and siRNA delivery. J Control Release Off J Control Release Soc 97(2):345–356

Karlsson H, Birch J, Halim J, Barsoum MW, Persson POÅ (2015) Atomically resolved structural and chemical investigation of single MXene sheets. Nano Lett 15(8):4955–4960

Kim H, Lee S, Chen X (2013) Nanotheranostics for personalized medicine. Expert Rev Mol Diagn 13(3):257–269

Kinsella JM et al (2011) X-ray computed tomography imaging of breast cancer by using targeted peptide-labeled bismuth sulfide nanoparticles. Angew Chem Int Ed 50(51):12308–12311

Kudr J et al (2017) Magnetic nanoparticles: from design and synthesis to real world applications. Nano 7(9)

Kunjachan S, Ehling J, Storm G, Kiessling F, Lammers T (2013) Noninvasive imaging of nanomedicines and nanotheranostics: principles, progress, and prospects. Chem Rev ACS Publications. Nanoparticles in medicines 115:10907–10937

Kunjachan S et al (2014) Passive versus active tumor targeting using RGD- and NGR-modified polymeric nanomedicines. Nano Lett 14(2):972–981

Lammers T, Kiessling F, Hennink WE, Storm G (2010) Nanotheranostics and image-guided drug delivery: current concepts and future directions. Mol Pharm Am Chem Soc 7(6):1899–1912

Li X et al (2017) Nanostructured phthalocyanine assemblies with protein-driven switchable photoactivities for biophotonic imaging and therapy. J Am Chem Soc 139(31):10880–10886

Li X, Lee S, Yoon J (2018) Supramolecular photosensitizers rejuvenate photodynamic therapy. Chem Soc Rev 47(4):1174–1188

Liang C et al (2014) Tumor metastasis inhibition by imaging-guided photothermal therapy with single-walled carbon nanotubes. Adv Mater Deerfield Beach Fla 26(32):5646–5652

Licha K, Olbrich C (2005) Optical imaging in drug discovery and diagnostic applications. Adv Drug Deliv Rev 57(8):1087–1108

Lin J et al (2016) Multimodal imaging guided cancer phototherapy by versatile biomimetic theranostics with UV and γ irradiation protection. Adv Mater Deerfield Beach Fla 28(17):3273–3279

Lin H, Wang X, Yu L, Chen Y, Shi J (2017) Two-dimensional ultrathin MXene ceramic nanosheets for photothermal conversion. Nano Lett 17(1):384–391

Locatelli E et al (2012) Biocompatible nanocomposite for PET/MRI hybrid imaging. Int J Nanomedicine 7:6021–6033

Long M, van Laarhoven HWM, Bulte JWM, Levitsky HI (2009) Magnetovaccination as a novel method to assess and quantify dendritic cell tumor antigen capture and delivery to lymph nodes. Cancer Res 69(7):3180–3187

Lorusso V et al (2014) Non-pegylated liposome-encapsulated doxorubicin citrate plus cyclophosphamide or vinorelbine in metastatic breast cancer not previously treated with chemotherapy: a multicenter phase III study. Int J Oncol 45(5):2137–2142

Lucky SS, Soo KC, Zhang Y (2015) Nanoparticles in photodynamic therapy. Chem Rev 115(4):1990–2042

Luo K et al (2009) Functional L-lysine dendritic macromolecules as liver-imaging probes. Macromol Biosci 9(12):1227–1236

Ma Y, Huang J, Song S, Chen H, Zhang Z (2013) Cancer-targeted nanotheranostics: recent advances. Small J.:1–19 Wiley

Medarova Z, Pham W, Kim Y, Dai G, Moore A (2006) In vivo imaging of tumor response to therapy using a dual-modality imaging strategy. Int J Cancer 118(11):2796–2802

Meng Z, Hou W, Zhou H, Zhou L, Chen H, Wu C (2018) Therapeutic considerations and conjugated polymer-based photosensitizers for photodynamic therapy. Macromol Rapid Commun 39(5)

Menon JU, Jadeja P, Tambe P, Vu K, Yuan B, Nguyen KT (2013) Nanomaterials for photo-based diagnostic and therapeutic applications. Theranostics 3(3):152–166. https://doi.org/10.7150/thno.5327

Mikhail S, Allen C (2010) Poly(ethylene glycol)-b-poly(epsilon-caprolactone) micelles containing chemically conjugated and physically entrapped docetaxel: synthesis, characterization, and the influence of the drug on micelle morphology. Biomacromolecules 11(5):1273–1280

Mikhaylov G et al (2011) Ferri-liposomes as an MRI-visible drug-delivery system for targeting tumours and their microenvironment. Nat Nanotechnol 6(9):594–602

Mountz JM, Alavi A, Mountz JD (2012) Emerging optical and nuclear medicine imaging methods in rheumatoid arthritis. Nat Rev Rheumatol 8(12):719–728

Müller C et al (2014) Promising prospects for 44Sc-/47Sc-based theragnostics: application of 47Sc for radionuclide tumor therapy in mice. J Nucl Med 55(10):1658–1664

Nahrendorf M et al (2009) Hybrid in vivo FMT-CT imaging of protease activity in atherosclerosis with customized nanosensors. Arterioscler Thromb Vasc Biol 29(10):1444–1451

Ntziachristos V, Weissleder R (2001) Experimental three-dimensional fluorescence reconstruction of diffuse media by use of a normalized born approximation. Opt Lett 26(12):893–895

Ntziachristos V, Tung C-H, Bremer C, Weissleder R (2002) Fluorescence molecular tomography resolves protease activity in vivo. Nat Med 8(7):757–760

Ntziachristos V, Bremer C, Weissleder R (2003) Fluorescence imaging with near-infrared light: new technological advances that enable in vivo molecular imaging. Eur Radiol 13(1):195–208

Nwe K et al (2009) A new approach in the preparation of dendrimer-based bifunctional diethylenetriaminepentaacetic acid MR contrast agent derivatives. Bioconjug Chem 20(7):1412–1418

Pedrosa P et al (2015) Gold nanotheranostics: proof-of-concept or clinical tool? Nano 5(4):1853–1879

Pérez-Campaña C et al (2013) Biodistribution of different sized nanoparticles assessed by positron emission tomography: a general strategy for direct activation of metal oxide particles. ACS Nano 7(4):3498–3505

Pérez-Medina C et al (2016) Nanoreporter PET predicts the efficacy of anti-cancer nanotherapy. Nat Commun 7

Petersen L et al (2011) 64Cu loaded liposomes as positron emission tomography imaging agents. Biomaterials 32(9):2334–2341

Pichler J, Wehrl HF, Kolb A, Judenhofer MS (2008) Positron emission tomography/magnetic resonance imaging: the next generation of multimodality imaging? Semin Nucl Med 38(3):199–208

Pressly ED et al (2007) Structural effects on the biodistribution and positron emission tomography (PET) imaging of well-defined (64)Cu-labeled nanoparticles comprised of amphiphilic block graft copolymers. Biomacromolecules 8(10):3126–3134

Roesch (2012) Scandium-44: benefits of a long-lived PET radionuclide available from the (44)Ti/(44)Sc generator system. Curr Radiopharm 5(3):187–201

Shanmugam V, Selvakumar S, Yeh C-S (2014) Near-infrared light-responsive nanomaterials in cancer therapeutics. Chem Soc Rev 43(17):6254–6287

Sharma R, Mody N, Vyas SP (2017) Bioinspired nanotheranostics. Bipolymer based composites, vol 609. Elsevier, pp 279–288.

Sohail A, Ahmad Z, Bég OA, Arshad S, Sherin L (2017) A review on hyperthermia via nanoparticle-mediated therapy. Bull Cancer (Paris) 104(5):452–461

Sonali, Viswanadh MK, Singh RP, Agrawal P, Mehata AK, Pawde DM, Narendra, Sonkar R, Muthu MS (2018) Nanotheranostics: emerging strategies for early diagnosis and therapy of brain cancer. Nanotheranostics Ivyspring 2:70–86

Spikes D (1985) The historical development of ideas on applications of photosensitized reactions in the health sciences. In: Bensasson RV, Jori G, Land EJ, Truscott TG (eds) Primary photo-processes in biology and medicine. Springer US, Boston, pp 209–227

Sun X et al (2018) Aggregation-induced emission nanoparticles encapsulated with PEGylated nano graphene oxide and their applications in two-photon fluorescence bioimaging and photodynamic therapy in vitro and in vivo. ACS Appl Mater Interfaces 10(30):25037–25046

Swanson D et al (2008) Targeted gadolinium-loaded dendrimer nanoparticles for tumor-specific magnetic resonance contrast enhancement. Int J Nanomedicine 3(2):201–210

Tang Y et al (2015) An aptamer-targeting photoresponsive drug delivery system using 'off–on' graphene oxide wrapped mesoporous silica nanoparticles. Nanoscale 7(14):6304–6310

Terreno E, Castelli DD, Viale A, Aime S (2010) Challenges for molecular magnetic resonance imaging. Chem Rev 110(5):3019–3042

Torchilin P (2006) Micellar nanocarriers: pharmaceutical perspectives. Pharm Res 24(1):1

Torchilin VP, Frank-Kamenetsky MD, Wolf GL (1999) CT visualization of blood pool in rats by using long-circulating, iodine-containing micelles. Acad Radiol 6(1):61–65

Trubetskoy VS, Gazelle GS, Wolf GL, Torchilin VP (1997) Block-copolymer of polyethylene glycol and polylysine as a carrier of organic iodine: design of long-circulating particulate contrast medium for X-ray computed tomography. J Drug Target 4(6):381–388

Urvashi S, Dar MM, Hashmi AA (2014) Dendrimers: synthetic strategies, properties and applications. Orient J Chem 30(3):911–922

van Schooneveld MM et al (2010) A fluorescent, paramagnetic and PEGylated gold/silica nanoparticle for MRI, CT and fluorescence imaging. Contrast Media Mol Imaging 5(4):231–236

Varela-Moreira AA, Shi Y, Fens MHAM, Lammers T, Hennink WE, Schiffelers RM (2017) Clinical application of polymeric micelles for the treatment of cancer. Mater Chem Front 1(8):1485–1501

Vats S, Singh M, Siraj S, Singh H, Tandon S (2017) Role of nanotechnology in theranostics and personalized medicines. J Health Res Rev Wolters kluwer 4(1):1–7

Von Tappeiner H (1903) Therapeutische Versuche mit fluoreszierenden Stoffen. Munch Med Wochenschr 1:2042–2044

Wang YX, Hussain SM, Krestin GP (2001) Superparamagnetic iron oxide contrast agents: physicochemical characteristics and applications in MR imaging. Eur Radiol 11(11):2319–2331

Watermann A, Brieger J (2017) Mesoporous silica nanoparticles as drug delivery vehicles in cancer. Nanomaterials 7(7)

Wehrl F, Judenhofer MS, Wiehr S, Pichler BJ (2009) Pre-clinical PET/MR: technological advances and new perspectives in biomedical research. Eur J Nucl Med Mol Imaging 36(Suppl 1):S56–S68

Wishart DS (2016) Emerging applications of metabolomics in drug discovery and precision medicine. Nat Rev Drug Des Discov 15:473–484

Working K, Dayan AD (1996) Pharmacological-toxicological expert report. CAELYX. (stealth liposomal doxorubicin HCl). Hum Exp Toxicol 15(9):751–785

Wu SY, McMillan NAJ (2009) Lipidic systems for in vivo siRNA delivery. AAPS J 11(4):639–652

Wu P, Deng D, Gao J, Cai C (2016) Tubelike gold sphere-attapulgite nanocomposites with a high photothermal conversion ability in the near-infrared region for enhanced cancer photothermal therapy. ACS Appl Mater Interfaces 8(16):10243–10252

Wunder A et al (2003) Albumin-based drug delivery as novel therapeutic approach for rheumatoid arthritis. J Immunol (Baltim) 170(9):4793–4801

Xiao Q et al (2013) A core/satellite multifunctional nanotheranostic for in vivo imaging and tumor eradication by radiation/photothermal synergistic therapy. J Am Chem Soc 135(35):13041–13048

Xie T, Jing C, Long Y-T (2017) Single plasmonic nanoparticles as ultrasensitive sensors. Analyst 142(3):409–420

Xu H et al (2007) Preparation and preliminary evaluation of a biotin-targeted, lectin-targeted dendrimer-based probe for dual-modality magnetic resonance and fluorescence imaging. Bioconjug Chem 18(5):1474–1482

Yan X et al (2015) Optical and photoacoustic dual-modality imaging guided synergistic photodynamic/photothermal therapies. Nanoscale 7(6):2520–2526

Yang Y (2011) cRGD-functionalized, DOX-conjugated, and 64Cu-labeled superparamagnetic iron oxide nanoparticles for targeted anticancer drug delivery and PET/MR imaging. Biomaterials 32(17):4151–4160

Zhang M, Kievit FM (2011) Cancer nanotheranostics: improving imaging and therapy. Adv Healthc Mater Wiley 23:217–247

Zhang H et al (2015) Graphene oxide-BaGdF5 nanocomposites for multi-modal imaging and photothermal therapy. Biomaterials 42:66–77

Zheng J, Jaffray D, Allen C (2009) Quantitative CT imaging of the spatial and temporal distribution of liposomes in a rabbit tumor model. Mol Pharm 6(2):571–580

Zhou Z (2014) Iron/iron oxide core/shell nanoparticles for magnetic targeting MRI and near-infrared photothermal therapy. Biomaterials 35(26):7470–7478

Part VI

Tools and Techniques in NanoBioMedicine

Nanodevices: The Future of Medical Diagnostics

15

Ahana Mukherjee, Jaydeep Bhattacharya, and Ranjita Ghosh Moulick

Abstract

The speedy lifestyle of the twenty-first century has increased the disease complexity into severalfold. In this regard, biomedical devices have influenced the modern society by providing all kinds of medical comfort and have enhanced life expectancy worldwide. The need for these devices is rising due to the fast life, where every one is busy in their own way and thus people have much less time for regular monitoring of their own health. This is the reason why personalized point-of-care medical devices are in high demand. These smart systems with optimal capabilities are enabling the society with physical, mental, and clinical support. In this chapter, we have discussed about the modernization of the devices and their significant advancement in various fields using nanotechnology. The ultrasmall size and sensitivity of materials in the nanoregime have helped in development of new tools and systems which have taken clinical diagnosis and treatment into the next level.

Keywords

Biomedical devices · Nanotechnology · Nanobiosensors · Fabrication · Medical diagnostics

A. Mukherjee · R. G. Moulick (✉)
Amity Institute of Integrative Sciences and Health, Amity University Haryana, Amity Education Valley, Gurugram, Haryana, India

J. Bhattacharya
School of Biotechnology, Jawaharlal University, New Delhi, India

S. K. Saxena, S. M. P. Khurana (eds.), *NanoBioMedicine*,
https://doi.org/10.1007/978-981-32-9898-9_15

15.1 Biomedical Devices: An Introduction

A biomedical device is a combined or single system used for diagnostics or therapeutic purposes to monitor, prevent, or treat any disease, injury, and physiological ailment. It could be a material, apparatus (Bradbury et al. 2004), an article, or an appliance for application either in disease detection, monitoring, prevention, treatment, or any investigation, modification, and replacement required for any kind of physiological process. Even assistance required to a human body through metabolic, pharmacological, or immunological means could be a biomedical device. They have proved to be the most essential part of modern medicine as they can provide immense support and automation. The control mechanism using software with variety of modelling platforms and computers has made it technologically smart and advanced. These devices vary according to their need. They can be as simple as disposable gloves or as complicated as chip implants. Few examples of these devices include general hospital devices, dental devices, cardiovascular devices, neurological devices (Aebischer and Winn 1992), in vitro diagnostics, surgery devices, reprocessing of reusable medical devices (Chin and Imran 1995), etc. Therefore, rules and regulations regarding the safety of such devices become the primary issue since its malfunctioning might affect the human health either by escalating the injury or by giving false measurement values leading to the wrong treatments. For this reason, FDA (Food and Drug Administration, USA) monitors the safety and effectiveness of the devices and has classified them into three categories: class I, class II, and class III based on the risk (Kramer et al. 2012) they posed to the patients or users. Biomedical devices like elastic bandages, gloves, and tongue depressors that provide the lowest level of risk and therefore require the lowest level of regulatory control fall under class I biomedical devices. On the other hand, class II devices in spite of being simple devices are more complicated than class I devices; they provide slightly more risk than class I devices. They require more care and regulatory control for safety assurance. Examples of such devices are condoms, pregnancy test kits, infusion pump, etc. Finally, devices like pacemakers which have the highest risk and need the most precise regulatory control fall under class III biomedical devices. Currently, Central Drugs Standard Control Organization (CDSCO) of India has also introduced classification of these devices based on the associated risk. Class A represents low risk, class B poses moderate low risk, whereas class C and class D are associated with moderate high risk and high risk, respectively; WHO has developed a guideline containing policies, strategies, and action plan for accessing medical devices in a safe and effective way to prevent risk but also to assist the patients (https://www.who.int/medical_devices/safety/en/). Designing medical devices and its application has not only captured a major segment in biomedical engineering but also has become multidisciplinary because application of such devices varied from basic biological sciences to clinical medicines enabling new discoveries, diagnosis, and treatments using knowledge of algorithms, sensors, electronics, photonics, or magnetics. Notable amendments from various disciplines have improved the functionality, quality, safety, and reliability of the existing and new devices. One such upgradation has happened to a very simple device that we use to measure body temperature at bedside – a mercury thermometer (Bolton 1900). Mercury thermometer was invented in 1714 by Daniel Gabriel Fahrenheit in Amsterdam. The clever idea of using expansion of mercury with temperature and easy operation made it a household item. Currently, its intelligent version contains a combination of thermocouple, battery, voltmeter, and electronic display. The resolution and noise are very low. It is commonly known as digital thermometer now. The devices are changing profoundly in recent years because of miniaturization and improved techniques. Ten such most innovative biomedical devices of 2018 are nominated in the 12th Annual Prix Galien USA Awards. Few of them are i) a glucose monitoring system which can provide real-time values, ii) a spyglass for enhanced visualization of digestive system cancer and other diseases, iii) a noninvasive handheld device for determining head injuries at point of care, iv) a dermal regeneration matrix (Yannas et al. 1989), etc. More than 10,000 medical devices are available in the market, so WHO has made a priority list for selection of proper device according to the need and understanding the technologies involved (https://www.who.int/medical_devices/priority/en/). These biomedical devices have become essential for modern medicine since they have ability to connect external devices wired or wireless and provide automation for monitoring and actuation to many patients. Several challenges are involved in the course of long running of these devices. As these devices collect personal data from many individuals and some of the arrangements are connected to the network interfaces, keeping all the information safe is prerequisite because network interfaces are the potential site for attacks and modified data would induce erroneous diagnosis in mass. Safety of users is another primary concern for medical devices. Avoiding any kind of accident should be the fundamental goal of medical device manufacturers. Heterogeneity (made up of different elements) and usability (minimum cognitive load on users) are also the other challenges (Lai et al. 2011).

15.2 Role of Nanotechnology in the Modernization of Biomedical Devices

The concept of nanotechnology started from the famous lecture "There is plenty of room at the bottom" by physicist Richard Feynman in 1959. Slowly nanotechnology emerged as a future research field, and the term nanotechnology was coined by Taniguchi. Buckyball (Chung and Sternberg 1993), the first nanostructure, was discovered and received the Nobel Prize in 1996. The era of nanotechnology started after 2010, capturing a significant part of modern technology and giving rise to the most exciting field of nanomedicine. It is the study of materials or structures between 1 and 100 nm sizes. In this range, the materials offer unique size-dependent properties that differ significantly from the bulk material and can be physically explained by quantum effect and that made the difference. The increased surface-to-volume ratio changed both physical property and chemical reactivity when bulk material is compared. These exceptional properties of materials in this range could be used to develop systems or devices or another new material that is not possible to achieve when made from bulk materials. The small size of nanostructures with special properties make them ultrasensitive. Nanotechnology is highly interdisciplinary as it combines physics, chemistry, material science, engineering, biology, and medicine. It was envisaged that sensors of nanoscale might offer significant advantages over conventional sensors because of similarities in sizes to the detectable molecules, e.g., a single molecule or an atom. For these reasons, nanomaterials exhibited a wide range of applications in diagnosis, drug delivery, imaging, prostheses, implants, sensors, etc. Nanoscale materials integrate well into the biological systems because of the similarities in sizes to the detectable molecules, e.g., a single molecule or an atom. The nanomaterials could be developed from a variety of materials, e.g., organic, inorganic, or metal substrates using chemical or physical methods. Taking advantage of surface chemistry reactions and conjugating biological molecules with nanomaterials, specific biological reactions like antigen-antibody interaction, ligand-receptor interaction, and DNA-DNA hybridization could be determined specifically.

All these have generated the new area named as nanomedicine which has become extremely useful for improving human health based on diagnostics, patient treatment, prevention, and curing disease or injury. Novel nanomaterials are being used for targeted delivery of drugs, pharmaceutical molecules, or imaging agents at the disease site and are endocytosed inside the cell. The size, shape, composition, surface chemistry, and charge, all these factors play important roles (Sahay et al. 2010) and investigation is still on for development of ideal vehicle in nanomedicine. A variety of nanocarriers made of polymers (Tong and Cheng 2007), liposomes (Fenske and Cullis 2008; Sen et al. 2018), bioinspired materials (Chen et al. 2019), or PLGA (Saneja et al. 2018) encapsulating therapeutic agents and formulations are presently in the market and in clinical trials. Since it has been demonstrated that nanotechnology has effectively reduced the toxicity of the anticancer compounds (Rodriguez-Nogales et al. 2018), a variety of tumor microenvironment is being tested (Marcazzan et al. 2018; Cong et al. 2018; Nicolas et al. 2018). Targeted delivery often requires complex surface modifications, supramolecular architecture presented as a strategy to achieve better responses aimed at a specific site (Alshaer et al. 2018). Additionally, RNAi-based nanomedicine (Liu et al. 2018a), dual nanomedicine (Zhu et al. 2018), microfluidic technology (Liu et al. 2018b), and phototheranostics (Vankayala and Hwang 2018) are the recent advances. Hematological (Deshantri et al. 2018) or nervous system pathologies (Mizrahy et al. 2019) are also being explored.

Clinical and healthcare system has been benefited by another emerging field, nanorobotics. It is the technology to create machines or robots of nanoscale. Nanorobots are being used in treating cancer (Li et al. 2018), dentistry (Kumar and Vijayalakshmi 2006), medicine (Freitas 2005), surgery (Li et al. 2017), etc.

The unique properties displayed by the nanomaterials helped in creating new systems or devices. Nanoscale devices have the ultrafast response time because the speed with which the species is detected depends on the sensor's dimension. Their small size, light weight, and high surface-to-volume ratio makes them ideal candidates for high signal amplification and trace amount detection that were initially thought to be unattainable.

15.3 Development of Biosensors for Medical Diagnostics

Biosensors are advanced analytical devices (Turner et al. 1987) which consist of a biological sensing element (Perumal and Hashim 2014a) that converts a biological response into an electrical, optical, or thermal indication. It finds huge application in medical diagnosis and a lot of other things like food analysis, study of biomolecules and their interactions, drug development, drug delivery, crime detection, environmental field examining, industrial process control, agricultural industries (Mohanty and Kougianos 2006), manufacturing of pharmaceuticals, and replacement of organs. Biosensors generally have three parts according to its working principle (Perumal and Hashim 2014b): The first component is bioreceptor which is a sensitive biological element (complex materials like tissue slices, microorganisms, cell organelles, and cell receptors, proteins including enzymes and antibodies, nucleic acids, plant proteins or lectins, etc.). The second component is transducer which is the detector part that changes the resulting signal due to contact of the analyte, and it displays the results in an accessible way. The final section comprises of an amplifier which generates the signal when the sensor interacts with the analyte (Fig. 15.1).

The sensitive biological material (usually microorganisms, antibodies, or enzymes) is immobilized on the sensor surface by covalent or noncovalent binding and stays in contact with the transducer. The substances whose chemical constituents have to be measured bind to the biological element which in turn generates electrical signal that can be measured. The signals can also be thermal or optical (Fig. 15.2).

There are several types of biosensors (Monošík et al. 2012):

(a) Electrochemical Biosensors: Electrochemical biosensors detect the binding of a target molecule by sensing a redox reaction (Pohanka and Skládal 2008; Ronkainen et al. 2010). These biosensors have been a topical area of research for nearly 50 years. Leland C. Clerk was the first to introduce an enzyme electrode with immobilized glucose oxidase in the year 1962. These biosensors basically convert biochemical data such as analyte concentrations into an electrical signal.

(b) Amperometric Biosensor: Amperometric biosensors are sensors that produce current when a potential is applied between the two microelectrodes (Ianniello

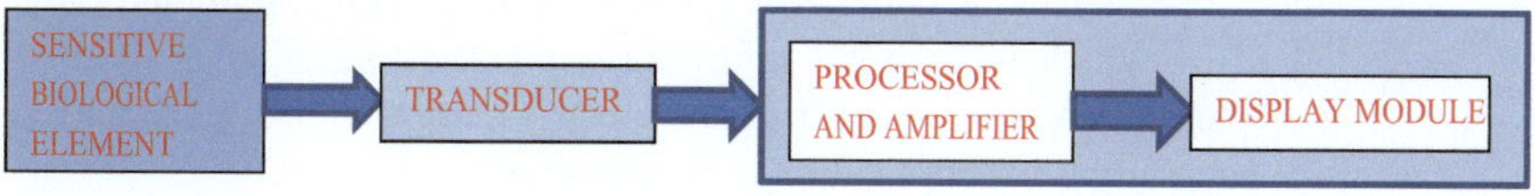

Fig. 15.1 The schematic represents the components of a biosensor

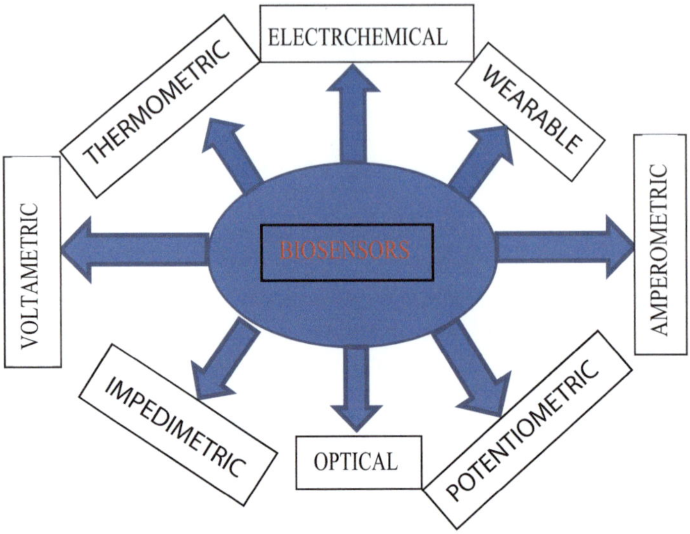

Fig. 15.2 Different kinds of biosensors

and Yacynych 1981). The common amperometric biosensor used involves the Clark oxygen electrode. They mainly find application in diagnosis of diseases; analysis of glucose, ethanol, and lactate in wine; and many more.

(c) Potentiometric Biosensors: Potentiometric biosensors are sensors that make use of ion-selective electrodes in order to convert a biological response into an electrical indication. They rely on indicating the variations in pH or potential, and these determinations are applicable in environmental analysis, food investigation, and clinical diagnostics.

(d) Impedimetric Biosensors: Impedance is defined as the resistance of a circuit to alternating current. It is a two-dimensional quantity arising from two one-dimensional quantities – reactance and resistance. Impedance which is denoted by the letter Z is different from resistance because it is dependent on frequency while resistance is not. Biosensors that convert a biological response to impedance are impedimetric biosensors. They are based on the sensitive technique termed as electrochemical impedance spectroscopy (EIS). EIS is a sensitive indicator of different chemical and physical properties. There are different types of impedimetric biosensors: (i) enzyme-based impedimetric biosensors, (ii) immunobinding-based impedimetric biosensors, (iii) nucleic acid-based impedimetric biosensors, (iv) cell- and microorganism-based impedimetric biosensors (Katz and Willner 2003).

(e) Voltammetric Biosensor: Voltammetric biosensors depend on voltammetry that is a method by which we can get information about a particular analyte by varying a potential and then measuring the electric current that is generated as a result. It generally finds application in food analysis.

(f) Piezoelectric Biosensors: Piezoelectric biosensors are biosensors which basically depend on mass (Babacan et al. 2000). They are based on the principle of sound vibrations and hence are also known as acoustic biosensors. In these types of sensors, a sensing molecule is attached to a mass-to-frequency transducer in which a mechanical vibration is generated when the sensing molecule and the analyte to be explored interact with each other. These mechanical vibrations that are generated due to interaction are then converted into an electrical signal that is proportional to the amount of the analyte. Quartz crystal micro- and nanobalance are examples of piezoelectric biosensors. They have wide level of technological applications, they are fitted in electronic devices, they are used in the construction of biosensors that are used in affinity interactions, etc.

(g) Thermometric Biosensor: Thermometric biosensors are also called calorimetric biosensors or thermal biosensors and are based on the principle of generation of heat during a chemical or biological reaction (Ramanathan and Danielsson 2001). These biosensors find application in estimation of serum cholesterol, glucose (Ramanathan et al. 2001), uric acid, penicillin G, and urea.

(h) Optical Biosensor: Optical biosensors are biosensors which are based on the principle of optical measurements like light scattering (Cooper 2002), fluorescence, chemiluminescence, or absorption (Borisov and Wolfbeis 2008). They use optical fibers to detect the properties of the analyte of importance. They allow safe non-electrical sensing of materials and do not require a reference sensor. Some important optical biosensors are (i) fiber optic lactate biosensor, (ii) optical biosensors for detecting urinary infections, (iii) optical biosensors for blood glucose, and (iv) optical sensors that detect changes in pH and pCO2 during critical surgeries.

(i) Wearable Biosensor: Wearable biosensors are biosensors which are wireless (Ajami and Teimouri 2015) and can be worn like bandages. They are gaining a lot of interest due to their ability to provide real-time information about a body area by measuring the biochemical markers in the biofluids such as saliva, tears, sweat, and interstitial fluid. Biosensors have come into play with a lot of advantages like they are cheap, small in size, and quick and easy to use, and their sensitivity and selectivity are much more than the conventional instruments used. They are used in a lot of things like clinical analysis, environmental monitoring, food analysis, industrial process control, agricultural industry, and many other things. The most important application of biosensors is its application in the field of clinical diagnostics. The electrochemical biosensors are used to estimate glucose and lactic acid in patient samples. Piezoelectric biosensors are fitted in electronic devices, whereas the optical biosensors detect changes in the pCO2 during important surgeries. Commercial biosensors are used for self-testing by self-monitoring of biofluids which has improved the efficiency of healthcare replacing the slow diagnostic methods. It has made clinical decision-making a fast process (Malhotra and Chaubey 2003). The most popular example of a clinically used biosensor is the glucose oxidase-based sensor which is used by diabetic patients in order to check their blood glucose levels. Biosensors are used for detection of the increasing number of pollutants finding their way into the groundwater systems and thereby into drinking water. The important targets for biosensors are nitrates and phosphates. Commercially available biosensors are used for the measurement of carbohydrates, acids, and alcohol. They are used for quality check in different laboratories to measure the carbohydrate content, gases, cofactors, amino acids, amides, amines, carboxylic acids, inorganic ions, phenols, and heterocyclic compounds. They are used in beer and soft drink companies. They are also used in poultry and fish in order to detect the presence of pathogenic microorganisms in them. New products are produced by a huge amount via cell culture along with the normal industrial fermentation products. Biosensors find application in assessing these products and measuring their generation (Alocilja and Radke 2003; Terry et al. 2005; Luong et al. 1997; Karube and Tamiya 1987; Scott 1998). The agricultural industry depends on biosensors for quality control. They are used for the measurement of various components of agricultural samples. Biosensors used for the detection of carbamates and organophosphates from pesticides are based on the cholinesterases inhibition. Besides, these sensors are also used for the measurement of ammonia and methane. The only commercially available biosensors used in agricultural industries for wastewater quality check are the biological oxygen demand (BOD) organisms which are based on *Rhodococcus erythropolis* immobilized on a polyacrylamide or collagen sensor surface. The most important diagnostic application of a biosensor is the disease-related biomarkers in the biofluids. Biomarker concentration is an important tool to determine the disease type and its stage as well as the patient's response to medication. This is a very important application as biomarkers are significant in certain diseases, for example, a rise in blood sugar level indicates diabetes. However, many biomarkers are still a topical part of research. Some biomarkers can be found to identify the subtypes of diseases which can help in providing efficient healthcare. Currently, testing of biomarkers in human fluids takes place in laboratories using analyzers based on DNA or protein microarrays including immunoassay methods. They require trained staff and increased time and effort. Hence, biosensors provide quick and self-testing of analytes in biofluids. For example, portable biosensors (Srinivasan and Tung 2015) are available to test blood glucose levels or blood coagulation which provide results within minutes.

15.4 Fabrication of Nanostructure-Based Highly Sensitive Platforms

Nanostructures are novel structures that have dimensions within 100 nm. The common nanostructures are cylindrical nanotubes, nanosurfaces, and nanospheres. There are two kinds of fabrication methods for the formation of nanostructures: (a) top-down method and (b) bottom-up method (Fig. 15.3).

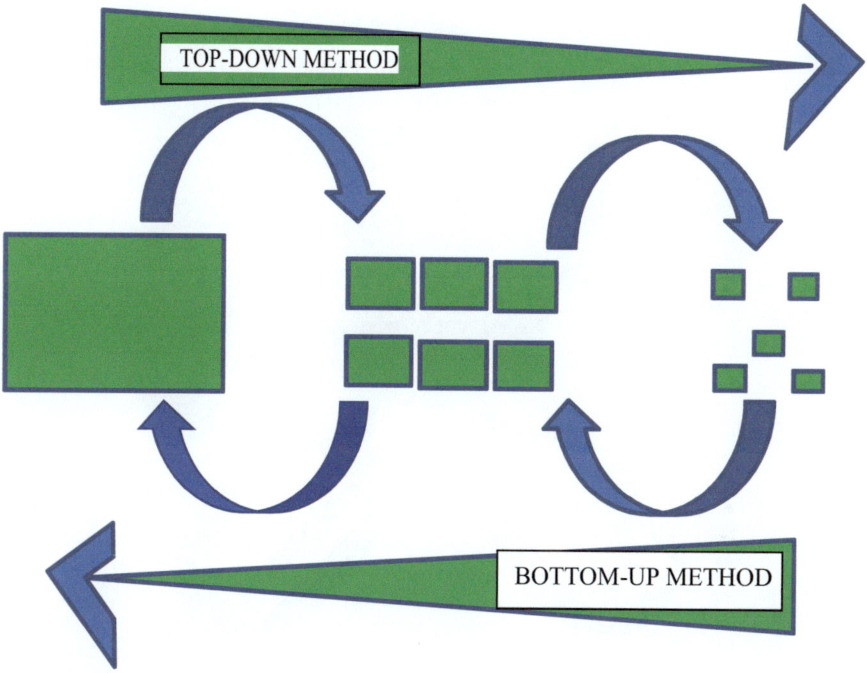

Fig. 15.3 The schematics represent the concept of both top-down and bottom-up approach

Top-down methods are generally known as lithographic methods in the formation of nanostructures. Semiconductor nanostructures are formed by these lithographic methods and have come into existence due to the advanced resolution of the microelectronics performing technologies. The methods involve the hiding of few selected nanoscale regions and removal of the surrounding nonhidden regions. The combined use of the lithography methods like photolithography and electron-beam lithography has led to the formation of various nanostructures below 100 nm. Yet, fabrication of quantum wires and quantum dots which have dimensions in the range of 10 nm is not possible by the use of these well-known conventionally used lithographic techniques. Bottom-up methods are also known as "self- assembly" methods. Extensive researches are going on in semiconductor nanostructure formations using these bottom-up methods. In the 1990s, quantum dots were made using these techniques. Since that time, clusters of nanoscale dimensions are being tried to be made. These methods are now used for making several varieties of quantum dots and quantum wires. In this chapter, we have mainly focused on the different top-down approaches (lithographic techniques) of formation of nanostructures. Lithography in nanotechnology is a technique which involves etching, writing, and printing of structures which have dimensions in the range of 100 nm. Nanostructure patterning becomes easier due to the lithographic techniques.

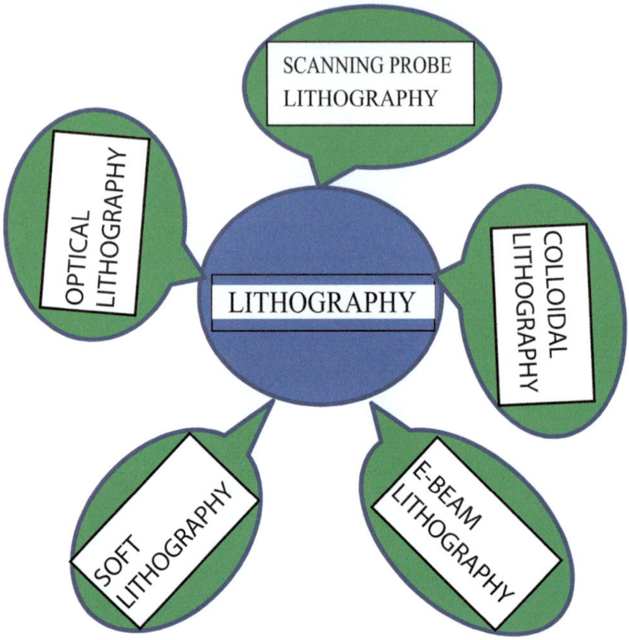

Fig. 15.4 The various lithography techniques are shown in this figure

The various lithography techniques are (a) colloidal lithography, (b) E-beam lithography, (c) optical lithography, (d) soft lithography, and (e) scanning probe lithography (Fig. 15.4).

Colloidal lithography is an important unconventional fabrication method that uses colloidal materials as a covering for engraving and leaving a layer of the desired material in a base. This technique includes ordered and disordered colloidal particles. For example, polystyrene beads are deposited in a Si surface, and then plasma O2 shrinks the size of the beads. C4F8 is used for etching of the remaining Si layer. Finally, the beads are removed by standardized methods to form only the Si nanoislands. Electron beam lithography generally known as E-beam lithography is a technique in which a beam of electrons is focussed on a resist (an electron-sensitive region) covered surface in order to pattern extremely small and fine structures. This method can form nanostructures of dimensions less than 10 nm. It is slower than the other lithographic methods and is more expensive than them. In Glasgow, E-beam lithography is being used to generate gaps of 1 nm. Their Department of Physics and Astronomy is trying to pattern magnetic materials with the help of E-beam lithography. Optical lithography also known as photolithography or UV lithography is a technique of biofabrication which transfers different patterns with the help of ultraviolet light to a photoresist (photosensitive area) before the following steps of doping, etching, and deposition. This is a conventionally used technique, but it faces a lot of difficulties. It is mainly used in the formation of microstructures. It is faster and cheaper than E-beam lithography. This lithography is the most commonly used method for tissue engineering as it can make the formation of microchips easier. This process is nowadays being used to make micro-electro-mechanical system (MEMS) devices. Chen et al. are using this method in generating patterns which are micron-sized on a flexible substrate. Soft lithography is an extension of photolithography that often used non-photolithographic techniques for patterning. It is termed as "soft" because it uses soft materials like polydimethylsiloxane (PDMS) for construction of patterns. PDMS is very common in this method because it is inexpensive and biocompatible. Photolithography has a lot of limitations which are overcome by soft lithography. It can make structures of dimensions between 10 and 100 nms which photolithography is incapable of. Soft lithography is used not only in cell biology industries but also in optics and microelectronic areas. The types of soft lithography are (a) microcontact printing which has a resolution of 35 nm, (b) replica molding which has a resolution of 30 nm, (c) microtransfer molding which has a resolution of 1 μm, (d) micromolding in capillaries having a resolution of 1 μm, and (e) solvent-assisted micromolding that has a resolution of 60 nm. Scanning probe lithography (SPL) is a nanofabrication technique that uses scanning probe as used in atomic force microscopy in order to generate nanoscale patterns. It is a one-step process and can make nanostructures of less than 10 nm dimensions. It is a "maskless" technique and is inexpensive. This technique makes 3D structures by a nanoscopic stylus used for transferring nanoscale patterns. This method includes nanoshaving, nanografting, and nanopen reader and writer (NPRW). Cheong et al. used thermal SPL (tSPL) to construct a binary hard covering on the polymer HM8006. This covering was further used for making nanopatterns in silicon.

15.5 Importance of Nanobiosensors in Clinical Diagnostics

Sensitive and low-cost biosensors in association with nanomaterials are termed as nanobiosensors (Touhami 2014). They serve as a bridge between fast detection and routine self-testing. Nowadays, medical diagnosis has become advanced due to fast detection of analytes and their concentration which tells us both the presence of the disease and its stage (Prasad 2014). Nanobiotechnology is being used in molecular diagnostics, and several nanotechnologies are in the processing. Here we will see some of the latest uses of nanobiosensors in medical diagnostics (Fig. 15.5).

Biochip based on nanotechnology is a completely new idea in the field of analyte recognition and quantification (Vo-Dinh et al. 2001; Vo-Dinh and Cullum 2000; Ghoshal et al. 2010; Vo-Dinh 2004). Some devices that include biochips and microarrays based on nanotechnology are biochips recognizing proteins and nanofluidic arrays. One of the promising uses of nanofluidic devices is isolation and analysis of biomolecules like DNA. This detection schemes can very well detect cancer. One similar device is used for the construction of silicon nanowires on a chip, using photolithographic and etching methods which is followed by an oxidation step that converts the wires into a bare nanotube. Using this process, the investigators can probably create nanotubes with diameters very small as 10 nm. Nanofluidic technology has broad applications in pathogen detection, drug development, personalized

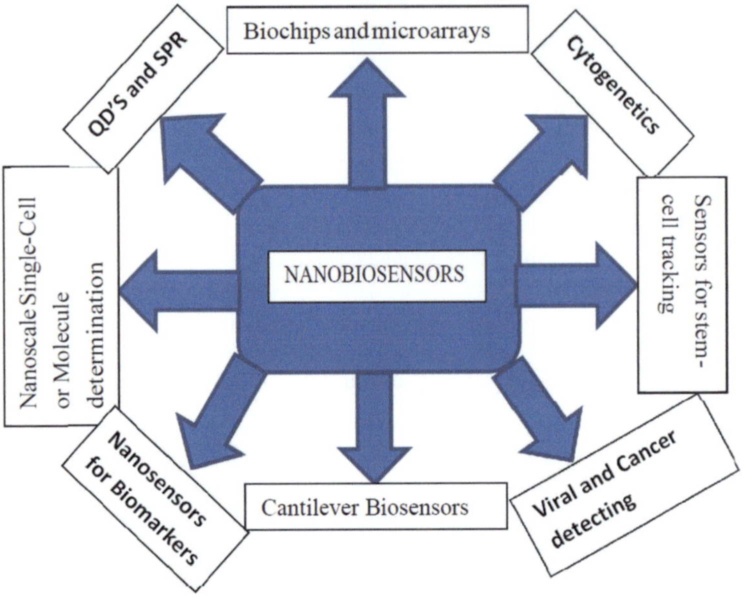

Fig. 15.5 A variety of nanobiosensors available are demonstrated in this picture

medicine, and clinical research. On the other hand, protein microarrays are used in efficient purifying of proteins without the problem of storing and give high protein yield for use in functional purposes. Cytogenetics is the part of genetics that deals with the behavior of chromosomes during cell division. Molecular cytogenetics is now increased by the use of atomic force microscopy and quantum dot (QD) (Zhang et al. 2005; Medintz et al. 2003) FISH. This process focuses on the combination of biochemical and nanomanipulation techniques, which allows both nanoextraction and nanodissection of DNA present in the chromosome. SPIONs (superparamagnetic iron oxide nanoparticles) (Ghoshal et al. 2011) are becoming very important for cell tracking. As it is not able to track intracellularly, hence a new particle named perfluorocarbon nanoparticles is primarily used for tracking endothelial progenitor cells in human umbilical cord. Many perfluorocarbon compounds exist in nature, and cell's different parts can be labeled. It has been seen that there is no change in cell markers after labeling activity, and they are seen to be able to get inside blood vessels forming tumors in mice. Hence, it can be concluded that perfluorocarbon-based biosensors can be used to detect tumors and cure cardiovascular diseases. Nanotechnology has developed the method for identification of single cells. Nanolaser scanning confocal spectroscopy has the capability of single-cell resolution and can be used to identify certain not known properties of many cancer cells that finds the difference between closely related noncancerous cells. The application of nanobiotechnology to proteomics is able to detect a single molecule of protein. Nanotechnology has created a new level for the detection of biomarkers. Some biomarkers (Arya and Bhansali 2011; Li et al. 2012; Tehrani et al. 2014) form the basis of molecular analytical tests. The high surface areas of nanoparticles make them ideal candidates for developing biomarker attachment platforms. The variety of nanoparticles that are available can be tailored to selectively bind a subset of biomarkers and make them ready for later study using the different proteomic tests. Functional polymer-coated nanoparticles can be used in the quick determination and quantification of biomarkers and separation of DNA. Gold particles of sizes not greater than 13 nm are attached to DNA surfaces (Ravalli and Marrazza 2015; Devi et al. 2015). The gold nanoparticles only assemble onto a sensor surface due to the presence of a complementary target. When a patterned sensor surface is made up of multiple DNA strands, the technique can detect millions of different DNA sequences at a given point of time. Inorganic fluorophores that have advantages over conventionally used fluorescent markers are termed as QDs (Huang and Zhu 2009; Hansen et al. 2006). They show high sensitivity and stable fluorescence without the use of lasers. QDs have a wide range of applications for molecular diagnostics and genotyping. QDs also allow multiplexed diagnostics, and they combine diagnostics with therapeutics. The most important use of QDs is for cancer detection. Stable QD bioconjugates help in the visualization of cancer cells in living animals. QD along with fluorescence microscopy can follow cells at high resolution in living organisms. QDs with a polyacrylate cap covalently linked to antibodies are used to detect Her2 which is a marker for breast cancer. Cantilevers generally convert a reaction into a mechanical motion on the nanometer scale, and it can be measured directly by a light beam deflection from its surface. Cantilever technology can be used as an alternative to PCR and corresponds to the new DNA and protein microarrays without the need to label the target molecules (Fritz 2008; Ziegler 2004). The advantages of cantilevers are that they provide fast, label-free determination of specific DNA sequences for oncogenes, SNPs, and genotyping (Raiteri et al. 2001). Nanocantilevers are crucial in designing an ultrasmall sensor for detecting bacteria, viruses, and other disease-causing organisms. A real-time cantilever biosensor provides regular assessment of pathological parameters. Surface plasmon resonance (SPR) (Gupta and Verma 2009; Hegnerová et al. 2009; Zhu et al. 2008; Anker et al. 2010) is generally used in optical biosensors. Tags can be optically detected by surface-enhanced Raman scattering (SERS). Each of its type uses Raman spectrum of different molecules. SERS have width which are 1/50 of that of normal fluorescent band. The spectral intensity of SERS-based tags is directly proportional to the number of particles, and that is why these tags facilitate multiplexed analyte quantification. SERS-based tags come with low-cost instrumentation. The particles can be checked for near-infrared range, which allows its detection in blood and in tissues. Another advantage of these particles is that they are stable and are not prone to photodegradation. The SERS technology can be used to directly detect biomarkers. DNA-based technique based on surface-enhanced Raman gene (SERGen) probes is used to detect gene targets by hybridization to DNA sequences corresponding to these probes. SERS nanoprobes (Jun et al. 2010; Chon et al. 2011) are nowadays being widely used for biological sensing. Virus particles are also known as biological nanoparticles. Herpes simplex virus and adenovirus are used to initiate the assembly of magnetic nanobeads as nanosensors for clinically important viruses. Detection of as few as five viral particles is possible in a 10 mL serum sample. This system is more sensitive, cheaper, and faster than ELISA-based and PCR-based analysis. Cancer is a leading cause of death in the world. It has been found out that early diagnosis favors fast treatment of this disease. Cancer biomarkers are being extensively used in the diagnosis of cancer. Electrochemical sensors are becoming important in the laboratories for the analysis of diverse biological elements. 2D is becoming very important like graphene, graphene oxide, and reduced graphene oxide, and graphene-like nanomaterials are being used in electrochemical sensors for cancer diagnosis. Nanoparticles are being used for dual-mode imaging of cancer. The best characteristics of QDs and magnetic iron oxide nanoparticles are combined to create a single nanoparticle probe that can yield clinically useful images of both tumors and the other molecules that are involved in cancer. Silica nanoparticles of 30 nm in size are saturated with rhodamine and 9 nm diameter iron oxide nanoparticles which are soluble in water. The resulting combination of nanoparticles that is formed is nearly 45 nm in diameter and results in a better means of detection for both fluorescent imaging tests and magnetic resonance imaging than any of the individual components. An antibody binding to polysialic acid molecules that are found on the surface of the lung is attached to the nanoparticles. They are taken up by tumor cells, and hence they can be checked under a fluorescent microscope. Antibiotics accumulate in the human body by food chain and can cause severe influence to human health. Hence, the development of simple cost-effective and cost-sensitive methods for detection of antibiotic levels is highly desirable. Nanomaterials with excellent electronic, optical, mechanical, and thermal properties are being used in making nanobiosensors which can detect the amount of antibiotics (Lan et al. 2017) present in the body. Nanoparticles are being used to develop MRI. A new MRI contrast agent which is using manganese oxide nanoparticles to show the anatomic structures of mouse brain is producing images that are very clear. The contrast agent which is newly made is used for better diagnosis of neurological disorders such as Parkinson's disease, Alzheimer disease (Hegnerová et al. 2009), and stroke. Antibodies attached to the manganese oxide nanoparticles are able to detect specific binding to receptors on the surface of breast cancer cells. This has been studied in mouse. The tumors become clearly visible by the antibody attached to the contrast agent. The same principle also allows to detect other diseases with the help of proper antibodies.

15.6 Future Aspects of Nanobiosensors

Within the next few years, nanobiosensors which can make ample number of measurements rapidly and at a very low cost will become available (Sagadevan and Periasamy 2014). The most important application of sensors based on nanotechology in the clinical field will be biomarker analysis. Biomarkers reflect the presence and different stages of disease in most organs. Therefore, detection of these markers will provide a sensitive analysis of health and disease (Malhotra et al. 2017; Gruhl et al. 2011). Molecular electronics and nanoscale chemical sensors can be used to build microscopic sensors which can detect arrangements of different types of chemicals in a fluid. Nanodevices which can detect the presence of different kinds of chemicals which are secreted into the tissues in response to injuries or infections can be developed. With the presently used procedures for blood analysis, such a chemical would be difficult to detect. Another motive is to replace fluorescent labeling by fluorescent labeling with nanoparticles as the signal in normal fluorescence labeling is less intense. Cantilever sensors will provide diagnosis without the use of PCRs. Nanotechnology can basically be applied for assessment of a single cell for diagnosis mainly related to the genes. Nanodiagnostics could reduce the time needed for the clinical results in the near future (Jain 2007; Demidov 2004). For example, patients with sexually transmitted diseases can provide their urine samples before they go in person to the clinic, so that their samples can be checked and the results can come by the time they visit the doctor. The doctor can then give the prescription hand in hand at that very moment, thereby reducing the time for diagnosis and medication and making the process less costly. In the next few years, nanobiotechnology will help in the development of personalized medicines. Another important area of application will be the diagnosis of cancer (Wang et al. 2017; Pathak et al. 2007). Cancer is detected so slowly by recently available procedures that sometimes it is often too late for cure. Nanorobotics method can be used in the future for early detection as well as for cancer treatment. Finally, nanotechnologies promise to develop the identification of the nature of illness extremely fast and enable point-of-care testing (POCT) which is useful in the short-time determination of analytes, linking diagnostics with medication. The most important applications of nanosensors are in the discovery of biomarkers, diagnosis of cancer, and testing the presence of pathogenic microorganisms. Nanobiosensors promise to play a leading role in the development of clinical diagnostics and curative methods (Chamorro-Garcia and Merkoçi 2016) in the fast life of the twenty-first century.

References

Aebischer P, Winn SR (1992) Neurological therapy devices. U.S. Patent No. 5,106,627. 21

Ajami S, Teimouri F (2015) Features and application of wearable biosensors in medical care. J Res Med Sci 20(12):1208

Alocilja EC, Radke SM (2003) Market analysis of biosensors for food safety. Biosens Bioelectron 18(5–6):841–846

Alshaer W, Hillaireau H, Fattal E (2018) Aptamer-guided nanomedicines for anticancer drug delivery. Adv Drug Deliv Rev 134:122–137

Anker JN et al (2010) Biosensing with plasmonic nanosensors. In: Nanoscience and technology: a collection of reviews from nature journals. Macmillan Publishers Ltd., London, pp 308–319

Arya SK, Bhansali S (2011) Lung cancer and its early detection using biomarker-based biosensors. Chem Rev 111(11):6783–6809

Babacan S et al (2000) Evaluation of antibody immobilization methods for piezoelectric biosensor application. Biosens Bioelectron 15(11–12):615–621

Bolton HC (1900) "Evolution of the thermometer", 1592–1743. The Chemical Publishing Co., Easton

Borisov SM, Wolfbeis OS (2008) Optical biosensors. Chem Rev 108(2):423–461

Bradbury TJ, et al (2004) System and method for rapidly customizing design, manufacture and/or selection of biomedical devices. U.S. Patent No. 6,772,026. 3
Chamorro-Garcia A, Merkoçi A (2016) Nanobiosensors in diagnostics. Nano 3:194–215. 1849543516663574
Chen Z, Wang Z, Zhen G (2019) Bioinspired and biomimetic Nanomedicines. Acc Chem Res 52:1255–1264
Chin D, Imran MA (1995) Reusable medical device with usage memory, system using same. U.S. Patent No. 5,425,375. 20
Chon H et al (2011) Simultaneous immunoassay for the detection of two lung cancer markers using functionalized SERS nanoprobes. Chem Commun 47(46):12515–12517
Chung F, Sternberg S (1993) Mathematics and the Buckyball. Am Sci 81(1):56–71
Cong Y et al (2018) Dual drug backboned shattering polymeric Theranostic Nanomedicine for synergistic eradication of patient-derived lung cancer. Adv Mater 30(11):1706220
Cooper MA (2002) Optical biosensors in drug discovery. Nat Rev Drug Discov 1(7):515
Demidov W (2004) Nanobiosensors and molecular diagnostics: a promising partnership. Expert Rev Mol Diagn 4:267–268
Deshantri AK et al (2018) Nanomedicines for the treatment of hematological malignancies. J Control Release 287:194–215
Devi RV, Doble M, Verma RS (2015) Nanomaterials for early detection of cancer biomarker with special emphasis on gold nanoparticles in immunoassays/sensors. Biosens Bioelectron 68:688–698
Fenske DB, Cullis PR (2008) Liposomal nanomedicines. Expert Opin Drug Deliv 5(1):25–44
Freitas RA (2005) Current status of nanomedicine and medical nanorobotics. J Comput Theor Nanosci 2(1):1–25
Fritz J (2008) Cantilever biosensors. Analyst 133(7):855–863
Ghoshal S et al (2010) Biosensors and biochips for nanomedical applications: a review. Sensors & Transducers 113(2):1
Ghoshal S et al (2011) Superparamagnetic iron oxide nanoparticle attachment on array of micro test tubes and microbeakers formed on p-type silicon substrate for biosensor applications. Nanoscale Res Lett 6(1):540
Gruhl FJ, Rapp BE, Länge K (2011) Biosensors for diagnostic applications. Adv Biochem Eng Biotechnol 133:115–148
Gupta BD, Verma RK (2009) Surface plasmon resonance-based fiber optic sensors: principle, probe designs, and some applications. J Sens 2009:194–215
Hansen JA et al (2006) Quantum-dot/aptamer-based ultrasensitive multi-analyte electrochemical biosensor. J Am Chem Soc 128(7):2228–2229
Hegnerová K et al (2009) Surface plasmon resonance biosensors for detection of Alzheimer disease biomarker. Sensors Actuators B Chem 139(1):69–73
Huang H, Zhu J-J (2009) DNA aptamer-based QDs electrochemiluminescence biosensor for the detection of thrombin. Biosens Bioelectron 25(4):927–930
Ianniello RM, Yacynych AM (1981) Immobilized enzyme chemically modified electrode as an amperometric sensor. Anal Chem 53(13):2090–2095
Jain KK (2007) Applications of Nanobiotechnology in clinical diagnostics. Clin Chem 53(11):2002–2009
Jun B-H et al (2010) Multifunctional silver-embedded magnetic nanoparticles as SERS nanoprobes and their applications. Small 6(1):119–125
Karube I, Tamiya E (1987) Biosensors fob food industry. Food Biotechnol 1(2):147–165
Katz E, Willner I (2003) Probing biomolecular interactions at conductive and semiconductive surfaces by impedance spectroscopy: routes to impedimetric immunosensors, DNA-sensors, and enzyme biosensors. Electroanalysis 15(11):913–947
Kramer DB, Xu S, Kesselheim AS (2012) How does medical device regulation perform in the United States and the European union? A systematic review. PLoS Med 9(7):e1001276
Kumar SR, Vijayalakshmi R (2006) Nanotechnology in dentistry. Indian J Dent Res 17(2):62–65

Lai DTH, Palaniswami M, Begg R (eds) (2011) Healthcare sensor networks: challenges toward practical implementation. CRC Press, Boca Raton
Lan L, Yao Y, Ping J, Ying Y (2017) Recent advances in nanomaterial- based biosensors for antibiotics detection. Biosens Bioelectron 91:504–514
Li J, Li S, Yang CF (2012) Electrochemical biosensors for cancer biomarker detection. Electroanalysis 24(12):2213–2229
Li J et al (2017) Micronanorobots for biomedicine: delivery, surgery, sensing, and detoxification. Sci Robot 2(4):194–215
Li S et al (2018) A DNA nanorobot functions as a cancer therapeutic in response to a molecular trigger in vivo. Nat Biotechnol 36(3):258
Liu Y et al (2018a) Engineering multifunctional RNAi nanomedicine to concurrently target cancer hallmarks for combinatorial therapy. Angew Chem Int Ed 57(6):1510–1513
Liu D et al (2018b) Current developments and applications of microfluidic technology toward clinical translation of nanomedicines. Adv Drug Deliv Rev 128:54–83
Luong JHT, Bouvrette P, Male KB (1997) Developments and applications of biosensors in food analysis. Trends Biotechnol 15(9):369–377
Malhotra BD, Chaubey A (2003) Biosensors for clinical diagnostics industry. Sensors Actuators B Chem 91(1–3):117–127
Malhotra S, Verma A, Tyagi N, Kumar V (2017) Biosensors: Principle, types and applications. IJARIIE-ISSN(O) 3(2):2395–4396
Marcazzan S et al (2018) Nanomedicine, an emerging therapeutic strategy for oral cancer therapy. Oral Oncol 76:1–7
Medintz IL et al (2003) Self-assembled nanoscale biosensors based on quantum dot FRET donors. Nat Mater 2(9):630
Mizrahy S et al (2019) Targeting central nervous system pathologies with nanomedicines. J Drug Target 27(5–6):542–554
Mohanty SP, Kougianos E (2006) Biosensors: a tutorial review. Ieee Potentials 25(2):35–40
Monošík R, Streďanský M, Šturdík E (2012) Biosensors- classification, characterization and new trends. Acta Chimica Slovaca 5(1):109–120
Nicolas S et al (2018) Polymeric nanocapsules as drug carriers for sustained anticancer activity of calcitriol in breast cancer cells. Int J Pharm 550(1–2):170–179
Pathak P, Katiyar VK, Giri S (2007) Cancer research-nanoparticles, nanobiosensors and their use in cancer research. J Nanotechnol Online:14
Perumal V, Hashim U (2014a) Advances in biosensors: principle, architecture and applications. J Appl Biomed 12(1):1–15
Perumal V, Hashim U (2014b) Advances in biosensors: principle, architecture and applications. J Appl Biomed 12(1):1–15
Pohanka M, Skládal P (2008) Electrochemical biosensors–principles and applications. J Appl Biomed (De Gruyter Open) 6:2
Prasad S (2014) Nanobiosensors: the future for diagnosis of disease? Configurations 8:9
Raiteri R et al (2001) Micromechanical cantilever-based biosensors. Sensors Actuators B Chem 79(2–3):115–126
Ramanathan K, Danielsson B (2001) Principles and applications of thermal biosensors. Biosens Bioelectron 16(6):417–423
Ramanathan K, Jönsson BR, Danielsson B (2001) Sol–gel based thermal biosensor for glucose. Anal Chim Acta 427(1):1–10
Ravalli A, Marrazza G (2015) Gold and magnetic nanoparticles-based electrochemical biosensors for cancer biomarker determination. J Nanosci Nanotechnol 15(5):3307–3319
Rodriguez-Nogales C et al (2018) Nanomedicines for pediatric cancers. ACS Nano 12(8):7482–7496
Ronkainen NJ, Halsall HB, Heineman WR (2010) Electrochemical biosensors. Chem Soc Rev 39(5):1747–1763
Sagadevan S, Periasamy M (2014) Recent trends in nanobiosensors and their applications-a review. Rev Adv Mater Sci 36:62–69

Sahay G, Alakhova DY, Kabanov AV (2010) Endocytosis of nanomedicines. J Control Release 145(3):182–195
Saneja A et al (2018) CD44 targeted PLGA nanomedicines for cancer chemotherapy. Eur J Pharm Sci 121:47–58
Scott AO (1998) Biosensors for food analysis. Elsevier, Amsterdam
Sen S et al (2018) Computational studies of micellar and nanoparticle nanomedicines. Chem Soc Rev 47(11):3849–3860
Srinivasan B, Tung S (2015) Development and applications of portable biosensors. J Lab Autom 20(4):365–389
Tehrani Z et al (2014) Generic epitaxial graphene biosensors for ultrasensitive detection of cancer risk biomarker. 2D Materials 1(2):025004
Terry LA, White SF, Tigwell LJ (2005) The application of biosensors to fresh produce and the wider food industry. J Agric Food Chem 53(5):1309–1316
Tong R, Cheng J (2007) Anticancer polymeric nanomedicines. J Macromol Sci Polym Rev 47(3):345–381
Touhami A (2014) Biosensors and nanobiosensors: design and applications. Nanomedicine 15:374–403
Turner A, Karube I, Wilson GS (1987) Biosensors: fundamentals and applications. Oxford University Press, Oxford
Vankayala R, Hwang KC (2018) Near-infrared-light-Activatable nanomaterial-mediated Phototheranostic Nanomedicines: an emerging paradigm for Cancer treatment. Adv Mater 30(23):1706320
Vo-Dinh T (2004) Biosensors, nanosensors and biochips: frontiers in environmental and medical diagnostics. In: Proceedings of the first international symposium on micro and nano technology, Hawaii. Vol. 2004
Vo-Dinh T, Cullum B (2000) Biosensors and biochips: advances in biological and medical diagnostics. Fresenius J Anal Chem 366(6–7):540–551
Vo-Dinh T, Cullum BM, Stokes DL (2001) Nanosensors and biochips: frontiers in biomolecular diagnostics. Sensors Actuators B Chem 74(1–3):2–11
Wang L, Xiong Q, Xiao F, Duan H (2017) 2D nanomaterials based electrochemical biosensors for cancer diagnosis. Biosens Bioelectron 89:136–151
Yannas IV et al (1989) Synthesis and characterization of a model extracellular matrix that induces partial regeneration of adult mammalian skin. Proc Natl Acad Sci 86(3):933–937
Zhang C-Y et al (2005) Single-quantum-dot-based DNA nanosensor. Nat Mater 4(11):826
Zhu S et al (2008) A localized surface plasmon resonance nanosensor based on rhombic Ag nanoparticle array. Sensors Actuators B Chem 134(1):193–198
Zhu Y, Feijen J, Zhong Z (2018) Dual-targeted nanomedicines for enhanced tumor treatment. Nano Today 18:65–85
Ziegler C (2004) Cantilever-based biosensors. Anal Bioanal Chem 379(7–8):946–959

Nanobiosensor: Current Trends and Applications

16

Nitai Debnath and Sumistha Das

Abstract

Biosensor is an analytical device which incorporates a biologically active element with an appropriate physical transducer to generate a measurable signal proportional to the concentration of analyte in any type of sample. Integration of nanomaterials in these biosensors can make them much portable, more sensitive and more economical. Moreover, as the sensitivity of detection in nanomaterial-based biosensors is enhanced, probe material is required in extremely low concentration, and by tuning the surface property of the nanomaterials, these sensors can easily be employed for selective and specialized detection. An efficient detection system can help in management of many a problem of our modern-day life. A nanobiosensor can be used for early diagnosis of deadly diseases like cancer, and early detection of cancer can drastically reduce cancer-related mortality. It also can detect presence of microbial pathogens in our body, food, drinking water, and the atmosphere. Nanobiosensor-based precision farming has the promise to reduce the excess usage of toxic agrochemicals. These sensors can be used for smart food packaging which can instantly make the consumer aware of any spoiled food. This detection system is much suitable for effectively detecting environmental pollutants. In this chapter, several nanomaterial-based fabrication techniques and their applications in different facets of lives are discussed.

Keywords

Surface-enhanced Raman scattering · Gold nanoparticle · Carbon nanotube · Nanobiochip · Biobarcode

N. Debnath · S. Das (✉)
Amity Institute of Biotechnology, Amity University Haryana, Gurgaon, India

S. K. Saxena, S. M. P. Khurana (eds.), *NanoBioMedicine*,
https://doi.org/10.1007/978-981-32-9898-9_16

16.1 Introduction

Biosensors are devices used to detect the presence or concentration of a biological analyte, which may be a biomolecule, a biological structure, or a microorganism (Wilson and Gifford 2005). Biosensors consist of three parts: a component that recognizes the analyte and produces a signal, a signal transducer, and a reader device. Biosensors have a wide range of applications starting from food quality monitoring, medical diagnosis to estimation of environmental pollution. The first biosensor was developed in the early 1960s by Led and Clark (Patel et al. 2016). This biosensor could only detect presence of oxygen. Biosensor technology has witnessed a paradigm shift from the days of Led and Clark. Biosensors are now more complex, sensitive and compact. Sensitivity and performance of biosensors can be much more enhanced (even to single molecule detection) by incorporating nanomaterials within these bioanalytical devices. This type of biosensor where nanomaterials constitute integral part of it is called as nanobiosensor (Fig. 16.1).

Nanomaterials show completely novel properties which cannot be explained by classical Newtonian mechanics. At nanodomain, most of the electrons become surface electrons and electromagnetic property become very dominant, while gravitational force become really negligible. Nanoscale materials have huge surface area to volume ratio, and they become very reactive. These materials show new optical, electrical and chemical properties in comparison with their bulk counterpart (Debnath et al. 2011). These novel properties of nanomaterials are exploited for building nanobiosensors. Different nanomaterials like gold, silver, graphene, carbon nano tube, silica etc. are being utilized in biosensors for faster detection and higher reproducibility.

Biosensors or nanobiosensors can be classified on the type of signal transduction they employ. They are grossly classified as (i) optical, (ii) electrochemical and (iii) piezoelectric biosensor. Optical biosensor detects change in optical property,

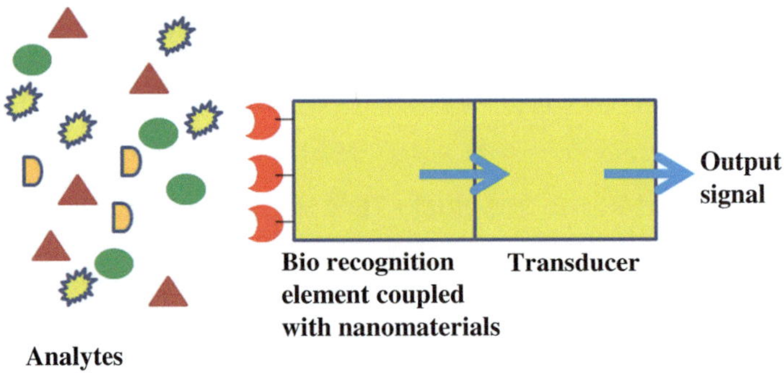

Fig. 16.1 Working principle of a nanobiosensor. Here, analytes (biological or non-biological) combine with a physicochemical detector in which nanomaterials and biological component constitute the integral components. The transducer or the detector element transforms the signal resulting from the interaction of the analyte with the biological element into another signal (i.e. transduces) that can be more easily measured and quantified

whereas electrochemical biosensor is equipped for detection of changes in electrical properties. Piezoelectric biosensor can detect minute change in mass.

(i) *Optical nanobiosensors*

Optical nanobiosensors are ultra small devices with unique optoelectronic properties to identify materials through different optical measurement such as absorption, refraction, dispersion, fluorescence, phosphorescence, Raman and interference spectroscopy etc. As the principle of this type of nanobiosensor is nonelectrical in nature, it has the advantage of in vivo detection over other types of nanobiosensors. Based on application optical nanobiosensors can further be classified into fluorescence, surface plasmon resonance (SPR), and optical fiber-based biosensors.

Fluorescence biosensing can directly detect a molecule before and after the excitation. Another type of fluorescence biosensor relies on the signal transmitted by a fluorophore (like green fluorescent protein or GFP) when a target molecule is present. A third type of nanobiosensor can detect fluorescence energy transfer (FRET) signal. In SPR nanobiosensors, the adsorption of targeted analyte by surface ligand is detected by measuring the change in condition of the resonance coupling of incident light. Fiber-optic-based biosensor is the most commonly used technique which has opened a novel area of research known as near-field optical scanning microscopy (NSOM). With respect to conventional larger fiber optical sensors, NSOMs are designed with one end tapered fibers in the range of 20–100 nm. Here the basic principle of excitation is also different where level of photon escape is minimized and hence sensitivity is much enhanced. Similarly, infrared spectroscopy-based optical nanobiosensors are also gaining much interest because of their unique vibrational fingerprint through which biomolecules can be identified unambiguously. To amplify the signal further, surface-enhanced infrared spectroscopy (SEIRS) was introduced around IR nanoantennas which exhibit better signal and referred as hot spots. Ceramic phosphorus-based optical nanosensors are even more advantageous because of their unique luminescence emission under incident light. Conversion of infrared response to visible range makes this phosphorous-based optical sensor more advantageous and more robust for sensing applications.

(ii) *Electrochemical nanobiosensors*

Enzyme-based biosensors are very common in the field of disease diagnostics and therapy. However, enzymes have their self-limitations like heat sensitivity, effect of natural inhibitor, slow reactions kinetics under different microenvironmental conditions, internal protein susceptibility, and effect of different denaturing agents. To overcome all these limitations, introduction of nanoelectrode-based sensors is gaining immense interest because of their contribution in increasing the stability and sensitivity of conventional enzyme-based sensors. Majorly because of chemical inertness and extreme biocompatibility, carbon and noble metal-based nanomaterials are of popular choice for construction of electrochemical nanosensors. Other materials such as copper, nickel, cadmium, silicon and metal oxides are of preference as nanoelectrodes.

Electron beam lithography is a commonly used method to fabricate such electrochemical nanobiosensors which ensures very minimal analysis time. Moreover, highly sensitive nanoelectrodes gives quick signal change in every minute applied potential. Laser pulling technology is another emerging science in the field where further improvisation in terms of miniaturization of disk-shaped nanoelectrodes has potential to amplify signal from neurotransmitters, thus enabling lesser rate of cerebral damage to drug delivery to ultra small target areas.

(iii) *Piezoelectric nanobiosensors*

As already discussed piezoelectric nanobiosensors are based on detection of a change of mass. The quartz crystal microbalance (QCM) is the most common piezoelectric biosensor used since 1959. The main components of a traditional QCM are an oscillator circuit containing quartz disc with circular electrodes on both sides of the quartz. A mechanical oscillation is generated due to the alternating voltage between these two electrodes. Any addition or adsorption of mass by the surface of the piezoelectric crystal brings changes in the frequency of oscillation which is again proportional to the amount of mass. When nanoparticles (NPs) are integrated in the QCM, it becomes sensitive to such a level that it can even detect nucleotide hybridization.

This chapter summarizes fabrication and application of nanostructure-based biosensors in all facets of biological domain.

16.2 Signal Amplification for Nanobiosensing

The major objective of developing nanobiosensor is to detect biomolecules with high sensitivity. Traditional physical and bioanalytical methods cannot be used to achieve this sensitivity because of their inherent background noise. Moreover, use of nanomaterials will be helpful to miniaturize the assay setup. The better efficiency of the nanobiosensor is achieved by enhancing the signal amplification in various ways. (i) Electron transfer rate can be enhanced by exploiting the optoelectronic properties of nanomaterials which make the optical signal more sensitive; (ii) optoelectrical and visual signals can be amplified by applying the catalytic and enzyme mimetic functions of nanomaterials for detecting biomolecules; (iii) nanomaterials like carbon quantum dot, CdSe nanoparticle which have excellent fluorescent property, can be used as label in fluorescence or chemiluminescence-based assays (Ding et al. 2008); (iv) otherwise nanomaterials can be used as carrier of signaling molecules which may be fluorescent or are optically active. Sometimes enzymes also can be tagged with nanomaterials as enzymes can produce molecules with redox or optical activity while catalyzing the reaction; (v) biofunctionalization of nanomaterials with covalent or noncovalent interactions has enormous potential for signal amplification in a nanobiosensor.

16.3 Nanobiosensors for Biomedical Applications

Probably the widest range of nanobiosensors is in the field of medical diagnostics. These ultra small, smart devices can enter very small organs, tissues and cells wherein through their unique properties can identify unique pathophysiological changes. These sensors are robust because of their stability and target specificity.

16.3.1 Identification of Bacterial Pathogens

Biofilm associated infections are very challenging because of their heterogeneous nature. A number of techniques are employed for identification of such deadly infections. Gold nanoparticle (GNP)-based multichannel optical nanosensor was developed by Li et al. (2014) which could simultaneously identify six bacterial biofilms including two uropathogenic bacteria. A fluorescent dye loaded micelle which was moreover equipped with bacterial cell wall biomarker was able to detect very minute number of bacteria. It was so sensitive that it could identify presence of 15 *Escherichia coli* cells in a 1 ml sample (Mouffouk et al. 2011). SPR and localized surface plasmon resonance (LSPR), a unique phenomenon shown by metal nanostructures, are often utilized in optical nanobiosensors for detecting a number of pathogenic bacteria. As for example, Charlermroj et al. (2013) reported immune detection of bacterial cell with the help of a modified optical biosensor. Similarly, pathogenic microbes could be identified by gold nanorod coupled optical sensor (Wang and Irudayaraj 2008). Not only SPR, but also SERS (surface enhanced Raman scattering)-based sensors can be developed from GNP and can be used for detection of infection in blood sample (Cheng et al. 2013).

Mechanical nanosensors based on QCM or cantilever technology are also quite popular for detection of whole cell bacteria like *E. coli* (Jiang et al. 2011), *Salmonella enterica* serovar Typhimurium (Salam et al. 2013), *Campylobacter jejuni*, *Bacillus anthracis* (Hao et al. 2009), etc. This is an emerging point of care technology which promises ultrasensitivity and reduced reaction time and miniaturization of overall process. Label-free potentiometric detection of *Staphylococcus aureus* was reported using electrochemical sensors of carbon nanotube aptamer (Zelada-Guillen et al. 2012). Similarly, graphene-based nanoimmunosensors were fabricated for non-labeled detection of bacteria (Wan et al. 2011). Nanoporous membrane based electrochemical immunosensors were also developed for rapid and ultrasensitive detection of *E. coli* (Chan et al. 2013). Electromagnetic nanosensor for detection of *S. aureus* was reported by Ávila et al. (2012).

Although nanosensor-based bacterial identification is paving newer ways day by day, the technology still suffers due to a number of limitations:

(a) Complex diversity of sample in nature
(b) Miniaturization is not easy all the time
(c) Reliability on the existing methodology
(d) False-positive results

(e) Very small sample size
(f) Lack of commercialization and investment

16.3.2 Identification of Viruses

Human immunodeficiency virus (HIV) is one of the mostly pathogenic viruses. Rate of patient survival in HIV infection is almost negligible. Detection of infection is very important in terms of disease prognosis as it is a multi symptomatic disease. In many of the cases, late recognition of the causing biomolecule/microorganism is actually the reason of late or no response to treatments. In this scenario, nanosensors are of immense interest for rapid and ultrafine detection. As, for example, streptavidin-coated NP-based biobarcode was developed for quick identification of HIV surface (capsid) antigen p24 through monoclonal antibody platform. Sensitivity of this technology reaches to as minimum as 1 pg/ml sample. In comparison to conventional ELISA-based detection, use of this antibody tagged nanobiosensor is more promising (Abraham et al. 2008; Tang et al. 2007).

Hepatitis is a disease that causes abnormal liver inflammation majorly driven by five types of hepatitis viruses (A, B, C, D, and E). All these viruses are responsible for liver disorder, but their mechanism of action varies. This infection not only leads to liver cirrhosis and carcinoma but also leads to chronic ailment to the patient with multiple disorders. Using Rayleigh light scattering (RLS) spectroscopy, gold/silver nano-oligo conjugates were used to identify HBV (hepatitis B virus) and HCV (hepatitis C virus) infection selectively with nano-amplification principle (Wang et al. 2003). GNP-labeled staphylococcal protein A (SPA)-based protein chip assay was found to be further effective where 3 ng/ml detection limit was achieved (Duan et al. 2005). Almost similar strategies were also developed for detection of HEV (hepatitis E virus) and HAV (hepatitis A virus) (Wan et al. 2005; Liu et al. 2006).

Herpes simplex virus (HSV) is responsible for cold sores or genital sores where the virus can enter into the epidermis level of skin and results in severe sores. To combat this recurrent and profound infection many antiviral strategies were developed. Nanosensor-based immune assays with antibody (against viral surface antigen) tagged gold-oligo nanospheres showed a much promising result (Jain 2005). Viral respiratory diseases are illnesses caused by a number of viruses that share common trait and generally infect upper respiratory tract. The symptoms of such infections become manifest within 1–10 days post exposure. Such conditions are spread rapidly through direct or indirect inhalation, infected cough droplets as well as contaminated handkerchiefs and toys. Influenza virus, adenoviruses and respiratory syncytial viruses are common causes of respiratory infections. Monoclonal antibody (anti-hexon) tagged highly fluorescent europium (III)-chelate-doped NP was reported for identification of adenovirus, and it had the sensitivity of detection at the level of 5000 virus particles/ml. Highly stable carbon quantum dots (CQDs) are extremely valued sensor material for their unique optoelectronic properties that encompass broad excitation spectra and narrow emission spectra. CQDs are now extensively utilized for identification of various viral infections (Chan et al. 2007).

16.3.3 Detection of Cancer

Cancer is a type of disease which can be caused by innumerable possibilities, and its symptoms are also diversified in nature. Over the years several diagnostic, drug tracking and therapeutic strategies are employed in the field of cancer management to combat this deadly condition. Yet cancer remains to be one of the major challenges in medical science. Immune detection techniques like radioimmunoassay (RIA), enzyme-linked immunosorbent assay (ELISA) etc., tissue histology analysis, cancer tissue imaging, cell signaling pathways tracking as well as tumor positive cell biomarker identifications are long being employed for cancer detection. Still much of the cancer patients die due to lack of robust cancer diagnosis method at very early stage of the ailment. So research articles related to nanomechanical sensor devices are getting much focus for clinical trial and marketability (Elisa et al. 2015).

For cancer detection nanomaterials are specifically designed for amalgamation with tumor specific markers. Tumor cell vesicular shreds or mobile tumor cells can now not only be rapidly detected with the help of nanobiosensors but also dynamic tracking of the anticancer drug is possible (Guosoong et al. 2011; Salvati et al. 2015). This novel biosensing system obviously has promise toward increasing survival rate of the cancer patients. GNPs, owing to their easy synthesis process, biocompatibility, less cytotoxicity, and resistance to photo bleaching (human cell experiments), are being exploited mostly in colorimetric nanobiosensors for cancer detection (Medly et al. 2008; Chen et al. 2009). GNPs have the added advantage of being Raman active for which they are widely used to detect very common cancer biomarker like epidermal growth factor receptor (EFGR), etc. (Qian et al. 2008). Another very useful nanotool in cancer detection is carbon-based nanostructures. Due to very unique optoelectronic properties, CQDs are highly popular contrast enhancers for imaging studies. As, for example, prostate-specific membrane antigen (PSMA) recognizing aptamer bioconjugate attached QDs are very popular cancer detecting sensors (Bagalkot et al. 2007). QD-based nanochips are ultrasensitive and capable of very fast cancer detection. These nanobiosensors are being utilized for detection of multiple cancer biomarkers (Jokerst et al. 2008).

Fluorescence resonance energy transfer (FRET)-associated aptamer-conjugated silica NPs are reported to detect cancers like T-cell leukemia and B-cell lymphoma with beautiful fluorescent emission spectra on single wavelength excitation (Chen et al. 2009). Apart from these, different nanopore-based sensors are also quite useful for detecting several cancer biomarkers for their useful features like synthetic protein-based material, easy fabrication, and ability to probe a wide array of samples like drug molecules, nucleic acids, proteins, etc. (Kang et al. 2006; Clarke et al. 2009; Kasianowicz et al. 2001).

Magnetic nanobiosensors, which principally work by measuring magnetism, can be used for detection of several pathophysiological conditions. Superparamagnetic nanomaterial-based sensors use the principle of magnetic relaxation switches (MRSws) which are now clinically approved MR (magnetic resonance) contrast agents. In contrast to optical sensors, these MRSws use radiofrequency radiation, readily penetrable to the biological sample of very inert nature. The characteristic inhomogeneity of such materials results in contrast enhancement in a given magnetic field and ultimately leads to ultrasensitive nanobiosensing system.

16.3.4 Nanosensor-Based Breath Analyzers

Breath analysis is a recent advancement of disease detection through identification of organic volatile substances from exhaled breath samples. Nanosensor-based breath analyzers are already in clinical trial stage. Spectroscopy-based volatile compound detection method has shown much potential to detect multiple abnormal situations ranging from carcinomas to tuberculosis, to Parkinson's disease and even to diabetes mellitus (Konvalina and Haick 2014). Kahn et al. (2015) developed GNP-based dynamic detection array where unique spatial organization results in sensing of volatile organic compounds (VOCs) from breath samples of ovarian carcinoma patients with accuracy of nearly 82%. Similarly a micro-nano-chip array sensor has been developed to detect and monitor regular glucose level in Type 1 diabetes mellitus patients (Mwakikunga 2018). Oxide nanomaterials-based gas sensors like Si-doped WO_3 (silicon doped tungsten oxide) gas sensors equipped with proton-transfer-reaction time of flight mass spectrometry (PTR-TOF-MS) are capable of monitoring acetone level present in healthy human breathing both online and offline (Righettoni et al. 2013). A good number of nano-nose or breath analyzers are developed for identification of VOCs, specifically for lung cancer detection (Thriumani et al. 2014) . As, for example, Cyranose 320 consisting of 32 carbon polymer sensors are capable of differentiating breath samples of a healthy individual and lung cancer affected patient (Ulanowska et al. 2008). A chemiresistor associated electronic nose composed of GNPs can accurately detect lung cancer with sensitivity to 88% (Folinsky et al. 2005. http://web.mit.edu/~kilroi/Public/text/ifprop.pdf).

16.4 Nanobiochip: A New Paradigm in Lab-on-a-Chip Technology

Miniaturization of hardware components is of paramount importance as all the modern-age gadgets are becoming smaller day by day. Moreover, smaller size ensures less energy utilization, provides portability and is economical in the long run. Nanochip is basically an integrated circuit where multiple molecules are mechanically assembled one at a time and nanobiochips are fabricated by the amalgamation of biology in existing chip technology. Several interactions can be performed in this single platform, and hence this type of array-based chips reduce wastage of energy and time. Introduction of NPs in biochips has made these sensors more sensitive and advanced. According to R. Bashir of Laboratory of Integrated Biomedical Micro/Nanotechnology and Applications (LIBNA), Purdue University, USA, biochips can be defined as "microelectronic-inspired devices that are used for delivery, processing, analysis, or detection of biological molecules and species."

Nanobiochips have a huge prospect in detection, therapy, analysis, designing and modeling biological networks, probing of multiple cell signaling pathways, identifying chromosomal aberration, neuronal signal amplification and analysis, immunotherapy, immune detection and editing, etc. More specifically nanobiochips are designed for detection of bio micro/nano molecules like nucleic acids (DNA/RNA), proteins, lipids and microorganisms (bacteria, virus and parasites etc.).

16.4.1 Salient Features of Nanobiochip

An ideal biochip should have minimum dimension, small volume, and high surface area to volume ratio. This is why nanomaterials, which are gifted with specialized features, are the ideal building blocks for nanobiochip construction. The following are few more reasons for which NPs are used in biochip fabrication:

- Requirement of probe material in extremely low concentration because of high surface area of nanomaterials
- Highly sensitive nanostructures can be engineered based on requirement
- Even tuning not only in size but also in surface properties can be employed for selective and specialized detection
- Small size enhances the portability and miniaturization of the entire system
- Reduction in effective reaction volume
- Multiple detection
- Because of wide array of existing natural and engineered sources of nanomaterials, the technology can be tuned for both in vitro and in vivo applications
- Opto-electronic properties of nanomaterials are an added feature for imaging-based bioassays

16.4.2 Silicon Nanobiochip

Silicon-based nanostructures are most commonly used for nanobiochip development. Traditional microarray platforms which suffer from poor accessibility of the target and low loading capacity can be overcome by creating a pseudo three-dimensional structure with rough polycrystalline silicon covered with a SiO_2 layer. This surface is capable of multiple functionalizations which ultimately help in covalent binding with a number of bio-molecules. Porous silicon biochips functionalized with photochemical passivation seem to have maximum specificity for DNA detection by means of optical reflectivity measurements (Stefano et al. 2006). Apart from this silicon-based single-stranded DNA detection, optical biochips are also very popular for their wide biomedical applications (Rendina et al. 2014). Programmable nanochips can be used to relieve the painful cancer diagnostic processes like tissue biopsy. The McDevitt Lab of Rice University has developed a Programmable Bio-Nano-Chip (PBNC), based on a small silicon chip attached with fluorophore-tagged antibodies to detect heart disease or cancer with just a small saliva sample. This particular device is at phase III clinical trial level and waiting for FDA approval.

16.4.3 Carbon Nanotube (CNT)-Based Biochips

Cylindrical hollow nanostructures of carbon allotropes with length to diameter ratio of 132,000,000:1, assembled in tubular fashion, where the walls are fabricated with one atom thin carbon sheets, are known as CNTs. Similar to other nanostructures, CNTs also have some unique features which make them very useful for a number of applications:

- Fabrication of light-weight spacecrafts
- Easy penetration of biological membranes due to ultra small size
- Development of solar cells
- Development of single component memory cell (Memristor)
- Creation of bicycle components
- Modeling of light-weight boat
- Development of high speed and capacity memory in exchange of lesser energy investment
- Fabrication of precise atomic level CNT-based circuits

CNT has also become very popular as biochip component. Specifically its ultra small size and emission of less heat in any circuit makes it one of the most suitable components for nanobiochip fabrication. As soon as any biomolecule gets attached to CNT, its electrical resistance changes dynamically. This property makes it ideal as bioanalytical material. Similarly, its unique optical features give added level of advantage in nanobiochips. This is why nowadays silicon-based chips are getting replaced with that of CNT. CNTs are also becoming popular for making nano-wires and nano-cables. CNT is being used to make a highgain single Lab-on-a-Chip (LOC) microarray unit which has high sensitivity and multiplex capability of rapidly analyzing biochemical samples like nucleic acids (DNA/RNA) without requirement of any further amplification (Sorgenfrei et al. 2011). This patented technology has already created a boom into nanochip industry as it provides multiple advantages over conventional systems (qPCR etc.) like:

- Better yield and sensitivity
- Reduced array timing
- Cost-efficiency
- More potable than qPCR units
- Real-time monitoring makes the technology a robust bioanalytical approach

Though nanoarray/nanochip systems have a huge potential to replace bulky instrumental arrangements and laborious and time-consuming sample amplification methods, the technology is still at its preliminary lab-application level and needs further characterization in terms of safety, stability and false-positive signaling problems before their full-fledged industrial scale application.

16.4.4 Nanolithography-Based Biochip Construction

Nanolithography is another emerging technology in chip development which provides accurate protein immobilization and array development (Bearinger et al. 2009). Similarly Dip-pen nanolithography (DPN)-based technology is being utilized in nanolithography for multiple chip fabrication (Salaita et al. 2007). Investigation of focal growth became much easier with nanochip-based protein array system fabricated with non contact-based polymer nanopillars (Kuo et al. 2011).

16.4.5 Nanoskiving

It is another novel approach for nanochip development. This is actually amalgamation of thin-film deposition of metal on a topographically scaffold substrate with sectioning using an ultramicrotome (Xu et al. 2008). Such arrays embedded in epoxy matrix provide unique application majorly in two systems, optics and electronics after removal of polymer matrix by plasma oxidation.

16.4.6 Nanofluidics-Based Biochips

Application of nanofluidic technology to devise silicon nanowire-based biochip, where the nanowire readily oxidizes to nanotubes, is able to detect DNA promptly. This nanobiochip paved a new detection scheme for cancer (Fan et al. 2005). LSPR-dependent ultrasensitive lab-on-a-chip platform based on metallic NPs was developed by a group of eminent researchers led by the Institute of Photonic Sciences (ICFO) in Castelldefels, which can detect cancer biomarkers in blood even at a very early stage within minutes. This nanobiochip is an ultrasensitive, state-of-the-art, powerful sensor (Aćimović et al. 2014) for cancer detection.

16.4.7 Nanonose for Detection of Human Ailment

Application of biochips for diagnostics can not only reduce the time of detection but also lessen dependence on heavy costly instruments. For example, scientists from the University of Massachusetts have developed a biochip that can mimic the functional mechanism of human nose referred to as "nanonose" to detect signs of illness of a body. The "nanonose" is composed of gold nanochore-fluorescent polymer conjugate with six receptors that can detect proteins at nano-molar concentration the way human nose detects olfactory molecules (You et al. 2007).

16.5 Nanobiosensors for Pollution Detection

Global urbanization and rapid deforestation has already made environment alarmingly unhealthy in terms of pollution load. In this scenario, environmentalists around the world is taking a number of approaches to identify the level of different air, water and soil pollutants. Nowadays several nanosensors have become popular for detecting environmental pollution. These nanosensors are rapid, ultrasensitive, portable and cheaper in comparison with traditional pollution monitoring devices.

16.5.1 Nanosensors for Air Pollution Detection

Nanomaterial-based electrochemical and gas sensors are gaining much interest because of their easy design, portable smart network, and cost-effective nature (Ampelli et al. 2015). Carbon-based nanomaterials are very promising due to their physical, optical, and resonating frequency capacities that can sense very minute amount of pollutants like NH_3 (Manzetti et al. 2015). Metal oxide nanomaterial like ZnO nanorods (thermolysis-assisted chemical solution) was reported to be instrumental in polyimide sensing (Ahn et al. 2010). Metal oxide NPs are considered as better pollution indicator than organic compounds because of their chemical, thermal, and mechanical stability. Similarly researchers have also developed low-cost black carbon sensor for detection of air quality with some additional advantages like weatherproof enclosure, solar-powered rechargeable battery, and cellular communication to enable long-term, remote operation (Julien et al 2018). Optically active palladium NP-based VOC detectors were designed using a single-mode tapered optical fiber by the flame-brushing technique. This sensor shows both enhanced sensing ability and quicker response time (González-Sierra et al. 2017). Another smart gas sensor was developed where nanomaterial-based sensors were amalgamated with GIS modeling. This gas sensor was incorporated in a Personal Digital Assistant (PDA) which was again linked with bluetooth communication tools and Global Positioning System (GPS). This smart nanosensor is capable of robust circulation of information on pollution levels at multiple geographical positions (Pummakarnchanaa et al. 2005). Another nanosensor composed of LiCl doped TiO_2 electrospun nano fiber was developed by researchers which can monitor air humidity very efficiently. This nanosensor is ultra responsive, requires minimum recovery time, and has good reproducibility, linearity, and environmental stability.

16.5.2 Nanomaterial-Based Sensors for Water Pollutant Detection

According to WHO, globally there are more than 790 million who are deprived of potable water resources. This extensive water scarcity is majorly due to lack of infrastructures required for detection of water contaminants which is required to differentiate potable and non-potable water. Moreover, drinking water should be under continuous monitoring and maintenance. Application of nanotechnology in detection and waste water management is creating high hope. As nanomaterials have high surface area to volume ratio, they are extremely good adsorbents. They also show photoluminescence, chemiluminescence, and increased oxidative-reductive reaction, which make them very much useful for devising nanobiosensors suitable for detection of water contaminants.

Optical sensor based on GNP can detect Hg (II) and organophosphate pesticides in aqueous solution very efficiently through fluorescence-based emission spectra. Using chemical vapor deposition method, porous CNT membranes can be fabricated for making an electrochemical sensor to detect waterborne microbial pathogens like *E. coli* (Cheng et al. 2008). Similarly silver-gold nano integrated electrochemical sensors are also very popular for their potential to detect aqueous nitrate level (Jingfang et al. 2012; Wang and Yu 2013). Based on ion beam chemical vapor deposition (FIB-CVD) method, a new generation of nano thermal sensors was also developed where calibration of nanosensor was carried out in water using hot stage of atomic force microscopy (AFM) cantilever (ElShimy et al. 2007). Turbidity of water is a very important parameter for water quality assessment. Nanomaterial-based nephelometric turbidity sensors are also reported (Lambrou et al. 2009). Another very sensitive nano-molybdenum-based optical sensor can detect aqueous phosphate level very effectively with a photo diode detector using 690 nm optical filters (AZTEC®).

16.5.3 Nanobiosensors for Detection of Soil Pollutants

Ever growing population and excessive need of food resources are directly correlated with overuse of chemical and natural fertilizers, pesticides, herbicides and industrial waste as soil seepage. Different pollutants from other sources are also making the earth contaminated with several toxic residues every day. Nanotechnology has already contributed a lot in soil pollution detection and management. From determination of microbial presence in soil to detection of different soil and food contaminants and pollutants are possible by nanobiosensors (Joyner and Kumar 2015). Similarly, identification of the pesticide level becomes easier based on nanosensors coated with markers that can attach selectively to the pollutant (Rathbun 2013). Apart from very basic applications of nanobiosensors in monitoring various soil parameters like moisture, soil pH, chemical pollutants (pesticides, herbicides, fertilizers, insecticides), these sensors can also be fabricated for detection of crop pathogens (García et al. 2010). Nanobiosensors are also fabricated in such smart way that it can also detect climate change in prior. CNT-based sensors can act as wireless nanoantenna having ability to sense and signal weather parameters (Garcia-Martinez 2016). Apart from this microcantilever-based sensors are also useful to determine total carbon level of soil (Plata et al. 2008). There are reports on array-based multisensory which can act as an electronic tongue and decipher food and soil contaminants (Ciosek and Wróblewski 2007).

16.5.4 Nanobiosensor in Agriculture Sector

The current agricultural practices need input from the modern technologies as human population is ever growing while there is continuous shrinkage of cultivable land. All agricultural scientists are now concerned with sustainable agriculture which aims at increasing agricultural produce without irreparably damaging the natural resources. Nanotechnology has enormous promise to make the agricultural system sustainable by minimizing application of agrochemicals and lowering nutrient losses. Moreover, with a network of nanosensors, the precise time and place for intervention with crop protection strategies can be applied. This concept is called as precision farming.

16.5.4.1 Nanobiosensor in Precision Farming

The agricultural output can be maximized while keeping the input cost very low if environmental variables are closely monitored, and the insecticide, herbicide and fertilizer are sprayed to targeted site only. Precision farming comprises sensors, computers, GPS (global satellite positioning system) and remote sensing which are used to measure environmental condition at micro scale and to determine whether any intervention is required to protect the crop from pests or phytopathogens. Here nanobiosensors play a crucial role in providing accurate data and thus help in minimizing unnecessary usage of agrochemicals (Cioffi et al. 2004) (Fig. 16.2).

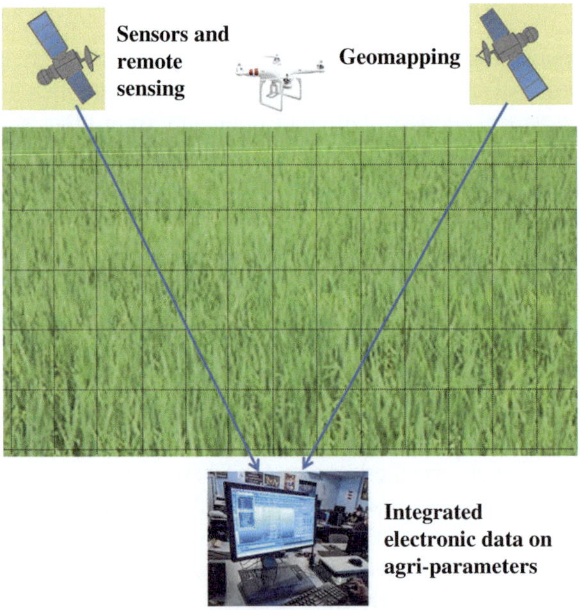

Fig. 16.2 Nanobiosensor-based precision farming has the potential to reduce the usage of natural resources and produce food in a sustainable way

Specifically nanosensors can be used for detection of plant pathogens, level of soil nutrients, and presence of plant viruses. An atomic force microscope (AFM)-based nanobiosensor can be used for detection of metsulfuronmethyl, a herbicide, as the AFM tip is functionalized with acetolactate synthase enzyme (Da Silva et al. 2013). Surface plasmon resonance (SPR) of gold nanoparticle (GNP) can be utilized for detection of organophosphorous pesticide (Fahimi-Kashani and Hormozi-Nezhad 2016). In this nanosensor, an acetylcholinesterase (AChE) layer is immobilized on GNP. In presence of suitable pesticides, AChE activity to hydrolyze acetylcholine chloride would be inhibited, and in turn lead to the change of the light attenuation due to a local increase of the refractive index. The correlation between inhibition rate and light attenuation can be utilized to determine the concentration of pesticide. A nanobiosensor made up of single-walled carbon nanotube and metal oxide NP can detect gases like ammonia, nitrogen oxides, SO_2, H_2S, volatile organics, etc., and this can be used for monitoring agricultural pollutants and also can be utilized to enhance agricultural productivity.

16.5.4.2 Nanobiosensor for Detection of Residual Agrochemicals and Pathogens in Food

Most of the food we consume is contaminated with agrochemical residue, naturally occurring toxicants like aflatoxin, microbial pathogens, etc. Detection of food contaminants is of paramount importance as it is directly correleated to human health. Nanobiosensors can be utilized to determine even trace amount of toxic chemicals in food. A multi-walled CNT-based sensor can be used for detection of organophosphate pesticides in vegetables (Yu et al. 2015). Carbofuran pesticide residue can be detected with electrochemical nanosensor made up with GNP (Sun et al. 2012; Zhu et al. 2013). An immunodipstick assay where GNP is the integral part can be utilized for detection of DDT (dichlorodiphenyltrichloroethane), one of the most common organochlorine pesticides worldwide (Lisa et al. 2009). Similarly, Zn-Se quantum dot immobilized acetylcholinesterase can detect organophosphate with the help of graphene-chitosan nanocomposite modified electrode (Dong et al. 2013). An antibody tagged GNP immunosensor can be utilized for detection of aflatoxin in the level of 10–100 ng/dL (Sharma et al. 2010). Ochratoxin, another naturally occurring toxin, can be detected by antibody modified magnetic NP or antibody functionalized rare earth (sodium yttrium fluoride) nanoimmunosensor. Heavy metal contamination in food product also presents a serious threat for human health. Multi-walled carbon nanotube or MWCNT (Bagheri et al. 2013), silver NP (Zhou et al. 2011), GNP (Zhou et al. 2014), graphene oxide (Zhu et al. 2015), mesoporous and nanoporous silica NP (Cheng et al. 2015) are now being used for detection of heavy metals in food product. Nanobiosensors also play a major role in detection of food borne pathogens. Jain et al. (2012) have reported that antibody against *Salmonella* can be tagged with CNT, and this can ultimately detect presence of *Salmonella* in food. A nanosensor assembly was developed by growing CNT on a graphite substrate. These CNTs were tagged with GNPs which were eventually added with *E. coli* specific thiolated RNA.

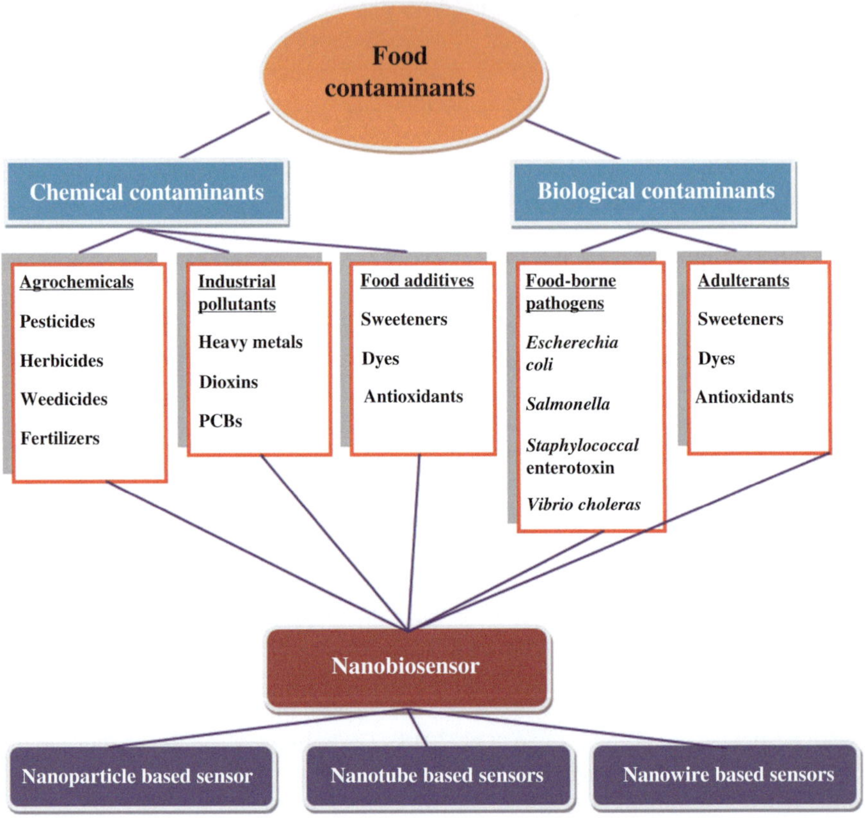

Fig. 16.3 Nanobiosensors for detection of different food contaminants. (Adapted from Kuswandi B., Futra D., Heng L.Y. (2017) Nanosensors for the detection of food contaminants in "Nanotechnology applications in food" Elsevier, pp. 307–333)

Even nanosensor-based smart packaging enables to maintain quality of food product during different phases of logistic process, and ultimately the consumer can remain ensured about the quality of the food material within the package. These nanosensors can detect not only presence of pathogenic microbes but also presence of gases, change in temperature, and humidity which are integral parameters related with freshness of food (Mills 2005). Electronic tongue which is specifically an array of nanosensors is extremely efficient for detecting foul gases released by spoiled food (Bowles and Lu 2014). Kraft is in the process of delivering personalized food to consumers by making food profile based on likes, dislikes, allergies and nutritional deficiencies of the consumers (Meetoo 2011). Application of nanobiosensors is summarized in Fig. 16.3.

16.6 Future Perspectives

Nanobiosensors have the promise to replace many traditional diagnostic devices as these are more sensitive and do not always rely on costly infrastructural setup. These ultrasensitive detection devices can recognize very minute signal coming out of any biological interaction. Modern-day sensors are gradually becoming smaller in size and highly capable of multitasking, which has practically miniaturized a whole laboratory setup in a chip. The enhanced efficiency and capability of early diagnosis of these sensors have the promise to lessen the mortality from deadly diseases like cancer. Precision farming can make use of ever depleting natural resources more judiciously. At the same time, the necessary precautionary measures should also be taken to avoid excess exposure to the nanomaterials as long-term nanomaterial-related toxicity profiling is yet to be done.

References

Abraham AM, Kannangai R, Sridharan G (2008) Nanotechnology: a new frontier in virus detection in clinical practice. Indian J Med Microbiol 26:297–301

Aćimović SS, Ortega MA, Sanz V, Berthelot J, Garcia-Cordero JL, Renger J, Maerkl SJ, Kreuzer MP, Quidant R (2014) LSPR chip for parallel, rapid, and sensitive detection of cancer markers in serum. Nano Lett 14(5):2636–2641

Ahn H, Park J-H, Kim S-B, Jee SH, Yoon YS, Kim D-J (2010) Vertically aligned ZnO nanorod sensor on flexible substrate for ethanol gas monitoring. Electrochem Solid-State Lett 13:J125–J128

Ampelli C, Leonardi SG, Bonavita A, Genovese C, Papanikolaou G, Perathoner S, Centi G, Neri G (2015) Electrochemical H_2O_2 sensors based on Au/CeO_2 nanoparticles for industrial applications. Chem Eng Trans 43:733–738

Ávila BEF, Pedrero M, Campuzano S, Escamilla-Gómez V, Pingarrón JM (2012) Sensitive and rapid amperometric magnetoimmunosensor for the determination of *Staphylococcus aureus*. Anal Bioanal Chem 403:917–925

Bagalkot V, Zhang L, Levy-Nissenbaum E, Jon S, Kantoff P, Langer R, Farokhzad O (2007) Quantum dot-aptamer conjugates for synchronous cancer imaging, therapy, and sensing of drug delivery based on bi-fluorescence resonance energy transfer. Nano Lett 7:3065–3070

Bagheri H, Afkhami A, Khoshsafar H, Rezaei M, Shizradmehr A (2013) Simultaneous electrochemical determination of heavy metals using a triphenylphosphine/MWCNTs composite carbon ionic liquid electrode. Sens Acutators B 186:451–460

Bowles M, Lu J (2014) Removing the blinders: a literature review on the potential of nanoscale technologies for the management of supply chains. Technol Forecast Soc Chang 82:190–198

Bearinger JP, Stone G, Dugan LC, E Dasher B, Stockton C, Conway JW, Kuenzler T, Hubbell JA (2009) Porphyrin-based photocatalytic nanolithography: a new fabrication tool for protein arrays. Mol Cell Proteomics 8:1823–1831

Chan CP, Tzang LC, Sin K, Ji S, Cheung K, Tam T (2007) Biofunctional organic nanocrystals for quantitative detection of pathogen deoxyribonucleic acid. Anal Chim Acta 584:7–11

Chan KY, Ye WW, Zhang Y, Xiao LD, Leung PHM, Li Y, Yang M (2013) Ultrasensitive detection of E. coli O157:H7 with biofunctional magnetic bead concentration via nanoporous membrane based electrochemical immunosensor. Biosens Bioelectron 41:532–537

Charlermroj R, Oplatowska M, Gajanandana O, Himananto O, Grant IR, Karoonuthaisiri N, Elliott CT (2013) Strategies to improve the surface plasmon resonance-based immmunodetection of bacterial cells. Microchim Acta 180:643–650

Chen Z, Luo S, Liu C, Cai Q (2009) Simple and sensitive colorimetric detection of cysteine based on ssDNA-stabilized gold nanoparticles. Anal Bioanal Chem 395:489–494

Cheng YX, Liu YY, Huang JJ, Man YZ, Zhang W, Zhang ZH, Jin LT (2008) Rapid amperometric detection of coliforms based on MWNTs/Nafion composite film modified glass carbon electrode. Talanta 75:167–171

Cheng ZH, Li G, Liu M (2015) Metal-enhanced fluorescence effect of Ag and Au nanoparticles modified with rhodamine derivative in detecting Hg^{2+}. Sen Actuators B 212:495–504

Cheng IF, Chang HC, Chen TY, Hu C, Yang FL (2013) Rapid (<5 min) identification of pathogen in human blood by electrokinetic concentration and surface-enhanced Raman spectroscopy. Sci Rep 3:2365

Cioffi N, Torsi L, Ditaranto N, Sabbatini L, Zambonin PG, Tantillo G, Ghibelli L, D'Alessio M, Bleve-Zacheo T, Traversa E (2004) Antifungal activity of polymer-based copper nanocomposite coatings. Appl Phys Lett 85:2417–2419

Ciosek P, Wróblewski W (2007) Sensor arrays for liquid sensing – electronic tongue systems. Analyst 132:963–978

Clarke J, Wu H, Jayasinghe L, Patel A, Reid S, Bayley H (2009) Continuous base identification for single-molecule nanopore DNA sequencing. Nat Nanotechnol 4:265–270

Da Silva ACN, Deda DK, da Róz AL, Prado RA, Carvalho CC, Viviani V, Leite FL (2013) Nanobiosensors based on chemically modified AFM probes: a useful tool for metsulfuron-methyl detection. Sensors (Basel) 13:1477–1489

Debnath N, Das S, Seth D, Chandra R, Bhattacharya SC, Goswami A (2011) Entomotoxic effect of silica nanoparticles against *Sitophilus oryzae* (L.). J Pest Sci 84(1):99–105

Ding L, Cheng W, Wang X, Ding S, Ju H (2008) Carbohydrate monolayer strategy for electrochemical assay of cell surface carbohydrate. J Am Chem Soc 130:7224–7225

Dong J, Zhao H, Qiao F, Liu P, Wang XD, Ai SY (2013) Quantum dot immobilized acetylcholinesterase for the determination of organophosphate pesticides using graphene-chitosan nanocomposite modified electrode. Anal Methods 5:2866–2872

Duan L, Wang Y, Li SS, Wan Z, Zhai J (2005) Rapid and simultaneous detection of human hepatitis B virus and hepatitis C virus antibodies based on a protein chip assay using nano-gold immunological amplification and silver staining method. BMC Infect Dis 5:53–58

Elisa S, Francesco S, Silke K (2015) Nanosensors for early cancer detection and for therapeutic drug monitoring. Nanomedicine (London) 10:3495–3512

ElShimy HM, Nakajima M, Arai F, Fukuda T (2007) Nano thermal sensors for sensing temperature in water environment. In: Proceedings of the 7th IEEE international conference on nanotechnology, August 2–5, 2007, Hong Kong, pp 1045–1049

Fahimi-Kashani N, Hormozi-Nezhad MR (2016) Gold-nanoparticle-based colorimetric sensor array for discrimination of organophosphate pesticides. Anal Chem 88:8099–8106

Fan R, Karnik R, Yue M, Li D, Majumdar A, Yang P (2005) DNA translocation in inorganic nanotubes. Nano Lett 5:1633–1637

García M, Forbe T, Gonzalez E (2010) Potential applications of nanotechnology in the agro-food sector. Ciênc Tecnol Aliment 30:571–581

Garcia-Martinez J (2016) https://www.scientificamerican.com/article/the-internet-of-things-goes-nano/

González-Sierra NE, Gómez-Pavón LC, Pérez-Sánchez GF, Luis-Ramos A, Zaca-Morán P, Muñoz-Pacheco JM, Chávez-Ramírez F (2017) Tapered optical fiber functionalized with Palladium nanoparticles by drop casting and laser radiation for H_2 and volatile organic compounds sensing purposes. Sensors 17(9):2039–2046

Guosoong L, Jie W, Huangxian J, Feng Y (2011) Streptavidin-functionalized silver-nanoparticle-enriched carbon nanotube tag for ultrasensitive multiplexed detection of tumor markers. Adv Funct Mater 21:2938–2943

Hao R, Wang D, Zhang X, Zuo G, Wei H, Yang R, Zhang Z, Cheng Z, Guo Y, Cui Z, Zhou Y (2009) Rapid detection of Bacillus anthracis using monoclonal antibody functionalized QCM sensor. Biosens Bioelectron 24:1330–1335

Jain KK (2005) Nanotechnology in clinical laboratory diagnostics. Clin Chim Acta 358:37–54
Jain S, Singh SR, Horn DW, Davis VA, Pillai S (2012) Development of an antibody functionalized carbon nanotube biosensor for foodborne bacterial pathogens. J Biosens Bioelectron 24:15–28
Jiang X, Wang R, Wang Y, Su X, Ying Y, Wang J, Li Y (2011) Evaluation of different micro/nanobeads used as amplifiers in QCM immunosensor for more sensitive detection of E. coli O157:H7. Biosens Bioelectron 29(1):23–28
Jingfang H, Jizhou S, Chao B, Jianhua T, Shanhong Z (2012) Three-dimensional nanostructured silver on gold interdigitated microband array electrode for nitrate determination, Ind. J Chem Tech 19:414–419
Jokerst J, Floriano P, Christodoulides N, Simmons G, McDevitt J (2008) Integration of semiconductor quantum dots into nano-bio-chip systems for enumeration of CD4+ T cell counts at the point-of-need. Lab Chip 8:2079–2090
Joyner JR, Kumar DV (2015) Nanosensors and their applications in food analysis: a review. Int J Sci Technol 3(4):80–90
Kahn N, Lavie O, Paz M, Segev Y, Haick H (2015) Dynamic nanoparticle-based flexible sensors: diagnosis of ovarian carcinoma from exhaled breath. Nano Lett 15:7023–7028
Kang X, Cheley S, Guan X, Bayley H (2006) Stochastic detection of enantiomers. J Am Chem Soc 128:10684–10685
Kasianowicz J, Henrickson S, Weetall H, Robertson B (2001) Simultaneous multi analyte detection with a nanometer-scale pore. Anal Chem 73:2268–2272
Konvalina G, Haick H (2014) Sensors for breath testing: from nanomaterials to comprehensive disease detection. Acc Chem Res 47:66–76
Kuo CW, Chien FC, Shiu JY, Tsai SM, Chueh DY, Hsiao YS, Yang ZH, Chen P (2011) Investigation of the growth of focal adhesions using protein nanoarrays fabricated by nanocontact printing using size tunable polymeric nanopillars. Nanotechnol J 22:265302
Lambrou TP, Anastasiou C, Panayiotou CG (2009) A nephelometric turbidity system for monitoring residential drinking water quality. In: Komninos N (ed) Sensor applications experimentation, and logistics, vol 29, pp 1–13
Li X, Kong H, Mout R, Saha K, Moyano DF, Robinson SM, Subinoy RS, Zhang X, Riley MA, Rotello VM (2014) Rapid identification of bacterial biofilms and biofilm wound models using a multichannel nanosensor. ACS Nano 8:12014–12019
Liu HH, Cao X, Yang Y, Liu MG, Wang YF (2006) Array-based nano-amplification technique was applied in detection of hepatitis E virus. J Bio Chem Mol Biol 39:247–252
Rathbun LC (2013) https://www.nanooze.org/author/lcr2-2/page/2/
Lisa M, Chouhan RS, Vinayaka AC, Manonmani HK, Thakur MS (2009) Old nanoparticles based dipstick immunoassay for the rapid detection of dichlorodiphenyltrichloroethane: an organochlorine pesticide. Biosen Bioelecton 25:224–227
Manzetti S, Vasilache D, Francesco E (2015) Emerging carbon-based nanosensor devices: structures, functions and applications. Adv Manuf 3:63–72
Medly C, Smith J, Tang Z, Wu Y, Bamrungsap W, Tan W (2008) Gold nanoparticle-based colorimetric assay for the direct detection of cancerous cells. Anal Chem 80:1067–1072
Meetoo DD (2011) Nanotechnology and the food sector: From the farm to the table. Emir J Food Agric 23:387–403
Mills A (2005) Oxygen indicators and intelligent inks for packaging food. Chem Soc Rev 34:1003–1011
Mouffouk F, da Costa AM, Martins J, Zourob M, Abu-Salah KM, Alrokayan SA (2011) Development of a highly sensitive bacteria detection assay using fluorescent pH-responsive polymeric micelles. Biosens Bioelectron 26:3517–3523
Mwakikunga BW (2018) A computerized breath analyzer device based on a micro-nano-chip of VO_2 nanoparticles for personal pain-free determination of blood glucose levels in a type 1 diabetes mellitus patient. J Nanomed Nanotechnol 9:232–241
Patel S, Nanda R, Sahoo S, Mohapatra E (2016) Biosensors in health care: the milestones achieved in their development towards Lab-on-Chip-Analysis. Biochem Res Int. https://doi.org/10.1155/2016/3130469

Plata MR, Hernando J, Zougagh M, Contento AM, Villasenor MJ, Sánchez-Rojas JL, Rios A (2008) Characterization and analytical validation of a microcantilever-based sensor for the determination of total carbonate in soil samples. Sens Actuators B 134:245–251

Pummakarnchanaa O, Tripathia N, Dutta J (2005) Air pollution monitoring and GIS modeling: a new use of nanotechnology based solid state gas sensors. Sci Tech Adv Mat 6:251–255

Qian X, Peng X, Ansari D, Yin-Goen Q, Chen G, Shin D, Yang L, Young A, Wang M, Nie S (2008) In vivo tumor targeting and spectroscopic detection with surface-enhanced Raman nanoparticle tags. Nat Biotechnol 26:83–90

Rendina I, Lamberti A, Rea I, Arcari P, De Stefano L (2014) Silicon based optical biochips for biomedical applications, Latin America optics and photonics conference, OSA Technical Digest, Optical Society of America, paper LF1D.3

Righettoni M, Schmid A, Amann A, Pratsinis SE (2013) Correlations between blood glucose and breath components from portable gas sensors and PTR-TOF-MS. J Breath Res 7:037110

Sorgenfrei S., Chiu C.Y., Gonzalez Jr. R. L., Yu Y.J., Kim P., Nuckolls C., . Shepard K.L. (2011) Label-free single-molecule detection of DNA-hybridization kinetics with a carbon nanotube field-effect transistor. Nat Nanotechnol, 6: 126-132.

Salaita K, Wang Y, Mirkin CA (2007) Applications of dip-pen nanolithography. Nat Nanotechnol 2:145–155

Salam F, Uludag Y, Tothill IE (2013) Real-time and sensitive detection of Salmonella Typhimurium using an automated quartz crystal microbalance (QCM) instrument with nanoparticles amplification. Talanta 115:761–767

Salvati E, Francesco S, Silke K (2015) Nanosensors for early cancer detection and for therapeutic drug monitoring. Nanomedicine. https://doi.org/10.2217/nnm.15.180. http://web.mit.edu/~kilroi/Public/text/ifprop.pdf

Sharma A, Mathar Z, Sumana G, Solanki PR, Kim CG, Malhotra BD (2010) Antibody immobilized cysteamine functionalized-gold nanoparticles for aflatoxin detection. Thin Solid Films 519:1213–1218

Stefano LD, Rotiroti L, Rea I, Moretti L, Francia GD, Massera E, Lamberti A, Arcari P, Sanges C, Rendina I (2006) Porous silicon-based optical biochips. J Opt A Pure App Opt 8:188–192

Sun X, Li Q, Wang X (2012) Amperometric immunosensor based on gold nanoparticles and saturated thiourea for carbofuran detection. IEEE Sensors J 12(6):2071–2076

Tang S, Zhao J, Storhoff JJ, Norris PJ, Little RF, Yarchoan R (2007) Nanoparticle-based biobarcode amplify cation assay (BCA) for sensitive and early detection of human immunodeficiency type 1 capsid (p24) antigen. J Acquir Immune Defic Syndr 46:231–237

Thriumani R, Zakaria A, Jeffree AI, Hishamuddin NA, Omar MI, Adom AH, Shakaff AYM (2014) Lung cancer detection using an electronic nose. California Institute of Technology: a preliminary study on detection and discrimination of cancerous cells, 2014 IEEE Conference on Biomedical Engineering and Sciences (IECBES), pp 752–756

Ulanowska A, Ligor M, Amann A, Buszewski B (2008) Determination of volatile organic compounds in exhaled breath by ion mobility spectrometry. Chem Anal 53:953–965

Wan Y, Lin Z, Zhang D, Wang Y, Hou B (2011) Impedimetric immunosensor doped with reduced graphene sheets fabricated by controllable electrodeposition for the non-labelled detection of bacteria. Biosens Bioelectron 26:1959–1964

Wan Z, Wang Y, Li SS, Duan L, Zhai J (2005) Development of array-based technology for detection of HAV using Gold-DNA probes. J Biochem Mol Biol 38:399–406

Wang C, Irudayaraj J (2008) Gold nanorod probes for the detection of multiple pathogens. Small 4:2204–2208

Wang C, Yu C (2013) Detection of chemical pollutants in water using gold nanoparticles as sensors: a review. Rev Anal Chem 32(1):1–14

Wang YF, Pang DW, Zhang ZL, Zheng HZ, Cao JP, Shen JT (2003), Visual gene diagnosis of HBV and HCV based on nanoparticle probe amplification and silver staining enhancement. *Journal of Medical Virology 70*(2):205–211

Wilson GS, Gifford R (2005) Biosensors for real-time in vivo measurements. Biosens Bioelectron 20(12):2388–2403

Xu Q, Rioux RM, Dickey MD, Whitesides GM (2008) Nanoskiving: a new method to produce arrays of nanostructures. Acc Chem Res 12:1566–1577

You C, Miranda OR, Gider B, Ghosh PS, Kim IB, Erdogan B (2007) Detection and identification of proteins using nanoparticle-fluorescent polymer 'chemical nose' sensors. Nat Nanotechnol 2:318–323

Yu GX, Wu WX, Zhao Q, Wei XY, Lu Q (2015) Efficient immobilization of acetylcholinesterase onto amino functionalized carbon nanotubes for the fabrication of high sensitive organophosphorus pesticides biosensors. Biosens Bioelectron 68:288–294

Zhou Y, Zhao H, He Y, Ding N, Cao Q (2011) Colorimetric detection of Cu^{2+} using 4-mercaptobenzoic acid modified silver nanoparticles. Colloids Surf A Physiochem Eng Asp 391:179–183

Zelada-Guillen GA, Sebastian-Avila JL, Blondeau P, Riu J, Rius FX (2012) Label-free detection of *Staphylococcus aureus* in skin using real-time potentiometric biosensors based on carbon nanotubes and aptamers. Biosens Bioelectron 31:226–232

Zhou Y, Hu D, Liu L, Miaomiao L, Xiao K, Xu M (2014) Selective and sensitive colorimetric sensor of mercury (II) based on gold nanoparticles and 4-mercaptophenylboronic acid. Sens Actuators B 196:106–111

Zhu H, Xu Y, Liu A, Kong N, Shan F, Wenrong Y, Barrow CJ, Liu J (2015) Graphene nanodots-encaged porous gold electrode fabricated via ion beam sputtering deposition for electrochemical analysis of heavy metal ions. Sens Actuators B 206:592–600

Zhu Y, Cao Y, Sun X, Wang X (2013) Amperometric immunosensor for carbofuran detection based on MWCNTs/GS-PEI-Au and AuNPS-antibody conjugate. Sensors 1:5286–5301

Nanoparticles as Potential Endocrine Disruptive Chemicals

17

Gunjan Dagar and Gargi Bagchi

Abstract

Nanoparticles are three-dimensional particulates different from their parent compound and possess novel electrical, magnetic, and optical properties. They have applications in a variety of areas like medicine, engineering, environmental remediation, etc. Small particle size and high surface area further increase the exposure of NPs to humans through inhalation and dermal absorption. Use of NPs in our daily life is increasing everyday as a part of cosmetics, food, etc. Exposure of humans to NPs can cause modulation or alterations in cell signaling process. These NPs penetrate into the cell and disrupt normal functions. In recent years, there is a lack of information about the possible hazardous effect of nanoparticles on human health and disruption of endocrine system. NPs cause hormonal disruption, neurological and immune disorders, impact fertility, and act as endocrine disruptive chemicals (EDCs). EDCs disrupt the body's normal function due to their ability to block or mimic a hormone's natural function. EDCs include industrial (bisphenol A, PCB, plasticizers, dioxins, etc.) and synthetic chemicals (pesticides, solvents, etc.). Exposure of humans to EDCs is unavoidable, so there is a serious need to identify the compounds that have high impact on human health especially by acting through the hormonal system to alter cell signaling and functions. This review summarizes different types of NPs and their potential impact on hormone signaling and functioning of major systems in the human body.

Keywords

Nanoparticles · Endocrine disruptive chemicals · Nuclear hormone receptors · Neurological disorders and cancer

G. Dagar · G. Bagchi (✉)
Amity Institute of Biotechnology, Amity University Gurgaon, Gurgaon, India
e-mail: gbagchi@ggn.amity.edu

S. K. Saxena, S. M. P. Khurana (eds.), *NanoBioMedicine*,
https://doi.org/10.1007/978-981-32-9898-9_17

17.1 Introduction

Nanotechnology as a concept has been attributed to the Nobel laureate Richard Feynman in 1959 (Feynman 1960). In 1974 Norio Taniguchi employed the word nanotechnology. The term "nano" comes from the Greek word "nanos" meaning "dwarf" (Taniguchi 1974). Nanoparticles (NPs) or nanomaterials are three-dimensional particulates with one dimension less than 100 nanometers (about the size of a virus or a smoke particle). They possess novel magnetic, electrical, and optical properties often different from those in the parent compound (National Institute of Environmental Health Sciences 2013). NPs may find practical applications in a variety of areas, including catalysis, engineering, medicine, and environmental remediation. Applications of NPs are likely to further increase human exposure through ingestion, inhalation, and dermal absorption, among other mechanisms, due to their relatively high surface area and small particle size (Buzea et al. 2007; Oberdorster et al. 2005). The potential hazards of exposure to the human body and environment to NPs cannot be ignored. Toxic effect of NPs on human health depends on individual factors such as genetics and existing disease as well as on the NPs' properties like size, shape, structure, and inorganic and organic coating (Nohynek and Dufour 2012). Many studies have reported the toxic effects of metal, carbon, and silica NPs. In recent years, the use of NPs has increased severalfold in industrial, medical, and consumer applications. NPs form the core of nano-biomaterials. Humans are exposed to NPs at certain workplaces in the ambient air (Nel et al. 2006).

There are two types of NPs: combustion-derived NPs like welding fumes, diesel exhaust particles, and particulate matters and manufactured NPs like palladium, silicon, copper oxide, carbon black, silver, zinc oxide, titanium dioxide, carbon nanotubes, etc. Although these two types of NPs are of similar size, they have some obvious differences. Engineered or manufactured NPs are monodispersed, have chemical nature, and have solid mainly spherical, tube, fiber, or wire form. Combustion-derived NPs are polydispersed, with chemically complex nature and soluble or poorly soluble form (Win-Shwe and Fujimaki 2011). Potential benefits of using nanoparticles and nano-devices include an expanded range of label multiplexing (Oberdorster et al. 2005). NPs are gradually being incorporated into our daily lives as part of medicine, food, clothing, electronic goods, or cosmetics, and a large sector of the economy depends on nanotechnology. However, NPs can non-specifically bind to several cellular components within the body such as enzymes, antibodies, and nucleotides, resulting in modulation of the cell signaling process (Jain 2011). NPs can penetrate cells and subsequently disrupt their biological structure and normal functions via generation of reactive oxygen species (ROS) or by increasing intracellular oxidative stress (Singh et al. 2016).

There is a serious lack of information on the hazards of potential nanoparticle to human health and their possible toxic effects on the endocrine system (Iavicoli et al. 2013). Since the disruption of hormonal functions is associated with severe effects on human health, this topic is of primary importance. Over the past 50 years, epidemiological data have revealed a significant increase in the incidence and prevalence of a number of adverse effects on human health such as reduced fertility and the onset of diseases; disorders of the immune and neurological systems; and alterations in the development and growth process, such as cancer in the breast, ovary, prostate, and testes, diabetes, and obesity (De Coster and van Larebeke 2012; World Health Organization (WHO) 2012). Increase in these diseases is due to the growing exposure of general population to contaminants, which may have adverse effect on their body's hormonal system or in other terms act as endocrine disruptive chemicals (EDCs). Some studies performed on EDCs have revealed that by altering hormonal and homeostatic systems, nanoparticles may play a role in the onset of the aforementioned diseases. They have significant impact on biological systems and can disrupt the normal pattern of development (Knez 2013). They disrupt the body's normal functions by their ability to mimic or block hormonal action (Diamanti-Kandarakis et al. 2009). EDCs are highly heterogeneous. They include synthetic chemicals like pesticides, solvents, industrial chemicals, or plasticizers. They also include industrial solvents and lubricants such as plastic compounds, dioxins, polychlorinated biphenyls (PCB), bisphenol A (BPA), persistent organic pollutants (POP), plasticizers, pesticides such as chlorinated insecticides, triazole and imidazole pharmaceutical agents, chemical compounds that are widely used in cosmetics such as phthalates, and heavy metals such as cadmium, mercury, zinc, arsenic, lead, and manganese (De Coster and van Larebeke 2012; Casals-Casas and Desvergne 2011).

Most EDCs interfere with steroids or thyroid hormones, while others disrupt the protein/peptide hormone signaling (Diamanti-Kandarakis et al. 2009). Since workers and general populations' exposure to EDCs is unavoidable, there is a need to increase the efforts to identify the chemicals that can act upon the hormonal system and behave as endocrine disruptors to study their molecular mechanism of action as the impact of these chemicals on human health is high.

Starting in the early 1900s, endocrine disruption was observed in the USA when a pig farmer complained of fertility problems in swineherds fed on moldy grain. From the last two decades, feminization of fishes was observed in the presence of estrogenic compounds present in sewage effluent (Okkerman and van der Putte 2002). The presence of these chemicals in water caused a change in the sex of fish and changed the egg production of female fish and mating behavior of male fish. These observations indicated that xenobiotic chemicals could disrupt the endocrine system and modulate the sex steroid hormones, thereby causing harmful effects on wildlife (Parks et al. 2001).

However, knowledge about EDCs is still very limited, even after the various studies carried out in recent years. Data on a number of xenobiotics is still incomplete, as the research issue has only concentrated on a few groups of chemicals such as pesticides or POP (Parrott et al. 2003). Therefor to address EDC issue, first thing should be to identify all the possible compounds, which may interfere and disrupt the regulatory mechanisms and homeostasis of the endocrine system. This provision is needed for those chemicals that are used for consumer products and in workplaces and also chemicals whose toxicological profile is not yet clear or defined including NPs. Current evidence shows that different types of NPs are capable of altering the normal and physiological activity of the endocrine system (Gracia et al. 2008). Therefore, considering the scarcity of knowledge regarding the toxicological impact of NPs and potential adverse effects on endocrine functions, further investigation is needed to elucidate the possible threats to mammalian reproduction and endocrine disruption posed by NPs (Kshirsagar et al. 2005).

17.2 Mechanism of Action of EDCs

An endocrine-disrupting substance or chemical is a compound that mimics the homeostasis and hormonal system of the human body. EDCs were originally thought to exert their actions through nuclear hormone receptors, including androgen receptor (AR), progesterone receptor (PR), thyroid receptor (TR), retinoid receptor, and estrogen receptor (ER), among other receptors. Today, basic scientific research shows that these chemicals often act via more than one mechanism. Some EDCs have mixed estrogenic and anti-androgenic properties (Rasier et al. 2007). EDCs may metabolize to generate by-products with different properties. Receptor-mediated mechanisms have received the most attention, but other mechanisms (e.g., hormone synthesis, transport, and metabolism) have been shown to be equally important. For most EDCs, the mechanism(s) of action are poorly understood. This makes it difficult to distinguish between direct and indirect effects and primary and secondary effects of exposure to EDCs (Diamanti-Kandarakis et al. 2009). Direct binding of EDCs to nuclear receptor leading to receptor activation or receptor antagonism is a well-known mechanism. However, indirect mechanisms of endocrine disruption have also been identified and are discussed below (Tabb and Blumberg 2006; Norris et al. 2009).

17.2.1 Hormone Sensitizer

Hormone-sensitizing EDCs affect nuclear receptor activity via non-genomic intracellular signaling pathways and cause an increase in the intrinsic transcriptional activity of the receptor without direct interactions between the EDC and the hormone or its receptor (Priyandoko et al. 2011). Methoxyacetic acid (MAA), the principle metabolite of a commonly used industrial solvent, ethylene glycol ether 2-methoxyethanol (2-ME), belongs to this class. MAA is a hormone sensitizer. MAA is known to exert transient but catastrophic effects on testicular histology and function. Exposure to MAA is associated with various developmental and reproductive toxicities. MAA exposure also causes neural toxicity, blood and immune disorders, and limb degeneration. In normal human fibroblast, MAA results in DNA damage and mitochondrial membrane loss by inducing production of reactive oxygen species (Bagchi et al. 2009). MAA treatment alters the expression of androgen receptor (AR) and androgen-binding protein (ABP) in a stage-specific manner in rat seminiferous tubules. MAA has been found to activate the tyrosine kinase-P13K pathway to enhance or antagonize androgen-induced gene expression.

MAA can inhibit histone deacetylases (HDACs) and can activate mitogen-activated protein kinase (MAPK), thus increasing the levels of acetylated histone H4. Induced hyperacetylation of histone H3 and H4 by MAA is associated with rapid spermatocyte death. Study performed by Bagchi et al. in 2010 demonstrated the changes in the expression of testicular Leydig cell genes that play a critical role in germ cell survival and male reproductive function (Tirado et al. 2003). Changes induced by MAA have a significant impact on germ cell behavior, cellular metabolism, gene expression, and reproductive functions (Bagchi et al. 2010).

17.2.2 Changes in DNA Methylation or Histone Modifications

Methylation of DNA occurs at cytosine residues in CpG dinucleotides and plays a role in genomic imprinting, X-chromosome inactivation, and suppression of retrotransposons, as well as gene expression regulation (Ehrlich 2003). Methylation of DNA can interfere with transcription factor binding, leading to a reduction of gene expression. EDCs can cause changes in DNA methylation and histone modifications. During the period of gonadal sex determination in gestating female, exposure to EDCs results in increased male fertility and decreased spermatogenic capacity. These effects were transferred through the male germ line to nearly all males of all subsequent generations examined (Anway and Skinner 2006).

The effects on reproduction correlate with altered DNA methylation patterns in the germ line. During the early life, exposures to EDCs may alter gene expression in hypothalamic nuclei via non-genomic signaling, histone acetylation, DNA methylation, and epigenetic mechanism.

17.2.3 EDCs That Cause Genomic Instability by Interfering with the Spindle Fiber

During the process of growth and development of an organism, the exposure to EDCs may activate or deactivate some part of the genome in specific time and locations, to cause epigenetic effects and genomic instability. An altered epigenome will transfer to the subsequent generations by germ line in the process called "epigenetic trans-generational inheritance." Reports show that in MCF-7 cells, estrogens like bisphenol A and E2 induce micronuclei which indicates that the exposure of xenohormones may lead to genomic instability (Nel et al. 2006). It was observed that ER antagonists did not prevent micronucleus formation by these estrogens.

17.3 Nanoparticles and Endocrine Disruption

With the development and wide use of nanomaterials, concerns and fears have been expressed regarding potential risks of nanoparticles to worker's safety, human health, and environment contamination (Mercer et al. 2008). Accumulating research data using animal and/or cultured cell models have shown that nanoparticles can cause unusual rapid lung injury and toxicity to various organs (Wennerberg et al. 2011). It is possible theoretically that long-term exposure to nanoparticles may cause some unexpected damage to humans in the same manner as seen in animals. Delivery of NP products to humans and the development of novel nanoparticle-based technologies are constantly expanding. For the treatment of human disease, nanoparticles are currently delivered by intravenous injection or implantation, oral administration, and transdermal delivery (Yah et al. 2012). Sometimes exposure of nanoparticles to human body may also occur accidently by dermal contact, inhalation, and swallowing. Nanoparticles undergo metabolism in human circulation and also excretion or retention in various body compartments. Distinct physiochemical properties of different nanoparticles (as they vary in shape, size, and surface property) determine the way they are absorbed, distributed, metabolized, and eliminated in the body and in the cell. This process is popularly referred to as the ADME (absorption, distribution, metabolism, and elimination) (Wu et al. 2009). The most important barrier of human body against natural and environmental hazards is the skin, by the contact of nanoparticles, which prevents absorption. Yet smaller nanoparticles can penetrate the skin barrier and enter human circulation (Bai et al. 2010). Upon entering the body, nanoparticles use the blood stream to reach vital organs such as the brain, testis (De Jong et al. 2008), liver, and spleen, and lastly these are captured by reticuloendothelial system (RES) cells, which lead to the foreign body elimination. Nevertheless, these vital organs are not fully protected, since some NPs can pass through their barriers (Ma et al. 2011) and it depends on their physical (e.g., shape, size, aspect ratio) (Herbert 1994) as well as their chemical properties (e.g., surface chemical, charge status, aggregation). Nanoparticles with positive charge can pass through the cell more effectively, since the cellular membrane (contains double layer phospholipids) has negative charge.

17.4 Commonly Used Nanoparticles and Their Impact on Health and Hormonal Functions

In this section, the commonly used nanoparticles and diseases associated with their exposure are summarized (Table 17.1).

17.4.1 Silica NPs

Silica nanoparticles (SNPs) occupy a prominent position in scientific research, because of their easy preparation and wide use in various industrial applications, such as catalysis, pigments, pharmacy, electronic and thin film substrates, electronic and thermal insulators, and humidity sensors (Yamashita et al. 2011). The quality of some of these products is highly dependent on the size and distribution of these particles. In 2011 Yamashita et al. performed an experiment where the toxic effects of silicon NPs were monitored, when injected into the pregnant mice. SNPs

Table 17.1 Commonly used nanoparticles and associated diseases

S. No.	Nanoparticles		Size	Disease	Industrial use	Effects and mechanism
1	Silicon NPs		70 nm	Lung inflammation	Electronic and pharmacy	Fetotoxicity, placental dysfunction (restriction of fetal resorption and fetal growth)
				Interstitial fibrosis		
				Industrial bronchitis		
				Small airway disease		
2	Palladium NPs			Kidney disease	Automobile catalysts	Renal tubular dysfunction
				Disease related to female reproductive system		
3	Gold NPs		10–20 nm	Fetal malformations	Chemical catalysis and electronics	Endocrine disruption
						Increase estrogen accumulation
4	Titanium NPs		26.4	Kidney disease, renal failure, cancers	Cosmetic	Induce insulin resistance in liver-derived cells directly via macrophage activation
5	Carbon-based NPs	MWCNT	20–30 nm	Visceral or skeletal malformations	Medical/environmental	Changes in sex hormone
		SWCNT	2.37	Fetus abnormalities	Agricultural applications	Fetal malformation in females
						High percentages of early miscarriages

produced complication in pregnancy, as female mice exhibited 20% lower uterine weight and approx. 10% smaller fetuses (Brandenberger et al. 2013). Christina Brandenberger in 2013 also studied the effect of engineered amorphous SNP on inhalation. These may act as adjuvants to enhance allergic airway disease. For the study, OVA (ovalbumin)-induced murine model of asthma was used to test the hypothesis. The exposure to SNP during OVA sensitization (SNP/OVA mice) resulted in an exacerbation of allergic airway disease after OVA challenge, implying an adjuvant effect of SNP in the development of asthma. The adjuvant effect observed in SNP/OVA mice increased with SNP dose (0, 10, 100, 400 μg SNP/animal) (Cristaudo et al. 2005). Co-exposure to SNP during OVA sensitization caused a dose-dependent enhancement of allergic airway disease compared to challenge with OVA alone.

17.4.2 Palladium (Pd) NPs

Palladium (Pd) is a metal that belongs to the platinum group elements (PGEs). Exposure to this metal can cause acute toxicity and hypersensitivity, within respiratory systems (Garau et al. 2005; Goossens et al. 2006; Iavicoli et al. 2010). Although this metal is used in many industrial sectors, its most extensive use is for the production of automobile catalysts (Iavicoli et al. 2011b). These devices help to reduce unburned hydrocarbon (HC), carbon monoxide (CO), and nitrogen oxide (NOx) emissions in the exhaust of lean-burn engines. However, surface abrasion and deterioration of catalytic converters release palladium into the environment, causing a constant and progressive increase of its level in road dust, soil, airborne particulate, and groundwater tables (Iavicoli et al. 2008). A number of studies have shown that palladium nanoparticles are able to exert adverse health effects, such as apoptosis and alterations of the release, and expression of numerous cytokines. Recent studies have shown that the exposure of palladium nanoparticles in female Wistar rats is able to induce a significant renal tubular dysfunction at concentration of 0.012, 0.12, 1.2, and 12 mg/kg. From the past 20 years, there has been a significant increase in the industrial use of Pd, and emission of this metal from automobile catalytic converters has resulted in higher environmental Pd levels (Iavicoli et al. 2011a; Leopold et al. 2008). Although the weight distribution of PGE emission is largest in the micrometer range, catalysts also produce a nanoparticulate form of these metals (Wilkinson et al. 2011). Therefore, it is clear that in recent years, there has been a notable increase in exposure to micrometer- and nanometer-sized Pd particles that has generated a growing concern regarding the possible adverse effects that Pd-UFPs might have on human health.

17.4.3 Gold NPs (AuNPs)

Gold nanoparticles exhibit unique chemical and physical properties, quite different from the properties of bulk gold. For example, gold particles that are 10–20 nanometers in diameter are red-colored, a fact exploited in a number of medical diagnostic products (Production and Uses of Gold Nanomaterials By Richard Holliday 2008). Gold nanoparticles (AuNPs) have also been investigated for many beneficial applications that include novel uses in targeted drug delivery, electronics, bioimaging, and industry (Dastjerdi and Montazer 2010). For example, CYT-6091, a novel nanomedicine composed of recombinant human tumor necrosis factor-alpha conjugated to 27 nm AuNPs, has shown promise in targeting malignant tumors in vivo and has exhibited less toxicity than recombinant human tumor necrosis factor-alpha alone. However, other studies have shown that in vivo exposure to AuNPs can induce pulmonary toxicity, neurotoxicity, cardiotoxicity, nephrotoxicity, and hepatotoxicity in rodents (Abdelhalim and Jarrar 2012).

17.4.4 Titanium Dioxide Nanoparticles (TiO2 NPs)

TiO2 NPs account for over 70% of the total production volume of nanoparticles worldwide. TiO2 nanoparticle has a wide range of application in industrial and consumer products especially in cosmetics such as high sun protection factor creams in order to protect the skin from UV light. TiO2 NPs are widely used in paints, plastics, inks, food colorants, toothpastes, papers, and cosmetic and skin care products. The TiO2 NPs are the earliest industrially produced nanomaterials and one of the most highly manufactured nanomaterials in the world (Liang et al. 2009). There are three different crystalline structures of TiO2 NPs: anatase, rutile, and brookite (Chen et al. 2014a). Rutile is the most natural form of TiO2 and is said to be chemically inert (Shi et al. 2013). TiO2 nanoparticles are produced abundantly and used extensively because of their high stability, anticorrosive, and photocatalytic properties. It is particularly used in sunblock creams in order to protect the skin from UV light and is under investigation as a novel treatment for acne vulgaris, recurrent condyloma acuminata, atopic dermatitis, hyperpigmented skin lesions, and other non-dermatologic diseases. Further TiO2 nanoparticles are potential photosensitizers, and in nanotherapeutics, they can be used for advanced imaging and photodynamic therapy (Chen et al. 2014b). The wide use of TiO2 NPs has raised concerns about its potential toxicity as it is demonstrated that these have cytotoxic, genotoxic, and embryotoxic effects (Di Bucchianico et al. 2016; Sadiq et al. 2012). TiO2 NPs can impose adverse effects on male and female reproductive system especially in high doses, but its effects are strongly related to duration and routes of administration. After entering the body, TiO2 NPs circulate in the blood and accumulate in reproductive tissues such as the testis and ovary. They come into close contact with different cells of the testis and cause apoptosis and necrosis of testicular tissue, causing disturbance in spermatogenesis and reducing the sperm's motility as well as morphology. In the ovary, it increases the number of atretic follicles and causes inflammation and necrosis. Adverse effects of TiO2 NPs on reproductive tissue may be related to lipid peroxidation of cell membranes as TiO2 enhances the production of reactive oxygen species. Endocrine system appears to be the target of TiO2 in both male and female because administration of TiO2 decreases the expression of genes related to sex hormone metabolism. TiO2 nanoparticle may be associated with alteration of cytokines and inflammatory factors after administration, since the nanoparticles aggregate in the spleen, fertility reduction, and reproductive tissue injury. Intravenous injection of TiO2 nanoparticle in pregnant mice has shown increase in fetal reabsorption rate and decreased uterine weight at a dose of 0.8mg/mouse, because it can penetrate the placental barrier (Yamashita et al. 2011).

17.4.5 Carbon-Based NPs

Carbon nanotubes, fluorescent carbon quantum dots (CQDs), and graphene pertain to carbon materials family. Due to the emerging applications of carbon NPs clinically and in the environment, the potential toxicity of carbon NPs has aroused much attention. For instance, pristine graphene was reported to trigger macrophage apoptosis and stimulate cytokine secretion (Zhou et al. 2012). Carbon nanotubes are tube-shaped carbon material and can be divided into two types: single-walled carbon nanotubes (SWCNTs) and multi-walled carbon nanotubes (MWCNTs). Bai in 2010 investigated the effect of water-soluble MWCNTs on the testis. MWCNTs accumulated in the testes 24 h after a single dose (5 mg/kg per dose). MWCNTs caused oxidative stress after 5 doses over 13 days and decreased germinative layer thickness at day 15, but not at days 60 and 90. Throughout the 90 days period, the serum levels of FSH, LH, and testosterone and spermatogenesis were not affected. The treated mice continued mating with healthy female mice to produce healthy offspring. Hence, it is concluded that repeated administration of CNTs in male mice produces reversible testis damage without affecting fertility. However, oxidative stress and alterations in the testes, such as decreased germinative layer thickness and vacuolization of Sertoli cells, have raised concerns (Bai et al. 2010).

In a separate study, the toxic effect of SWCNTs in mammals provided evidence that CNTs may possess endocrine-disrupting effects. SWCNTs could be pristine (pSWCNTs) and oxidized SWCNTs (oSWCNTs), and the effects of these two were examined by A. Pietroiusti, in 2011, on the development of the mouse embryo. SWNCT-exposed female mice showed early fetal resorption and presented swollen uteri with no developed embryos at a relatively high dose (30 μg/mouse) at GD 5.5. Early resorption in females was decreased at lower dose of SWCNTs, but the percentage of mothers with malformed fetuses increased. Increased ROS levels were likewise detected in malformed fetuses. The results showed that oxidized SWCNTs are more toxic to embryonic stem cell than pristine SWCNTs (Pietroiusti et al. 2011).

17.5 Diseases Associated with Nanoparticle Exposure

17.5.1 Male and Female Reproduction

Nanoparticles have many unique properties like magnetic, mechanical, thermal, and optical properties that allow their widespread applications in many industrial sectors and also in biomedicine (Zhang et al. 2007). There are more than 1310 marketed consumer products based on nanoparticles (Consumer Products Inventory 2013), and the numbers are increasing rapidly. Nanoparticles cause many diseases like pulmonary injury, hepatotoxicity (Bartneck et al. 2012), renal toxicity, reversible testis damage (Bai et al. 2010), immunotoxicity, and neurotoxicity (Wu et al. 2011) in both male and female. Recently, polymer nanoparticles caused severe pulmonary fibrosis in seven young female workers, providing new evidence for nanotoxicity in humans. Consequently, the increasing public exposure to nanoparticles is a call for concern with respect to endocrine disruption.

17.5.2 Female Reproductive System

In human females, the two primary sex hormones estrogen and progesterone are synthesized in the placenta or in the ovaries during pregnancy. There is some evidence to suggest that different NPs can alter the expression gene that is involved in steroidogenesis, including ovarian genes crucial to the synthesis of estrogens and progesterone. Normal female reproduction and fetal development are essential for the perpetuation of the species. However, the female reproductive system is considerably more fragile than the male systems. Compared with the reproductive male gametes, female gametes are rather limited. During a woman's lifetime, only about 400 follicles sequentially mature and ovulate (Hillier 1994). The female reproductive organs, ovary and uterus, exhibit periodic growth and regression, which is strictly regulated by hormones. Its dynamic activity and rigorous hormonal control make this system more sensitive to foreign bodies and physiological stress compared to other physiological processes (Warren and Perlroth 2001; Armenti et al. 2005). And the disturbance of female reproduction inevitably leads to abnormal fetal development. Many environmental chemicals have already demonstrated detrimental effects on the female reproductive system and embryonic development (Wang et al. 2011).

The emergence of nanoparticles has added a new threat to the vulnerable female population. The toxicity of nanoparticles to female reproductive and developmental health has been studied in various models (Shoults-Wilson et al. 2011; Zhao et al. 2013). The interference of xenobiotics with the female reproductive system may impair normal gonadal processes, such as oogenesis, ovulation, hormone production by granulosa cells, and the structure or function of the accessory reproductive structures.

Zhao et al. in 2013 exposed female mice to 2.5, 5, and 10 mg/kg TiO2 nanoparticle by intragastric administration for 90 consecutive days. Significant reduction of body weight, relative weight of ovary, and a decline in fertility were observed after the experiment's duration. Altering hematological parameters, TiO2 was deposited in the ovary. Additionally, atretic follicle count, sex hormone levels, and serum parameters increased, and inflammation and necrosis were observed. Fertility reduction and ovary injury of mice were seen as the inflammation related or follicular atresia related cytokines expressions were altered (Gao et al. 2012).

Jianling Sun et al. in 2013 studied the long-term exposure of TiO2 NPs (5–6 nm, intragastric administration) on nonpregnant mice; a concentration of 10mg/kg TiO2 NPs caused ovarian dysfunction and alterations in functional gene expression levels. Changes in the expression of genes regulating immune and inflammatory responses, oxidative stress, ion transport, cell proliferation, transcription, and oxidoreductase activity of the ovary were also observed. In the ovarian cells of these mice, TiO2 nanoparticles were detected, and the resultant cellular damage led to an imbalance in sex hormones and decreased fertility (Martino-Andrade and Chahoud 2009). In another study by Blum et al. in 2012, an increase in the uterine weight and altered placental weight of pregnant CD-1 mice were observed with daily inhalation of CdO NPs (cadmium oxide NPs) at 230 μg/mice dose. Furthermore, reduced levels of 17β-estradiol and altered expression levels of estrogen receptor α and β (ERα and ERβ) in the uterus eventually led to decreased implantation. Cd ions that are released from the CdO nanoparticles may act as an endocrine disruptor. These studies demonstrate that nanoparticles may adversely impact the female reproductive system and fertility, as has been shown for other toxic chemicals (Alaee and Ilani 2017).

17.5.3 Male Reproductive System

Females are particularly susceptible to nanoparticle toxicity. However, NPs have negative impacts on male germ cells too and development of male child. These impacts are associated with nanoparticle modification, composition, concentration, and route of administration and the species of the animal. Therefore, understanding the impacts of nanoparticles on animal growth and reproduction is essential. Exposure to nanoparticles also affects the male reproductive system, including an impact on spermatogenesis when it begins in the seminiferous tubules of the testes (Boisen et al. 2012).

In males, there is evidence to suggest that NPs accumulate in the testes. For example, Ag- and TiO2-based NPs may be more dangerous, while Si-based NPs appear to have few toxic effects, with an impact on cells in the seminiferous tubules, immune and inflammatory reactions, and sperm motility and morphology. All of these present a potential risk to male fertility if safe exposure levels are not defined and applied, especially among men working in the NP industry (Bakare et al. 2016).

Bakare et al. investigated the potential genotoxic effect of TiO2 nanoparticles at five different concentrations 9.38, 18.75, 37.50, 75.00, and 150.00 mg/kg for 5 consecutive days in the somatic and germ tissues of mice using the mouse bone marrow micronucleus and sperm morphology assays. This study showed the significant increase in abnormal sperm cells at tested concentrations after 5 and 10 weeks from the first day of exposure. Disrupted cellular architecture, vacuolation, and necrosis of testicular tissue were also observed (Boisen et al. 2013) (Fig. 17.1).

17.5.4 Fetal Development

NP exposure by inhalation is one of the most likely environmental exposure routes causing damage to fetal organs. The overall pathological aspects associated with these deleterious effects on the offspring still need to be deciphered. NPs may cause changes in embryogenesis and anomalies in the fetal reproductive system point. Further, NP exposure reduces sperm production (Gao et al. 2013). However, the impact varies from species to species (Yamashita et al. 2011). Decreased sperm production is associated with several molecular changes, thus altering the overall expression of genes involved in spermatogenesis. Yamashita in 2011 observed the accumulation of TiO2 and SiO2 nanoparticles in rat fetus that resulted in increased fetal resorption rate and smaller fetuses at gestational days 16 and 17, with a high

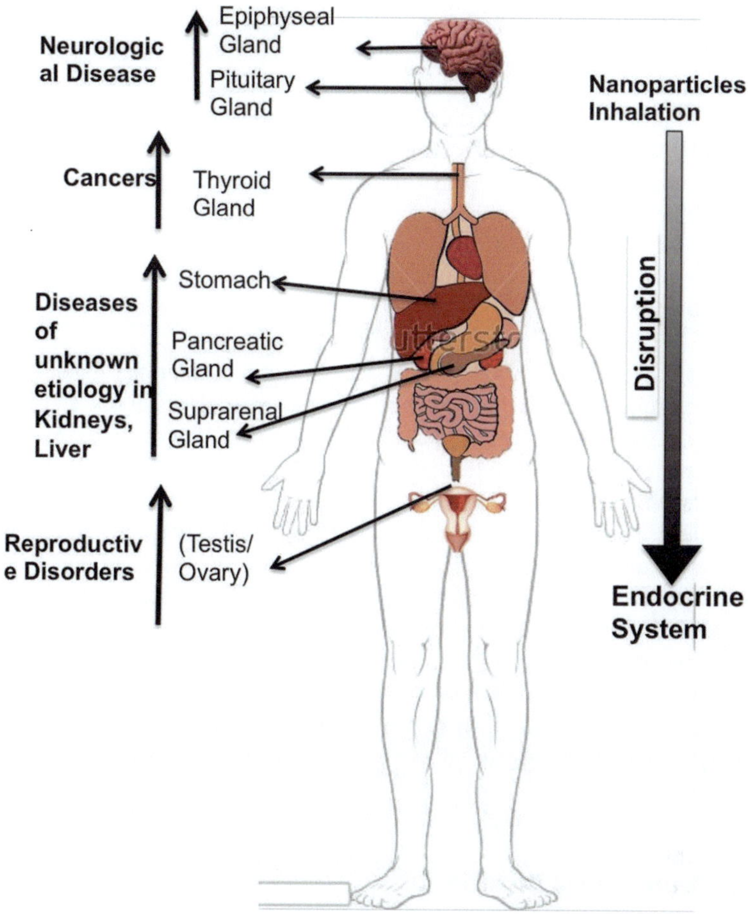

Fig. 17.1 Impact of NPs on male and female reproductive system and function

IV dosage of 800 μg/mouse but not at lower dosages (Rattanapinyopituk et al. 2014). Other studies demonstrated that gold nanoparticle (AuNP) transport with similar size as TiO2 and SiO2 across the placenta into the uterus did not lead to the accumulation and fetal toxicity (Semmler-Behnke et al. 2014; Tsuchiya et al. 1996). Dosages of the NPs were 100–1000 times less than TiO2 and SiO2 dosages employed by Yamashita et al. in 2011. These types of anomalies highlight the differences between different types of NPs and difficulties in determining the dosages of NPs that pose a threat to reproductive and fetal health, the uncertainties in determining what levels individuals are likely to be exposed to, especially pregnant women and fetuses. Toxic effects of nanoparticles are also associated with the fetal morphological development and organogenesis at different gestational periods (Park et al. 2010). High single oral dose of TiO2 NPs (100 or 1000 mg/kg) to pregnant dams causes a significant increase in fetal deformities and mortality (Pietroiusti et al. 2011). It would be of interest to determine what the effects would be over a range of doses in order to more accurately monitor the likely range of TiO2 doses arising from environmental, occupational, and therapeutic exposures. In other studies, the mortality of pups increased during the period of lactation, and decreased growth without deformities is detected after an oral administration of platinum (Pt) NPs at concentrations 0.25, 0.5, and 1 mg/kg, 14 days before and 4 days after mating in ICR mice (American Association for Cancer Research 2007).

17.5.5 Cancer

In 2007, a study presented at the Annual Meeting of the American Association for Cancer Research stated that the engineered nanomaterials of about a meter size could damage the DNA and lead to cancer (Bergin and Witzmann 2013). NPs are small enough to penetrate into the cell membranes and defenses, yet large enough to interfere with normal cells. Such NPs are used in electronics, cosmetics, chemical manufacturing, and among other industries. They can be very difficult to isolate from the larger environment because of their extremely small size, as they are much too small for removal by conventional filtering techniques. A study by Pacheco et al. looked at how these different types of NPs could cause DNA damage in MCF-7 line of breast cancer cells. They observe that NPs can cause both dose-dependent and time-dependent increases in the risk of DNA damage in breast cancer cells exposed to either aqueous colloidal silica or C60 fullerenes. The damage in DNA could potentially lead to mutations and ultimately increase the risk of cancer (Freire et al. 2010).

17.5.6 Neurological Disorders

The understanding of mechanisms accounting for neurological diseases has extended opportunities for NP technology to treat these diseases.

Traffic-related air pollution may cause adverse effects on neurodevelopment in children (156). Prenatal exposure of pregnant mice to nanoparticles caused neurological disorders in their offspring (Sugamata et al. 2012). In an inhalation study, prenatal exposure to DE m (0.3, 1, and 3 mg/ ^{3}) resulted in various types of damage, including caspase-3-positive cells in the cerebral cortex and hippocampus and crescent-shaped spaces in some cells. Furthermore, the granular epithelial cells and scavenger cells that constitute the blood-brain barrier (BBB) underwent apoptosis (Sugamata et al. 2006). Maternal exposure to TiO2 (0.1 mg/mouse) by subcutaneous injection also resulted in the apoptosis of endothelial cells, capillary stenosis, and degenerative changes in the neighboring parenchyma (Takeda et al. 2009). The nanoparticle-induced reduction of dopamine (DA) turnover in the nucleus accumbens and striatum induced a decrease in spontaneous motor activity, thereby emphasizing the adverse effects of TiO2 nanoparticles on the central dopaminergic system. Analysis of the gene expression in the brain of the offspring indicated that the alterations are related to inflammation, oxidative stress, and neurotransmitters (Shimizu et al. 2009).

17.6 Conclusion

With the increase in applications of nanoparticles in industries, cosmetics, medicine, and environmental remediation together with the inhalation of NPs polluting the air, it is necessary to understand the transferable adverse effects of these NPs to next generation and reproductive toxicity. NPs, on one hand owing to their physical and chemical properties, can be used in drug delivery systems, textiles, and clinical therapy and can provide great advantage and effectiveness in these fields. On the other hand, NPs have many potential adverse health effects especially in humans coming in contact with them due to their nondegradable and as yet unidentified properties. However, the studies on the toxicity of NPs have been far fewer than those on the positive application of NPs. Many of these concerns have been highlighted from both in vivo and in vitro studies. Subsequently, these NPs can enter systemic circulation and cells and organelles as well as the reproductive organs, particularly via the skin, eyes, gastrointestinal tract, and nasal olfactory structures, at which point many potential harmful effects may occur. While the use of NPs has significant medical benefits, one should not overlook their potential as hazardous and toxic molecules especially with respect to endocrine disruption.

References

Abdelhalim MA, Jarrar BM (2012) Histological alterations in the liver of rats induced by different gold nanoparticle sizes, doses and exposure duration. J Nanobiotechnol 10:5

Alaee S, Ilani M (2017) Effect of titanium dioxide nanoparticles on male and female reproductive systems. J Adv Med Sci Appl Technol 3(1):3

American Association for Cancer Research (2007) Nanoparticles can damage DNA, increase cancer risk. ScienceDaily. ScienceDaily

Anway MD, Skinner MK (2006) Epigenetic transgenerational actions of endocrine disruptors. Endocrinology 147(6):S43–S49

Armenti AE, Zama AM, Passantino L, Uzumcu M (2005) Developmental methoxychlor exposure affects multiple reproductive parameters and ovarian folliculogenesis and gene expression in adult rats. *Toxicol. Appl. Pharm.* 2008, *233*, 286–296. Anway, M.D.; Cupp, A.S.; Uzumcu, M.; Skinner, M.K. Epigenetic transgenerational actions of endocrine disruptors and male fertility. Science 308:1466–1469

Bagchi G, Hurst CH, Waxman DJ (2009) Interactions of methoxyacetic acid with androgen receptor. Toxicol Appl Pharmacol 238:101–110. https://doi.org/10.1016/j.taap.2008.03.015

Bagchi G, Zhang Y, Waxman DJ (2010) Impact of methoxyacetic acid on mouse Leydig cell gene expression Bagchi et al. Reprod Biol Endocrinol 8:65. http://www.rbej.com/content/8/1/65

Bai Y, Zhang Y, Zhang J, Mu Q, Zhang W, Butch ER et al (2010) Repeated administrations of carbon nanotubes in male mice cause reversible testis damage without affecting fertility. Nat Nanotechnol 5:683–689. https://doi.org/10.1038/nnano.2010.153

Bakare AA, Udoakang AJ, Anifowoshe AT, Fadoju OM, Ogunsuyi OI, Alabi OA et al (2016) Genotoxicity of titanium dioxide nanoparticles using the mouse bone marrow

micronucleus and sperm morphology assays. J Pollu Effects Control 04(02). https://doi.org/10.4172/2375-4397.1000156
Bartneck M, Ritz T, Keul HA, Wambach M, Bornemann J, Gbureck U, Ehling J, Lammers T, Heymann F, Gassler N (2012) Peptide-functionalized gold nanorods increase liver injury in hepatitis. ACS Nano 6:8767–8777
Bergin IL, Witzmann FA (2013) Nanoparticle toxicity by the gastrointestinal route: evidence and knowledge gaps. Int J Biomed Nanosci Nanotechnol 3(1–2). https://doi.org/10.1504/IJBNN.2013.054515
Boisen AMZ, Shipley T, Jackson P, Hougaard KS, Wallin H, Yauk CL et al (2012) NanoTIO 2 (UV-Titan) does not induce ESTR mutations in the germline of prenatally exposed female mice. Part Fibre Toxicol 9:19. https://doi.org/10.1186/1743-8977-9-19
Boisen AMZ, Shipley T, Jackson P, Wallin H, Nellemann C, Vogel U et al (2013) *In utero* exposure to nanosized carbon black (Printex90) does not induce tandem repeat mutations in female murine germ cells. Reprod Toxicol 41:45–48. https://doi.org/10.1016/j.reprotox.2013.06.068
Brandenberger C, Rowley NL, Jackson-Humbles DN, Zhang Q, Bramble LA, Lewandowski RP, Wagner JG, Chen W, Kaplan BL, Kaminski NE, Baker GL, Worden RM, Harkema JR (2013) Engineered silica nanoparticles act as adjuvants to enhance allergic airway disease in mice. Part Fibre Toxicol 10:26
Buzea C, Pacheco II, Robbie K (2007) Nanomaterials and nanoparticles: sources and toxicity. Biointerphases 2:MR17–MR71
Casals-Casas C, Desvergne B (2011) Endocrine disruptors: from endocrine to metabolic disruption. Annu Rev Physiol 73:135–162
Chen T, Yan J, Li Y (2014a) Genotoxicity of titanium dioxide nanoparticles. J Food Drug Anal 22:95–104
Chen Z, Wang Y, Ba T, Li Y, Pu J, Chen T et al (2014b) Genotoxic evaluation of titanium dioxide nanoparticles in vivo and in vitro. Toxicol Lett 226(3):314–319. https://doi.org/10.1016/j.toxlet.2014.02.020
Consumer Products Inventory (2013) An inventory of nanotechnology-based consumer products currently on the market. Available online: http://www.nanotechproject.org/inventories/consumer/ (accessed on 23 April 2013). Int J Mol Sci 14: 9331
Cristaudo A, Sera F, Severino V, De Rocco M, Di Lella E, Picaro M (2005) Occupational hypersensitivity to metal salts, including platinum, in the secondary industry. Allergy 60:159–164
Dastjerdi R, Montazer M (2010) A review on the application of inorganic nano- structured materials in the modification of textiles: focus on anti- microbial properties. Colloids Surf B Biointerfaces 79:5–18
De Coster S, van Larebeke N (2012) Endocrine-disrupting chemicals: associated disorders and mechanisms of action. J Environ Public Health. https://doi.org/10.1155/2012/713696
De Jong WH, Hagens WI, Krystek P, Burger MC, Sips AJ, Geertsma RE (2008) Particle size-dependent organ distribution of gold nanoparticles after intravenous administration. Biomaterials 29:1912–1919. https://doi.org/10.1016/j.biomaterials.2007.12.037
Di Bucchianico S, Cappellini F, Le Bihanic F, Zhang Y, Dreij K, Karlsson HL (2016) Genotoxicity of TiO2 nanoparticles assessed by mini-gel comet assay and micronucleus scoring with flow cytometry. Mutagenesis 32(1):127–137. https://doi.org/10.1093/mutage/gew030
Diamanti-Kandarakis E, Bourguignon JP, Giudice LC, Hauser R, Prins GS, Soto AM, Zoeller RT, Gore AC (2009) Endocrine-disrupting chemicals: an Endocrine Society scientific statement. Endocr Rev 30(4):293–342. [PubMed: 19502515]
Ehrlich M (2003) Expression of various genes is controlled by DNA methylation during mammalian development. J Cell Biochem 88:899–910
Feynman RP (1960) There's plenty of room at the bottom. Eng Sci 23(5):22–36
Freire C, Ramos R, Puertas R, Lopez-Espinosa MJ, Julvez J, Aguilera I, Cruz F, Fernandez MF, Sunyer J, Olea N (2010) Association of traffic-related air pollution with cognitive development in children. J Epidemiol Commun Health 64:223–228

Gao G, Ze Y, Li B, Zhao X, Zhang T, Sheng L, Hu R, Gui S, Sang X, Sun Q (2012) Ovarian dysfunction and gene-expressed characteristics of female mice caused by long-term exposure to titanium dioxide nanoparticles. J Hazard Mater 243:19–27

Gao G, Ze Y, Zhao X, Sang X, Zheng L, Ze X et al (2013) Titanium dioxide nanoparticle-induced testicular damage, spermatogenesis suppression, and gene expression alterations in male mice. J Hazard Mater 258:133–143. https://doi.org/10.1016/j.jhazmat.2013.04.046

Garau V, Masala MG, Cortis MC, Pittau R (2005) Contact stomatitis due to palladium in dental alloys: a clinical report. J Prosthet Dent 93:318–320

Goossens A, De Swerdt A, De Coninck K, Snauwaert JE, Dedeurwaerder M, De Bonte M (2006) Allergic contact granuloma due to palladium following ear piercing. Contact Dermatitis 55:338–341

Gracia T, Jones PD, Higley EB et al (2008) Modulation of steroidogenesis by coastal waters and sewage effluents of Hong Kong, China, using the H295R assay. Environ Sci Pollut Res 15(4):332–343

Herbert G (1994) Synthesis of monodispersed silica powders I. Particle properties and reaction kinetics. J Eur Ceram Soc 14:205

Hillier S (1994) Current concepts of the roles of follicle stimulating hormone and luteinizing hormone in folliculogenesis. Hum Reprod 9:188–191

Iavicoli I, Bocca B, Caroli S, Caimi S, Alimonti A, Carelli G et al (2008) Exposure of Rome city tram drivers to airborne platinum, rhodium, and palladium. J Occup Environ Med 50:1158–1166

Iavicoli I, Bocca B, Fontana L, Caimi S, Bergamaschi A, Alimonti A (2010) Distribution and elimination of palladium in rats after 90-day oral administration. Toxicol Ind Health 26:183–189

Iavicoli I, Fontana L, Bergamaschi A (2011a) Palladium: exposure, uses and human health effects. In: Nriagu JO (ed) Encyclopedia of environmental health, vol 4. Elsevier, Burlington, pp 307–314

Iavicoli I, Fontana L, Marinaccio A, Alimonti A, Pino A, Bergamaschi A et al (2011b) The effects of iridium on the renal function of female Wistar rats. Ecotoxicol Environ Saf 74:1795–1799

Iavicoli I, Fontana L, Leso V, Bergamaschi A (2013) The effects of nanomaterials as endocrine disruptors. Int J Mol Sci 14:16734

Jain KK (2011) Nanobiotechnology. In: Comprehensive biotechnology, vol 1. Elsevier, New York, pp 599–614

Knez J (2013) Endocrine-disrupting chemicals and male reproductive health. Reprod BioMed Online 26:440–448

Kshirsagar NA, Pandya SK, Kirodian GB, Sanath S (2005) Liposomal drug delivery system from laboratory to clinic. J Postgrad Med 51(Suppl. 1):S5–S15

Leopold K, Maier M, Weber S, Schuster M (2008) Long-term study of palladium in road tunnel dust and sewage sludge ash. Environ Pollut 156:341–347

Liang G, Pu Y, Yin L (2009) Influence of different sizes of titanium dioxide nanoparticles on hepatic and renal functions in rats with correlation to oxidative stress. J Toxicol Environ Hlth 72:740–745

Ma X, Wu Y, Jin S, Tian Y, Zhang X, Zhao Y et al (2011) Gold nanoparticles induce autophagosome accumulation through size-dependent nanoparticle uptake and lysosome impairment. ACS Nano 5:8629–8639. https://doi.org/10.1021/nn202155y

Martino-Andrade AJ, Chahoud I (2009) Reproductive toxicity of phthalate esters. Mol Nutr Food Res 54:148–157

Mercer RR, Scabilloni J, Wang L et al (2008) Alteration of deposition pattern and pulmonary response as a result of improved dispersion of aspirated single-walled carbon nanotubes in a mouse model. Am J Physiol—Lung Cell Mol Physiol 294(1):L87–L97

National Institute of Environmental Health Sciences (2013) Since you asked: nanotechnology and NIEHS [Inter- net]. Research Triangle Park (NC): NIEHS; [cited 2010 Dec 13]. From: http://www.niehs.nih.gov/news/media/questions/sya-nano.cfm

Nel A, Xia T, Mädler L, Li N (2006) Toxic potential of materials at the nanolevel. Science 311(5761):622–627

Nohynek GJ, Dufour EK (2012) Nano-sized cosmetic formulations or solid nanoparticles in sunscreens: a risk to human health? Arch Toxicol 86:1063–1075

Norris JD, Chang C-Y, Wittmann BM, Kunder RS, Cui H, Fan D, Joseph JD, McDonnell DP (2009) The homeodomain protein HOXB13 regulates the cellular response to androgens. Mol Cell 36:405–416

Oberdorster G, Oberdorster E, Oberdorster J (2005) Nanotoxicology: an emerging discipline evolving from studies of ultrafine particles. Environ Health Perspect 113:823–839

Okkerman PC, van der Putte I (2002) Endocrine disrupters: study on gathering information on 435 substances with insufficient data. RPS BKH Consulting Engineers for European Commission Directorate-General for the Environment, Delft, pp 1–131

Park EJ, Kim H, Kim Y, Park K (2010) Effects of platinum nanoparticles on the postnatal development of mouse pups by maternal exposure. Environ Health Toxicol 25:279–286. Available online at: https://www.e-eht.org/journal/view.php?number_699

Parks LG, Lambright CS, Orlando EF, Guillette LJ Jr, Ankley GT, Gray LE Jr (2001) Masculinization of female mosquitofish in kraft mill effluent- contaminated fenholloway river water is associated with androgen receptor agonist activity. Toxicol Sci 62(2):257–267

Parrott JL, Wood CS, Boutot P, Dunn S (2003) Changes in growth and secondary sex characteristics of fathead minnows exposed to bleached sulfite mill effluent. Environ Toxicol Chem 22(12):2908–2915

Pietroiusti A, Massimiani M, Fenoglio I, Colonna M, Valentini F, Palleschi G et al (2011) Low doses of pristine and oxidized single-wall carbon nanotubes affect mammalian embryonic development. ACS Nano 5:4624–4633. https://doi.org/10.1021/nn200372g

Priyandoko D, Ishii T, Kaul SC, Wadhwa R (2011) Ashwagandha leaf derived withanone protects normal human cells against the toxicity of methoxyacetic acid, a major industrial metabolite. PLoS ONE 6:e19552. https://doi.org/10.1371/journal.pone.0019552

Production and Uses of Gold Nanomaterials By Richard Holliday, Head of Industrial Applications (2008), World Gold Council | Sep 05, 2008

Rasier G, Parent AS, Gérard A, Lebrethon MC, Bourguignon JP (2007) Early maturation of gonadotropin-releasing hormone secretion and sexual precocity after exposure of infantile female rats to estradiol or dichlorodiphenyltrichloroethane. Biol Reprod 77:734–742

Rattanapinyopituk K, Shimada A, Morita T, Sakurai M, Asano A, Hasegawa T et al (2014) Demonstration of the clathrin- and caveolin- mediated endocytosis at the maternal–fetal barrier in mouse placenta after intravenous administration of gold nanoparticles. J Vet Med Sci 76:377–387. https://doi.org/10.1292/jvms.13-0512

Sadiq R, Bhalli JA, Yan J, Woodruff RS, Pearce MG, Li Y et al (2012) Genotoxicity of TiO2 anatase nanoparticles in B6C3F1 male mice evaluated using Pig-a and flow cytometric micronucleus assays. Mutat Res/Genetic Toxicol Environ Mutagen 745(1–2):65–72. https://doi.org/10.1016/j.mrgentox.2012.02.002

Semmler-Behnke M, Lipka J, Wenk A, Hirn S, Schäffler M, Tian F et al (2014) Size dependent translocation and fetal accumulation of gold nanoparticles from maternal blood in the rat. Part Fibre Toxicol 11:33. https://doi.org/10.1186/s12989-014-0033-9

Shi H, Magaye R, Castranova V, Zhao J (2013) Titanium dioxide nanoparticles: a review of current toxicological data. Particle Fibre Toxicol 10(1):15. https://doi.org/10.1186/1743-8977-10-15

Shimizu M, Tainaka H, Oba T, Mizuo K, Umezawa M, Takeda K (2009) Maternal exposure to nanoparticulate titanium dioxide during the prenatal period alters gene expression related to brain development in the mouse. Part Fiber Toxicol 6. https://doi.org/10.1186/1743–8977–6-20

Shoults-Wilson WA, Reinsch BC, Tsyusko OV, Bertsch PM, Lowry GV, Unrine JM (2011) Effect of silver nanoparticle surface coating on bioaccumulation and reproductive toxicity in earthworms (Eisenia fetida). Nanotoxiclogy 5(3):432–444

Singh et al (2016) Structure, synthesis, and application of nanoparticles. In: Engineered nanoparticles. https://doi.org/10.1016/B978-0-12-801406-6.00002-9

Sugamata M, Ihara T, Takano H, Oshio S, Takeda K (2006) Maternal diesel exhaust exposure damages newborn murine brains. J Health Sci 52:82–84

Sugamata M, Ihara T, Umezawa M, Takeda K (2012) P-999-Maternal exposure to nanoparticles enhances the risk of mental neurological disorders in offspring. Eur Psychiatr. https://doi.org/10.1016/S0924-9338(12)75166-0

Tabb MM, Blumberg B (2006) New modes of action for endocrine-disrupting chemicals. Mol Endocrinol 20:475–482. [PubMed: 16037129] Anway MD, Cupp AS, Uzumcu M, Skinner MK. Epigenetic transgenerational actions of endocrine disruptors and male fertility. Science 2005;308:1466– 1469[PubMed: 15933200]

Takeda K, Suzuki K, Ishihara A, Kubo-Irie M, Fujimoto R, Tabata M, Oshio S, Nihei Y, Ihara T, Sugamata M (2009) Nanoparticles transferred from pregnant mice to their offspring can damage the genital and cranial nerve systems. J Health Sci 55:95–102

Taniguchi N (1974) On the basic concept of "nano-technology". In: Proceedings of International Conference on Production Engineering. Tokyo, Part II, Japan Society of Precision Engineering

Tirado OM, Martinez ED, Rodriguez OC, Danielsen M, Selva DM, Reventos J, Munell F, Suarez-Quian CA (2003) Methoxyacetic acid disregulation of androgen receptor and androgen-binding protein expression in adult rat testis. Biol Reprod 68:1437–1446

Tsuchiya T, Oguri I, Yamakoshi YN, Miyata N (1996) Novel harmful effects of [60] fullerene on mouse embryos *in vitro* and *in vivo*. FEBS Lett 393:139–145. https://doi.org/10.1016/0014-5793(96)00812-5

Wang J, Zhu X, Zhang X, Zhao Z, Liu H, George R, Wilson-Rawls J, Chang Y, Chen Y (2011) Disruption of zebrafish (Danio rerio) reproduction upon chronic exposure to TiO2 nanoparticles. Chemosphere 83:461–467

Warren M, Perlroth N (2001) The effects of intense exercise on the female reproductive system. J Endocrinol 170:3–11

Wennerberg A, Jimbo R, Allard S, Skarnemark G, Andersson M (2011) *In vivo* stability of hydroxyapatite nanoparticles coated on titanium implant surfaces. Int J Oral Maxillofac Implants 26:1161–1166

Wilkinson KE, Palmberg L, Witasp E, Kupczyk M, Feliu N, Gerde P et al (2011) Solution-engineered palladium nanoparticles: model for health effect studies of automotive particulate pollution. ACS Nano 5:5312–5324

Win-Shwe T-T, Fujimaki H (2011) Nanoparticles and Neurotoxicity. Int J Mol Sci 12:6267–6280

World Health Organization (WHO) (2012) State of the science of endocrine disrupting chemicals–2012. World Health Organization, Geneva, pp 23–237

Wu J, Liu W, Xue C, Zhou S, Lan F, Bi L et al (2009) Toxicity and penetration of TiO2 nanoparticles in hairless mice and porcine skin after subchronic dermal exposure. Toxicol Lett 191:1–8. https://doi.org/10.1016/j.toxlet.2009.05.020

Wu J, Wang C, Sun J, Xue Y (2011) Neurotoxicity of silica nanoparticles: brain localization and dopaminergic neurons damage pathways. ACS Nano 5:4476–4489

Yah CS, Simate GS, Iyuke SE (2012) Nanoparticles toxicity and their routes of exposures. Pak J Pharm Sci 25:477–491

Yamashita K, Yoshioka Y, Higashisaka K, Mimura K, Morishita Y, Nozaki M et al (2011) Silica and titanium dioxide nanoparticles cause pregnancy complications in mice. Nat Nanotechnol 6(5):321–328. https://doi.org/10.1038/nnano.2011.41

Zhang L, Gu F, Chan J, Wang A, Langer R, Farokhzad O (2007) Nanoparticles in medicine: therapeutic applications and developments. Clin Pharmacol Ther 83:761–769

Zhao X, Ze Y, Gao G, Sang X, Li B, Gui S et al (2013) Nano- sized TiO2-induced reproductive system dysfunction and its mechanism in female mice. PLoS ONE 8(4):e59378. https://doi.org/10.1371/journal.pone.0059378

Zhou H, Zhao K, Li W, Yang N, Liu Y, Chen C, Wei T (2012) The interactions between pristine graphene and macrophages and the production of cytokines/chemokines via TLR- and NF-κB-related signaling pathways. Biomaterials 33:6933

Part VII

NanoBioMedicine: Advanced Medical Devices

Techniques to Understand Mycobacterial Lipids and Use of Lipid-Based Nanoformulations for Tuberculosis Management

18

Saif Hameed, Sharda Sharma, and Zeeshan Fatima

Abstract

Tuberculosis (TB) is an airborne infectious disease caused by *Mycobacterium tuberculosis* (MTB) that claims 2 million lives every year around the globe. In the era of increasing burden of multidrug resistance (MDR), emergence of drug resistance has become a worrying cause of concern toward present therapeutics. Under such circumstances, in order to pave way for developing better therapeutic strategies, a deeper understanding to dissect the MDR mechanisms needs immediate attention given the persistent global burden of TB. Nowadays, understanding lipid biology of MTB has gained prominence due to the newly assigned roles of MTB lipids in pathogenesis particularly due to the functional interactions between lipids and MDR determinants. MTB possesses unique cell wall rich in lipids which are coded by almost 30% of the genome. The exclusive mycobacterial cell wall lipids such as trehalose monomycolate and dimycolate (TMM, TDM), phthiocerol dimycocerosate (PDIM), sulpholipid-1 (SL-1), diacyl trehalose (DAT), and pentacyl trehalose (PAT) among others are known to play a significant role in pathogens. Commensurate with this only limited resources are available which deals with the methods to study MTB lipids. The scarcity of relevant tools has led to investments in programs to develop new approaches for lipid research. This chapter describes the extraction methods to study lipids in *Mycobacterium* which are known to play crucial roles in MDR along with various lipid-based nanoformulations designed for TB management.

Keywords

Lipid · Mycolic acid · TDM · Prenol · Fatty acids · PIM · Nanoformulations

Saif Hameed and Sharda Sharma have contributed equally to this chapter.

S. Hameed · S. Sharma · Z. Fatima (✉)
Amity Institute of Biotechnology, Amity University Haryana, Gurugram (Manesar), India

S. K. Saxena, S. M. P. Khurana (eds.), *NanoBioMedicine*,
https://doi.org/10.1007/978-981-32-9898-9_18

18.1 Introduction

During the past decades, the emergence of multidrug resistance (MDR) among various microorganisms (infectious agents) against antimicrobial drugs has been a major problem globally. Tuberculosis (TB) is one of the dreadful diseases which is caused by *Mycobacterium tuberculosis* (MTB), an intracellular pathogen that claims millions of deaths annually. Due to the occurrence of MDR against anti-TB drugs, this problem is compounded by prolonged illness, higher medical cost for health care, and an extensive risk of death (Pal et al. 2017). Thus, there is a pressing need to study in depth the mechanisms contributing to MDR. Nowadays, a spurt in the research related to lipid biology of MTB has attracted considerable interest due to the emerging roles of MTB lipids in MTB pathogenesis particularly due to the functional interactions between lipids and MDR determinants (Hameed et al. 2012; Hameed and Fatima 2014). This chemical structure and organization of the mycobacterial cell envelope are responsible for persistence, resistance, and an unusual protective barrier that is vital for the survival of the bacteria inside the human host and thereby represent an attractive target to anti-tubercular drugs. Nanoformulation has received a significant attention as a drug delivery system because of its size-dependent properties. The lipid-based nanoformulation showed advantage of high degree of biocompatibility and versatility. The purposes of the lipid-based nanoformulation are to improve higher patient compliance, lessen toxicity, and reduce therapy time for TB (Matougui et al. 2016).

18.2 MTB Lipids

MTB cell wall possesses complex lipids that are responsible to confer many unique properties to MTB (Pal et al. 2017). One of the distinguishing features of the mycobacterial cell wall is the presence of mycolic acid which is the beta hydroxyl fatty acid with an alpha alkyl side chain covalently bound to arabinogalactan-peptidoglycan layer forming inner leaflet of bacterial inner surface of cell (Sharma et al. 2016, 2017). On the other hand, outer leaflets are composed of non-covalently associated several lipids such as trehalose monomycolate and dimycolate (TMM, TDM), phthiocerol dimycocerosate (PDIM), sulfolipid-1 (SL-1), diacyl trehalose (DAT), and pentacyl trehalose (PAT) that are involved in pathogenesis (Bailo et al. 2015). Although there are several useful studies depicting the role of lipids in MDR-TB, very little information is available which deals with the methodologies to study various *Mycobacterium* lipid class isolations. This chapter reviews the protocols which will be helpful to extract the different classes of mycobacterial lipids.

18.3 Total Lipid Extraction

The total cellular lipid isolation of MTB is routinely employed to study MTB lipids. There are many methods for total lipid extraction that has been developed over the years.

18.3.1 Folch Method

The Folch method is the most popular method for the extraction of total lipid from tissue samples and widely used for MTB (Folch et al. 1957). Firstly, the tissues are homogenized by the addition of chloroform/methanol in the ratio of 2:1 according to the sample (the final volume is 20 times volume of tissue sample). Then the mixture is kept on shaking for 15–20 min at RT, and the extract is either filtered with Whatman filter paper or centrifuged. The crude extract is mixed with 0.2 volumes (4 ml of 20 ml) of water and/or adequate amount of saline solution, and the mixture is allowed to separate into two phases: the lower phase consists of chloroform-methanol-water in the ratio of 86:14:1 (by volume) and contains lipids, which are evaporated under N_2 gas, whereas the upper phase contains the same solvents in the ratio of 3:48:47 (by volume), respectively, and contains a large amount of non-lipid contaminants.

18.3.2 Modified Folch Method

This method was adapted from the Folch method which was described by Pal (Pal et al. 2015, 2016). *Mycobacterium* cultures are harvested at 10000 rpm for 10 min and homogenized in distilled water for 3 min. The homogenate is suspended in chloroform and methanol in the ratio of 1:2 and vortexed for 5 min. The homogenate is centrifuged at 2000 rpm at 4 °C for 10–15 min. The supernatant is collected into another glass vial, and then the remaining chloroform is added to make the final ratio (1,1, v/v) and filtered through Whatman No. 1 filter paper. The extract is then washed with 0.88 % potassium chloride (KCl) to remove the non-lipid contamination and allowed to separate into two phases. The upper phase contains non-lipid contaminants along with methanol and water. The lower dense phase of chloroform containing lipid is transferred into another glass vial and resolved by thin-layer chromatography (TLC) (Fig. 18.1a) and stored at −20 °C until further use.

18.3.3 Bligh and Dyer Method

The Bligh and Dyer method is applied to tissues like cod muscle (Bligh and Dyer 1959). Homogenized 100 g of fresh or frozen tissues with blender for 2 min is added with 100 ml of chloroform and 100 ml of methanol in the ratio of 1:2. Additionally 100 ml of chloroform is added to the homogenate to make the ratio 1:1, and the mixture is blended for 30 s. Again, 100 ml of distilled water is added and blended for 30 s. Then the homogenate is filtered through Whatman filter paper and the sample is centrifuged, which forms two layers. The upper phase is removed which contains non-lipid contaminants and the lower phase collected containing lipid part which is concentrated under N_2 gas and resolved by TLC (Fig. 18.1b) and stored at −20 °C.

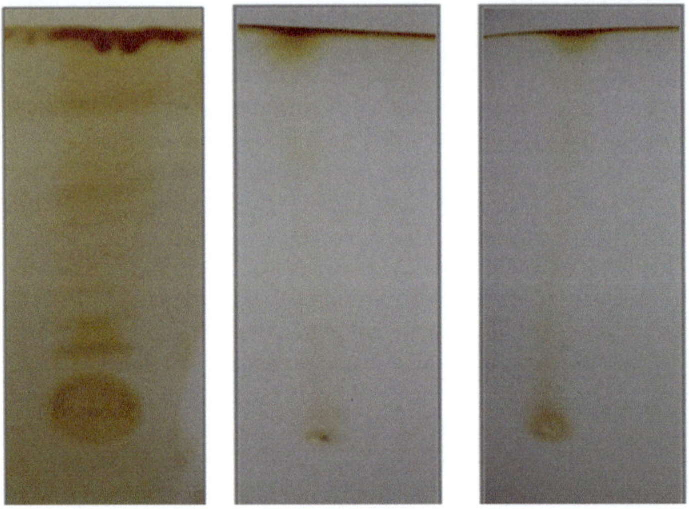

Fig. 18.1 TLC showing total lipid profile in *M. smegmatis* by various methods. (**a**) Modified method of Folch et al. (Image adapted from Pal et al. 2016). (**b**) Bligh and Dyer method (this study). (**c**) Chandramouli and Venkitasubramanian method (this study). Total lipids are resolved in chloroform/methanol/water (65:25:4) and visualized by iodine fumes

18.3.4 Chandramouli Method

The Chandramouli and Venkitasubramanian method is another method by which the total lipid extraction is performed (Chandramouli and Venkitasubramanian 1974). The MTB cells are harvested by centrifugation and suspended in chloroform and methanol in the ratio of 2:1(v/v) followed by stirring. After filtering the homogenate with Whatman filter paper, the extract is washed with 0.02% $CaCl_3$/0.017%M gC12/0.29%NaCl/0.37% KCl. The organic phase containing the lipid is collected, dried under N_2 gas, and resolved by TLC (Fig. 18.1c).

There are conflicting views related to total lipid isolation from MTB. For instance, the Bligh and Dyer (1959) procedure is considered to be more widely known and recommended for extraction of lipids from vascular plant tissues (Iverson et al. 2001; Axelsson and Gentili 2014). According to another study by Singh et al., either the Bligh and Dyer or Folch method provides low amount of lipid when applied to MTB. Moreover, their study reported Chandramouli and Venkitasubramanian's method, a less known method which gives considerably large concentration of mycobacterial lipids (Singh et al. 2014). However, in our studies, out of all the three methods, we have explored that modified Folch method is good for the recovery of total lipids (Pal et al. 2015, 2016, 2017).

18.4 Lipid Isolation Methodologies

18.4.1 Fatty Acids

Fatty acids are the long chains of hydrocarbon containing carboxylic acid which is present at the terminal in either saturated or unsaturated form. Saturated fatty acids have no double bond, whereas unsaturated fatty acids have double bond between their carbon atoms. Most naturally occurring fatty acids contains unbranched chain in the range from 4 to 28 carbon atoms (Nelson and Cox 2005).

18.4.1.1 Role in Pathogenesis

The outer layer of mycobacterium membrane contains glucan and proteins along with little amount of lipids, while the inner leaflet is made up of mycolic acid chain that is attached to arabinogalactan that is directly linked to peptidoglycan (Alderwick et al. 2015). The inner leaflet of mycomembrane is presumably composed of some free lipids such as TDM, TMM, and various glycolipids and phospholipids. Mycolic acids are the long fatty acid, present on the cell envelope of mycobacterium and known to be involved in pathogenesis (Marrakchi et al. 2014). Alpha mycolic acid contains 70% mycolic acid of the organism and several cyclopropane rings, whereas methoxy mycolic acid contains many methoxy groups that contribute between 10% and 15%, while the remaining 10% to 15% of mycolic acid are keto mycolic acids that contain keto groups.

Thus, MTB produces three main types of mycolic acids, viz., alpha, methoxy, and keto. Then another major component of cell envelope is TDM known as cord factor, which can elicit hypersensitive as well as T-cell-independent foreign body responses (Yamagami et al. 2001).

18.4.1.2 Mycolic Acid Isolation Protocols

The mycolic acid extraction method adapted from Daffé et al. (1983) method is described by Leite et al. (1998). The MTB cells are harvested (20–25 mg) by centrifugation at 10000 rpm for 10 min at 4 °C, and the cells are washed with PBS two to three times. The cells are dispersed into 5% potassium hydroxide solution which is prepared in 2-methoxyethanol, kept at 110 °C for 2 h, allowed to cool, and then acidified with 1 ml of sulfuric acid solution (20%). For the mycolic acid extraction, the above mixtures are washed two times with 5 ml diethyl ether. The phase containing ether is decanted and washed with 2 ml of water three times. Ether is removed on a water bath and the mycolic acid is methylated with the addition of 1 ml of diazomethane ether. Mycolic acids were analyzed through TLC by resolving in diethyl ether/petroleum ether in the ratio of 12:88 (v/v). The bands are visualized by spraying with 0.01% rhodamine in phosphate buffer (Fig. 18.2a).

Cantrell et al. have developed another method for mycolic acid extraction which is carried out by alkaline hydrolysis (Parish and Stoker 1998; Cantrell et al. 2013). Briefly, the *Mycobacterium* pellets are harvested by centrifugation, and chloroform and methanol are added in the ratio of 2:1 to extract the free total lipids, leaving behind the bound mycolyl arabinogalactan in cell pellets. 2 ml of 15% tetrabutyl ammonium hydroxide (TBAH) is added to dry cell mass to remove the bound mycolates from arabinogalactan and heated overnight at 100 °C followed by cooling the mixture. Then water is added to the TBAH solution to make the final ratio of 1:1. In the mixture, chloroform is added and vortexed for a minute. The bottom layer containing released mycolic acid could be dried under N_2 gas. The mycolic acid methyl esters are then resolved in petroleum ether/diethyl ether (85,15, v/v) solvent, and TLC plates were developed with phosphorimager (Fig. 18.2b).

Recently, Llorens-Fons et al. (2017) carried out mycolic acid extraction by Minnikin et al. method with little modification (Minnikin et al. 1980). Approximately 50 mg pellet is harvested from exponential phase of *Mycobacterium* culture. 2 ml of methanol, toluene, and sulfuric acid (30:15:1, v/v/v) is added to the pellet and kept overnight for heating at 80 °C. The sample is extracted twice with n-hexane, which yields methyl mycolates, and then dried under N_2 gas. The mycolate extract in n-hexane is concentrated by precipitation in cold methanol (4 °C, overnight). The sample is loaded to the TLC plates and resolved in n-hexane/diethyl ether (85:15; v/v, three runs). The mycolates are visualized after spraying with phosphomolybdic acid (10% in ethanol) and charring at 120 °C for 1 h (Fig. 18.2c).

Another variant for mycolic acid extraction can be performed by Singh et al. by using n-hexane solvent (Singh et al. 2012). Firstly, *Mycobacterium* cells are

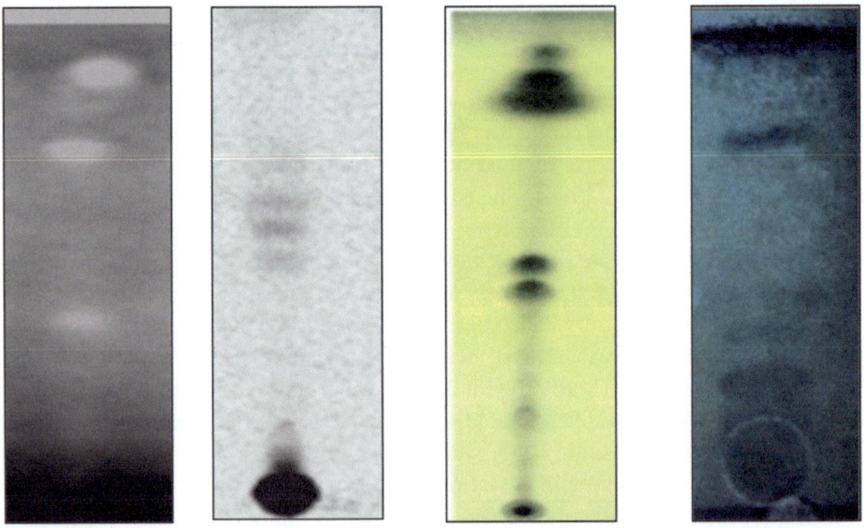

(a) (b) (c) (d)

Fig. 18.2 TLC showing mycolic acid profile of *Mycobacterium*. (**a**) Leite et al. method in MTB. (Image adapted from Leite et al. 1998). (**b**) Cantrell et al. method in MTB. (Image adapted from Cantrell et al. 2013). (**c**) Llorens-Fons et al. method in *M. abscessus*. (Image adapted from Llorens-Fons et al. 2017). (**d**) Mycolic acid extraction by Singh et al. from *M. smegmatis* (this study)

harvested, and about 2 gm of *Mycobacterium* pellet is suspended in 1 ml of hexane, and the mixture is incubated overnight at RT. The mixture is pelleted by centrifugation and the supernatant is transferred in a separate vial. An equal volume of chloroform and methanol (1:1,v/v) is added to the supernatant, and mixtures are then incubated until two phases are visible. The hexane layer containing mycolic acid is removed and analyzed for free mycolic acid. The sample is loaded to the TLC plates and developed in n-hexane/ethyl acetate (95:5, v/v, three runs). The mycolates are visualized after spraying with phosphomolybdic acid (10% in ethanol) and charring at 120 °C for 1 h (Fig. 18.2d). In our study, we have followed the Singh et al. method for extraction of mycolic acid from *Mycobacterium* cell wall and observed good separation of free mycolic acid on TLC plates.

18.4.1.3 TDM Isolation Protocols

Trehalose dimycolate is present on the cell wall of MTB and involved in pathogenesis (Jackson 2014). TDM extraction method described by Silva et al. (1985) was adapted by Indrigo et al. (2002). *Mycobacterium* cells are harvested by centrifugation, and the TDM is extracted with the addition of 1 ml petroleum ether followed by vigorous stirring for 2 min and allowing the mixture to incubate at RT. The mixture is centrifuged at 500 g for 10–15 min, and the supernatant is collected in another tube containing extracted material, and this extraction process is repeated twice more. All the supernatant is combined and dried under N_2 gas. The TLC plate is developed in chloroform and methanol in the ratio of 9:1 (v/v) which is visualized by spraying with 5% sulfuric acid containing 10% of α-naphthol followed by charring for 10 min (Fig. 18.3).

Slayden et al. performed another method for extraction of mycolic acid containing glycolipids TDM, TM, and Myc-PL (Slayden and Barry 3rd. 2001). The exponential phase of *Mycobacterium* culture (0.5–1.0 l) is harvested, and the pellets are dispersed in 10 ml of chloroform/methanol in the ratio of 2:1 (v/v) and shaken vigorously for 2–16 h (overnight). The mixture is centrifuged and the organic extract is removed in a glass vial. 500 ml acetone is added slowly (acetone should be in 50–100 times excess) in the organic extract, and the mixture is kept at −20 °C. For complete flocculation of material, the extract is kept at −20 °C for 12 h. Later the sample is centrifuged at 20,000 rpm for 15 min to recover the precipitated mycolate-containing material, and 1–2 ml of chloroform and methanol (2:1, v/v) is added in the recovered material and stored at −20 °C. The separation of the mycolate glycolipids (TDM, TMM, and Myc-PL) is performed by thin-layer chromatography and run in solvent system chloroform/methanol/ammonium hydroxide (80:20:2, v/v/v) where the bands are visualized by DPHT. Thus, the extraction of mycolic acid and TDM is apparent from a wide range of protocols.

18.4.2 Glycerolipids

Glycerolipids are composed of mono-, di-, and tri-substituted glycerols commonly called as triglycerides or triacyglycerides. It plays a very important role in the

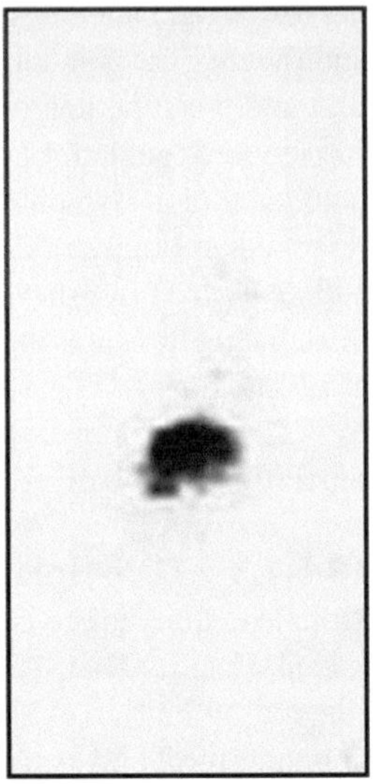

Fig. 18.3 TLC showing TDM in MTB by Indrigo et al. method. (Image adapted from Indrigo et al. 2002)

storage of metabolic energy. The hydrolysis of triglycerides releases glycerol and fatty acids from adipose tissues which is the initial step of fat metabolism (Coleman and Lee 2004; Ahmadian et al. 2007).

18.4.2.1 Role in Pathogenesis

TB infection starts when MTB enters into the lungs and infects the alveolar macrophages and eludes host defense resulting in immune response. The pathogens are present within granuloma and surrounded by foamy lipid-loaded macrophages (Stehr et al. 2013). During the formation phase of granuloma, MTB goes into a dormant state, in which the bacteria accumulate lipids in the form of lipid inclusion bodies from which many lipids including triacylglycerols (TAG) originate from host lipid degradation and/or fatty acid absorption. TAG plays a vital role in MTB metabolism, and it is hydrolyzed when MTB goes under the nutrient starvation condition (Daniel et al. 2011).

18.4.2.2 TAG Isolation Protocol

The apolar lipids are extracted as described by Slayden et al. (Slayden and Barry 3rd. 2001; Chauhan et al. 2013). The cells are pelleted, and apolar lipids are extracted with the addition of 2 ml of methanolic solution of 0.3% NaCl (100/10, v/v) and 1 ml petroleum ether to the cell pellet. The cell suspension is left on magnetic stirrer for 30 min, and then the mixture is centrifuged at 2500 rpm for 10 min. The upper layer containing apolar lipids is transferred into a separate vial. After adding 1 ml of petroleum ether in the lower phase, the mixture is kept on stirrer for 15 min. The cell suspension is again centrifuged as described above to recollect the supernatant. The supernatant from these extractions is combined and dried under N_2 gas. Apolar lipid

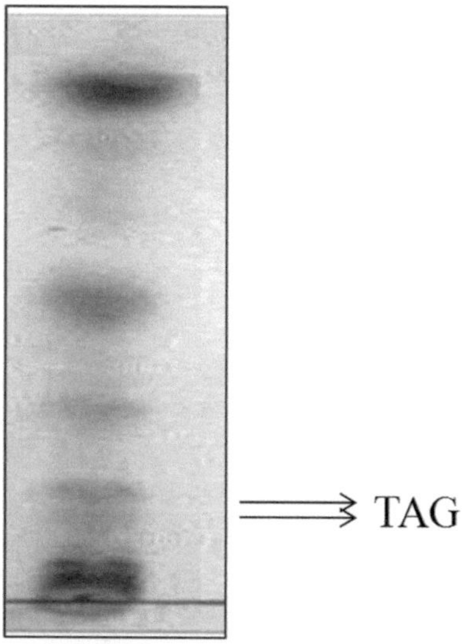

Fig. 18.4 TLC showing TAG in MTB by Chauhan et al. method. (Image adapted from Chauhan et al. 2013)

extracted is spotted on silica TLC plate and run in petroleum ether/diethyl ether (90:10 v/v) and visualized by spraying with 10% phosphomolybdic acid in ethanol and charred at 100 °C showing TAG (Fig. 18.4) and other apolar lipids (PDIM).

18.4.3 Glycerophospholipids

Glycerophospholipids are the major component of biological membrane of cells and are involved in metabolism and signaling. These are the membrane lipids in which two fatty acids are attached in ester linkage to the first and second carbons of glycerol and a highly polar or charged group is attached at third carbon through a phosphodiester bond. Glycerophospholipids include phosphatidylserine (PS), phosphatidylinositol (PI), phosphatidylglycerol (PG), phosphatidylethanolamine (PE), cardiolipin (CL), and phosphatidylinositol mannosides (PIM) which are the major lipid constituents present in the plasma membrane of MTB. Additionally, MTB plasma membrane consists of major lipoglycans referred to as lipomannan (LM) and Man LAM (Kaur et al. 2009; Guerin et al. 2010).

18.4.3.1 Role in Pathogenesis

The role of lipids such as PG, CL, PS, and PE is presently not well established in growth of MTB, but according to previous in vitro studies reported, PI, PIM, LM, and Man LAM are all essential for the viability of MTB (Kaur et al. 2009; Guerin et al. 2010). Previous data also showed that PIM is involved in the permeability of the cell envelope, inner membrane integrity, and regulation of cell septation and division (Parish et al. 1997; Kordulakova et al. 2002; Patterson et al. 2003; Morita et al. 2005, 2006).

18.4.3.2 PIM Isolation Protocols

The extractions of PIMs are performed as described previously (Besra 1998; Torrelles et al. 2009). Briefly the mid log phase culture of *Mycobacterium is harvested* by centrifugation, and the cells are washed with PBS. The pellets are suspended in methanol/0.3% sodium chloride (NaCl) (2 ml, 100:10, v/v) and 1 ml of petroleum to extract the lipids. The cell suspensions are centrifuged and the upper layer consisting of petroleum ether is removed. The lower phase consisting of methanolic saline fraction is heated at 65 °C for 5 min, and the lower phase mixture is mixed with 2.3 ml of chloroform:methanol:0.3% NaCl (9:10:3, v/v/v). Later, the mixture is centrifuged and supernatant is collected in a glass vial. The bacterial residue is further extracted by adding 0.75 ml of chloroform:methanol:0.3% NaCl in the ratio of 5:10:4 (v/v/v), respectively, to obtain all polar lipids. Then the supernatants from above extractions are pooled and further extracted with 1.3 ml of chloroform and 1.3 ml of 0.3% NaCl which form two phases. The lower phase is retained and dried under N_2 gas. Before spotting on TLC plate, the organic extract is resuspended in chloroform/methanol/water in the proportion 10:10:3 (v/v/v). The TLC plate is resolved in the first dimension solvent chloroform/methanol/water (60:30:6, v/v/v) and the second dimension solvent chloroform/acetic acid/methanol/water (40:25:3:6, v/v/v/v). PIMs are visualized by spraying α-naphthol or 10% sulfuric acid in ethanol and charred at 120 °C for 5 min (Fig. 18.5).

PIM extraction can be performed by another method by *Slayden and Barry* (Slayden and Barry 3rd. 2001). An overnight grown culture is harvested by centrifugation, and the pellet is suspended in 10 ml chloroform/methanol/water in the ratio of 10:10:3 (v/v/v) and the mixture kept on stirring for 15 min at 55 °C. The mixture is centrifuged at 750 g for 30 min, and the organic extract is decanted and dried under steam of air. The organic residues are hydrolyzed by adding 3 ml of 0.01 M hydrochloric acid for 5 min at 100 °C. Chloroform/methanol (2:1,v/v) is added in organic extract to separate the intact PIM from saccharide constituents of polyprenols and the final ratio of chloroform/methanol/water (8:4:3). The lower phase is

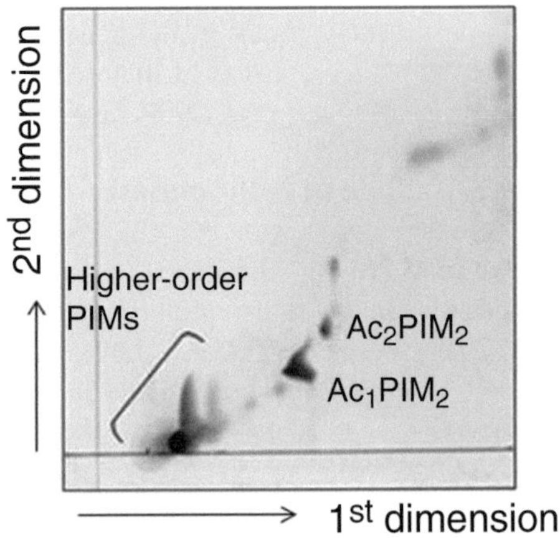

Fig. 18.5 TLC showing PIM in MTB by Torrelles et al. method. (Image adapted from Torrelles et al. 2009)

retained containing crude PIM and dried under N_2 gas. TLC analysis of PIM and polyprenol glycolipids is resolved in solvent systems of chloroform/methanol/water (65:25:4) or (60:30:6).

18.4.4 Saccharolipids

Saccharolipids are the class of bacterial membrane lipids in which fatty acids are linked with sugar backbone, therefore forming structures that are compatible with membrane bilayers.

18.4.4.1 Role in Pathogenesis

It includes sulfolipids, DAT, and PAT which play very crucial roles in pathogenesis (Bailo et al. 2015). DAT and PAT not only are involved in the structural role in the cell envelope of MTB but also are crucial for MTB replication and persistence in host, support the intracellular survival of MTB, and are involved in host immune response (Belardinelli et al. 2014). Sulfolipids contain sulfur functional group. It is the sulfated trehalose ester which is acylated with three or four acyl groups that contain one saturated fatty acid. Mmpl8 are responsible for the synthesis of sulfolipids in MTB which play a crucial role in the initial stage of MTB infection (Converse et al. 2003).

The extraction of sulfolipids is performed by using hexane-decylamine, or chloroform/methanol (2:1; v:v) solvents (100 g of pellets was extracted three times from 100 ml of solvent) are adapted from Goren (Goren 1970; Rhoades et al. 2011). Briefly, the cells are harvested and dispersed in 0.1% decylamine and incubated for 1–4 weeks at 4 °C and subsequently extracted in 0.5% decylamine with sonication of cells at RT for 30 min. For chloroform/methanol extraction, firstly the pellet is washed with PBS and resuspended in chloroform and methanol in the ratio of 2:1 (v/v). Then the pellet is extracted at 4 °C for 1–4 weeks for the first time, and subsequently extraction is done with sonication for 30 min at RT. The extract is pooled, filtered with PFTE filter, and dried by the help of rotary evaporator. To remove decylamine, residues are dissolved in hexane and equal volume of citric acid is added and it is mixed vigorously and allowed the separation of the mixture into two phases. The top layer is dried after percolation with the help of column of anhydrous sulfuric acid following with 1 column volume of hexane. Then total crude extract of lipid is dried by N_2 gas. The plates are firstly resolved in chloroform/methanol/acetic acid/water (95:1:5:0.3; v/v/v/v) to 20 cm, dried, and then again resolved in chloroform/acetone/methanol/water/acetic acid (158:83:1:6:32; v/v/v/v/v) till 15 cm. Plates are sprayed with ethanolic sulfuric acid (50%) followed by charring with a heat gun (Fig. 18.6a).

Another method for extraction is performed by Seeliger et al. method (Seeliger et al. 2012). The MTB cells are harvested and the pellet is suspended in 1 ml of hexane. The upper organic phase is removed by centrifugation, and equal volume of chloroform and methanol in the ratio of 1:1(v/v) is added to remove the surface

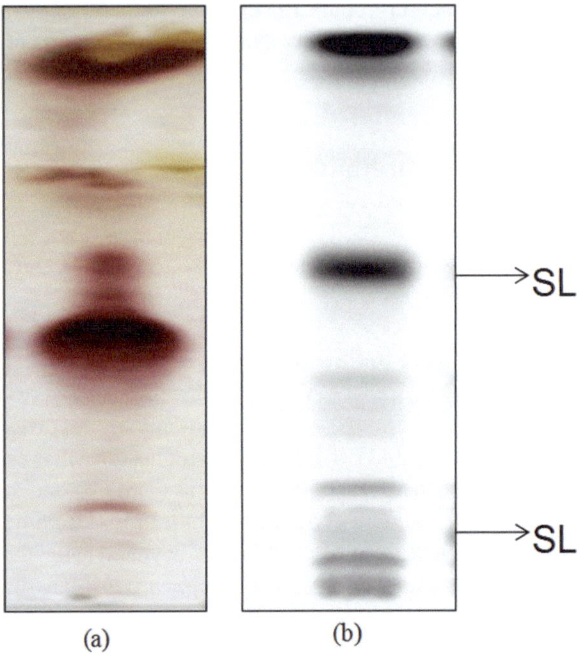

Fig. 18.6 TLC showing sulfolipids in MTB. (**a**) Rhoades et al. method. (Image adapted from Rhoades et al. 2011). (**b**) Seeliger et al. method. (Image adapted from Seeliger et al. 2012)

lipids. The remaining cell pellet and aqueous phase are extracted by adding 4 ml of chloroform/methanol (1:1, v/v) and incubated overnight at RT. The mixture is centrifuged and supernatant is collected in a glass vial. The TLC are resolved in chloroform/methanol/water (60:12:1, v/v/v), and the bands are visualized by phosphorimaging (Fig 18.6b).

18.4.5 Polyketides

Polyketides (PKs) are another class of complex lipids from MTB. Polyketides are biosynthesized by the decarboxylative condensation of malonyl-COA which is similar to fatty acid synthesis process.

18.4.5.1 Role in Pathogenesis

MTB produce two structurally similar methyl-branched fatty acids consisting of lipid, known as phenolic glycolipids and PDIM. Various secreted polyketides and polyketide-containing compounds are important virulence factors (Ferreras et al. 2008). Previous studies showed that PKs are present on the surface of mycobacterium which is linked with outer leaflets of cell wall which is a crucial factor responsible for pathogenesis (Cox et al. 1999; Camacho et al. 1999a; Domenech et al. 2004). Moreover these lipids showed a physical protection against host-induced damage (Camacho et al. 2001). Additionally it also modulates response by altering the cytokine profile in macrophages against infection (Rao et al. 2005). The synthesis of PKs takes place in cytoplasmic membrane of mycobacterium which requires transport of PKs from the cytoplasmic membrane to cell surface.

18.4.5.2 PDIM Extraction

PDIM is extracted from homogenate of *M. leprae* which is described as follows (Draper and Payne,1983). The tissues are homogenized in 0.15 M-NaCl/2 M-Tris base (9:1, v/v) and centrifuged at 1000 g for 10 min. The alkaline supernatant is decanted which is neutralized by the addition of hydrochloric acid. The extract is then washed with 0.37% (w/v) KCl resulting in the formation of two phases as described by Folch et al. KCl is present in the aqueous layer to reduce the accumulation of material at the interface (0.37% w/v). The lower phase is collected in a separate vial and evaporated for viscosity, and TLC is resolved in petroleum ether/acetone, in the ratio of 85:15 (v/v, three runs) (Fig. 18.7).

18.4.6 Prenol Lipids

Prenol is the class of lipid which is synthesized from isopentenyl diphosphate (5-carbon precursor) and dimethylallyl diphosphate, produced through mevalonic acid pathway (Kuzuyama and Seto 2003). Polyprenol phosphate is a key carrier lipid involved in the synthesis of the core structures of the mycobacterial cell wall, including peptidoglycan and arabinogalactan. The isoprenoids are formed by the successive addition of C5 unit (also known in the class of terpene) but the polyterpene are the subclass that contains more than 40 carbon (Rodriguez-Concepcion 2004; Porterand and Spurgeon 1981). Bacteria synthesized polyprenol in which isoprenoid unit is linked with oxygen that remains unsaturated typically containing

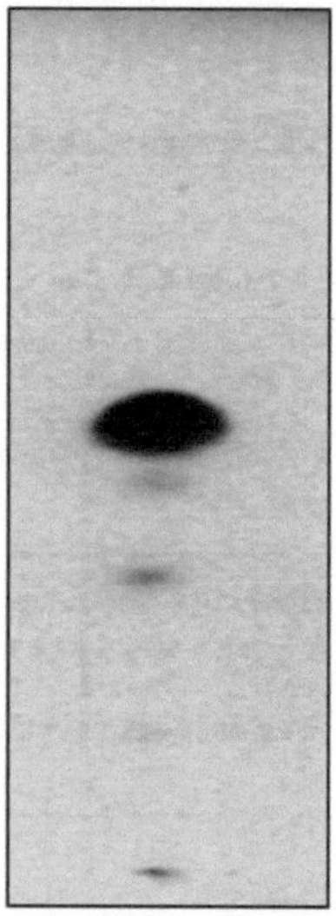

Fig. 18.7 TLC showing PDIM in MTB by Draper and Payne method. (Image adapted from Draper and Payne 1983)

10–12 unit long, whereas dolichols contain 18–22 isoprene units. In some bacteria, isoprenoid precursors are formed by methylerythritol phosphate pathway.

18.4.6.1 Role in Pathogenesis

Polyprenol and its phosphorylated derivates have an important role in the transport of oligosaccharide across the membrane. Polyprenol and phosphate sugar involve in the cytoplasmic glycosylation reaction in polysaccharide biosynthesis (Raetz and Whitfield 2002; Lazar and Walker 2002).

18.4.6.2 Prenol Isolation Protocol

The *Mycobacterium* cultures are harvested by centrifugation at 10,000 rpm for 15 min, and the pellets are dispersed in 10 ml of ethanol and kept on the heating plate at 70 °C for 20–30 min. The homogenate is centrifuged at 750 g for 15 min, and the supernatant is decanted containing the organic extract and air-dried. The residues are resuspended by the addition of 12 ml of chloroform/methanol in the ratio of 2:1 (v/v) and partition followed by the addition of 3 ml water; therefore, the final ratio of chloroform/methanol/water becomes 8:4:3, and organic phase containing glycolipids is removed. After evaporating the resultant to dryness, the residue is subjected to 1 ml of 0.1 M NaOH in ethanol for 45 min at 37 °C to separate the mixture into two phases. The upper layer is collected and performed for TLC, which is resolved in chloroform/methanol/distilled water (65:25:4, v/v/v) or (60:30:6, v/v/v) solvent system, and the bands are visualized by iodine (Slayden and Barry 3rd. 2001).

18.5 Lipid-Based Nanoformulations

In the era of nanotechnology, polymeric delivery as an alternative carrier for long-term delivery of therapeutic agents has suffered from limitation due to polymeric toxicity, high cost of polymers, solvent residue left after production, and the lack of feasibility for plant scale-up method. To overcome this problem, lipid-based nanoformulations as an alternative option to polymeric formulation in the management of TB due to inherited properties, biocompatibility and biodegradability, physical diversity, lower toxicity, high incorporation efficiency for lipophilic drug, and improved bioavailability stabilization of drug have emerged as success (Bibhas et al. 2017). Lipid-based nanoformulation consists of nanoemulsion, solid lipid nanoparticles, nanostructured lipid carriers, liposomes, and niosomes.

18.5.1 Nanoemulsion

It is kinetically stable, in which one phase is usually dispersed in another phase where it is usually immiscible covering the size ranging from 50 to 1000 nm (Devarajan and Ravichandran 2011). The nanosized particle of nanoemulsion shows various properties like surface area per unit volume, optically transparent appearance, tunable rheology, and robust stability. It plays a very important role in the field of antitubercular drug delivery. In a study, Nikonenko et al. developed phospholipid-based nanoemulsion system of SQ-64-NE which was effective against MTB (Nikonenko et al. 2014).

18.5.2 Solid Lipid Microparticles (SLMs)/Solid Lipid Nanoparticles (SLNs)

SLMs are a promising drug carrier system in which RIF-loaded lipid microsphere delivers the drug to alveolar macrophages through intranasal route for acquiring improved therapeutic efficacy in TB and TB/HIV patients. SLNs are the rapidly developing field of nanotechnology with potential applications in various fields such as in effective drug delivery. It consists of solid physiological lipids generally dispersed in water or an aqueous surfactant solution. The matrix of SLNs is composed of closely packed perfect crystalline solid lipid with very few empty spaces resulting in poor drug loading. Moreover, it provides unique characteristics such as good tolerability, freedom from organic solvent, pilot plant scale-up feasibility, and control and/or target drug release and has the ability to incorporate both lipophilic and hydrophilic drugs (Muller and Keck 2004). In a study, it has been reported that SLNs were prepared with an encapsulation efficiency for rifampicin (51%), for isoniazid (45%), and for pyrazinamide (41%) after a single oral administration to mice (Pandey et al. 2005).

18.5.3 Nanostructured Lipid Carrier (NLCs)

NLCs are second generation of colloidal lipid nanoparticles composed of both solid and liquid lipids as a core matrix. It provides some advantages over conventional carriers for drug therapy such as increased solubility, the ability to enhance storage stability, reduced adverse effect, prolonged half-life, improved permeability and bioavailability, and tissue-targeted delivery (Fang et al. 2013). NLCs are utilized by several researchers as committed for successful delivery of anti-TB drugs due to its unique properties (Bibhas et al. 2017).

18.5.4 Liposomes

Liposomes are simple microscopic vesicles in which an aqueous volume is enclosed by a lipid bilayer membrane. They are able to encapsulate both hydrophilic and lipophilic drug substances (Utreja et al. 2011). Moreover it can also be used as a nontoxic vehicle to encapsulate insoluble drugs size ranging from 0.05 to 5.0 mm in diameter (Mansoori et al. 2012). It possesses many advantages such as selective passive drug targeting, better therapeutic efficacy, and flexibility to couple with site-specific ligands to achieve active targeting, increased drug stability, and reduced toxicity of encapsulating agents with improved pharmacokinetic effects. Liposomes are readily taken up by the macrophages, and the contents are released intracellular that immediately acts against MTB (Bibhas et al. 2017).

18.5.5 Niosomes

Niosomes are one of the promising drug carrier systems for the delivery of different types of drugs, antigens, and hormones (Gopalkrishnan and Chenthilnathan 2012). It is a non-ionic surfactant-based vesicle structurally resembling to liposomes. It is a better lipid-based carrier over liposomes owing to its better chemical stability, easier pilot plant scale-up feasibility, greater osmotic activeness, and cheaper cost of production along with longer storage time. In niosome, hydrophobic drug molecules can be encapsulated inside surfactant bilayer and hydrophilic drug molecules inside the lipidic vesicle (Pravinagurjar and Chouksey 2014).

18.6 Conclusions

Although not very comprehensive, this chapter has recapitulated some of the most commonly used methods to isolate mycobacterial lipids and use of lipid-based nanoformulations in the management of TB. This information will help the researchers dealing with lipids of *Mycobacterium*, given the emerging roles of lipid in governing drug resistance in MTB.

References

Ahmadian M, Duncan RE, Jaworski K, Sarkadi-Nagy E, Sul HS (2007) Triacylglycerol metabolism in adipose tissue. Future Lipidol 2:229–223

Alderwick LJ, Harrison J, Lloyd GS, Birch HL (2015) The mycobacterial cell wall--peptidoglycan and arabinogalactan. Cold Spring Harb Perspect Med 5:a021113

Axelsson M, Gentili F (2014) A single –step method for rapid extraction of total lipids from green microalgae. PLoS One 9:e89643

Bailo R, Bhatt A, Ainsa JA (2015) Lipid transport in *Mycobacterium tuberculosis* and its implications in virulence and drug development. Biochem Pharmacol 96:159–167

Belardinelli JM, Larrouy-Maumus G, Jones V, Sorio de Carvalho LP, McNeil MR, Jackson M (2014) Biosynthesis and translocation of unsulfated acyltrehaloses in *Mycobacterium tuberculosis*. J Biol Chem 289:27952–27965

Besra GS (1998) Preparation of cell wall fraction from mycobacteria. Methods Mol Biol 101:91–107

Bibhas CM, Subas CD, Gitanjali M, Narahari NP (2017) Exploring the use of lipid based nanoformulations for the management of tuberculosis. J Nanosci Curr Res 2:1–15

Bligh EG, Dyer WJ (1959) A rapid method of total lipid extraction and purification. Can J Biochem Physiol 37:911–917

Camacho LR, Ensergueix D, Perez E, Gicquel B, Guilhot C (1999a) Identification of a virulence gene cluster of *Mycobacterium tuberculosis* by signature-tagged transposon mutagenesis. Mol Microbiol 34:257–267

Camacho LR, Constant P, Raynaud C, Laneelle MA, Triccas JA, Gicquel B, Daffe M, Guilhot C (2001) Analysis of the phthiocerol dimycocerosate locus of *Mycobacterium tuberculosis*. Evidence that this lipid is involved in the cell wall permeability barrier. J Biol Chem 276:19845–19854

Camacho LR, Ensergueix D, Perez E, Gicquel B, Guilhot C (1999b) Identification of a virulence gene cluster of *Mycobacterium tuberculosis* by signature-tagged transposon mutagenesis. Mol Microbiol 34:257–267

Cantrell SA, Leavell MD, Marjanovic O, Lavarone AT, Leary JA, Riley LW (2013) Free mycolic acid accumulation in the cell wall of the mce1 operon mutant strain of *Mycobacterium tuberculosis*. J Microbiol 51:619–626

Chandramouli V, Venkitasubramanian TA (1974) Effect of age on the lipids of mycobacteria. Indian J Chest Dis 16:199–207

Chauhan P, Reddy PV, Singh R, Jaisinghani N, Gandotra S, Tyagi AK (2013) Secretory phosphatases deficient mutant of *Mycobacterium tuberculosis* imparts protection at the primary site of infection in guinea pigs. PLoS One 8:e77930

Coleman RA, Lee DP (2004) Enzymes of triacylglycerol synthesis and their regulation. Prog Lipid Res 43:134–176

Converse SE, Mougous JD, Leavell MD, Leary JA, Bertozzi CR, Cox JS (2003) MmpL8 is required for sulfolipid-1 biosynthesis and *Mycobacterium tuberculosis* virulence. Proc Natl Acad Sci U S A 100:6121–6126

Cox JS, Chen B, McNeil M, Jacobs WR Jr (1999) Complex lipid determines tissue-specific replication of *Mycobacterium tuberculosis* in mice. Nature 402:79–83

Daffé M, Lanéelle MA, Asselineau C, Lévy-Frébault V, David H (1983) Intérêt taxonomique des acides gras des mycobactéries; proposition d'une méthode d'analyse. Ann Microbiol (Paris) 134B:241–256

Daniel J, Maamar H, Deb C, Sirakova TD, Kolattukudy PE *(2011) Mycobacterium tuberculosis* uses host triacylglycerol to accumulate lipid droplets and acquires a dormancy-like phenotype in lipid-loaded macrophages. PLoS Pathog 7:e1002093

Devarajan V, Ravichandran V (2011) Nanoemulsions: as modified drug delivery tool. Int J Compr Pharm 2:1–6

Domenech P, Reed MB, Dowd CS, Manca C, Kaplan G, Barry CE (2004) The role of MmpL8 in sulfatide biogenesis and virulence of *Mycobacterium tuberculosis*. J Biol Chem 279:21257–21265

Draper P, Payne SN (1983) Isolation of a Characteristic Phthiocerol Dimycocerosate from *Mycobacterium leprae*. J Gen Microbiol 129:859–863

Ferreras JA, Stirrett KL, Lu X, Ryu J-S, Soll CE, Tan DS, Quadri EN (2008) Mycobacterial phenolic glycolipid virulence factor biosynthesis: mechanism and small-molecule inhibition of polyketide chain initiation. Chem Biol 15:51–61

Fang CL, Al-Suwayeh SA, Fang JY (2013) Nanostructured lipid carriers (NLCs) for drug delivery and targeting. Recent Pat Nanotechnol 7:41–55

Folch J, Lees M, Sloane Stanley GH (1957) A simple method for the isolation and purification of total lipides from animal tissues. J Biol Chem 226:497–509

Gopalkrishnan S, Chenthilnathan AN (2012) A novel drug delivery device. Res J Pharm Biol Chem Sci 3:1090–1098

Goren MB (1970) Sulfolipid I of Mycobacterium tuberculosis, strain H37Rv I. Purification and properties. Biochim Biophys Acta 210:116–126

Guerin ME, Korrdulakova J, Alzari PM, Brennan PJ, Jackson MJ (2010) Molecular basis of phosphatidyl-myo-inositol mannoside. J Biol Chem 285:33577–33583

Hameed S, Fatima Z, Khurana SMP (2012) Lipid molecule: a key player in human pathophysiology. In: Behera KK (ed) Advance frontier on biotechnology. Jaya Publishing House, Delhi, pp 213–229

Hameed S, Fatima Z (2014) Lipidomics: tool to redefine lipids. In: Biotechnology—progress and prospects. Studium Press LLC, Houston, pp 302–312

Indrigo J, Hunter RL Jr, Actor JK (2002) Influence of trehalose 6,6H-dimycolate (TDM) during mycobacterial infection of bone marrow macrophages. Microbiology 148:1991–1998

Iverson SJ, Lang SL, Cooper MH (2001) Comparison of the bligh and dyer and folch methods for total lipid determination in a broad range of marine tissue. Lipids 36:1283–1287

Jackson M (2014) The Mycobacterial cell envelope- Lipids. Cold Spring Harb Perspect Med 4:a021105

Kaur D, Guerin ME, Skovierova H, Brennan PJ, Jackson M (2009) Biogenesis of the cell wall and other glycoconjugates of *Mycobacterium tuberculosis*. Adv Appl Microbiol 69:23–78

Kordulakova J, Gilleron M, Mikusova K, Puzo G, Brennan PJ, Gicquel B, Jackson M (2002) Definition of the first mannosylation step in phosphatidylinositol mannoside synthesis. PimA is essential for growth of mycobacteria. J Biol Chem 277:31335–31344

Kuzuyama T, Seto H (2003) Diversity of the biosynthesis of the isoprene units. Nat Prod Rep 20:171–183

Lazar K, Walker S (2002) Substrate analogues to study cell-wall biosynthesis and its inhibition. Curr Opin Chem Biol 6:786–793

Leite CQ, de Souza CW, Leite SR (1998) Identification of mycobacteria by thin layer chromatographic analysis of mycolic acids and conventional biochemical method: four years of experience. Mem Inst Oswaldo Cruz 93:801–805

Llorens-Fons M, Perez-Trujillo M, Jualian E, Brambilla C, Alcaide F, Byrd TF, Luquin M (2017) Trehalose polyphleates, external cell wall lipids in *Mycobacterium abscessus*, are associated with the formation of clumps with cording morphology, which have been associated with virulence. Front Microbiol *8*:1402

Marrakchi H, Laneelle MA, Daffe M (2014) Mycolic acids: structure, biosynthesis, and beyond. Chembiol 21:67–85

Mansoori M, Agrawal S, Jawade S, Khan M (2012) A review on liposome. Int J Adv Res Pharm Biosci 2:453–464

Matougui N, Boge L, Groo AC, Umerska A, Ringstad L, Bysell H, Saulnier P (2016) Lipid-based nanoformulations for peptide delivery. Int J Pharm 502:80–97

Minnikin DE, Hutchinson IG, Caldicott AB, Goodfellow M (1980) Thin-layer chromatography of methanolysates of mycolic acid-containing bacteria. J Chromatogr A 188:221–233

Morita YS, Velasquz R, Taig E, Waller RF, Patterson JH, Tull D, Williams SJ, Billman-Jacobe H, McConville MJ (2005) Compartmentalization of lipid biosynthesis in mycobacteria. J Biol Chem 280:21645–21652

Morita YS, Sena CB, Waller RF, Kurokawa K, Sernee MF, Nakatani F, Haites RE, Billman Jacobe H, Mc Conville MJ, Maeda Y, Kinoshita T (2006) PimE is a polyprenol-phosphate-mannose-dependent mannosyltransferase that transfers the fifth mannose of phosphatidylinositol mannoside in mycobacteria. J Biol Chem 281:25143–25155

Muller RH, Keck CM (2004) Challenges and solutions for the delivery of biotech drugs--a review of drug nanocrystal technology and lipid nanoparticles. J Biotechnol 113:151–170

Nelson DL, Cox MM (2005) Chapter 10: Lipids. In: Lehninger principles of biochemistry, 4th edn. W.H Freeman and Company, New York

Nikonenko B, Reddy VM, Bogatcheva E, Protopopoya M, Einck L, Nacv CA (2014) Therapeutic efficacy of SQ641-NE against *Mycobacterium tuberculosis*. Antimicrob Agents Chemother 58:587–589

Pal R, Hameed S, Kumar P, Singh S, Fatima Z (2015) Comparative lipidome profile of sensitive and resistant. *Mycobacterium tuberculosis* strain. Int J Curr Microbiol App Sci 2015:189–197

Pal R, Hameed S, Sharma S, Fatima Z (2016) Influence of iron deprivation on virulence traits of mycobacteria. Braz J Infect Dis 20:585–591

Pal R, Hameed S, Kumar P, Singh S, Fatima Z (2017) Comparative lipidomics of drug sensitive and resistant *Mycobacterium tuberculosis* reveals altered lipid imprints. 3 Biotech 7:325

Pandey R, Sharma S, Khuller G (2005) Oral solid-lipid nanoparticle based antitubercular chemotherapy. Tuberculosis 85:415–420

Parish T, Stoker NG (1998) Mycobacteria protocols, vol 101. Humana Press, Totowa. (Chapter 8)

Parish T, Liu J, Nikaido H, Stoker NG (1997) A *Mycobacterium smegmatis* mutant with a defective inositol monophosphate phosphatase gene homolog has altered cell envelope permeability. J Bacteriol 179:7827–7833

Patterson JH, Waller RF, Jeevarajah D, Billman-Jacobe H, McConville MJ (2003) Mannose metabolism is required for mycobacterial growth. Biochem J 372:77–86

Porterand JW, Spurgeon SL (1981) Biosynthesis of isoprenoid compounds, vol 1. Wiley, New York

Pravinagurjar N, Chouksey S (2014) Niosome: a promising pharmaceutical drug delivery. Int J Pharm Drug Anal 2:425–431

Raetz CRH, Whitfield C (2002) Lipopolysaccharides endotoxins. Annu Rev Biochem 71:635–700

Rao V, Fujiwara N, Porcelli SA, Glickman MS (2005) *Mycobacterium tuberculosis* controls host innate immune activation through cyclopropane modification of a glycolipid effector molecule. J Exp Med 201:535–543

Rhoades ER, Streeter C, Turk J, Hsu FF (2011) Characterization of sulfolipids of *Mycobacterium tuberculosis* H37Rv by multiple-stage linear ion-trap high resolution mass spectrometry with electrospray ionization reveals that family of sulfolipid II predominates. Biochemistry 50:9135–9147

Rodriguez-Concepcion M (2004) The MEP pathway: a new target for the development of herbicides, antibiotics and antimalarial drugs. Curr Pharm Res 10:2391–2400

Seeliger JC, Holsclaw CM, Schelle MW, Botyanszki Z, Gilmore SA, Tully SE, Niederweis M, Cravatt BF, Leary JA, Bertozzi CR (2012) Elucidation and chemical modulation of sulfolipid-1 biosynthesis in *Mycobacterium tuberculosis*. J Biol Chem 287:7990–8000

Sharma S, Pal R, Hameed S, Fatima Z (2016) Antimycobacterial mechanism of vanillin involves disruption of cell-surface integrity, virulence attributes, and iron homeostasis. Int J Mycobacteriol 5:460–468

Sharma S, Hans S, Hameed S, Fatima Z (2017) Elucidating the mechanism of vanillin induced mycobacterial membrane disruption: implications of lipid alteration. SOJ Microbiol Infect Dis 5:1–7

Silva CL, Ekizlerian SM, Fazioli RA (1985) Role of cord factor in the modulation of infection caused by mycobacteria. Am J Pathol 118:238–247

Singh P, Sinha R, Tandon R, Tyagi G, Khatri P, Chandra Shekhar Reddy L, Saini NK, Pathak R, Varma-Basil M, Prasad AK, Bose M (2014) Revisiting a protocol for extraction of mycobacterial lipids. Int J Mycobacteriol 3:168–172

Singh V, Jamwal S, Jain R, Verma P, Gokhale R, Rao KV (2012) Mycobacterium tuberculosis-driven targeted recalibration of macrophage lipid homeostasis promotes the foamy phenotype. Cell Host Microbe 12:669–681

Slayden RA, Barry CE 3rd. (2001) Analysis of the lipid of lipids of *Mycobacterium tuberculosis*. Analysis of the Lipids of *Mycobacterium tuberculosis*. Methods Mol Med 54:229–245

Stehr M, Elamin AA, Singh M (2013) Lipid inclusion in mycobacterial infections. In: Mahboub BH, Mayank G (eds) Tuberculosis –Current issues in diagnosis and management. (Chapter 3)

Torrelles JB, DesJardin LE, MacNeil J, Kaufman TM, Kutzbach B, Knaup R, McCarthy TR, Gurcha SS, Besra GS, Clegg S, Schlesinger LS (2009) Inactivation of *Mycobacterium tuberculosis* mannosyltransferase *pim B* reduces the cell wall lipoarabinomannan and lipomannan content and increases the rate of bacterial-induced human macrophage cell death. Glycobiology 19:743–755

Utreja P, Jain S, Tiwary A (2011) Localized delivery of paclitaxel using elastic liposomes: formulation development and evaluation. Drug Deliv 18:367–376

Yamagami H, Matsumoto T, Fujiwara N, Arakawa T, Kaneda K, Yano I, Kobayashi K (2001) Trehalose 6, 6′-dimycolate (cord factor) of *Mycobacterium tuberculosis* induces foreign-body- and hypersensitivity-type granulomas in mice. Infect Immun 69:810–815

Nanomaterial-Assisted Mass Spectrometry: An Evolving Cutting-Edge Technique

19

Ashutosh Singh, Nitin Bhardwaj, and Rajendra Prasad

Abstract

In the last decade, the area of "omics research" has received tremendous recognition and found significances in the field of biomedicine. And so did develop the technologies for analyzing various kinds of biomolecules. The advances made in omics tools are quite diverse and advanced. In this chapter, we give an overview of two most common approaches used for the analysis of biomolecules, namely, electrospray ionization and matrix-assisted laser desorption/ionization (MALDI) mass spectrometry. The conventional MALDI approach for biomolecular analysis relies on organic matrices for ionization of analytes, which have several disadvantages in analysis of small molecules. Here we discuss the types and application of nanomaterials in laser desorption/ionization mass spectrometry in the analysis of biomolecules. Additionally, examples of nanomaterial-assisted mass spectrometry imaging are discussed. Together this chapter provides insights into mass spectrometry and significance of nanomaterials in analysis of biomolecules, which have large-scale implications in the field of biomedicine.

Keywords

Electrospray ionization · Matrix-assisted laser desorption/ionization · Biomolecules · Mass spectrometry · Nanoparticles

A. Singh (✉)
Department of Biochemistry, University of Lucknow, Lucknow, Uttar Pradesh, India
e-mail: singh_ashutosh@lkouniv.ac.in

N. Bhardwaj
Department of Zoology and Environmental Science, Gurukul Kangri University, Haridwar, India

R. Prasad
Amity Institute of Integrative Sciences and Health, Amity University, Gurgaon, Haryana, India

Amity Institute of Biotechnology, Amity University, Gurgaon, Haryana, India

S. K. Saxena, S. M. P. Khurana (eds.), *NanoBioMedicine*,
https://doi.org/10.1007/978-981-32-9898-9_19

Abbreviations

CHCA	α-cyano-4-hydroxycinnamic acid
CNT	carbon nanotubes
DHBA	2,5-dihydroxybenzoic acid
DWCNTs	double-walled carbon nanotubes
EI	electron impact
ESI	electrospray ionization
FCCA	α-cyano-4-fluorocinnamic acid
GC	gas chromatography
HPA	3-hydroxypicolinic acid
HPLC	high-performance liquid chromatography
LDI	laser desorption ionization
MALDI	matrix-assisted laser desorption/ionization
MRM	multiple reaction monitoring
MS/MS	triple quadrupole
MSI	mass spectrometry imaging
MS^n	tandem quadrupole linear ion trap
MWCNTs	multi-walled carbon nanotubes
NALDI	nanoparticle-assisted laser desorption/ionization
PC	phosphatidylcholine
PREIS	precursor ion scanning
Q-TOF	quadrupole time-of-flight
SRM	single reaction monitoring
SWCNTs	single-walled carbon nanotubes
TOF	time-of-flight

19.1 Introduction to Mass Spectrometry

The last decade has seen a tremendous advancement in the area of structure determination using mass spectrometry approaches. These mass spectrometry-based methods have allowed sensitive and accurate analysis of several classes of biomolecules. These range from drugs, simple and complex lipids, peptides, proteins, glycans, oligonucleotides, and oligosaccharides among others (Fenn et al. 1989; Bakry et al. 2011; Cho et al. 2013; Sandrin et al. 2013; Bergman et al. 2014; Brügger 2014; Görgens et al. 2018). In order to understand how this biomolecular analysis is achieved, we need to understand the basic workflow and various components of a mass spectrometer.

19.1.1 Chromatographic Separation

Chromatographic separation is a necessity for analysis of molecules that have similar masses. In majority of mass spectrometry analysis, samples are pre-resolved on a high-performance liquid chromatography (HPLC) column, where the sample separation is dependent on factors like surface characteristics of column matrix, flow rate of mobile phase, temperature, pH, etc. (Torres-Lapasió and García-Alvarez-Coque 2006). It is expected that HPLC will provide sufficient resolution between analytes prior to their introduction into the mass spectrometer (Wong et al. 2018). Therefore the correct choice of column is absolutely critical to achieve proper separation (Torres-Lapasió and García-Alvarez-Coque 2006; Wong et al. 2018). Recently, nano-HPLC columns have come into practice with several thousand-fold higher sensitivity compared to the conventional HPLC columns. In a nano-HPLC column, a low inner diameter reduces the sample dilution factor which in turn results in an increased sensitivity (Gaspari and Cuda 2011). Of note HPLC is preferred for nonvolatile biomolecules, and for volatile biomolecules, gas chromatography (GC) is a preferred separation technique. Both these techniques have been extensively reviewed in literature (Cha et al. 2015).

19.1.2 Ionization Methods

Post-HPLC separation, the analytes are subjected to ionization to give them charge and bring them into gaseous phase. The fundamental principal underlying the detection of biomolecules on a mass spectrometer is that the molecule should be ionized into the gaseous phase without fragmentation prior to its delivery into the mass analyzer. The conventional ionization methods like electron impact (EI) have a problem of in-source fragmentation and therefore are not preferred for analyzing biomolecules of over 1000 Da. EI however remains the ionization method of choice for GC (Cha et al. 2015). This problem has been successfully resolved by using very soft ionization techniques like electrospray ionization (ESI) and matrix-assisted laser desorption/ionization (MALDI), which allow maximum ionization and minimal in-source fragmentation of the analyte, called as molecular ion once ionized (Fenn et al. 1989; Bergman et al. 2014).

19.1.2.1 ESI

ESI is the most frequently used ionization method because it is easily interfaced with HPLC and mass analyzers and can ionize samples of very low to remarkably high mass range (Fenn et al. 1989; Cha et al. 2015). Matthias Wilm has reviewed the principle and existing models involved in ESI mechanism in detail (Wilm 2011). Briefly, a high voltage is applied across the heated fine metal capillary through which the analytes in liquid phase flow continuously toward a counter electrode. Due to narrow tip of the capillary, solvent droplets carrying the analytes are formed. These droplets rapidly evaporate into smaller droplets till they no longer can carry the charge called as the Rayleigh limit, resulting in electrified droplet fission called as Coulomb explosion which in turn yields charged analyte ions that are attracted to counter electrode and into the mass analyzer (Wilm 2011). ESI instruments can be coupled with several different kinds of mass analyzers like tandem quadrupole linear ion trap (MS^n) or triple quadrupole (MS/MS) or quadrupole time-of-flight (Q-TOF) (Köfeler et al. 2012). The applications and flexibility of analysis of an ESI-based instrument are huge. For example, an ESI-MS/MS instrument can be used to record simply the parent ion mass (in Q1-MS), precursors of select parent ions (in precursor ion scanning, PREIS), single parent and daughter ion reactions (in single reaction monitoring, SRM), multiple parent and daughter ion reactions (in multiple reaction monitoring, MRM), and the neutral loss of specific fragment from the parent ion (Singh and Del Poeta 2016). Additionally, ESI instruments can be operated in negative and positive ion modes depending upon the ionization efficiency of the analyte. Due to this flexibility in operating modes, ESI-MS/MS is quite suitable for "omics" studies. A layout of a HPLC-coupled ESI-MS/MS instrument is shown in Fig. 19.1. Q-TOF-based mass spectrometers on the other hand provide high mass resolution and provide accurate mass of the analyte. However, Q-TOFs are expensive, and data acquisition and analysis require time and expertise (Köfeler et al. 2012).

ESI however has some limitations. We often see production ions with multiple charges especially in case of large peptides. Since principally mass analyzers detect *m/z* ratios, large interference is seen between ions coming from large peptide fragments to those of small peptides in the mass spectrum. Additionally, multiplicity of charge carried by ions also causes severe interference (Wilm 2011). Because ESI is often coupled with HPLC, salt contamination is another issue which results in lower sensitivity and additional peaks in the mass spectrum (Metwally et al. 2015). Therefore, sample preparation process requires extensive care. Nonetheless, ESI-based instruments remain the method of choice for high-throughput screenings (Smith et al. 1990).

19.1.2.2 MALDI

MALDI is another preferred technique for soft ionization of large synthetic polymers and biomolecules (Tanaka et al. 1988; Costello 1997), mostly because MALDI provides incredible sensitivity and low fragmentation and requires minimal sample preparation. MALDI ionization requires a matrix to facilitate the ionization process by absorbing the UV laser radiation and transferring the energy to analytes leading to generation of ions. The nitrogen UV laser of wavelength 337 nm is most common

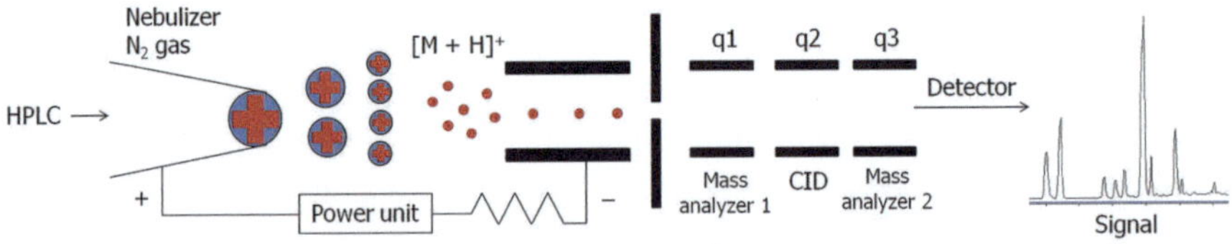

Fig. 19.1 Workflow of ESI-MS/MS. Ion detection as $[M + H]^+$ in positive ion mode has been depicted. Ions are shown in small red circles. q1, q2, and q3 are three quadrupoles where q1 and q3 are mass analyzers and q2 is the collision cell

and is applied in pulses of 0.5–20 ns. In MALDI process, about 1 μL sample in solvent is applied to the sample spot on MALDI plate (stainless steel or gold), and to this equal amount of matrix in solvent is added, gently mixed, and allowed to crystallize on the plate itself. Other approaches include an analyte-matrix mixture which is made and applied to the MALDI plate, spray-based application of the sample, and the more automated plate loaders. The idea here is to get uniform, thin crystallized layer of analyte-matrix mixture. This is important to achieve uniform signal throughout the sample spot allowing to record uniform mass spectrum for multiple pulses at the same or different spots. It has been observed that analyte-matrix ratios are important determinants for MALDI ionization of various molecules (Roepstorff 2000; Lei et al. 2013; Nadler et al. 2017). Therefore, optimization of exact analyte-matrix ratio for proper ionization is a necessary step.

The ionization step involves focusing a laser beam onto the sample matrix, which results in extirpation of sample and matrix molecules from the plate surface. Subsequent desorption and desolvation result in ionization of the sample usually by transfer of protons resulting in positively charged ions, in positive ion mode (Fig. 19.2). Ions generated in MALDI are usually singly charged. Next, a constant external electric field helps accelerate these ions into the time-of-flight (TOF) tube toward the detector, where the ions with low *m/z* ratio travel at a faster rate compared to those with high *m/z* ratio. The detector records the time taken by each ion to reach the detector and calculates the *m/z* ratio. In a linear ion MALDI-TOF instrument, the ion path in TOF tube is linear (Fig. 19.2a) (Jiang et al. 2012). Although

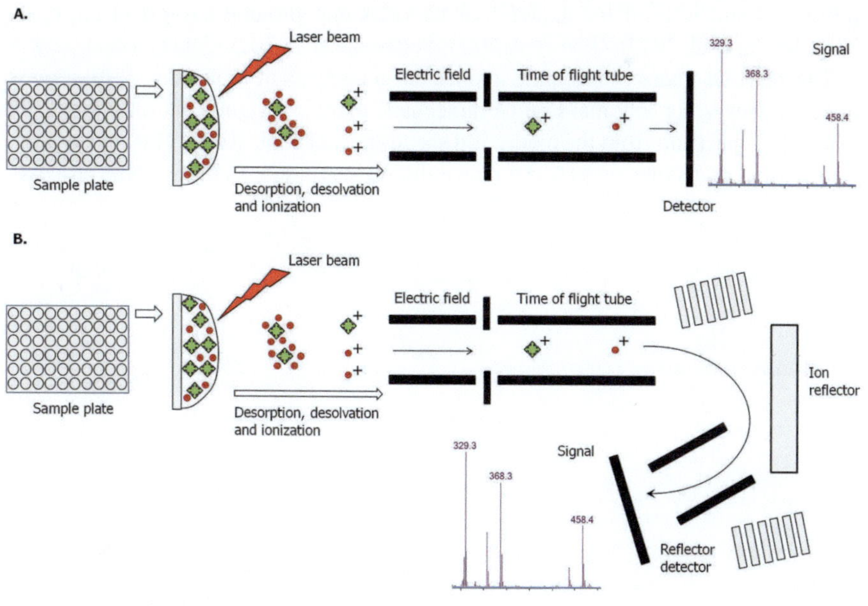

Fig. 19.2 Workflow of MALDI-TOF. (**a**) Depiction of MALDI-TOF in linear ion mode. (**b**) Depiction of MALDI-TOF in reflecting ion mode. Ions are shown in small red circles, and the MALDI matrix is represented with green symbols

these analyses are usually robust and sensitive, there are limitations. Irregularities in sample surface or in ionization step may lead to different ions of same analyte hit the detector at slightly different intervals. This results in broadening of peaks in the spectrum and causes interference with other analyte ions (Köfeler et al. 2012). This issue can be much resolved by using a nonlinear MALDI-TOF instrument (Fig. 19.2b) (Jiang et al. 2012). Here the ionized ions in TOF tube are deflected using an electrical field-based ion deflector toward a separate deflector detector. This results in two things: (i) it allows more time for different analyte ions to resolve, thereby improving resolution, and (ii) it allows ions of same analyte more time to catch up, thereby resulting in tighter peaks in the spectrum (Jiang et al. 2012).

19.1.3 MALDI Matrices

Traditional MALDI approach is based on small molecule organic matrices. In the past two decades, these matrices have extensively used to achieve ionization in MALDI which enabled the analysis of biomolecules. Some common examples of MALDI matrices are shown in Fig. 19.3. For example, 2,5-dihydroxybenzoic acid (DHBA) and α-cyano-4-hydroxycinnamic acid (CHCA) are suitable for analysis of peptides, proteins, and carbohydrates (glycans). CHCA is also suited for lipids and organic molecules. The 3-hydroxypicolinic acid (HPA) matrix is specific for oligonucleotides. The α-cyano-4-fluorocinnamic acid (FCCA) is a broad-spectrum matrix and is used to analyze lipids, drugs, peptides, and phosphopeptides. Sinapic acid on the other hand is used for analysis of peptides and proteins (Karas et al. 1993; Wiegelmann et al. 2013; Zhou et al. 2004; Soltwisch et al. 2012; Jackson et al. 2005).

The MALDI matrices are quite applicable in analysis of molecules of high mass range. However, for low mass range molecules, there is a significant interference in analyte ion spectrum from the matrix ion spectrum (Lei et al. 2013). This makes analysis of small molecules on MALDI matrix-based instrument a cumbersome process.

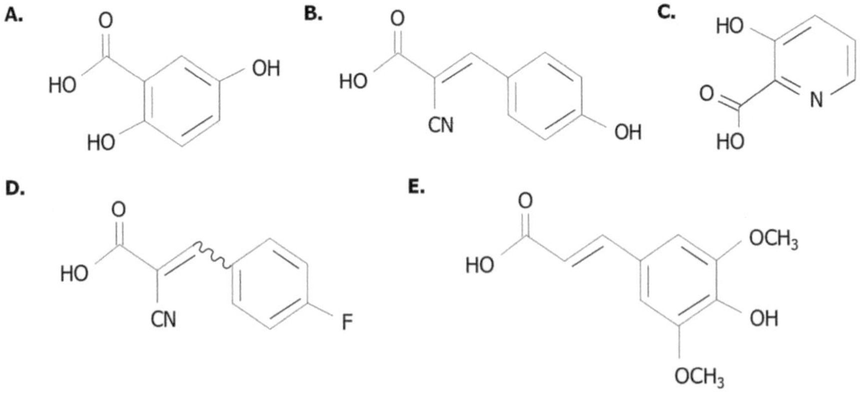

Fig. 19.3 **Some common examples of MALDI matrices.** (**a**) 2,5-Dihydroxybenzoic acid. (**b**) α-Cyano-4-hydroxycinnamic acid. (**c**) 3-Hydroxypicolinic acid. (**d**) α-Cyano-4-fluorocinnamic acid. (**e**) Sinapic acid

19.2 Nanomaterials and Laser Desorption Ionization (LDI)

Nanomaterials have provided a suitable alterative to MALDI matrices for detection of biomolecules, especially small mass molecules, in matrix-free LDI mode (Lei et al. 2013; Yagnik et al. 2016). Nanosized materials or nanoparticles (size range 10–100 nm) have wide biological applications in biosensing and drug delivery mechanisms (Baetke et al. 2015; Yohan and Chithrani 2014). In recent years, nanoparticle-assisted LDI (NALDI) technique has evolved as a major approach to analyze small molecules (Chiang et al. 2011; Tata et al. 2012a). Nanoparticles are well suited for LDI as they have (i) high laser absorption and heat capacity and (ii) low thermal conductivity. Nanoparticles have few major advantages over the conventional MALDI matrices:

(i) High chemical affinity toward select biomolecules. For example, TiO_2 and ZrO_2 can more specifically adsorb the phosphate groups of peptides via coordination (Fíla and Honys 2012; Zhou et al. 2007). Similarly boronate monoliths showed specific enrichment of neutral sugar-containing glycoproteins (Zhou et al. 2017).
(ii) Enrichment by size exclusion and physical interaction. For example, hierarchical ordered silica monoliths can extract peptides by size exclusion (Grimes et al. 2007). Zeolite nanocrystals and oxidized diamond nanoparticles have hydrophobic and hydrophilic characteristics, respectively (Zhang et al. 2005; Li et al. 2013). Several studies have also implicated electrostatic interaction-based enrichment for nanoparticles (Giri et al. 2014).
(iii) Different nanoparticle surfaces may allow different rates of proteolysis, which can be advantageous during the analysis of select peptides (Lei et al. 2013; Ju and Yeo 2012).
(iv) They show no interference from matrix ions of low mass range (Yagnik et al. 2016).

A wide variety of nanoparticles have been implicated in LDI studies. These include metal nanoparticles of gold and silver (used usually with a capping) (Ju and Yeo 2012; Ding et al. 2018); metal oxides like TiO_2, WO_3, and Fe_3O_4 (Castro et al. 2008; Ausekar et al. 2018; Reddy et al. 2014); and carbon-based nanoparticles like diamond and functionalized C_{60} $[C_{60}((CH_2)_2COOH)_n]$ (Li et al. 2013; Ausekar et al. 2018; Ugarov et al. 2004). A recent study exhaustively compared the thermal desorption efficiencies of these nanoparticles, providing guidelines for their usage in LDI (Yagnik et al. 2016).

Carbon nanotubes (CNT) are one of the most commonly used nanoparticles and are divided into three major categories: single-walled carbon nanotubes (SWCNTs) with single layer of carbon, double-walled carbon nanotubes (DWCNTs), and multi-walled carbon nanotubes (MWCNTs) with multiple layers of concentric hollow carbon cylinders. SWCNT can be used as ionization matrix. This however requires acid functionalization (AF) of SWCNT attaching sulfonate or carboxyl groups resulting in AF-SWCNT that easily disperses in solution and may be used for LDI. Naked unfunctionalized SWCNTs show a strong tendency to aggregate and are not recommended for LDI (Ugarov et al. 2004; Pan et al. 2005).

19.3 Mass Spectrometry Imaging (MSI)

The ability of MALDI to ionize molecules from solid surface coupled with high-end mass analyzers like Q-TOF has been useful in imaging the sample surface based on the ion intensity mapping, referred to as MALDI mass spectrometry imaging (MALDI MSI). The approach is quite simple. A thin slice of sample tissue is placed directly on matrix-coated MALDI sample plate. Sequential ion masses representing a diverse set of biomolecules are recorded for the entire sample section. This allows us to follow spatial changes on the biological surface providing a more in vivo like image of the cellular surfaces. MSI has been used to monitor both plant and animal origin samples. Animal tissues such as the eye, liver, kidney, and brain are most studied (Prideaux et al. 2010; Hankin et al. 2011; Beine et al. 2016; Guran et al. 2017; Hart and Clench 1618; Qin et al. 2018). The technique has tremendous applications in biomedical research. Since most biological surfaces comprise lipids as an abundant component, it is not surprising that *m/z* of most ions in the range of 400–1000 correspond to lipids, especially phosphatidylcholine (PC).

In an imaging study on rat brain model, comparison of ion spectrum coming from sections of hippocampus region of the control and ischemic rat brain shows signals for PC species ranging from *m/z* 700 to 850 (Fig. 19.4a). Specifically PC species 16:0/16:0 and 16:0/18:1 were detected. The *m/z* of 734.5, 756.5, and 772.5 and *m/z* of 760.5, 782.5, and 798.5 represented $[M + H]^+$, $[M + Na]^+$, and $[M + K]^+$ species for PC16:0/16:0 and PC16:0/18:1, respectively. Inset images in Fig. 19.4 show MSI image obtained for *m/z* 760.5 ion which represents PC16:0/18:1 as $[M + H]^+$. The analyses showed loss of *m/z* 798.5, potassium adduct of PC16:0/18:1, in ischemic sample suggestive of specific loss of Na/K-ATPase activity (Fig. 19.4a). The result also showed an accumulation of Cer(d18:1/18:0), *m/z* 548.5 detected as $[M + H - H_2O]^+$, in ischemic hemisphere of the rat (Fig. 19.4b). It was proposed that Cer(d18:1/18:0) are accumulated due to onset of apoptotic pathways upon tissue injury (Hankin et al. 2011). This is a good example to understand how MSI can be used to understand tissue damage and disease.

Due to lack of homogeneity during application of MALDI matrix, more often than not the MSI data lack resolution and reproducibility. This issue was resolved by using nanowires made up of silicon, zinc oxide, tin oxide, etc. to generate nano-assisted LDI surfaces, resulting in a new ionization approach called as nanoparticle-assisted laser desorption/ionization (NALDI) (Tata et al. 2012b; Chiu 2014; Tata et al. 2014). This matrix-free approach shows no interference in spectrum in low mass range, which is essential for mapping lipids. The NALDI approach has been successfully used for imaging kidney tissue and melanoma sections in mice, direct surface detection of secondary metabolites, and affinity probes for microbial

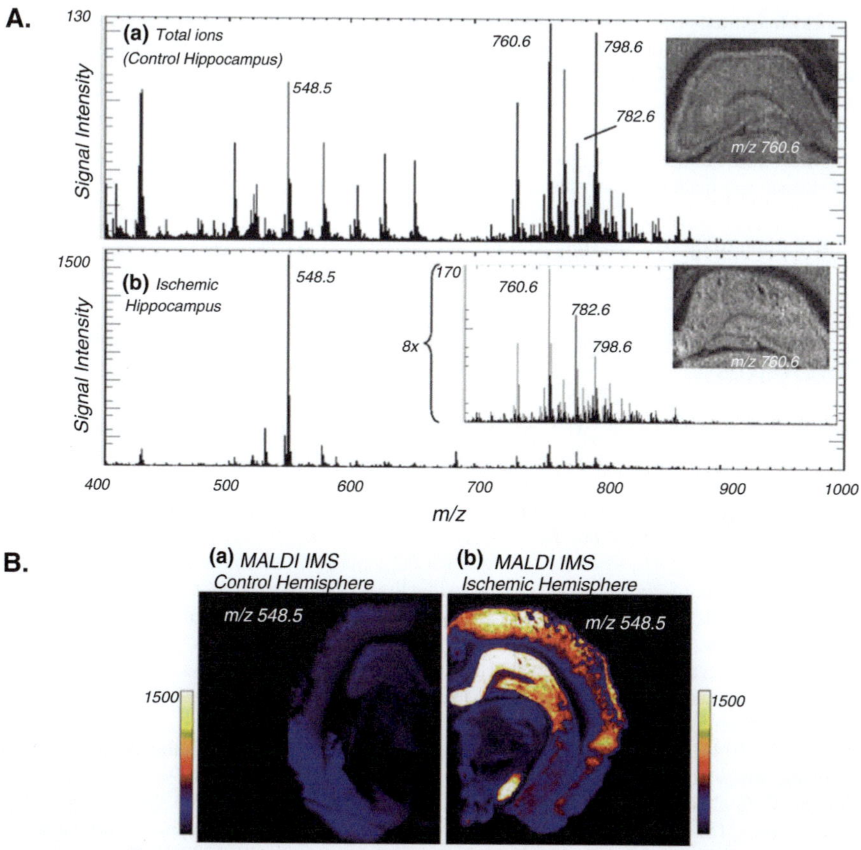

Fig. 19.4 **Example of MALDI-MSI of lipids in rat brain injury model.** (**a**) Averaged MALDI-TOF mass spectra (a) control hippocampus and (b) ischemic hippocampus. (**b**) MALDI IMS representing m/z 548.5 ([M + H – H2O]$^+$, Cer [d18:0/18:1]) from (a) control hippocampus and (b) ischemic hippocampus. (Images reprinted with permission from Ref. (Hankin et al. 2011))

detection, among others. NALDI coupled with Q-TOF is no doubt the future of MSI (Tata et al. 2012b, 2014; Chiu 2014).

19.4 Concluding Remarks

In this chapter, we have given a brief overview of mass spectrometry and the two most commonly used soft ionization methods, ESI and MALDI. We discussed the advantages and disadvantages of these methods over the other. We also discussed the applications, advantages, and disadvantages of organic MALDI matrices compared to nanomaterial-based matrices. Lastly we discuss the MSI approach with examples. Together this chapter provides insights into some of the key aspects of mass spectrometry and nanomaterial research.

Acknowledgments We thank grants to AS from ICMR No. 52/08/2019-BMS and University of Lucknow, Lucknow, grants to NB from ICMR No. 56/2/2019-Hae/BMS, and grants to RP from DBT No. BT/01/CEIB/10/III/02, BT/PR7392/MED/29/652/2012 and BT/PR14117/BRB/10/1420/2015.

Financial and Competing Interest Disclosure Author has no financial and competing interests with the subject matter or materials discussed in the manuscript.

Contribution to the Manuscript AS, RP and NB wrote the manuscript.

References

Fenn JB, Mann M, Meng CK, Wong SF, Whitehouse CM (1989) Electrospray ionization for mass spectrometry of large biomolecules. Science 246(4926):64–71

Bakry R, Rainer M, Huck CW, Bonn GK (2011) Protein profiling for cancer biomarker discovery using matrix-assisted laser desorption/ionization time-of-flight mass spectrometry and infrared imaging: a review. Anal Chim Acta 690(1):26–34

Cho YT, Su H, Huang TL, Chen HC, Wu WJ, Wu PC et al (2013) Matrix-assisted laser desorption ionization/time-of-flight mass spectrometry for clinical diagnosis. Clin Chim Acta 415:266–275

Sandrin TR, Goldstein JE, Schumaker S (2013) MALDI TOF MS profiling of bacteria at the strain level: a review. Mass Spectrom Rev 32(3):188–217

Bergman N, Shevchenko D, Bergquist J (2014) Approaches for the analysis of low molecular weight compounds with laser desorption/ionization techniques and mass spectrometry. Anal Bioanal Chem 406(1):49–61

Brügger B (2014) Lipidomics: analysis of the lipid composition of cells and subcellular organelles by electrospray ionization mass spectrometry. Annu Rev Biochem 83:79–98

Görgens C, Guddat S, Thomas A, Thevis M (2018) Recent improvements in sports drug testing concerning the initial testing for peptidic drugs (< 2 kDa) – sample preparation, mass spectrometric detection, and data review. Drug Test Anal. https://doi.org/10.1002/dta.2503

Torres-Lapasió JR, García-Alvarez-Coque MC (2006) Levels in the interpretive optimisation of selectivity in high-performance liquid chromatography: a magical mystery tour. J Chromatogr A 1120(1–2):308–321

Wong JW, Wang J, Chow W, Carlson R, Jia Z, Zhang K et al (2018) Perspectives on liquid chromatography-high-resolution mass spectrometry for pesticide screening in foods. J Agric Food Chem 66(37):9573–9581

Gaspari M, Cuda G (2011) Nano LC-MS/MS: a robust setup for proteomic analysis. Methods Mol Biol 790:115–126

Cha E, Kim S, Kim HJ, Lee KM, Kim KH, Kwon OS et al (2015) Sensitivity of GC-EI/MS, GC-EI/MS/MS, LC-ESI/MS/MS, LC-Ag(+) CIS/MS/MS, and GC-ESI/MS/MS for analysis of anabolic steroids in doping control. Drug Test Anal 7(11–12):1040–1049

Wilm M (2011) Principles of electrospray ionization. Mol Cell Proteomics 10(7):M111.009407

Köfeler HC, Fauland A, Rechberger GN, Trötzmüller M (2012) Mass spectrometry based lipidomics: an overview of technological platforms. Meta 2(1):19–38

Singh A, Del Poeta M (2016) Sphingolipidomics: an important mechanistic tool for studying fungal pathogens. Front Microbiol 7:501

Metwally H, McAllister RG, Konermann L (2015) Exploring the mechanism of salt-induced signal suppression in protein electrospray mass spectrometry using experiments and molecular dynamics simulations. Anal Chem 87(4):2434–2442

Smith RD, Loo JA, Edmonds CG, Barinaga CJ, Udseth HR (1990) New developments in biochemical mass spectrometry: electrospray ionization. Anal Chem 62(9):882–899

Tanaka K, Waki H, Ido Y, Akita S, Yoshida Y, Yoshida T (1988) Protein and polymer analyses up to m/z 100 000 by laser ionization time-of-flight mass spectrometry. Rapid Commun Mass spectrum 2(8):151–153

Costello CE (1997) Time, life … and mass spectrometry. New techniques to address biological questions. Biophys Chem 68(1–3):173–188

Roepstorff P (2000) MALDI-TOF mass spectrometry in protein chemistry. EXS 88:81–97

Lei C, Qian K, Noonan O, Nouwens A, Yu C (2013) Applications of nanomaterials in mass spectrometry analysis. Nanoscale 5(24):12033–12042

Nadler WM, Waidelich D, Kerner A, Hanke S, Berg R, Trumpp A et al (2017) MALDI versus ESI: the impact of the ion source on peptide identification. J Proteome Res 16(3):1207–1215

Jiang SS, Liang XN, Li X, Wang SL, Lv DW, Ma CY et al (2012) Wheat drought-responsive grain proteome analysis by linear and nonlinear 2-DE and MALDI-TOF mass spectrometry. Int J Mol Sci 13(12):16065–16083

Karas M, Ehring H, Nordhoff E, Stahl B, Strupat K, Hillenkamp F et al (1993) Matrix-assisted laser desorption/ionization mass spectrometry with additives to 2,5-dihydroxybenzoic acid. Org Mass Spectrom 28:1476–1481

Wiegelmann M, Soltwisch J, Jaskolla TW, Dreisewerd K (2013) Matching the laser wavelength to the absorption properties of matrices increases the ion yield in UV-MALDI mass spectrometry. Anal Bioanal Chem 405(22):6925–6932

Zhou L, Deng H, Deng Q, Zhao S (2004) A mixed matrix of 3-hydroxypicolinic acid and pyrazinecarboxylic acid for matrix-assisted laser desorption/ionization time-of-flight mass spectrometry of oligodeoxynucleotides. Rapid Commun Mass Spectrom 18(7):787–794

Soltwisch J, Jaskolla TW, Hillenkamp F, Karas M, Dreisewerd K (2012) Ion yields in UV-MALDI mass spectrometry as a function of excitation laser wavelength and optical and physicochemical properties of classical and halogen-substituted MALDI matrixes. Anal Chem 84(15):6567–6576

Jackson SN, Wang HY, Woods AS, Ugarov M, Egan T, Schultz JA (2005) Direct tissue analysis of phospholipids in rat brain using MALDI-TOFMS and MALDI-ion mobility-TOFMS. J Am Soc Mass Spectrom 16(2):133–138

Yagnik GB, Hansen RL, Korte AR, Reichert MD, Vela J, Lee YJ (2016) Large scale nanoparticle screening for small molecule analysis in laser desorption ionization mass spectrometry. Anal Chem 88(18):8926–8930

Baetke SC, Lammers T, Kiessling F (2015) Applications of nanoparticles for diagnosis and therapy of cancer. Br J Radiol 88(1054):20150207

Yohan D, Chithrani BD (2014) Applications of nanoparticles in nanomedicine. J Biomed Nanotechnol 10(9):2371–2392

Chiang CK, Chen WT, Chang HT (2011) Nanoparticle-based mass spectrometry for the analysis of biomolecules. Chem Soc Rev 40(3):1269–1281

Tata A, Fernandes AM, Santos VG, Alberici RM, Araldi D, Parada CA et al (2012a) Nanoassisted laser desorption-ionization-MS imaging of tumors. Anal Chem 84(15):6341–6345

Fíla J, Honys D (2012) Enrichment techniques employed in phosphoproteomics. Amino Acids 43(3):1025–1047

Zhou H, Tian R, Ye M, Xu S, Feng S, Pan C et al (2007) Highly specific enrichment of phosphopeptides by zirconium dioxide nanoparticles for phosphoproteome analysis. Electrophoresis 28(13):2201–2215

Zhou C, Chen X, Du Z, Li G, Xiao X, Cai Z (2017) A hybrid monolithic column based on boronate-functionalized graphene oxide nanosheets for online specific enrichment of glycoproteins. J Chromatogr A 1498:90–98

Grimes BA, Skudas R, Unger KK, Lubda D (2007) Pore structural characterization of monolithic silica columns by inverse size-exclusion chromatography. J Chromatogr A 1144(1):14–29

Zhang Y, Wang X, Shan W, Wu B, Fan H, Yu X et al (2005) Enrichment of low-abundance peptides and proteins on zeolite nanocrystals for direct MALDI-TOF MS analysis. Angew Chem Int Ed Engl 44(4):615–617

Li HC, Hsieh FJ, Chen CP, Chang MY, Hsieh PC, Chen CC et al (2013) The hemocompatibility of oxidized diamond nanocrystals for biomedical applications. Sci Rep 3:3044

Giri K, Shameer K, Zimmermann MT, Saha S, Chakraborty PK, Sharma A et al (2014) Understanding protein-nanoparticle interaction: a new gateway to disease therapeutics. Bioconjug Chem 25(6):1078–1090

Ju S, Yeo WS (2012) Quantification of proteins on gold nanoparticles by combining MALDI-TOF MS and proteolysis. Nanotechnology 23(13):135701

Ding F, Qian Y, Deng Z, Zhang J, Zhou Y, Yang L et al (2018) Size-selected silver nanoparticles for MALDI-TOF mass spectrometry of amyloid-beta peptides. Nanoscale 10(46):22044–22054

Castro AL, Madeira PJ, Nunes MR, Costa FM, Florêncio MH (2008) Titanium dioxide anatase as matrix for matrix-assisted laser desorption/ionization analysis of small molecules. Rapid Commun Mass Spectrom 22(23):3761–3766

Ausekar MV, Mawale RM, Pazdera P, Havel J (2018) Matrix assisted and/or laser desorption ionization Quadrupole ion trap time-of-flight mass spectrometry of WO3 clusters formation in gas phase. Nanodiamonds, fullerene, and Graphene oxide matrices. J Am Soc Mass Spectrom 29(3):581–587

Reddy PM, Chang KC, Liu ZJ, Chen CT, Ho YP (2014) Functionalized magnetic iron oxide (Fe_3O_4) nanoparticles for capturing gram-positive and gram-negative bacteria. J Biomed Nanotechnol 10(8):1429–1439

Ugarov MV, Egan T, Khabashesku DV, Schultz JA, Peng H, Khabashesku VN et al (2004) MALDI matrices for biomolecular analysis based on functionalized carbon nanomaterials. Anal Chem 76(22):6734–6742

Pan C, Xu S, Hu L, Su X, Ou J, Zou H et al (2005) Using oxidized carbon nanotubes as matrix for analysis of small molecules by MALDI-TOF MS. J Am Soc Mass Spectrom 16(6):883–892

Prideaux B, Staab D, Stoeckli M (2010) Applications of MALDI-MSI to pharmaceutical research. Methods Mol Biol 656:405–413

Hankin JA, Farias SE, Barkley RM, Heidenreich K, Frey LC, Hamazaki K et al (2011) MALDI mass spectrometric imaging of lipids in rat brain injury models. J Am Soc Mass Spectrom 22(6):1014–1021

Beine B, Diehl HC, Meyer HE, Henkel C (2016) Tissue MALDI mass spectrometry imaging (MALDI MSI) of peptides. Methods Mol Biol 1394:129–150

Guran R, Vanickova L, Horak V, Krizkova S, Michalek P, Heger Z et al (2017) MALDI MSI of MeLiM melanoma: searching for differences in protein profiles. PLoS One 12(12):e0189305

Hart PJ, Clench MR (1618) MALDI-MSI of lipids in human skin. Methods Mol Biol 2017:29–36

Qin L, Zhang Y, Liu Y, He H, Han M, Li Y et al (2018) Recent advances in matrix-assisted laser desorption/ionisation mass spectrometry imaging (MALDI-MSI) for in situ analysis of endogenous molecules in plants. Phytochem Anal 29(4):351–364

Tata A, Fernandes AM, Santos VG, Alberici RM, Araldi D, Parada CA et al (2012b) Nanoassisted laser desorption-ionization-MS imaging of tumors. Anal Chem 84(15):6341–6345

Chiu TC (2014) Recent advances in bacteria identification by matrix-assisted laser desorption/ionization mass spectrometry using nanomaterials as affinity probes. Int J Mol Sci 15(5):7266–7280

Tata A, Montemurro C, Porcari AM, Silva KC (2014) Lopes de Faria JB, Eberlin MN. Spatial distribution of theobromine–a low MW drug–in tissues via matrix-free NALDI-MS imaging. Drug Test Anal 6(9):949–952

Part VIII

NanoBioMedicine: Risk Assessment and Management

Nanotoxicology in Medicine

20

Shailendra K. Saxena, Rajni Nyodu, Swatantra Kumar, and Vimal K. Maurya

Abstract

With the advancement of use of nanoparticles in the field of medicine in the recent times, its production and formulations have also risen to a steep. However, with the use of a particle as minute as less than the size of 100 nm, the chances of having its adverse effect have also been seen to be the on the rise. The nanoparticles are made to be administered into the human body as a drug-delivery system or mediator, at many instances, to blood or to CNS to deliver drugs for multiple diagnostic purposes, and hence, these nanoparticles have a prospective of effecting functional organs in a dreadful manner along with possible healing purposes, which is why study of its toxicology is very much advantageous and incumbent in order to prevent any of its dreadful effects. Study of toxicology in nanoparticles is termed as nanotoxicology. In this chapter, we will be discussing on nanotoxicology in the field of medicine, its mechanism, and its prevalence in the biological forms.

Keywords

Nanotoxicology · Nanomedicine · Nanoparticles · Liposomes · Particle toxicology · Magnetic nanoparticles · Quantum dots · Therapeutics · Cytotoxicity · Biodistribution of particles Regenerative medicine

S. K. Saxena (✉) · R. Nyodu · S. Kumar · V. K. Maurya
Centre for Advanced Research, Faculty of Medicine,
King George's Medical University (KGMU), Lucknow, India
e-mail: shailen@kgmcindia.edu

S. K. Saxena, S. M. P. Khurana (eds.), *NanoBioMedicine*,
https://doi.org/10.1007/978-981-32-9898-9_20

20.1 Introduction

In order to understand the subcategory of toxicology study, which is nanotoxicology, the outlook of nanoparticles and the field of nanotechnology have to be understood in the frontline. Nanotechnology being the field of science where the particles are engineered, manipulated, and manufactured at nanosize (1–100 nm) and the application of this technology in the field of medicine are what nanomedicine is all about. Nanoparticles as the name suggests are particles with a size ranging at a nano (submicroscopic) level, which is less than 100 nm (Singh et al. 2019). Such nanosized particles are used as a mediator or delivery system for the transport of various therapeutic and imaging agents (e.g., proteins, peptides, and nucleic acids), which are otherwise difficult to translocate to the target sites. Consisting of myriad of construct material, nanoparticles are constructed of lipids, metals, silicon, silica, polymers, proteins, and one of the most important nonmetals, carbon (Almeida et al. 2011a). Nanoparticles are shown to be efficient in reducing side effects since it refines the accumulation of drugs in infected tissue by decreasing the clearance of drugs. Certain nanoparticles have electrical properties, which are efficient features to be employed in therapeutic purposes. The energy available externally in combination with metal nanoparticles (AuNP and FeONP) is used to thermally ablate the infected tissue. So far, there are several nano-based drugs, which are designed to combat various deadly diseases, such as cancer, infectious diseases, neurological diseases, and many more (Pelaz et al. 2017) (Fig. 20.1).

Alongside of their efficient therapeutic effects, certain class of nanoparticles has shown to have toxic/adverse effects due to their size, which on the other hand has proved to be a utility factor. In nature, nanosized particles are released in all combustion processes including numerous natural processes that are the matter of concern since ages. One of the daily cases of particle toxicology is the case with TiO_2 particles (25 nm), which are the ingredients of cosmetic sunscreens to shield from UV rays (Bierkandt et al. 2018). Upon application of sunscreen onto the skin, TiO_2 exposure takes place. Studies have shown that the penetration of TiO_2 is relatively low as it is supposed to be, and hence, the biological effects are generally small, and exposure of TiO_2 into the environment is significantly high (Ruszkiewicz et al. 2017).

Particle toxicology in medicine is seen in the carbon-based nanoparticles, which has been shown in many studies, although many more extensive studies are needed. These carbon-based nanoparticles especially carbon nanotubes (CNTs) have been reported to induce mesothelioma and have been claimed that this might be due to its shape. Hence, drawing a conclusion, the deciding factors for the toxicology of any nanoparticle are size and morphology of nanoparticles (Lohcharoenkal et al. 2013). However, at the current scenario in medicine, such toxicity-inducing properties of these nanoparticles are not an element to terminate them as a potential candidate for drug delivery vehicle, although enough care and approaches have to be taken while incorporating these toxicity-inducing nanoparticles so as to prevent them from any unspecific targeting of healthy tissue.

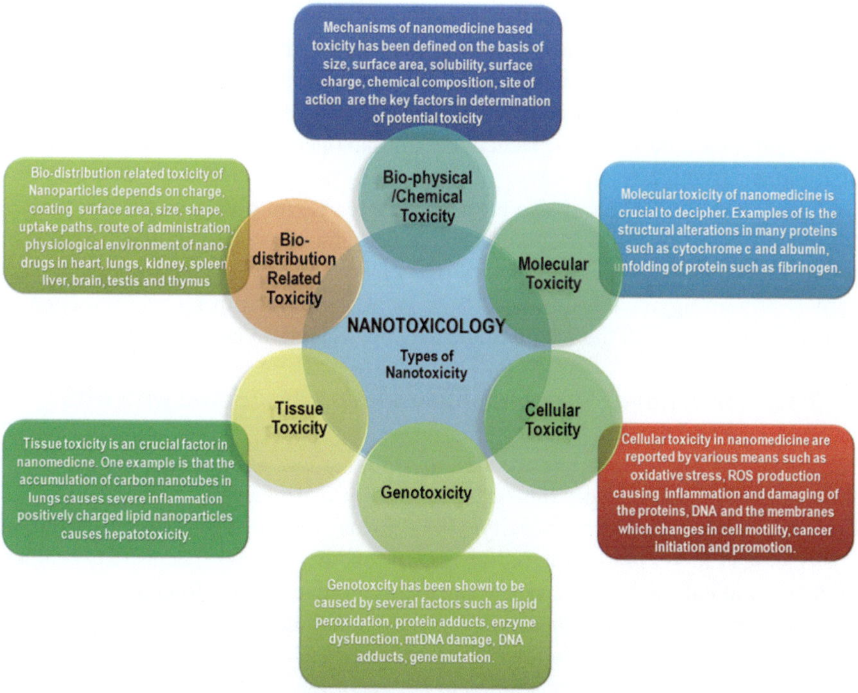

Fig. 20.1 Toxicity associated with nanomedicine. This figure depicts the various forms of toxicity associated with nanomedicine such as biodistribution-related toxicity, tissue toxicity, gene toxicity, cellular toxicity, molecular toxicity, and biophysical/chemical toxicity of nanomedicines

20.2 Mechanism of Nanotoxicity

The precise mechanism of toxicity caused by nanosized particulates is not yet fully understood. However, considering the size of nanosized particles, it is quite generalized that the surface area of nanoparticles could be one of the possible discussions. Since the size of the nanoparticles is submicroscopic, its surface area would be large, and this surface area-to-volume ratio of the nanoparticles is the most triggering spot to emphasize (Fu et al. 2014). Since these nanoparticles are targeted and delivered into various different organs and tissues, the toxicity would also be affecting the human body at multilevel which includes molecular, cellular, and at tissue level.

20.2.1 Molecular Toxicity of Nano-Based Drugs

Post-entry, the nanoparticles interact with the biomolecules present in the biological fluids. These biomolecules form a coating around the surface of the particle, which is termed as protein corona consisting of different proteins. Eventually, the presence of the protein corona may alter the properties of nanoparticles such as shape, size, etc. For instances, the interaction among the nanoparticles and the protein corona can decrease the size of the nanoparticles and lead it to become more anionic. Alongside, even the interior contents of biomolecules may also undergo some functional and structural changes. Hence, certain nanoparticles are reported to induce alterations structurally in many proteins such as cytochrome c, albumin, etc. (Sukhanova et al. 2018) and also have found to cause unfolding of protein such as fibrinogen. Subsequently, such conformational changes may alter the function of proteins and stimulating inflammatory pathways. Additionally, another major concern is protein aggregation induced by nanoparticle. To sum up, the presence of protein layer decreases the nanoparticle toxicity (Saptarshi et al. 2013).

20.2.1.1 Oxidative Stress and Inflammation Associated with ROS Production

Nanoparticles have a large surface area-to-volume ratio, and in the case of some of the nanoparticles, the larger the surface area, the more elevated their reactivity to the chemicals and biological activity will be. Thus, this elevated reactivity to chemicals can lead to yet increasing production of reactive oxygen species (ROS) along with free radicals and subsequently causing oxidative stress, alongside inflammation and damaging of the proteins, DNA, and membranes and also causing changes in cell motility, cancer initiation, and promotion (Khanna et al. 2015). Notably, ROS overproduction has been studied to be one of the most pivotal mechanisms of nanotoxicity. This mechanism has its base upon the chemical characterization of the nanoparticles, e.g., metals used in nanoparticles such as Cu, silica, ZnO, etc. and on the nanofibers such as CNTs, C60 fullerenes, etc. When inside the cell, such metals react with the cellular components such as DNA, proteins, and lipids leading to damage (Manke et al. 2013). The use of one of the most sought-after outcomes of nanomedicine, quantum dots, has shown to be producing oxidative stress causing damage to cell that was due to production of ROS. Silver-based nanoparticles have been shown to induce oxidative stress and apoptosis in the cells (Kim and Ryu 2013). CNTs in bronchial epithelial cells have also been studied to be producing ROS which causes oxidative stress. When not shielded by the antioxidative enzymes, the free radicals are reported to be triggering reactions with regard to inflammation. In the case of ROS overproduction, it is anticipated that these free radicals react with macromolecules in the cell resulting in the contrasting results (Shvedova et al. 2003).

20.2.1.2 ROS Production Leading to Cyto- and Genetoxicity

The generation of reactive oxygen species has been dreadful because of its subsequent consequences leading to cytotoxicity and genetoxicity. Cytotoxicity induced by silica nanoparticles is responsible for ROS production in the cell membrane. The same group came up with the report showing the induction of cytotoxicity by CuO-based nanoparticles, which was also due to the ROS production (Gallo et al. 2018). In another metal oxide, ZnO- and TiO2-based nanoparticles have also been reported to be inducing cytotoxicity resulting in the alteration of BEAS-2B cells, human bronchial epithelial cells. In vitro study on NIH3T3 cells showed that the Ag-based nanoparticle-induced apoptosis was mediated by the production of ROS (Vandebriel and De Jong 2012). When compared the cytotoxicity level caused by different metal oxides, CuO-based nanoparticles topped the lists based on the reports so far. Hence, many studies are still ongoing and yet to report. Taking into account the reports, ROS production induced by the nanoparticles is predominantly the causes of oxidative stress resulting in the mitochondrial damage, damage to DNA, and protein modification, leading to cytotoxicity and cancer and also gene toxicity (Karlsson et al. 2008).

20.2.2 Tissue Toxicity of Nano-Based Drugs

Toxicity caused by nanoparticles in the cellular or molecular level eventually affects the whole function of the organs, and the organs which are most affected are the ones which has got high level of accumulation with nanoparticles. One of the most prominent examples is lungs getting deposited with CNTs causing its severe inflammation. Another example is blood that has been found to be induced with hepatotoxicity by the injection of positively charged lipid nanoparticles (Wolfram et al. 2015).

20.3 Biodistribution of Nano-Based Drugs in Terms of Toxicity

Biodistribution of nanoparticles has been one of the major deciding factors about efficacy and safety of drugs administered. Since nanoparticles consist of metal, their biodistribution is one of the major concerns because of the fact that they can get cleared very rapidly from the blood and can stay in organs for a longer duration of time. Studies have been reported about the inability to prevent unspecific distribution of nanoparticles, which is one of the prevalent causes of nanotoxicity. The size of the particles plays a pivotal role in the study of biodistribution. Biodistribution based on size has not been fully explained yet, but more profoundly, it has to do with the filtering effect such as the particles which are larger in size and are removed by the liver and spleen more rapidly, and the particles lesser in size are diverted to bone marrow (Moghimi et al. 2012). In vivo distribution of nano drugs are studied categorically based upon many aspects. These include its surface, its size and shape, its characteristics (charge and coating etc.), the uptake paths or route of administration (for synthetic and free nanoparticles), properties of nanoparticles, and also the physiological environment where the nanoparticles were being induced. This is evident from the report of biodistribution in vivo study of various Ag-based nanoparticles and Au-based nanoparticles on rodents (Almeida et al. 2011b). The report also shows dose-dependent toxicity when these nanoparticles were being administered intravenous. Their major accumulation organs were found to be in the spleen and liver. Evidently, the smaller nanoparticles were studied to be distributed widely to multiple organs which include the heart, lungs, kidney, spleen, liver, brain, and testis and also in the thymus. In addition, the FeO-based nanoparticles also have the same model of distribution, as Ag- and Au-based nanoparticles (Ruttkay-Nedecky et al. 2017).

One of the most sought-after nanoparticles, quantum dots, coated with PEG, applied in the imaging techniques has been studied by a group based on their coating. Quick and prolonged uptake of quantum dots (QDs) has shown into RES organs including prolonged QD blood circulation upon PEGylation. Report claiming that though QDs are not yet the candidate to induce toxicity, an intensive in vivo study is needed to conclude the report precisely (Du et al. 2019). One of the clinically approved nano-based drug, Doxil, a carrier for doxorubicin is reported to be delivered to the tumor tissue, which results in moderate increase in antitumor activity though it is has been reported to be causing major cases of cytotoxicity yet traces of it. There are different methods to study biodistribution of the nanoparticles. In vitro methods include cell-associated fluorescence assays and flow cytometry on various cell lines (murine macrophages, human monocytes, and neutrophilic granulocytes), whereas in vivo methods include encapsulated histology studies and tracking the uptake of particle with the use of radiolabeled ^{14}C (Kanwal et al. 2018).

20.4 Safety Assessment for Nano-Based Drugs

In order to assess the safety concerns of nano-based drugs, it is very crucial to look for assays which precisely target the nanoparticles by considering its distinct size and surface properties as report has shown that many of the available safety assessment criteria have failed to target the nanoparticles. Another point to consider while assessing the safety of nano-based drugs is their concentration. Inappropriate concentrations of nanoparticles have been a crucial point of conflict of interest due to the same concentration failed to present a potential toxicity level in the in vivo studies (Accomasso et al. 2018). Heterogeneity in case of 3D in vivo culture whereas the conventional 2D models displayed homogenous and uniform distribution has been reported. Another report has shown the toxicity level of nanoparticles to be quite at low level in 3D cultures than in 2D monolayer. Therefore, one more approach for the improvement in the toxicology assays in vitro is including the 3D culture study which enables the co-culturing of different cell types and mimic tissue architecture very closely (Jones and Grainger 2009). A study group from International Life Sciences has categorized the assessment of nanoparticle toxicity into several steps, which includes physiochemical characterization, cell-free assays, cellular assays, and in vivo assays (Rushton et al. 2010). The basic approaches to assess toxicity are by dose-dependent responses at the steepest slope or point to be the cut off for comparison. To assure more efficacies in the safety assessment of nanoparticles, high-throughput screening platforms have been proposed to develop in order to yield high predictive information about the structure (Oberdörster 2010). In order to understand the overall physiochemical properties of nanoparticles prior to design of the nano-based drugs or product nanoparticles, safety libraries have been set up. Additionally, it has been contemplated that the precision medicine will be approving the therapeutics apart from the category of genome, proteome, epigenome, or metabolome and many more (Wick 2011).

20.5 Conclusions

So far, very compact size study has been conducted in the field of nanotoxicology, which, in comparison to nanotechnology studies, is far lagging behind. However, the information collected based on the available data concludes that the increased toxicity caused by nanoparticles is due to several factors such as size, shape, charges, their surface area-to-volume ratio, their ability to penetrate through the cellular barriers, and many other factors. Evidently, the toxicity of nanoparticles has been studied extensively in the systems of cardiovascular and neurology. However, many nanoparticles have been engineered to be studied in order to come up with a combating resolution. Nanoparticles have been shown to induce toxicity at both the molecular and cellular level prominently due to its size which is the resulting factor for the production of ROS and subsequently causing oxidative stress of the cells and their components. Research and study have been conducted, and report has been published on many approaches to combat this toxicity parameter and to prevent the cellular and organelle damage. The cloud showing all these potential risks and hazards of nano-based drugs can be combated only when there are enough information and data on the toxicological front. Exhilarating conclusions regarding nanomedicine and nanotoxicology can be seen in the future; yet there would be enough challenges due to the properties of nanosized particles, but detailed study could help the biologist and toxicologists to prove their prospective hypothesis to recognize and avoid potential risks associated with these toxicities. Much more extensive studies have to be conducted in order to provide for assessing safety measures of nano-based drugs/products.

20.6 Future Perspectives

Substantial amount of studies have to be conducted, and data has to be collected in order to understand the basicity of the toxicology caused by nanoparticles. In vivo studies that could incorporate long-term nanoparticle toxicity analyses have to be carried out, and these studies should be categorized in terms of time frame and dose dependency of toxicity of various nanoparticles. The effects of nanoparticles on biological distribution must be further studied to give a direction to the in vivo studies and, hence, help in concluding the data as more comparable. One of the major and critical disadvantages of nanotoxicology being the prolonged circulation in the blood which means more slow accumulation of tissue of the nanoparticles and very slow drug release. Henceforth, the toxicity induced by nanoparticles has been more proving to be more and more dreadful. For developing an efficient and successful formulation of nanoparticle in order to prevent the healthy cells and organs from getting affected adversely, many efforts have been put forward by various biologists and toxicologists. One of the strategies for the future direction has been about targeting system which delivers the particles into the target tissues from the blood or which keeps it in the target tissues (microvessels). As the accumulation of the nanoparticle in the target tissue would be faster and swift, the circulation time would be short and, hence, would result in fast drug release. Basically, this strategy is based upon rapid release formulation. The contemplation regarding the future of nanomedicine is that the inception of new technologies shall bring forth more advanced delivery system which would be efficient in not just the targeting and delivering precisely into the target area but also release it efficiently. Additionally, the strategy to come up with inexpensive and safe techniques and materials is at the frontline to be considered.

Acknowledgments The authors are grateful to the Vice Chancellor, King George's Medical University (KGMU), Lucknow for the encouragement and support for this work. SK Saxena is also supported by CCRH, Government of India, and US NIH grants: R37DA025576 and R01MH085259. The authors have no other relevant affiliations or financial involvement with any organization or entity with a financial interest in or financial conflict with the subject matter or materials discussed in the manuscript apart from those disclosed.

References

Accomasso L, Cristallini C, Giachino C (2018) Risk assessment and risk minimization in nanomedicine: a need for predictive, alternative, and 3Rs strategies. Front Pharmacol 9:228

Almeida JP, Chen AL, Foster A, Drezek R (2011a) In vivo biodistribution of nanoparticles. Nanomedicine (London) 6(5):815–835

Almeida JP, Chen AL, Foster A, Drezek R (2011b) In vivo bio distribution of nanoparticles. Nanomedicine 6(5):815–835

Bierkandt FS, Leibrock L, Wagener S, Laux P, Luch A (2018) The impact of nanomaterial characteristics on inhalation toxicity. Toxicol Res (Cambridge) 7(3):321–346

Du Y, Zhong Y, Dong J, Qian C, Sun S, Gao L, Yang D (2019) The effect of PEG functionalization on the in vivo behavior and toxicity of CdTe quantum dots. RSC Adv 9(22):12218–12225

Fu PP, Xia Q, Hwang HM, Ray PC, Yu H (2014) Mechanisms of nanotoxicity: generation of reactive oxygen species. J Food Drug Anal 22(1):64–75

Gallo A, Manfra L, Boni R, Rotini A, Migliore L, Tosti E (2018) Cytotoxicity and genotoxicity of CuO nanoparticles in sea urchin spermatozoa through oxidative stress. Environ Int 118:325–333

Jones CF, Grainger DW (2009) In vitro assessments of nanomaterial toxicity. Adv Drug Deliv Rev 61(6):438–456

Kanwal U, Irfan Bukhari N, Ovais M, Abass N, Hussain K, Raza A (2018) Advances in nanodelivery systems for doxorubicin: an updated insight. J Drug Target 26(4):296–310

Karlsson HL, Cronholm P, Gustafsson J, Möller L (2008) Copper oxide nanoparticles are highly toxic: a comparison between metal oxide nanoparticles and carbon nanotubes. Chem Res Toxicol 21(9):1726–1732

Khanna P, Ong C, Bay BH, Baeg GH (2015) Nanotoxicity: an interplay of oxidative stress, inflammation and cell death. Nanomaterials (Basel) 5(3):1163–1180

Kim S, Ryu DY (2013) Silver nanoparticle-induced oxidative stress, genotoxicity and apoptosis in cultured cells and animal tissues. J Appl Toxicol 33(2):78–89

Krug HF, Wick P (2011) Nanotoxicology: an interdisciplinary challenge. Angew Chem Int Ed 50(6):1260–1278

Lohcharoenkal W, Wang L, Stueckle TA, Dinu CZ, Castranova V, Liu Y, Rojanasakul Y (2013) Chronic exposure to carbon nanotubes induces invasion of human mesothelial cells through matrix metalloproteinase-2. ACS Nano 7(9):7711–7723

Manke A, Wang L, Rojanasakul Y (2013) Mechanisms of nanoparticle-induced oxidative stress and toxicity. Biomed Res Int 2013:942916

Moghimi SM, Parhamifar L, Ahmadvand D, Wibroe PP, Andresen TL, Farhangrazi ZS, Hunter AC (2012) Particulate systems for targeting of macrophages: basic and therapeutic concepts. J Innate Immun 4(5–6):509–528

Oberdörster G (2010) Safety assessment for nanotechnology and nanomedicine: concepts of nanotoxicology. J Intern Med 267(1):89–105

Pelaz B, Alexiou C, Alvarez-Puebla RA, Alves F, Andrews AM, Ashraf S, Balogh LP, Ballerini L, Bestetti A, Brendel C, Bosi S, Carril M, Chan WC, Chen C, Chen X, Chen X, Cheng Z, Cui D, Du J, Dullin C, Escudero A, Feliu N, Gao M, George M, Gogotsi Y, Grünweller A, Gu Z, Halas NJ, Hampp N, Hartmann RK, Hersam MC, Hunziker P, Jian J, Jiang X, Jungebluth P, Kadhiresan P, Kataoka K, Khademhosseini A, Kopeček J, Kotov NA, Krug HF, Lee DS, Lehr CM, Leong KW, Liang XJ, Ling Lim M, Liz-Marzán LM, Ma X, Macchiarini P, Meng H, Möhwald H, Mulvaney P, Nel AE, Nie S, Nordlander P, Okano T, Oliveira J, Park TH, Penner RM, Prato M, Puntes V, Rotello VM, Samarakoon A, Schaak RE, Shen Y, Sjöqvist S, Skirtach AG, Soliman MG, Stevens MM, Sung HW, Tang BZ, Tietze R, Udugama BN, VanEpps JS, Weil T, Weiss PS, Willner I, Wu Y, Yang L, Yue Z, Zhang Q, Zhang Q, Zhang XE, Zhao Y, Zhou X, Parak WJ (2017) Diverse applications of nanomedicine. ACS Nano 11(3):2313–2381

Rushton EK, Jiang J, Leonard SS, Eberly S, Castranova V, Biswas P, Elder A, Han X, Gelein R, Finkelstein J, Oberdörster G (2010) Concept of assessing nanoparticle hazards considering nanoparticle dosemetric and chemical/biological response metrics. J Toxicol Environ Health A 73(5):445–461

Ruszkiewicz JA, Pinkas A, Ferrer B, Peres TV, Tsatsakis A, Aschner M (2017) Neurotoxic effect of active ingredients in sunscreen products, a contemporary review. Toxicol Rep 4:245–259

Ruttkay-Nedecky B, Krystofova O, Nejdl L, Adam V (2017) Nanoparticles based on essential metals and their phytotoxicity. J Nanobiotechnol 15(1):33

Saptarshi SR, Duschl A, Lopata AL (2013) Interaction of nanoparticles with proteins: relation to bio-reactivity of the nanoparticle. J Nanobiotechnol 11:26

Shvedova AA, Castranova V, Kisin ER, Schwegler-Berry D, Murray AR, Gandelsman VZ, Maynard A, Baron P (2003) Exposure to carbon nanotube material: assessment of nanotube cytotoxicity using human keratinocyte cells. J Toxicol Environ Health A 66(20):1909–1926

Singh AV, Laux P, Luch A, Sudrik C, Wiehr S, Wild AM, Santomauro G, Bill J, Sitti M (2019) Review of emerging concepts in nanotoxicology: opportunities and challenges for safer nanomaterial design. Toxicol Mech Methods 29(5):378–387

Sukhanova A, Bozrova S, Sokolov P, Berestovoy M, Karaulov A, Nabiev I (2018) Dependence of nanoparticle toxicity on their physical and chemical properties. Nanoscale Res Lett 13(1):44

Vandebriel RJ, De Jong WH (2012) A review of mammalian toxicity of ZnO nanoparticles. Nanotechnol Sci Appl 5:61–71

Wolfram J, Zhu M, Yang Y, Shen J, Gentile E, Paolino D, Fresta M, Nie G, Chen C, Shen H, Ferrari M (2015) Safety of nanoparticles in medicine. Curr Drug Targets 16(14):1671–1681

Cellular and Organismal Toxicity of Nanoparticles and Its Associated Health Concerns

21

Sudarshan Kini, Anjana Kaveri Badekila, Debmalya Barh, and Anurag Sharma

Abstract

The demand for nanotechnology in biomedical science is escalating rapidly as novel nanomaterials help in rebuilding the life of patients suffering from serious health conditions. Nanomaterials are widely used for biomedical applications such as drug delivery carriers, diagnostic agents, image-contrasting agents, tissue engineering, targeted cancer therapy, and so on. However, due to poor understanding of mechanisms at the nanoscale, nature had to deal with the negative face of the nanotechnology broadly called as nanotoxicity. Nanotoxicology is therefore the study of the toxicity of nanomaterials at the cellular, organism, and environmental levels. Variety of nanoparticles (NPs) prepared from sources like metals, semiconductors, polymers, and lipids behave differently in cells due to the difference in their surface functionality, size and shape anisotropy, charge and dispersity in polar or nonpolar solvents, etc. Therefore, since the last decade, the scientific community has shown keen interest to understand the NPs toxicity at different biological levels of the organization. Cellular toxicity is mainly due to the intervention of NPs in cellular processes leading to oxidative stress, altered signaling, proliferation, and death pathways. Nanotoxicity in organism level causes defects in physiological functioning, behavior, and reproduction. Herein, this chapter enlightens various effects of commonly used NPs at cellular level as well as in organisms that may have implications linked to serious abnormal conditions such as cancer, diabetes, neurodisorders, cardiovascular, and hepatotoxicity.

Keywords

Nanoparticles · Nanotoxicity · Cellular and organismal toxicity · Disease

S. Kini · A. K. Badekila · D. Barh · A. Sharma (✉)
Nitte University Centre for Science Education and Research (NUCSER), Mangaluru, India
e-mail: anuragsharma@nitte.edu.in

S. K. Saxena, S. M. P. Khurana (eds.), *NanoBioMedicine*,
https://doi.org/10.1007/978-981-32-9898-9_21

Abbreviations

Ag	Silver
Al_2O_3	Aluminum oxide
ALCL	Anaplastic large cell lymphoma
Au	Gold
Bi_2O_3	Bismuth oxide
Ca	Calcium
CAT	Catalase
CdSe	Cadmium selenide
CeO_2	Cerium oxide
CNP	Carbon nanoparticles
CNT	Carbon nanotube
Cu	Copper
CuO	Copper oxide
DNA	Deoxyribonucleic acid
Fe_3O_4	Iron oxide
GPx	Glutathione peroxidise
GSH	Glutathione
HaCaT	Human keratinocyte cell line
Hb	Hemoglobin
HepG2	Liver hepatocellular carcinoma cells
Hsps	Heat shock proteins
IL-6	Interleukin 6
JNK	c-Jun N-terminal kinase
MCN	Mesoporous carbon nanoparticles
miRNA	MicroRNA
MRC-5	Medical Research Council cell strain-5
mRNA	Messenger ribonucleic acid
MSN	Mesoporous silica nanoparticles
MT	Metallothionein
NPs	Nanoparticles
p38MAPK	p38 mitogen-activated protein kinase
pax	Paired box gene
PCL	Polycaprolactone
PEG	Poly ethylene glycol
pH	Potential of hydrogen
PLA	Polylactic acid
PLGA	Polylactic-co-glycolic acid
PNIPAM	Poly(N-isopropylacrylamide)
QD	Quantum dot
RNA	Ribonucleic acid
ROS	Reactive oxygen species
SOD	Superoxide dismutase
TiO_2	Titanium dioxide

TLRs	Toll-like receptor signaling
TNFα	Tumor necrosis factor α
WC-Co	Tungsten carbide cobalt
ZnO	Zinc oxide
ZrO_2	Zirconium dioxide

21.1 Introduction

Today with emerging trends, the transformation of technology overlooks harmful effects aggravated by the exposure of nanotoxicants that normal eyes fail to pick in a day's routine. Advances in nanotechnology in the field of health and biomedical science have improved the quality of life of patients suffering from various dreadful diseases. Molecules in the nanodimension interact with cells and their microenvironment and induce a therapeutic effect. Currently, there is no concurrence among the scientific community regarding what feature is most important in highlighting response for each type of nanomaterial even though a range of findings is available. Various nanoparticles (NPs) have been successfully used for several biomedical applications such as drug delivery carriers (Koo et al. 2005), tissue engineering (Gorain et al. 2017), cancer therapy (Peer et al. 2007), contrasting imaging agents (Koo et al. 2005), gene delivery (Xiao et al. 2019), biosensing, and environmental applications (Sanvicens et al. 2009). Using the bottom-up approach, bulk materials are chemically reduced into smaller nuclei followed by growth results in nanomaterials. Due to the high surface to volume ratio, individual NPs can able to interact with each biomolecule independently which aids to enhancement in functional attributes essential for biomedical applications. Further, various material properties including solubility, scattering, fluorescence, magnetization, reflectance, drug targeting, and thermal properties are significantly affected. Biomedical applications entail interaction of nanodrug carriers with target tissues. Therefore, desirable properties of nanocarriers should oblige a) encapsulation or entrapment of drug molecules in a nanosystem, b) biocompatibility and biodegradability of NPs, c) long circulating drug delivery with minimal burst effect, d) site-specific and selective tumor targeting, e) stability of NPs in aqueous solvents, f) high surface area for enhanced accumulation into the tissue site, and g) controlled drug delivery which includes ultrasound, pH, magnetic hyperthermia, and photothermal and enzyme-triggered drug release (Brigger et al. 2012; Cao and Wang 2011; Chen et al. 2018; Cuenca et al. 2006; Koo et al. 2005; Logothetidis 2006; Mohanraj and Chen 2006; Patravale et al. 2004; Peer et al. 2007).

The general synthesis of NPs involves two approaches: the top-down approach and the bottom-up approach. Generally, the top-down approach is a technique where large materials are broken down by external forces into small nanoscale structure in a precise pattern, whereas in bottom-up approach, the force of chemical oxidation or reduction is in total control with the formation of nanoparticle. Here crystal growth takes place, growth species like atoms, ions, and molecules impinging on the surface assemble into crystal structures. Hence this approach is building up of materials from the bottom. In general bottom-up approach involves the principle of supersaturation and nucleation followed by growth. Supersaturation involves an increase in concentration as a function of time. Nucleation does not occur even above equilibrium. Nucleation would start when the minimum supersaturated concentration is attained that overcomes the critical energy barrier. Uniform monodispersed nuclei formed grows subsequently. When the solute concentration decreased below supersaturated concentration, no new nuclei would form, and growth proceeds further (Cao and Wang 2011). Thus, NPs formed in solution are monodispersed. However, size, shape, and dispersion in aqueous or organic solvents can be controlled by varying the concentration of solvents, temperature, types of surfactants, and pH. Major challenges in synthesis are huge surface energy, acquiring mono-dispersed particle, stabilization, and prevention of agglomeration.

Depending on the specific applications, NPs are fabricated into metallic, semiconductor, lipid, protein, and polymeric structures. For example, metallic gold (Au), silver (Ag), and copper (Cu) NPs have plasmonic characteristics which upon excitation with laser induce surface plasmon resonance (SPR). Therefore, surface functionalization with monoclonal or polyclonal antibodies can be probed for the detection of biomolecules such as cells, proteins, enzymes, growth factors, and so on. Further, the bactericidal activity of AgNPs makes them a prominent player in developing antimicrobial surfaces. The semiconductor quantum dots (QDs), also known as "zero-dimensional" NPs, display inherent fluorescence with high photostability and high quantum yield. The photoluminescent QDs encapsulated in biocompatible polymeric NPs have been widely used for bioimaging and fluorescent resonance energy transfer (FRET) applications (Beija et al. 2012; Kini et al. 2018, 2019; Van Vlerken and Amiji 2006). Use of biocompatible polymers like polylactic acid (PLA), poly(N-isopropylacrylamide) (PNIPAM), chitosan, alginate, polylactic-co-glycolic acid (PLGA), polyethylene glycol (PEG), polycaprolactone (PCL), and respective copolymers is highly promising and demanding for therapeutic applications (Uhrich et al. 1999). Nanocarriers fabricated from such polymeric materials for drug delivery not only induce the therapeutic efficacy but also enable selective targeting and enhance the cellular uptake efficiency. Other nanocarriers including liposomes (lipid-based NPs), dendrimers (hyperbranched polymeric macromolecules with the central core from where polymeric branches originated for drug entrapment), polymeric micelle formed by oil in water or water in oil emulsions, and fullerenes (spheroidal carbon nanostructures) are used as drug carriers for drug delivery applications. NPs such as QDs (semiconductor-based fluorescent nanocrystals), iron oxide-based superparamagnetic NPs (IO for magnetic hyperthermia), Au NPs (for photothermal therapy), AgNPs (for water disinfectant), and titanium oxide (TiO_2) and zinc oxide (ZnO) NPs are mainly used for imaging and diagnostic purposes (Beija et al. 2012; Van Vlerken and Amiji 2006), and carbon-based NPs like graphene and carbon nanotubes (CNTs) are extensively used for drug delivery and biosensing applications. Liposome-based doxorubicin (Doxil) (Mousa and Bharali 2011) and albumin-based paclitaxel (Abraxane) are clinically approved and marketed for cancer treatment (Hawkins et al. 2008; Li and Wallace 2008). Some of the PLGA NPs for the delivery of anticancer agents such as leuprolide acetate, buserelin acetate, and triptorelin pamoate were successfully used for prostate cancer therapy (Mundargi et al. 2008).

Nanotoxicology deals with the study of the toxicity of nanomaterials at the cellular, organism, and environmental levels. So the serious concern is about the effects of NPs exposure because individual NPs are highly reactive due to high surface energy which means a significant mass of NPs can pose a serious threat to health compared to the equivalent mass of bulk material. Earlier evidence and current knowledge on NPs suggest that one of the main reasons for nanotoxicity in cells is oxidative stress (Aguilar 2012; Stone and Donaldson 2006). Tiny NPs such as carbon black, fullerenes, CNTs, asbestos, etc. causing ambient pollution in the atmosphere generate radicals with unpaired electrons which makes them highly reactive species (Elsaesser and Howard 2012; Oberdörster 2010). Therefore, the interaction of such free radicals with cells disturbs the balance maintained by antioxidants such as vitamins, glutathiones (GSH), and peroxidases. Industrial production of NPs derived from metals and semiconductors like silicon, titanium, gold, zinc, silver, and their respective oxides has shown a link between oxidative damage and diseases like asthma, cancer, cardio- and hepatotoxicity, and immune-related disorders (Baky et al. 2013; Chen et al. 2018; Gaiser et al. 2013; Khanna et al. 2015; Lanone and Boczkowski 2011).

Toxicological aspects of NPs are chasing technological developments like a shadow as by-products of nanosynthesis process, tissue accumulation, unpredictable and adverse consequences after exposure, systemic toxicity, enhanced reactivity due to high surface area, and toxic degradation products cause adverse health effects. Therefore, toxicological data of NPs at cellular as well as organ level is a prerequisite before taking into the field and also useful for safety and risk assessment. In vitro studies are conducted in the presence of antioxidants present in the serum which may neutralize the oxidative effect of NPs leading to false positive estimation of biocompatibility. Nevertheless, there are several physical, chemical, and biological factors of NPs that influence the toxicity. For example, ultrafine particles with size <100 nm can penetrate the skin; deposit in the lungs, liver, and kidney; and cause chemical and physical effects in cells (Borm et al. 2006; Buzea et al. 2007). Chemical effects include solubility, reactive oxygen species (ROS) generation, lipid peroxidation, catalytic oxidation and reduction of functional proteins, ionic imbalance, and change in intracellular pH. Physically NPs can cause disruption of the membrane, protein aggregation and DNA damage, and barrier formation for cellular communications. Though these effects are dependent on the size, shape, charge, surface energy, and ligand functionalization, there is no accurate pattern in physicochemical parameters which could predict the toxicity of neither NPs nor epidemiological data available. For example, different NPs having the same structural backbone may cause variable toxicity due to different surface functionalization. Sometimes lack of awareness about handling prepared NPs may cause direct exposure of NPs to human body or environment as most of the bottom-up methods do not require well-equipped clean rooms and aseptic conditions. Similarly, by-products of purified nanomaterials also contain nanomolecules which contaminate water and air resources. Apart from physicochemical and cellular factors, assessment of doses; systemic administration including nanodrug delivery to intravenous, transdermal, and blood-brain barrier; and pulmonary routes and biodistribution are contributing to systemic toxicity. To counter such problems, internationally standardized protocols for handling NPs and safety protocols need to be developed and validated. However, monitoring long-term effects of engineered NPs which are already commercialized for applications like cosmetics, electronics, food additives, clothing, sports materials, etc. becomes challenging. Therefore, it requires strong earnestness to understand the basic scientific mechanism of toxicity of various anisotropic NPs at cellular levels both in vitro and in vivo before toxicants reach the alarming level. Further, accurately determining the toxicity limit of physicochemical parameters of NPs used for clinical studies could allow researchers to design nanomaterials in the future.

Considering the above facts, assessment of various NPs toxicity at cellular as well as organism level is indispensable. This knowledge of nanotoxicology would help the researchers and clinicians to set safety parameters for the usage of NPs in various biomedical applications. Herein, this chapter describes the toxic effects of various NPs on cellular processes involving signaling, proliferation, growth, division, and death pathways. Similarly, the toxicity of NPs is associated with the physiological functioning of the organisms such as development, reproduction, and behavior. Furthermore, it is well documented that cellular and organism toxicity leads to various lifestyle diseases. Henceforth, this study focuses on the potential dangers of diverse NPs and their associated health risks like cancer and other problems as described in the following (Fig. 21.1).

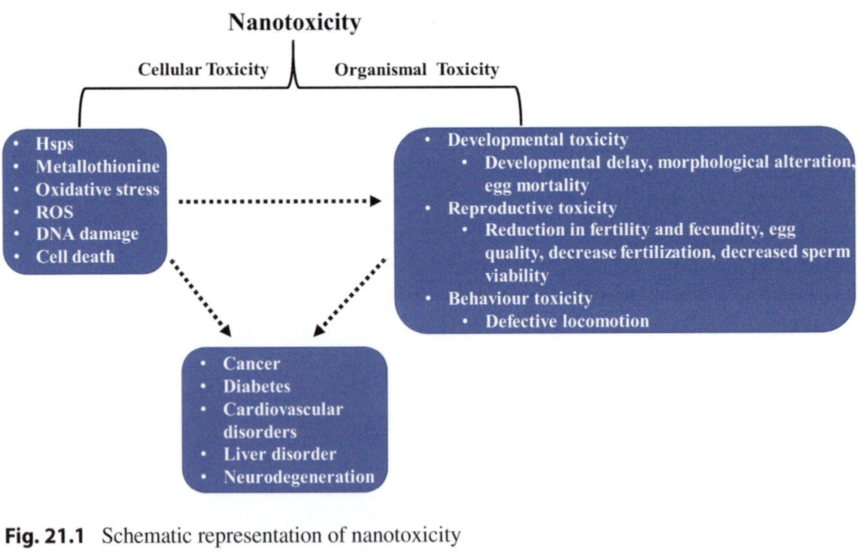

Fig. 21.1 Schematic representation of nanotoxicity

21.2 Cellular Nanotoxicity

Responses at the cellular level are one of the earliest reactions which help the cells to defend and recover from stressful events. The interaction of NPs with cells could trigger various cytotoxic effects (Table 21.1). Therefore, illustrating the effect of NPs at the cellular level is a critical aspect of human risk assessment and has been receiving great interest in the process of implementation of NPs. NPs are different in nature due to their composition and conformation; therefore cellular toxicity could be assessed at the different level of regulations. Among the multiple cellular responses, upregulation of heat shock proteins (Hsps) is considered as an early toxicity biomarker (Gupta et al. 2010). Increased Hsps level in the cell prevents protein aggregation and other protein modifications (Chatterjee and Burns 2017; Takalo et al. 2013). Since the last decade, modulation in Hsps has been considered as an early sensor of NPs toxicity. Recently, Masouleh et al. (2017) and Krishnaraj et al. (2016) observed an increase in *hsp70* mRNA levels in Ag NPs-exposed juvenile *Caspian kutum* and zebrafish, respectively. Similar upregulated levels of Hsp70 were observed with ZnO and silica NPs exposure in *Drosophila*, TiO_2 NPs in Caribbean reef-building coral, Au NPs in *Daphnia magna*, Cu NPs in *Takifugu fasciatus*, and IONPs in mice (Dominguez et al. 2015; Jovanovic and Guzman 2014;

Table 21.1 Nanoparticles toxicity at cellular and organismal levels

Nanoparticles	Cellular effect	Organismal effect
Ag NPs	Hsp70, metallothionein, oxidative stress and ROS	Decrease body proportion and depigmentation
MSN and silica NPs	Hsp70, Hsp22, Hsp27, Hsp60, Hsp90, metallothionein, oxidative stress and ROS	Sperm abnormalities, malformation, and impairment in swimming activity
ZnO NPs	Hsp70, metallothionein, ROS, DNA, and protein damage	Decreases the developmental stage, increases egg mortality, and small size of organism
TiO_2 NPs	Hsp70, apoptosis and TLR signaling, ROS, and JNK activation	Defective larval crawling and climbing behavior
Au NPs	Hsp70, Hsp90, apoptosis, inflammatory response	Inhibition of ectodermal differentiation, abnormal embryonic development and abortion, abnormal reproduction, reduced swimming activity
Fe_3O_4NPs	Hsp 70, cytokine production, oxidative stress	Embryo mortality, delay in hatching and malformation
QD NPs	Metallothionein, oxidative stress, and ROS	Decrease heart and hatching rate, pericardial and yolk sac edema
CuO NPs	Hsp70, metallothionein, and oxidative stress	Developmental delay and structural changes such as skeletal rods and shorter arms
Bi_2O_3	Hsp70, metallothionein, and oxidative stress	
ZrO_2	Glutathione peroxidase and oxidative stress	Reduction in the climbing activity

Pandey et al. 2013; Siddique et al. 2014; Sundarraj et al. 2017; Wang et al. 2018b). Such upregulation of Hsp70 has been proposed as an early bioindicator of cellular stress. Nevertheless, researchers raised the concern that one class of gene/protein cannot work as universal bioindicator as chemicals are different in nature (de Pomerai 1996; Gupta et al. 2010). In accordance with this concern, significant induction in the other class of *hsps* such as *hsp22* in *Drosophila* larvae after exposure to silica NPs has been observed (Pandey et al. 2013). Similarly, increased Hsp90 expression along with Hsp70 has been recorded after polystyrene nanoplastic and Au NPs exposure in *Daphnia pulex* and *Sparus aurata*, respectively (Liu et al. 2019; Teles et al. 2018). However, Petrache Voicu et al. (2015) observed downregulation in Hsp27, Hsp60, and Hsp90 protein expression in silica NPs-treated MRC-5 cell lines. Along with Hsps, metallothionein is also considered as a prominent biomarker of cellular toxicity. Metallothioneins are intracellular cysteine-rich, metal-binding proteins found from bacteria to human (Ruttkay-Nedecky et al. 2013). Increased metallothionein levels in the cell enhance detoxification, scavenge free radicals, and inhibit pro-apoptotic mechanisms (Thirumoorthy et al. 2007). Thus, the expression of metallothionein was examined in the metal NPs. Horie et al. (2018) and Miyayama and Matsuoka (2016) showed upregulation of metallothionein 2A (MT2A) in ZnO, copper oxide (CuO), bismuth oxide (Bi_2O_3), and AgNPs-exposed A549 cells due to release of metal ions in the cell. Rocha et al. (2018) investigated the effect of QDs on metallothioneins (MTs) isoforms (mt10IIIa and mt20IV) expression using mussel *Mytilus galloprovincialis*. Same group observed concentration and time-dependent changes in mt20IV mRNA levels and suggested its role in QDs metabolism. Parallel to Hsps and metallothionein induction, antioxidant defense mechanism and generation of ROS of an organism have been also used as an indicator of cellular toxicity (Auten and Davis 2009; Fu et al. 2014). ROS are unstable molecules containing unpaired oxygen (superoxide anion, hydroxyl radicals, peroxynitrite) atom. This unstable form of oxygen is collectively called free radicals. Excess formation of ROS due to the stressor toxicants may cause protein, lipid, and DNA damage in the cell that may lead to many disease conditions (Schieber and Chandel 2014; Sharma et al. 2012). As a protective mechanism, cells have universal conserved enzymatic (e.g., superoxide dismutase (SOD), catalase (cat), glutathione reductase, glutathione peroxidase (GPx)) and nonenzymatic (e.g., glutathione, vitamins C and D) antioxidant defense mechanism that helps in ROS detoxification. In this line, an ample number of reports highlight oxidative stress due to NPs exposure. Zhang et al. (2018b) studied the effect of AgNPs on soil nitrogen-fixing *Azotobacter vinelandii* bacteria. AgNPs exposure caused oxidative cellular damage to bacteria due to excess ROS and hydroxyl radical generation. Rossner Jr. et al. (2018) studied the effect of inhalation of acute and sub-chronic ZnO NPs by mice using next-generation sequencing and found modulation of splice junction genes associated with oxidative stress, immunity, and DNA repair. Similarly, increased ROS level was observed in ZnO NPs-exposed human umbilical vein endothelial cell line (Qiao et al. 2018b). Mesoporous silica NPs (MSNs) are extensively used as a drug delivery carrier. Hozayen et al. (2019) found that MSNs exposure to rats for 30 days causes cardiac and pulmonary toxicity due to increase in ROS generation and malondialdehyde and decline in antioxidant defense mechanism in the heart and lung of rats. Likewise, several other NPs have been shown oxidative stress (changes in antioxidant enzyme activities) and ROS generation in various in vivo and in vitro model systems (Abdelhalim et al. 2018; Gallo et al. 2018; Kong et al. 2019; Soltani et al. 2018; Yue et al. 2018). The immune system provides the first-line defense against infection. In addition immune toxicity is the sensitive area under chemical toxicants exposure (Vos et al. 1989). Dhupal et al. (2018) exposed RAW 264.7 cells (macrophage) to TiO_2 NPs and observed concurrent induction of macrophage-mediated apoptosis and Toll-like receptor (TLR) signaling through ROS-mediated JNK and p38MAPK pathways. Similarly, apoptosis and inhibition of TNF-α and IL-6 were found in T lymphocytes after exposure to mesoporous carbon NPs (MCN) (Li et al. 2018). Another study by Shah et al. (2018) showed human T lymphocytes exposed to IONPs decreased the cytokine production and proliferation of mitogen-activated T cells due to a redox imbalance. The effect of ZnO NPs was assessed by Abass et al. (2017) on albino mice spleen and thymus. The group observed an increase in total leucocytic count and decrease in RBCs, platelet counts, and Hb % due to the oxidative or inflammatory pathway. Recently, Manzo et al. (2017) exposed sea urchin with nanosized ZnO via food and observed DNA damage (through comet assay) in their coelomocytes (immune effector cells). Alaraby et al. (2015) examined the effect of cadmium selenide (CdSe) QDson *Drosophila* hemocytes. They found CdSeQDs cross the intestinal barrier and cause oxidative stress and DNA damage in hemocytes. Asadpour et al. (2014) found reduction in cell viability and glutathione peroxidase activity in ZrO2-exposed N2a and PC12 cells. Similar to immunotoxicity, NPs also interfere with hematopoiesis that leads to several blood disorders. Due to high metabolic activity, the hematopoietic system is highly prone to biological and physical stress. In this line, several reports underline the negative impact of NPs on hematopoiesis process such as inhibition of erythrogenesis process in zebrafish after exposure to AgNPs (Cui et al. 2016). Liu et al. (2014) demonstrated the shrinkage and apoptosis of hematopoietic organs via an increase in ROS due to cadmium QDs exposure to *Bombyx mori*. However, melanin NPs exposure restores hematopoietic homeostasis in γ-radiation-treated mice (Rageh et al. 2015). Toxicomics study gives a broad idea to understand the toxicity in totality and to unravel the associated molecular mechanism. Analysis of global gene, protein, and metabolic changes associated with NPs helped environmental researchers to predict adverse responses to NPs. In this context, transcriptomic data identified differentially expressed genes associated with oxidative stress, detoxification, endocytosis, intestinal integrity, and iron homeostasis in IONPs-exposed *Caenorhabditis elegans* (Gonzalez-Moragas et al. 2017). In another toxicogenomic study, AgNPs exposed HepG2 cells revealed misregulation of genes related to metabolism, stress response, cell differentiation, cell death, and development (Sahu et al. 2015). Kumar Babele (2019) did the total protein and metabolite profiling in ZnO NPs-exposed budding yeast and found changes in 40% of total proteins and metabolites involved in energy metabolism, oxidative stress, DNA and protein damage, and membrane integrity. Mirzajani et al. (2014) studied AgNPs toxicity in *Oryza sativa* L. through proteomic approach. The results revealed that

AgNPs exposure to *O. sativa L.* root affects oxidative stress tolerance, Ca^{2+} regulation and signaling, cell division, apoptosis, and nucleic acid damage. Yang et al. (2010) identified 16 differentially expressed proteins in silicon dioxide NPs-exposed HaCaT cells that were linked to oxidative stress, cytoskeleton, energy metabolism, apoptosis, and tumor-associated proteins. In the past decade, miRNA has been established as a promising biomarker for toxicological studies. In this line, Qiao et al. (2018a) identified miRNAs associated with inflammation and vesicle-mediated transport, oxidative stress, apoptosis, and autophagy in ZnO NPs-exposed rats. Similarly, 202 microRNAs were differentially expressed in Au NPs-exposed human dermal fibroblast cells. These microRNAs are involved in 71 different biological pathways such as metabolic process, cell-cell communication, cell cycle, apoptosis, and inflammatory response (Huang et al. 2015). In conclusion, risk assessment of NPs has been found to be crucial as they can affect the cellular homeostasis in different ways.

21.2.1 The Organismal Nanotoxicity

Toxicity of acute exposure at the cellular level is usually associated with cells/tissue and used to establish the safety levels of toxicants (Gormley and Teather 2003). Moreover, accumulation of damage at cellular and tissue levels may pose a negative impact at the organismal level. However, sometimes due to physiological complexity, changes at cellular do not reflect at organismal levels. Thus, evaluating long-term/chronic effects on the basis of these acute tests may be multifaceted. Therefore, systemic studies of NPs at organismal levels are essential to deciphering the long-term impact on the organisms fitness such as reproduction, development, and behavior (Table 21.1). Abnormal development due to chemical exposure such as malformation, growth retardation, low birth weight, and behavior deficit are the important defects in biomedical research which needs to be explored. In this line, Maisano et al. (2015) investigated the effect of CuO NPs on the development of sea urchin embryo. They observed that CuO NPs exposure causes developmental delay, morphological alteration such as absent of skeletal rods, and shorter arms in the exposed organism. Embryonic exposure of CuO NPs to zebrafish showed shorter body axis, smaller eyes, underdeveloped liver, and a delayed retinal neurodifferentiation along with reduced locomotory ability. A similar study in zebrafish decreased the expression of *pax2* and *pax6* genes which are involved in neural differentiation and decreased sizes of neural structures. Studies involving cerium oxide (CeO_2) NPs exposure in zebrafish larvae showed growth inhibition, decreased body weight, and delayed vertebral calcification (Lin et al. 2014). The effect of ZnO NPs on development was investigated in amphibians by Spence et al. (2016). The result showed that ZnO NPs decrease the developmental stages, increase egg mortality, and reduce the body size of an organism. In another case, Hao et al. (2017) studied the effect on offspring of ZnO-exposed hens and found that ZnO NPs cause liver dysfunction due to inadequate lipid synthesis (15 genes were found downregulated after ZnO exposure). Developmental exposure of Ag NPs to Drosophila larvae resulted in delayed and reduced developmental success and decreased body proportion (Panacek et al. 2011; Posgai et al. 2011). Phenotypic defects such as depigmentation and soft cuticle was also found when AgNPs were exposed in *Drosophila* (Gorth et al. 2011; Posgai et al. 2011). Administration of Au NPs to pregnant mice for over 3 days resulted in inhibition of ectodermal differentiation, uncharacteristic embryonic development, and abortion (Yang et al. 2018). Likewise, Hong et al. (2016) observed that TiO_2 exposure to mice during pregnancy/lactation time poses negative effect on the development of the central nervous system (diminishing of cerebral and cerebellar cortex, reduction in neurons, edema, nuclear condensation, and decrease in learning and memory capacity) in mice offsprings. Another report by P et al. (2018) showed ZrO_2NPs induced embryonic mortality, delay in hatching, axis and tail bent, and other malformation in zebrafish. In summary, these results show that exposure to NPs cause severe developmental problems, which might be responsible for various health adversities such as reproduction, neurological, and behavior abnormalities. In the last decade, several publications have shown the reproductive adversities of NPs. These include the effect on fertility and fecundity, fertilization, and egg quality, disrupting the balance of sex hormones and many more. Preaubert et al. (2016) investigated the effect of low CeO_2NPs concentration on in vitro fertilization in mice and found decrease in fertilization along with oxidative stress and DNA damage to spermatozoa and oocytes. However, supplementation of CeNPs at low concentration enhanced in vitro embryo production of prepubertal ovine oocytes (Ariu et al. 2017). A study involving female mice exposed to Cu NPs showed adverse changes in the reproductive biology of the organism (Zhang et al. 2018a). Researchers exposed human extravillous trophoblast cells and mice to Cu NPs and found an imbalance in sex hormones and also induced apoptosis and cell cycle arrest at cellular levels. In another study CuO NPs exposure to sea urchin showed sperm toxicity. This effect could be linked to decreased sperm viability, defective mitochondrial activity and increased ROS levels, lipid peroxidation, and DNA damage (Gallo et al. 2018). In *Paracentrotus lividus*, ZnO NPs exposure causes morphological alteration in the offspring, which may be due to sperm DNA damage. Wang et al. (2018a) observed lesser reproduction phenotype in Ag NPs-exposed *Daphnia similis*. This could be associated with downregulation of fatty acid contents after Ag NPs exposure to the organism. Hong et al. (2015) demonstrated that TiO_2 NPs cross the blood-testis barrier, accumulate in the testis, and negatively affect spermatogenesis process in mice. The authors found downregulation in the expression of several testis-specific genes (*Cdc2*, *Cyclin B1*, *Dmcl*, *TERT*, *Tesmin*, *TESP-1*, *XPD*, and *XRCCI*), which may be responsible for the reduced spermatogenesis process in TiO_2-exposed mice. In addition to that, TiO_2 NPs exposure (5–30μg/mL) in Sertoli cells showed cell inhibition, lactate dehydrogenase release, and induction of apoptosis (Hong et al. 2016). Kim et al. (2013) evaluated a multi-generational transfer effect of Au NPs using *C. elegans* and observed no signification effect on the survival rate of the organism. However, their reproduction rate was significantly decreased and caused abnormalities in the bag of worms. Evaluation of an organismal behavior is an important rule to determine physiological homeostasis which is crucial for proper body functioning. Among various behavior, locomotion is a vital behavior of an organism which affects various physiological processes such as reproduction, food intake, and predation. When the two *Daphnia* species *Daphnia similis and Daphnia pulex* were exposed to CeO_2 NPs, they exhibited a decreased swimming velocity by 30% and 40%, respectively (Artells et al. 2013). The author also found some morphological changes like presence of reliefs on the cuticle and longer distal spine in *D. similis*, which may be the cause of CeO_2 aggregation. Michalec et al. (2017) observed a decreased swimming activity and lower velocity in Au NPs-exposed calanoid copepod. Sabat et al. (2016) demonstrated that TiO_2 exposure to Drosophila larvae resulted in defective larval crawling and climbing behavior due to impaired brain physiology. Similar reduced (25%) climbing activity was recoded in zirconium dioxide (ZrO_2) NPs-exposed *Drosophila* (Mishra et al. 2017). Administration of silica NPs to mice showed negative effect on male reproductive biology such as damaged seminiferous epithelium, decreased sperm quality, and sperm abnormality (Zhang et al. 2016). Another study on zebrafish demonstrated that silica and Fe_3O_4 NPs exposure causes tail and head malformation and delayed hatching along with impaired swimming behavior (Duan et al. 2013; Zhu et al. 2012). The effect of carbon QD was investigated by Xiao et al. (2016). They exposed rare minnow embryos/larvae to carbon QD and found decreased movement in minnow embryo, increased heart rate, decreased hatching rate, pericardial and yolk sac edema, and malformation. The group suggested that these phenotypes might be mediated by increased oxidative stress and misregulation in development-associated genes.

21.3 Nanoparticles Affecting the Health

Recently emerging studies specifically concerning the behavior and toxicity of NPs that mediates health complications are gravitating. Moreover, not all toxicological studies to date deal with NPs. Hence, the pessimistic side of NPs is overlooked due to its huge impact on improving the technology. Health challenges are many; however, difficulties with inhalation, carcinogenicity, cardiovascular, neurodegenerative, and hepatotoxicity are the main problems associated with toxic NPs. Further, the specificity of nanotoxicants that eventually effect health is described.

21.3.1 Nanoparticles Linked to Cancer Development

According to a report submitted by the American Association for Cancer Research, it is prudent to limit the introduction of NPs into the environment until we understand which NPs are potentially harmful. For instance, Au NPs at an optimum nanosize range is pertained to photochemically damage tumor cells (Khanna et al. 2015). However, the rogue size of Au NPs can cause adverse effects at the cellular level for normal cells by interacting with cellular components and damaging the DNA (Alkilany and Murphy 2010). Additionally, exposure and inhalation of tungsten carbide cobalt (WC-Co) dust composed of NPs in metal manufacturing, drilling, and mining facilities can cause an increased risk of lung cancer (Armstead and Li 2016). Further, polyurethrane NPs-based breast implants are associated with causing anaplastic large cell lymphoma (ALCL), a rare form of cancer (Keith et al. 2017). Epidemiological research concerning leukemia-specific cancer and bladder cancer reports that low-cost commercial hair dyes slog by the formation of nanocrystals from lead sulfide (Towle et al. 2017; Turati et al. 2014). With emerging concerns over NPs safety, the application of inorganic ceramic NPs such as silica, titanium, and alumina is not being used in cancer therapy due to their non-biodegradable nature. Hence, the inability to molecularly decompose ceramic NPs in the environment limits its extent of application.

21.3.2 Nanoparticles Linked to Diabetes

Recently, CeO_2NPs have been implicated in diabetes-induced testicular sperm damage by attenuating hyperglycemia oxidative damage in different organs (Artimani et al. 2018). In contrast to the results of the study, it is reported that the administration of different doses of CeO_2NPs in healthy individuals causes oxidative damage in testes resulting in diminutive sperm quality, disruption of the endocrine system, and inflammation (Adebayo et al. 2018). In another approach, insulin-loaded aquasomes are fabricated with self-assembled nanocrystalline carbohydrate-coated calcium phosphate dihydrate ceramic core to optimize blood glucose in the targeted site using parenteral delivery system (Cherian et al. 2000; Umashankar et al. 2010). However, intrinsic biophysical constraints of the three layered conformations of aquasome can lead to an adverse or allergic reaction with suboptimal pharmacological activity (Collen et al. 2010).

21.3.3 Nanoparticles Linked to Cardiovascular Diseases

A leading cardiovascular disease, atherosclerosis is causing an increase in the mortality rate worldwide. This is also promoted by certain calcifying NPs (calcium phosphate). These mineral chaperones augment calcification of arterial vascular smooth muscle cells in vitro as suggested by many studies (Barba et al. 2012; Hunter et al. 2014). Another in vitro study has also revealed that engineered carbon NPs (CNPs) and single-walled nanotubes induce the aggregation of platelets, thus enhancing vascular thrombosis in rat carotenoid artery (Radomski et al. 2005). Recently Zhou et al. (2018) found the accumulation of CNPs in zebrafish after a long exposure. The study exhibited that the accumulation of CNP is responsible for structural changes in myocardial tissue and expression of inflammatory cytokines. Another study by El-Hussainy et al. (2016) showed that Al_2O_3 NPs exposure to rat for 14 days (30 mg/kg) altered ECG, disturbed serum markers, and enhanced inflammation and oxidative stress in myocardium. Altogether all these changes lead to myocardial dysfunction in the organism. Further, ceramic NPs used commonly for drug delivery exhibit carcinogenic effect as well as oxidative stress or cytotoxic activity in the heart, lungs, liver, and brain (Grundmann et al. 1989; Singh et al. 2016).

21.3.4 Nanoparticles Linked to Liver Diseases

The effect of ZnO NPs has been studied on a specific population of mice with inflammatory bowel disease induced by indomethacin. Histopathological examination shows high hepatic zinc detection postexposure that causes punitive lesions in the liver (Du et al. 2018). Jia et al. (2017b) investigated the effect of TiO_2 NPs on mice liver tissue and found adverse effects such as bulgy hepatocytes along with nuclear condensation and apoptosis. This could be associated with increased ROS levels and decreased expression of cytoprotective genes. Yu et al. (2017) demonstrated that SiO_2 NPs exposure in mice causes liver fibrosis. Further they explained SiO_2 NPs cause oxidative stress and activate TGF-β1/Smad3 signaling, which promotes liver fibrosis. Oral exposure of nontoxic doses of Ag NPs to normal and obese mice was studied by Jia et al. (2017a). The group found that AgNPs deposits only in the liver of obese mice which induce liver inflammation and suppresses fatty acid oxidation lead to steatohepatitis. Moreover, an evaluation of the toxic potential of AgNPs shows significant endoplasmic stress response in the liver, kidney, and lungs that can be avoided by rational use within safe dose (Huo et al. 2015).

21.3.5 Nanoparticles Linked to Neurodegeneration

The most common neurotoxic element on earth is aluminum with a plausible link to Alzheimer's disease (Tomljenovic 2011). Incremental acquisition of aluminum NPs (Al_2O_3) over a lifetime favors selective accumulation in sufficient amounts to cause brain damage (Krewski et al. 2007). In another study, a single-dose of oral ingestion of TiO_2, ZnO, and Al_2O_3NPs shows translocation of NPs to the central nervous system. The accumulation of NPs leads to axillary toxicity, subsequently damaging the normal metabolism of the brain (Shrivastava et al. 2014). Pathological effects such as destruction of blood-brain barrier, cellular edema, and brain tissue necrosis are observed in the presence of differently sized TiO_2 NPs in rat astrocytes (Liu et al. 2013). Moreover, the learning abilities of rats are affected due to dopaminergic neuronal dysfunction in the presence of manganese dioxide (MnO_2) NPs (Li et al. 2014). A study reveals that susceptibility of pregnant mice to neuroendocrine changes intensifies twofold in the presence of NPs compared to a nonpregnant female. Here, QDs are transferred across the placental barrier with increasing dose, suggesting the transplacental transfer potential of NPs (Chu et al. 2010; Wick et al. 2010). Further, neuropsychiatric complication was observed among patients due to flaking off of the metal shavings of the faulty hip implant followed by the release of chromium and cobalt toxicants into blood streams indicating possible dementia (Green et al. 2017).

21.4 Conclusions

The use and production of NPs have been grown worldwide from the last decade, and their exposure to human beings and other organisms has created an alarming situation. Moreover, there are no accepted occupational and environmental levels of NPs causing toxicity. Therefore, the rapid risk assessment of NPs is very much essential in the present scenario. In this context, researchers have explored various biomarkers or readouts for early and rapid risk assessment of NPs exposure. These include induction of Hsps, metallothionein, ROS generation, DNA damage, and different developmental, behavioral, and reproductive parameters (Table 21.1). Moreover, all these changes at the cellular and organismal levels might be responsible for various health emergencies. However, the mechanistic insight of NPs-induced health adversities is still open for an investigation. We hope that the ongoing studies across the world might be helpful to decipher the molecular mechanisms associated with NPs-induced toxicity. Apart from this, the promotion and execution of nanosafety programs at social, academic, and economic levels might be helpful to render the NPs-associated health risks.

References

Abass MA, Selim SA, Selim AO, El-Shal AS, Gouda ZA (2017) Effect of orally administered zinc oxide nanoparticles on albino rat thymus and spleen. IUBMB Life 69:528–539

Abdelhalim MAK, Moussa SAA, Qaid HA, Al-Ayed MS (2018) Potential effects of different natural antioxidants on inflammatory damage and oxidative-mediated hepatotoxicity induced by gold nanoparticles. Int J Nanomedicine 13:7931–7938

Adebayo OA, Akinloye O, Adaramoye OA (2018) Cerium oxide nanoparticle elicits oxidative stress, endocrine imbalance and lowers sperm characteristics in testes of balb/c mice. Andrologia 50. https://doi.org/10.1111/and.12920

Aguilar Z (2012) Nanomaterials for medical applications. Newnes, Boston

Alaraby M, Demir E, Hernandez A, Marcos R (2015) Assessing potential harmful effects of CdSe quantum dots by using Drosophila melanogaster as in vivo model. Sci Total Environ 530–531:66–75

Alkilany AM, Murphy CJ (2010) Toxicity and cellular uptake of gold nanoparticles: what we have learned so far? J Nanoparticle Res Interdisciplinary Forum Nanoscale Sci Technol 12:2313–2333

Ariu F, Bogliolo L, Pinna A, Malfatti L, Innocenzi P, Falchi L, Bebbere D, Ledda S (2017) Cerium oxide nanoparticles (CeO2 NPs) improve the developmental competence of in vitro-matured prepubertal ovine oocytes. Reprod Fertil Dev 29:1046–1056

Armstead AL, Li B (2016) Nanotoxicity: emerging concerns regarding nanomaterial safety and occupational hard metal (WC-Co) nanoparticle exposure. Int J Nanomedicine 11:6421–6433

Artells E, Issartel J, Auffan M, Borschneck D, Thill A, Tella M, Brousset L, Rose J, Bottero JY, Thiery A (2013) Exposure to cerium dioxide nanoparticles differently affect swimming performance and survival in two daphnid species. PLoS One 8:e71260

Artimani T, Amiri I, Soleimani Asl S, Saidijam M, Hasanvand D, Afshar S (2018) Amelioration of diabetes-induced testicular and sperm damage in rats by cerium oxide nanoparticle treatment. Andrologia 50:e13089

Asadpour E, Sadeghnia HR, Ghorbani A, Boroushaki MT (2014) Effect of zirconium dioxide nanoparticles on glutathione peroxidase enzyme in PC12 and n2a cell lines. Iranian J Pharm Res 13:1141–1148

Auten RL, Davis JM (2009) Oxygen toxicity and reactive oxygen species: the devil is in the details. Pediatr Res 66:121–127

Baky NA, Faddah LM, Al-Rasheed NM, Al-Rasheed NM, Fatani AJ (2013) Induction of inflammation, DNA damage and apoptosis in rat heart after oral exposure to zinc oxide nanoparticles and the cardioprotective role of alpha-lipoic acid and vitamin E. Drug Res 63:228–236

Barba I, Villacorta E, Bratos-Perez MA, Antolin M, Varela E, Sanchez PL, Tornos P, Garcia-Dorado D (2012) Aortic valve-derived calcifying nanoparticles: no evidence of life. Rev Esp Cardiol 65:813–818

Beija M, Salvayre R, Lauth-de Viguerie N, Marty J-D (2012) Colloidal systems for drug delivery: from design to therapy. Trends Biotechnol 30:485–496

Borm PJ, Robbins D, Haubold S, Kuhlbusch T, Fissan H, Donaldson K, Schins R, Stone V, Kreyling W, Lademann J et al (2006) The potential risks of nanomaterials: a review carried out for ECETOC. Part Fibre Toxicol 3:11

Brigger I, Dubernet C, Couvreur P (2012) Nanoparticles in cancer therapy and diagnosis. Adv Drug Deliv Rev 64:24–36

Buzea C, Pacheco II, Robbie K (2007) Nanomaterials and nanoparticles: sources and toxicity. Biointerphases 2:MR17–MR71

Cao G, Wang Y (2011) Nanostructures and nanomaterials: synthesis, properties, and applications. World Scientific, London

Chatterjee S, Burns TF (2017) Targeting heat shock proteins in cancer: a promising therapeutic approach. Int J Mol Sci 18. https://doi.org/10.3390/ijms18091978

Chen L, Liu J, Zhang Y, Zhang G, Kang Y, Chen A, Feng X, Shao L (2018) The toxicity of silica nanoparticles to the immune system. Nanomedicine (Lond) 13:1939–1962

Cherian AK, Rana AC, Jain SK (2000) Self-assembled carbohydrate-stabilized ceramic nanoparticles for the parenteral delivery of insulin. Drug Dev Ind Pharm 26:459–463

Chu M, Wu Q, Yang H, Yuan R, Hou S, Yang Y, Zou Y, Xu S, Xu K, Ji A et al (2010) Transfer of quantum dots from pregnant mice to pups across the placental barrier. Small 6:670–678

Collen J, Greenburg D, Holley A, King C, Roop S, Hnatiuk O (2010) Racial discordance in spirometry comparing four commonly used reference equations to the National Health and Nutrition Examination Study III. Respir Med 104:705–711

Cuenca AG, Jiang H, Hochwald SN, Delano M, Cance WG, Grobmyer SR (2006) Emerging implications of nanotechnology on cancer diagnostics and therapeutics. Cancer 107:459–466

Cui B, Ren L, Xu QH, Yin LY, Zhou XY, Liu JX (2016) Silver_ nanoparticles inhibited erythrogenesis during zebrafish embryogenesis. Aquat Toxicol 177:295–305

de Pomerai D (1996) Heat-shock proteins as biomarkers of pollution. Hum Exp Toxicol 15:279–285

Dhupal M, Oh JM, Tripathy DR, Kim SK, Koh SB, Park KS (2018) Immunotoxicity of titanium dioxide nanoparticles via simultaneous induction of apoptosis and multiple toll-like receptors signaling through ROS-dependent SAPK/JNK and p 38 MAPK activation. Int J Nanomedicine 13:6735–6750

Dominguez GA, Lohse SE, Torelli MD, Murphy CJ, Hamers RJ, Orr G, Klaper RD (2015) Effects of charge and surface ligand properties of nanoparticles on oxidative stress and gene expression within the gut of Daphnia magna. Aquat Toxicol 162:1–9

Du LJ, Xiang K, Liu JH, Song ZM, Liu Y, Cao A, Wang H (2018) Intestinal injury alters tissue distribution and toxicity of ZnO nanoparticles in mice. Toxicol Lett 295:74–85

Duan J, Yu Y, Shi H, Tian L, Guo C, Huang P, Zhou X, Peng S, Sun Z (2013) Toxic effects of silica nanoparticles on zebrafish embryos and larvae. PLoS One 8:e74606

El-Hussainy e-HM, Hussein AM, Abdel-Aziz A, El-Mehasseb I (2016) Effects of aluminum oxide (Al2O3) nanoparticles on ECG, myocardial inflammatory cytokines, redox state, and connexin 43 and lipid profile in rats: possible cardioprotective effect of gallic acid. Can J Physiol Pharmacol 94:868–878

Elsaesser A, Howard CV (2012) Toxicology of nanoparticles. Adv Drug Deliv Rev 64:129–137

Fu PP, Xia Q, Hwang HM, Ray PC, Yu H (2014) Mechanisms of nanotoxicity: generation of reactive oxygen species. J Food Drug Anal 22:64–75

Gaiser BK, Hirn S, Kermanizadeh A, Kanase N, Fytianos K, Wenk A, Haberl N, Brunelli A, Kreyling WG, Stone V (2013) Effects of silver nanoparticles on the liver and hepatocytes in vitro. Toxicol Sci, Off J Soc Toxicol 131:537–547

Gallo A, Manfra L, Boni R, Rotini A, Migliore L, Tosti E (2018) Cytotoxicity and genotoxicity of CuO nanoparticles in sea urchin spermatozoa through oxidative stress. Environ Int 118:325–333

Gonzalez-Moragas L, Yu SM, Benseny-Cases N, Sturzenbaum S, Roig A, Laromaine A (2017) Toxicogenomics of iron oxide nanoparticles in the nematode C. elegans. Nanotoxicology 11:647–657

Gorain B, Tekade M, Kesharwani P, Iyer AK, Kalia K, Tekade RK (2017) The use of nanoscaffolds and dendrimers in tissue engineering. Drug Discov Today 22:652–664

Gormley KL, Teather KL (2003) Developmental, behavioral, and reproductive effects experienced by Japanese medaka (Oryzias latipes) in response to short-term exposure to endosulfan. Ecotoxicol Environ Saf 54:330–338

Gorth DJ, Rand DM, Webster TJ (2011) Silver nanoparticle toxicity in Drosophila: size does matter. Int J Nanomedicine 6:343–350

Green B, Griffiths E, Almond S (2017) Neuropsychiatric symptoms following metal-on-metal implant failure with cobalt and chromium toxicity. BMC Psychiatry 17:33

Grundmann R, Said S, Krinke S (1989) Quality of life after rectal resection or extirpation. A comparison using different measurement parameters. Dtsch Med Wochenschr 114:453–457

Gupta SC, Sharma A, Mishra M, Mishra RK, Chowdhuri DK (2010) Heat shock proteins in toxicology: how close and how far? Life Sci 86:377–384

Hao Y, Liu J, Feng Y, Yu S, Zhang W, Li L, Min L, Zhang H, Shen W, Zhao Y (2017) Molecular evidence of offspring liver dysfunction after maternal exposure to zinc oxide nanoparticles. Toxicol Appl Pharmacol 329:318–325

Hawkins MJ, Soon-Shiong P, Desai N (2008) Protein nanoparticles as drug carriers in clinical medicine. Adv Drug Deliv Rev 60:876–885

Hong F, Zhao X, Si W, Ze Y, Wang L, Zhou Y, Hong J, Yu X, Sheng L, Liu D et al (2015) Decreased spermatogenesis led to alterations of testis-specific gene expression in male mice following nano-TiO2 exposure. J Hazard Mater 300:718–728

Hong F, Zhao X, Chen M, Zhou Y, Ze Y, Wang L, Wang Y, Ge Y, Zhang Q, Ye L (2016) TiO2 nanoparticles-induced apoptosis of primary cultured Sertoli cells of mice. J Biomed Mater Res A 104:124–135

Horie M, Shimizu K, Tabei Y (2018) Validation of metallothionein, interleukin-8, and heme oxygenase-1 as markers for the evaluation of cytotoxicity caused by metal oxide nanoparticles. Toxicol Mech Methods 28:630–638

Hozayen WG, Mahmoud AM, Desouky EM, El-Nahass ES, Soliman HA, Farghali AA (2019) Cardiac and pulmonary toxicity of mesoporous silica nanoparticles is associated with excessive ROS production and redox imbalance in Wistar rats. Biomed Pharmacother 109:2527–2538

Huang Y, Lu X, Qu Y, Yang Y, Wu S (2015) MicroRNA sequencing and molecular mechanisms analysis of the effects of gold nanoparticles on human dermal fibroblasts. Biomaterials 37:13–24

Hunter LW, Charlesworth JE, Yu S, Lieske JC, Miller VM (2014) Calcifying nanoparticles promote mineralization in vascular smooth muscle cells: implications for atherosclerosis. Int J Nanomedicine 9:2689–2698

Huo L, Chen R, Zhao L, Shi X, Bai R, Long D, Chen F, Zhao Y, Chang YZ, Chen C (2015) Silver nanoparticles activate endoplasmic reticulum stress signaling pathway in cell and mouse models: the role in toxicity evaluation. Biomaterials 61:307–315

Jia J, Li F, Zhou H, Bai Y, Liu S, Jiang Y, Jiang G, Yan B (2017a) Oral exposure to silver nanoparticles or silver ions may aggravate fatty liver disease in overweight mice. Environ Sci Technol 51:9334–9343

Jia X, Wang S, Zhou L, Sun L (2017b) The potential liver, brain, and embryo toxicity of titanium dioxide nanoparticles on mice. Nanoscale Res Lett 12:478
Jovanovic B, Guzman HM (2014) Effects of titanium dioxide (TiO2) nanoparticles on caribbean reef-building coral (Montastraea faveolata). Environ Toxicol Chem 33:1346–1353
Keith L, Herlihy W, Holmes H, Pin P (2017) Breast implant-associated anaplastic large cell lymphoma. Proc (Bayl Univ Med Cent) 30:441–442
Khanna P, Ong C, Bay BH, Baeg GH (2015) Nanotoxicity: an interplay of oxidative stress, inflammation and cell death. Nanomaterials 5:1163–1180
Kim SW, Kwak JI, An YJ (2013) Multigenerational study of gold nanoparticles in Caenorhabditis elegans: transgenerational effect of maternal exposure. Environ Sci Technol 47:5393–5399
Kini S, Ganiga V, Kulkarni SD, Chidangil S, George SD (2018) Sensitive detection of mercury using the fluorescence resonance energy transfer between CdTe/CdS quantum dots and Rhodamine 6G. J Nanopart Res 20:232
Kini S, Kulkarni SD, Ganiga V, Nagarakshit T, Chidangil S (2019) Dual functionalized, stable and water dispersible CdTe quantum dots: facile, one-pot aqueous synthesis, optical tuning and energy transfer applications. Mater Res Bull 110:57–66
Kong L, Hu W, Lu C, Cheng K, Tang M (2019) Mechanisms underlying nickel nanoparticle induced reproductive toxicity and chemo-protective effects of vitamin C in male rats. Chemosphere 218:259–265
Koo OM, Rubinstein I, Onyuksel H (2005) Role of nanotechnology in targeted drug delivery and imaging: a concise review. Nanomedicine 1:193–212
Krewski D, Yokel RA, Nieboer E, Borchelt D, Cohen J, Harry J, Kacew S, Lindsay J, Mahfouz AM, Rondeau V (2007) Human health risk assessment for aluminium, aluminium oxide, and aluminium hydroxide. J Toxicol Environ Health Part B Crit Rev 10(Suppl 1):1–269
Krishnaraj C, Harper SL, Yun SI (2016) In vivo toxicological assessment of biologically synthesized silver nanoparticles in adult Zebrafish (Danio rerio). J Hazard Mater 301:480–491
Kumar Babele P (2019) Zinc oxide nanoparticles impose metabolic toxicity by de-regulating proteome and metabolome in Saccharomyces cerevisiae. Toxicol Rep 6:64–73
Lanone S, Boczkowski J (2011) Titanium and gold nanoparticles in asthma: the bad and the ugly. Eur Respir J 37:225–227
Li C, Wallace S (2008) Polymer-drug conjugates: recent development in clinical oncology. Adv Drug Deliv Rev 60:886–898
Li T, Shi T, Li X, Zeng S, Yin L, Pu Y (2014) Effects of Nano-MnO2 on dopaminergic neurons and the spatial learning capability of rats. Int J Environ Res Public Health 11:7918–7930
Li X, Wang L, She L, Sun L, Ma Z, Chen M, Hu P, Wang D, Yang F (2018) Immunotoxicity assessment of ordered mesoporous carbon nanoparticles modified with PVP/PEG. Colloids Surf B Biointerfaces 171:485–493
Lin S, Wang X, Ji Z, Chang CH, Dong Y, Meng H, Liao YP, Wang M, Song TB, Kohan S et al (2014) Aspect ratio plays a role in the hazard potential of CeO2 nanoparticles in mouse lung and zebrafish gastrointestinal tract. ACS Nano 8:4450–4464
Liu Y, Xu Z, Li X (2013) Cytotoxicity of titanium dioxide nanoparticles in rat neuroglia cells. Brain Inj 27:934–939
Liu T, Xing R, Zhou YF, Zhang J, Su YY, Zhang KQ, He Y, Sima YH, Xu SQ (2014) Hematopoiesis toxicity induced by CdTe quantum dots determined in an invertebrate model organism. Biomaterials 35:2942–2951
Liu Z, Yu P, Cai M, Wu D, Zhang M, Huang Y, Zhao Y (2019) Polystyrene nanoplastic exposure induces immobilization, reproduction, and stress defense in the freshwater cladoceran Daphnia pulex. Chemosphere 215:74–81
Logothetidis S (2006) Nanotechnology in medicine: the medicine of tomorrow and nanomedicine. Hippokratia 10:7–21
Maisano M, Cappello T, Catanese E, Vitale V, Natalotto A, Giannetto A, Barreca D, Brunelli E, Mauceri A, Fasulo S (2015) Developmental abnormalities and neurotoxicological effects of CuO NPs on the black sea urchin Arbacia lixula by embryotoxicity assay. Mar Environ Res 111:121–127

Manzo S, Schiavo S, Oliviero M, Toscano A, Ciaravolo M, Cirino P (2017) Immune and reproductive system impairment in adult sea urchin exposed to nanosized ZnO via food. Sci Total Environ 599–600:9–13

Masouleh FF, Amiri BM, Mirvaghefi A, Ghafoori H, Madsen SS (2017) Silver nanoparticles cause osmoregulatory impairment and oxidative stress in Caspian kutum (Rutilus kutum, Kamensky 1901). Environ Monit Assess 189:448

Michalec FG, Holzner M, Barras A, Lacoste AS, Brunet L, Lee JS, Slomianny C, Boukherroub R, Souissi S (2017) Short-term exposure to gold nanoparticle suspension impairs swimming behavior in a widespread calanoid copepod. Environ Pollut 228:102–110

Mirzajani F, Askari H, Hamzelou S, Schober Y, Rompp A, Ghassempour A, Spengler B (2014) Proteomics study of silver nanoparticles toxicity on Oryza sativa L. Ecotoxicol Environ Saf 108:335–339

Mishra M, Sabat D, Sahu S, Unnikannan P, Dash P (2017) Oral intake of zirconia nanoparticle alters neuronal development and behaviour of *Drosophila melanogaster*. J Nanopart Res 19:282

Miyayama T, Matsuoka M (2016) Involvement of lysosomal dysfunction in silver nanoparticle-induced cellular damage in A549 human lung alveolar epithelial cells. J Occ Med Toxicol 11:1

Mohanraj V, Chen Y (2006) Nanoparticles-a review. Trop J Pharm Res 5:561–573

Mousa SA, Bharali DJ (2011) Nanotechnology-based detection and targeted therapy in cancer: nano-bio paradigms and applications. Cancers 3:2888–2903

Mundargi RC, Babu VR, Rangaswamy V, Patel P, Aminabhavi TM (2008) Nano/micro technologies for delivering macromolecular therapeutics using poly (D, L-lactide-co-glycolide) and its derivatives. J Control Release 125:193–209

Oberdörster G (2010) Safety assessment for nanotechnology and nanomedicine: concepts of nanotoxicology. J Intern Med 267:89–105

P K, M P, Samuel Rajendran R, Annadurai G, Rajeshkumar S (2018) Characterization and toxicology evaluation of zirconium oxide nanoparticles on the embryonic development of zebrafish, Danio rerio. Drug Chem Toxicol 42:1–8

Panacek A, Prucek R, Safarova D, Dittrich M, Richtrova J, Benickova K, Zboril R, Kvitek L (2011) Acute and chronic toxicity effects of silver nanoparticles (NPs) on Drosophila melanogaster. Environ Sci Technol 45:4974–4979

Pandey A, Chandra S, Chauhan LK, Narayan G, Chowdhuri DK (2013) Cellular internalization and stress response of ingested amorphous silica nanoparticles in the midgut of Drosophila melanogaster. Biochim Biophys Acta 1830:2256–2266

Patravale V, Date AA, Kulkarni R (2004) Nanosuspensions: a promising drug delivery strategy. J Pharm Pharmacol 56:827–840

Peer D, Karp JM, Hong S, Farokhzad OC, Margalit R, Langer R (2007) Nanocarriers as an emerging platform for cancer therapy. Nat Nanotechnol 2:751

Petrache Voicu SN, Dinu D, Sima C, Hermenean A, Ardelean A, Codrici E, Stan MS, Zarnescu O, Dinischiotu A (2015) Silica nanoparticles induce oxidative stress and autophagy but not apoptosis in the MRC-5 cell line. Int J Mol Sci 16:29398–29416

Posgai R, Cipolla-McCulloch CB, Murphy KR, Hussain SM, Rowe JJ, Nielsen MG (2011) Differential toxicity of silver and titanium dioxide nanoparticles on Drosophila melanogaster development, reproductive effort, and viability: size, coatings and antioxidants matter. Chemosphere 85:34–42

Preaubert L, Courbiere B, Achard V, Tassistro V, Greco F, Orsiere T, Bottero JY, Rose J, Auffan M, Perrin J (2016) Cerium dioxide nanoparticles affect in vitro fertilization in mice. Nanotoxicology 10:111–117

Qiao Y, Liang X, Yan Y, Lu Y, Zhang D, Yao W, Wu W, Yan Z (2018a) Identification of exosomal miRNAs in rats with pulmonary neutrophilic inflammation induced by zinc oxide nanoparticles. Front Physiol 9:217

Qiao YM, Liang X, Lu YK, Zhuo LB, Wu JJ, Wang HX, Yao W, Yan Z (2018b) The role of heme oxygenase-1 on oxidative stress injury induced by zinc oxide nanoparticles in human umbilical vein endothelial cells line EA.hy926 cells. Chin J Prev Med 52:1177–1181

Radomski A, Jurasz P, Alonso-Escolano D, Drews M, Morandi M, Malinski T, Radomski MW (2005) Nanoparticle-induced platelet aggregation and vascular thrombosis. Br J Pharmacol 146:882–893

Rageh MM, El-Gebaly RH, Abou-Shady H, Amin DG (2015) Melanin nanoparticles (MNPs) provide protection against whole-body -irradiation in mice via restoration of hematopoietic tissues. Mol Cell Biochem 399:59–69

Rocha TL, Bilbao E, Cardoso C, Soto M, Bebianno MJ (2018) Changes in metallothionein transcription levels in the mussel Mytilus galloprovincialis exposed to CdTe quantum dots. Ecotoxicology 27:402–410

Rossner P Jr, Vrbova K, Strapacova S, Rossnerova A, Ambroz A, Brzicova T, Libalova H, Javorkova E, Kulich P, Vecera Z et al (2018) Inhalation of ZnO nanoparticles: splice junction expression and alternative splicing in mice. Toxicol Sci, Off J Soc Toxicol 68(1):190–200

Ruttkay-Nedecky B, Nejdl L, Gumulec J, Zitka O, Masarik M, Eckschlager T, Stiborova M, Adam V, Kizek R (2013) The role of metallothionein in oxidative stress. Int J Mol Sci 14:6044–6066

Sabat D, Patnaik A, Ekka B, Dash P, Mishra M (2016) Investigation of titania nanoparticles on behaviour and mechanosensory organ of Drosophila melanogaster. Physiol Behav 167:76–85

Sahu SC, Zheng J, Yourick JJ, Sprando RL, Gao X (2015) Toxicogenomic responses of human liver HepG2 cells to silver nanoparticles. J Appl Toxicol 35:1160–1168

Sanvicens N, Pastells C, Pascual N, Marco M-P (2009) Nanoparticle-based biosensors for detection of pathogenic bacteria. TrAC Trends Anal Chem 28:1243–1252

Schieber M, Chandel NS (2014) ROS function in redox signaling and oxidative stress. Curr Biol 24:R453–R462

Shah A, Mankus CI, Vermilya AM, Soheilian F, Clogston JD, Dobrovolskaia MA (2018) Feraheme(R) suppresses immune function of human T lymphocytes through mitochondrial damage and mitoROS production. Toxicol Appl Pharmacol 350:52–63

Sharma A, Mishra M, Shukla AK, Kumar R, Abdin MZ, Chowdhuri DK (2012) Organochlorine pesticide, endosulfan induced cellular and organismal response in Drosophila melanogaster. J Hazard Mater 221-222:275–287

Shrivastava R, Raza S, Yadav A, Kushwaha P, Flora SJ (2014) Effects of sub-acute exposure to TiO2, ZnO and Al2O3 nanoparticles on oxidative stress and histological changes in mouse liver and brain. Drug Chem Toxicol 37:336–347

Siddique YH, Khan W, Khanam S, Jyoti S, Naz F, Rahul, Singh BR, Naqvi AH (2014) Toxic potential of synthesized graphene zinc oxide nanocomposite in the third instar larvae of transgenic Drosophila melanogaster (hsp70-lacZ)Bg9. BioMed Res Int 2014:382124

Singh D, Singh S, Sahu J, Srivastava S, Singh MR (2016) Ceramic nanoparticles: recompense, cellular uptake and toxicity concerns. Artifi Cells Nanomed Biotechnol 44:401–409

Soltani A, Kahkhaie KR, Haftcheshmeh SM, Jalali Nezhad AA, Akbar Boojar MM (2018) The comparative study of the effects of Fe2 O3 and TiO2 micro- and nanoparticles on oxidative states of lung and bone marrow tissues and colony stimulating factor secretion. J Cell Biochem. https://doi.org/10.1002/jcb.28031

Spence AR, Hopkins GR, Neuman-Lee LA, Smith GD, Brodie ED Jr, French SS (2016) Detrimental effects of zinc oxide nanoparticles on amphibian life stages. J Exp Zool A Ecol Genet Physiol 325:415–424

Stone V, Donaldson K (2006) Nanotoxicology: signs of stress. Nat Nanotechnol 1:23

Sundarraj K, Raghunath A, Panneerselvam L, Perumal E (2017) Iron oxide nanoparticles modulate heat shock proteins and organ specific markers expression in mice male accessory organs. Toxicol Appl Pharmacol 317:12–24

Takalo M, Salminen A, Soininen H, Hiltunen M, Haapasalo A (2013) Protein aggregation and degradation mechanisms in neurodegenerative diseases. Am J Neurodegener Dis 2:1–14

Teles M, Reyes-Lopez FE, Fierro-Castro C, Tort L, Soares A, Oliveira M (2018) Modulation of immune genes mRNA levels in mucosal tissues and DNA damage in red blood cells of Sparus aurata by gold nanoparticles. Mar Pollut Bull 133:428–435

Thirumoorthy N, Manisenthil Kumar KT, Shyam Sundar A, Panayappan L, Chatterjee M (2007) Metallothionein: an overview. World J Gastroenterol 13:993–996

Tomljenovic L (2011) Aluminum and Alzheimer's disease: after a century of controversy, is there a plausible link? J Alzheimer's Dis 23:567–598
Towle KM, Grespin ME, Monnot AD (2017) Personal use of hair dyes and risk of leukemia: a systematic literature review and meta-analysis. Cancer Med 6:2471–2486
Turati F, Pelucchi C, Galeone C, Decarli A, La Vecchia C (2014) Personal hair dye use and bladder cancer: a meta-analysis. Ann Epidemiol 24:151–159
Uhrich KE, Cannizzaro SM, Langer RS, Shakesheff KM (1999) Polymeric systems for controlled drug release. Chem Rev 99:3181–3198
Umashankar MS, Sachdeva RK, Gulati M (2010) Aquasomes: a promising carrier for peptides and protein delivery. Nanomedicine 6:419–426
Van Vlerken LE, Amiji MM (2006) Multi-functional polymeric nanoparticles for tumour-targeted drug delivery. Expert Opin Drug Deliv 3:205–216
Vos J, Van Loveren H, Wester P, Vethaak D (1989) Toxic effects of environmental chemicals on the immune system. Trends Pharmacol Sci 10:289–292
Wang P, Ng QX, Zhang H, Zhang B, Ong CN, He Y (2018a) Metabolite changes behind faster growth and less reproduction of Daphnia similis exposed to low-dose silver nanoparticles. Ecotoxicol Environ Saf 163:266–273
Wang T, Wen X, Hu Y, Zhang X, Wang D, Yin S (2018b) Copper nanoparticles induced oxidation stress, cell apoptosis and immune response in the liver of juvenile Takifugu fasciatus. Fish Shellfish Immunol 84:648–655
Wick P, Malek A, Manser P, Meili D, Maeder-Althaus X, Diener L, Diener PA, Zisch A, Krug HF, von Mandach U (2010) Barrier capacity of human placenta for nanosized materials. Environ Health Perspect 118:432–436
Xiao YY, Liu L, Chen Y, Zeng YL, Liu MZ, Jin L (2016) Developmental toxicity of carbon quantum dots to the embryos/larvae of rare minnow (Gobiocypris rarus). Biomed Res Int 2016:4016402
Xiao Y, Shi K, Qu Y, Chu B, Qian Z (2019) Engineering nanoparticles for targeted delivery of nucleic acid therapeutics in tumor. Mol Ther Methods Clin Dev 12:1–18
Yang X, Liu J, He H, Zhou L, Gong C, Wang X, Yang L, Yuan J, Huang H, He L et al (2010) SiO2 nanoparticles induce cytotoxicity and protein expression alteration in HaCaT cells. Part Fibre Toxicol 7:1
Yang H, Du L, Wu G, Wu Z, Keelan JA (2018) Murine exposure to gold nanoparticles during early pregnancy promotes abortion by inhibiting ectodermal differentiation. Mol Med 24:62
Yu Y, Duan J, Li Y, Li Y, Jing L, Yang M, Wang J, Sun Z (2017) Silica nanoparticles induce liver fibrosis via TGF-beta1/Smad3 pathway in ICR mice. Int J Nanomedicine 12:6045–6057
Yue L, Lian F, Han Y, Bao Q, Wang Z, Xing B (2018) The effect of biochar nanoparticles on rice plant growth and the uptake of heavy metals: implications for agronomic benefits and potential risk. Sci Total Environ 656:9–18
Zhang J, Ren L, Zou Y, Zhang L, Wei J, Li Y, Wang J, Sun Z, Zhou X (2016) Silica nanoparticles induce start inhibition of meiosis and cell cycle arrest via down-regulating meiotic relevant factors. Toxicol Res 5:1453–1464
Zhang CH, Wang Y, Sun QQ, Xia LL, Hu JJ, Cheng K, Wang X, Fu XX, Gu H (2018a) Copper nanoparticles show obvious in vitro and in vivo reproductive toxicity via ERK mediated signaling pathway in female mice. Int J Biol Sci 14:1834–1844
Zhang L, Wu L, Si Y, Shu K (2018b) Size-dependent cytotoxicity of silver nanoparticles to Azotobacter vinelandii: growth inhibition, cell injury, oxidative stress and internalization. PLoS One 13:e0209020
Zhou W, Tian D, He J, Yan X, Zhao J, Yuan X, Peng S (2018) Prolonged exposure to carbon nanoparticles induced methylome remodeling and gene expression in zebrafish heart. J Appl Toxicol 39(2):322–332
Zhu X, Tian S, Cai Z (2012) Toxicity assessment of iron oxide nanoparticles in zebrafish (Danio rerio) early life stages. PLoS One 7:e46286

Part IX

NanoBioMedicine: Clinical Translation and Commercialization

Opportunities in Clinical Translation and Commercialization of Nanomedicine

22

Nishant Srivastava and Shailendra K. Saxena

Abstract

Nanomedicine is emerging as a potential solution for many medical science problems and will drastically change the face of diagnostics, imaging, therapeutics, and drug delivery in the near future. The elevated use of inorganic and organic nanomaterials in medicinal science leads toward the development of potentially advanced and successful technologies. Nanomaterials are proven to be efficient drug carriers for the delivery of drug to the target site as well as for diagnostics of unnatural events in body. The conjugation of nanoparticles such as gold nanoparticles with antibiotics is found to accelerate the response of drugs severalfold. Similarly, in cancer imaging and therapy, the application of nanomaterials such as gold and iron oxide opens new dimensions of opportunities. There are certainly many challenges in front of researchers which need to be addressed such as compatibility, specificity, and toxicity of nanomaterials. The major pharmaceutical industries around the globe are presently more focused on the scientific research and developmental aspects of nanomedicine which is one of the major reasons of delayed or slow commercialization of nanomedicine in the market. It is believed that these under-research and under-trial drugs and technologies will soon get translated and be available in the market with the bright face of medicinal science.

N. Srivastava (✉)
Department of Biotechnology (DBT), Meerut Institute of Engineering and Technology (MIET), Meerut, India

S. K. Saxena
Centre for Advanced Research, Faculty of Medicine,
King George's Medical University (KGMU), Lucknow, India
e-mail: shailen@kgmcindia.edu

S. K. Saxena, S. M. P. Khurana (eds.), *NanoBioMedicine*,
https://doi.org/10.1007/978-981-32-9898-9_22

Keywords
Nanomaterials · Nanomedicine · Clinical translation · Specificity · Challenges

22.1 Introduction

Recent development and advancements in nanotechnology bring tremendous opportunity for medicinal and pharmaceutical scientists for the development of smart and highly effective drugs, therapies, and diagnostic tools. A majority of nanomaterials have been explored for their efficacy as nanomedicine. The advancements in medical sciences and technology make people's life happier and healthier with longevity (Meeto 2013). The use of nanotechnology in medicine opens up a wide door for the development of medicines that are smart, fast, accurate, and highly specific. Nanomedicine has a high potential to find solution for many untreatable diseases like cancer, viral infections, and genetic disorders. The small size and the high surface-to-volume ratio of nanomaterial lead to its appropriate applications in medicine. The small size of nanomaterial also facilitates smooth and specific action of drug molecule toward the target site. These nanosized materials can easily penetrate the cell and reach its DNA. Such properties of nanomaterials provide exciting opportunity for the treatment of many life-threatening diseases caused by bacterial-viral infection, genetic disorders, and metabolic abnormalities. The use of nanomaterials in medicine provides the opportunity to encapsulate non-soluble or badly soluble drugs, protection to therapeutic drug molecules, and improved specific distribution throughout the system (Bertrand et al. 2014).

The practice of nanomedicine is not new; in Indian Ayurveda the medical practitioner of ancient times was very well aware of the importance of Swarna Bhasma (globular gold particles or gold ash of 56–57 nm), Rajat Bhasma (silver particles or silver ash), Tamra Bhasma (copper particles), Abharak (Mika), Rasa (mercury), etc. The toxic materials will be converted into nontoxic ones by several stages of heating and burning and finally utilized as medicine for the treatment of a variety of diseases like anemia, tuberculosis, arthritis, stroke (ischemia), leukemia, and bacterial and fungal infections (Sanjoy Kumar 2015; Srivastava and Mukhopadhyay 2015a, b).

The continuing research and advancements in nanomedicines certainly come as an excellent opportunity for many problems related to medicine. From the diagnosis of cancer cells to their treatment, sensing of little changes at cellular or subcellular level and early detection with timely cure of illnesses are few advantages of nanotechnology. There are some nanomedicines that are already under trial in mice and further await human trials. Nanomedicines, for example gold nano shells for diagnosis of cancer and its treatment, application of liposome as vaccine adjuvants and as carrier for drug passage are under first phase of trial and successful results will lead these drugs for human trials (Nikalje 2015). The present chapter provides an insight and brief overview of nanomedicine and its challenges and scope of medical translation and commercialization. Lastly the ethical and social issues related to nanomedicine are discussed.

22.2 Nanomaterials for Nanomedicine

The term nanomedicine can be defined as the diagnosis, treatment, repair, monitoring, development, and control of human system at cellular, subcellular, and molecular levels utilizing engineered nanostructure materials and devices (Meeto 2013; Emerich 2005). Nanotechnology touches several dimensions of medical science and technology; the present and future applications of nanostructured materials are listed in Table 22.1.

There are several types of nanomaterials, including nanostructured crystalline and non-crystalline materials. Nanomaterials like quantum dots, polymers, silica nanoparticles, dendrimers, carbon nanotubes, micelles, metallic nanoparticles, etc. are some of the examples of pharmaceutical nanomaterials (Nikalje 2015).

22.3 Metallic Nanoparticles

Metallic nanoparticles like gold, silver, zinc, copper, nickel, etc. are widely explored for their applications in medical sciences. Metals like silver, gold, and copper have been used by mankind from the rise of civilization. In India, copper vessels were used to store water and in preparation of food. It is believed that copper has medicinal properties and it can destroy microbes and other pathogens from water. Similarly, silver and gold were utilized in many ways like for serving of food and for decoration and preservation of sweets and in Ayurvedic medicine. The utilization of these metals was not limited to India; in several parts of the world, people used them from thousands of years. In ancient civilizations Greeks and Romans used silver-coated pots to store and preserve water and wine. The antimicrobial property of silver

Table 22.1 Utilization of various types of engineered nanoparticles

Nanomaterial	Application	References
Gold	Antimicrobial, drug delivery, sensors, diagnostics, cancer treatment (gold nanorods), therapeutics, tissue engineering	Boisseau and Loubaton (2011), Srivastava and Mukhopadhyay (2015a, b) and Shamaila et al. (2016)
Silver	Antimicrobial, dermatology, therapeutics	Jyoti et al. (2016) and Buszewski et al. (2018)
Copper	Antimicrobial	Godymchuk et al. (2015) and Holubnycha et al. (2017)
Iron	Cellular imaging and targeted therapy	Wang (2016a, b)
Selenium	Cancer therapeutics, micronutrient, antimicrobial	Srivastava and Mukhopadhyay (2015a, b)
Quantum dots	Biomedical imaging, nano-biosensors	Wang (2016a, b)
Carbon (dots, wires, and tubes)	Cancer treatment, diagnostics, drug delivery, antibacterial	Markman et al. (2013) and Song et al. (2018)
ZnO	Antibacterial, antifungal	Gunalan et al. (2012)
Polymeric nanomaterials	Bandage, drug delivery, cosmetics	Banik et al. (2016)

makes it an important candidate in medicine. Silver compounds were utilized in the First World War for the treatment of infections caused by injury. Another compound of silver, silver sulfadiazine, is currently used for the treatment of burns (Solano-Umaña and Roberto 2015). These nanomaterials are found to be highly useful and effective in medicine and diagnostic applications. The brief applications and effect of metallic nanoparticles are discussed below.

22.3.1 Nanoparticles for Diagnosis and Imaging

Nanotechnology had incredible influence on medical diagnosis and imaging techniques and has played an important role in the technological advancement of presently available techniques. The exciting physicochemical properties of materials at the nanoscale open wide doors for revolutionary changes and fast-track development in the diagnosis of diseases. The SPR (surface plasmon resonance) of nanoparticles is a unique feature and is utilized for real-time, noninvasive, and label-free biosensing technique for monitoring of noncovalent molecular interactions (Tang et al. 2010; Bakhtiar 2013). SPR may be considered as a fingerprint of a particular nanomaterial. The noncovalent interactions of DNA-protein, DNA-DNA, RNA-DNA, cell-protein, protein-protein, protein-carbohydrate, small biomolecule-macromolecule (e.g., receptor-inhibitor complex), and peptide-protein and self-assembly of monolayers are examples of SPR applications in diagnosis (Bakhtiar 2013). The Lab on Chip technology constitutes a new paradigm for complete biochemical analysis in real time monitoring of metabolic disorders and diseases by analysis of cells, subcellualr components, fluids and even at molecular level monitoring for prediction of disease state (Meeto 2013).

The identification of cancer at the primary phase is one of the difficult tasks. The conventional method for identifying cancer and its type (malignant or benign) known as biopsy is one of the most reliable methods. Biopsy is an invasive and painful process of getting tissue from the affected site, and it becomes more complex when the site is a vital organ like the brain or liver. Nanotechnology provides an option in the form of nano-biopsy, a technique which is noninvasive or less invasive. The researchers developed a nanopattern pen device to collect cell or molecule from the target site without causing any harm to the tissues of the affected organ.

The application of nanoparticles like magnetic nanoparticles, supramagnetic nanoparticles, quantum dots, and gold nanoparticles found a great place in MRI as contrast agents. The desire of high contrast to view minute detailing is one of the major requirements of medicinal scientists. The application of nanoscale materials in MRI facilitates the detection of a solitary molecule or a solo cell in an intricate biological environment. The early detection of anomaly can help to restrict the progress of disease (Turkin 2010). Presently available imaging techniques can detect the abnormalities at tissue level, and there is need of early prognosis at molecular or cellular stage. The nanotechnology-based imaging provides specific identification and screening at cellular level by coating of nanomaterial with specific antibodies for receptors present on the targeted cell. The antibody-coated nanoparticle will bind with receptors on the diseased cell, and the operative illumination can be visualized by MRI scan. Furthermore, gold nanoparticles, quantum dots, carbon nanotubes, and nanoparticle coated with molecular markers may be used for enhancing light scattering in endoscopic and sonographic techniques that facilitate the observation of changes at cellular and molecular level which cannot be seen via conventional techniques (Turkin 2010; Boisseau and Loubaton 2011). The application of nanotechnology in medical diagnostics leads toward advancements of technologies for timely detection and cure of several life-threatening diseases like cancer, AIDS, cirrhosis, fibrosis, genetic disorders, etc. Some of the commercially available and US FDA- or Europe-approved nanoparticle-based products for imaging (especially for liver/spleen lesion) are Resovist (FeO nanoparticle layered with carboxydextran) and Feridex and Endoderm (FeO nanoparticles layered with dextran) (Wang et al. 2013).

22.3.2 Nanoparticles for Cancer Therapy

Cancer is one of the deadliest diseases claiming millions of world population every year. Worldwide extensive research is going on for the effective treatment of cancer. The presently available therapies for the treatment of cancer are chemotherapy, surgery, and radiation which have high side effects like bone marrow repression, cardiomyopathy, neurotoxicity, etc. and are not found to be much effective in the advanced stage of disease. Additionally, therapeutic molecules for the treatment of cancer poorly distinguish between cancerous cells and healthy cells present in the body. Due to these properties of cancer therapeutic drugs, the immune system of patient gets weak, making it more vulnerable toward infections. The utilization of nanomaterials for targeted cancer therapy is found to have exciting results. The application of nanotechnology in medicine provides effective and safe treatment of cancer without side effects. Their unusual physicochemical and optoelectronic properties, facile surface modification, high surface-to-volume ratio, and higher availability make these nanoscale materials excellent candidates for cancer therapy.

A group of researchers developed a gold and silver nanoparticle-based system of cancer drug delivery [b-Au-500-DOX and b-Ag-750-DOX] in the presence of doxorubicin (DOX) as an anti-cancer drug. They reported that the administrations of these nanoparticle-based drug delivery systems to cancer cells show significant inhibition of cancer cell proliferation (B16F10, MCF-7) (Patra et al. 2015).

In photodynamic therapy of cancer, nanoparticles (e.g., gold nanorods, quantum dots, magnetic nanoparticles) are inserted within the tumor and are irradiated out of the body by light source. The metallic nanoparticles absorb photo-light from the light source and get heat up due to photonic energy. This will lead to the formation of high-energy oxygen molecules which counter and finally destroy tumor cells without causing any damage to the normal cells of the body (Nikalje 2015).

The application of magnetic nanoparticles (e.g., superparamagnetic (Fe_3O_4) nanoparticles) for cancer therapy is also underway. The magnetic nanoparticles carrying therapeutic drug molecule are guided to the tumor site using magnets/

magnetic field and delivered to the target only (Markman et al. 2013). These magnetic nanoparticles are covered with tumor-specific receptors which specifically bind with tumor cells in the tumor tissues and deliver the drug molecule to targeted cells which prevent side effects of drug to normal cells. The nanotechnology-based therapies are highly specific and minimize or strike out risk of side effects associated with cancer treatments. The use of metallic nanoparticles, magnetic nanoparticles, quantum dots, and carbon-based and carbon nanomaterials for the treatment of cancer gives very exciting results and has the potential to overcome hitches correlated with presently available medicines.

22.3.3 Nanoparticles as Antimicrobial Agent

There are a number of studies available in the form of research literatures related to the antimicrobial activity of nanoparticles. Most of the available research articles are of silver nanoparticles followed by gold, copper, zinc oxide, and selenium nanoparticles. The antimicrobial efficacy of metal and their compounds is known from thousands of years. In the old medicinal patterns like Indian Ayurveda and Greek, Roman, and Unani systems, use of metals and its bhasma (ash) for the treatment of infections and purification of water and preservation of food was well known and properly documented. The elevated risk of multidrug-resistance microbes and depleting effect of antibiotics motivate researchers to develop an alternative, efficient, specific, and sustainable medicine for microbial infections. Table 22.2 summarizes a variety of metallic nanoparticles with their antimicrobial properties and efficiency.

Several studies demonstrated the excellent antimicrobial potential of silver nanoparticles. The effect of silver nanoparticles was studied on Vibrio cholerae and Escherichia coli and minimal inhibitory concentrations (MIC) found to be between 5×10^5 and 10^7 particles/ml respectively for both the microbes. Additionally, it was also observed that the single dose of silver nanoparticles to infant mice infected by these pathogens shows a significant depletion of 75–100-fold in colonization (Salem et al. 2015). Similarly, the mode of action of silver nanoparticles was observed against clinically isolated pathogenic bacteria *Bacillus cereus*, *Staphylococcus aureus*, *Helicobacter pylori*, *Pseudomonas aeruginosa*, and *Escherichia coli* in terms of membrane integrity and protein outflow. Interestingly, silver nanoparticles exhibit negligible cytotoxicity lower than 25 lg/mL, and it was found toxic at and/or higher than 50 lg/mL on human epidermoid larynx carcinoma (HEp-2). This research work proved that silver nanoparticles could inhibit bacterial replication and ultimately lead to cell death. In another study on different strains of gram-negative and gram-positive bacteria, it was observed that the combination of silver nanoparticles with antibiotics was more effective in comparison with silver nanoparticles alone (Jyoti et al. 2016; Buszewski et al. 2018).

Likewise, gold, copper, and zinc nanoparticles are also found to have exciting antimicrobial activity. The researchers found that nano-gold having a size below 20 nm is the best candidate for antimicrobial use. It was observed via an experiment

Table 22.2 Antimicrobial response of various types of engineered nanoparticles

Nanomaterials	Target microorganisms	References
Silver	*Vibrio cholerae* *Escherichia coli*	Salem et al. (2015)
	Escherichia coli *Pseudomonas aeruginosa* *Bacillus cereus* *Helicobacter pylori* *Staphylococcus aureus*	Gopinath et al. (2017)
	Escherichia coli	Raja et al. (2017)
	Staphylococcus aureus *Escherichia coli*	de Aragão et al. (2016)
Silver and silver conjugate with commercial antibiotics	*Staphylococcus aureus* *Pseudomonas aeruginosa* *Escherichia coli* *Bacillus subtilis* *Proteus mirabilis* *Klebsiella pneumoniae*	Buszewski et al. (2018)
	Salmonella typhimurium *Bacillus cereus* *Staphylococcus aureus* *Staphylococcus epidermidis* *Bacillus subtilis* *Klebsiella pneumoniae* *Escherichia coli* *Serratia marcescens*	Jyoti et al. (2016)
Gold	*Staphylococcus aureus* *Pseudomonas aeruginosa* *Streptococcus pyogenes* *Escherichia coli*	Srivastava and Mukhopadhyay (2015a, b)
	Staphylococcus aureus *Klebsiella pneumoniae* *Escherichia coli* *Pseudomonas aeruginosa*	Abdel-Raouf et al. (2017)
	Staphylococcus aureus *Escherichia coli*	Ahmad et al. (2013)
	Escherichia coli *Bacillus subtilis* *Klebsiella pneumoniae* *Staphylococcus aureus*	Shamaila et al. (2016)

(continued)

Table 22.2 (continued)

Nanomaterials	Target microorganisms	References
Copper	*Staphylococcus aureus* *Escherichia coli* *Candida albicans*	Bogdanović et al. (2014)
	Bacillus subtilis *Staphylococcus aureus* *Pseudomonas aeruginosa* *Candida albicans* *Salmonella choleraesuis*	Usman et al. (2013)
	Bacillus cereus *Staphylococcus aureus* *Escherichia coli* *Pseudomonas aeruginosa*	Godymchuk et al. (2015)
ZnO	*Staphylococcus aureus* *Escherichia coli* *Pseudomonas aeruginosa*	Dobrucka and Długaszewska (2016)
	Serratia marcescens *Staphylococcus aureus* *Proteus mirabilis* *Citrobacter freundii* *Rhizopus stolonifer* *Aspergillus flavus* *Aspergillus nidulans* *Trichoderma harzianum*	Gunalan et al. (2012)
Selenium	*Staphylococcus aureus* *Escherichia coli* *Streptococcus pyogenes* *Pseudomonas aeruginosa* *Aspergillus clavatus*	Srivastava and Mukhopadhyay (2015a, b)

that the gold nanoparticles (>20 nm) easily cross the cell membrane via porin channel and interacted with bacterial DNA. This interaction further restricts the transcription and unwinding of DNA followed by growth inhibition and lysis of bacterial cell (Srivastava and Mukhopadhyay 2015a, b).

In most of the studies, it was observed that nanoparticles cross the bacterial membrane and disrupt the normal function of cell by disrupting the metabolic pathways, damaging the DNA and causing the formation of membrane vesicles, leakage of intracellular material, and finally cell death.

22.3.4 Nanoparticles for Tissue Engineering and Regenerative Medicine

Tissue engineering and regenerative medicine is an interdisciplinary field of study comprised of engineering, biotechnology, medicine, pharmaceuticals, and biology to develop tissue and organ transplantation technologies. The organ and tissues are made up of cells which grow and differentiate into specific cell types on 3D scaffold. The growth of artificial tissues is still a big challenge with so many limitations and difficult and sophisticated protocols. The bad cell adhesive and mechanical properties of the scaffold, differentiation at the defect site, and ineffective cell growth and unstable growth factors are some of the limitations of tissue growth and regeneration (Vial et al. 2017). With the incredible advancements in nanotechnology and nanomedicine, the reproduction and regeneration of tissues and artificial stimulation for cell proliferation may be possibly achieved in the future (Nikalje 2015). The nanomaterial-based scaffolds are very helpful in tissue regeneration via stimulating cell proliferation, leading to organ development, growth, or regeneration. The researchers use gold nanorods for guiding orientation and variation of myoblast cells. The use of nano-gold in tissue scaffolds for wound curing, gold nanorods with gellan gum in bone tissue engineering, nano gold hydrogel complex for enhancement of bone rejuvenation are few examples of gold nanoparticles applications in tissue engineering (Cozad et al. 2011; Goreham et al. 2013; Zan et al. 2013; Vieira et al. 2015). Additionally, the researchers also reported that the particle size of gold nanoparticle plays a significant part and promotes osteogenic differentiation of adipose-derived stem cells; as an antioxidant, it inhibited the receptor activator of nuclear factor-κB ligand (RANKL) which induced osteoclast development. The size and spatial distance of gold nanoparticles also affect cell behavior (Sul et al. 2010; Goreham et al. 2013; Heo et al. 2014; Ko et al. 2015; Vial et al. 2017). The utilization and application of nanoparticles in the near future have potential answers for tissue engineering and regenerative medicine-related problems and diseases.

22.4 Nanoparticles for Drug Delivery

Delivery of drugs at the specific target with minimal loss and least side effects has been an utmost challenge for a long time in medicine. The main intention of targeted drug delivery is to deliver the drug at precise site where the problem is sited and maximize the drug concentration at diseased location. Another objective of drug delivery is to minimize drug side effect and increase specificity and decrease toxicity. Additionally, optimizing the minimum dose requirement with highest efficacy and the cost-effective availability of drug are additional important points for consideration. Nanotechnology emerging as solutions for these challenges as well as it have ability to carry molecules having minimal bioavailability and instability more efficiently (Emerich 2005). The incorporation of the main drug molecule with nanomaterials exerts a significant impact on drug pharmacokinetics. It is depending upon specific properties of drug molecules and carrier nanomaterial. The application of nanomaterial for drug delivery is carried out by the immobilization of drug molecule on nanomaterial surface in two ways, attachment and encapsulation, respectively. Currently the major concentration of nanomedicine researchers seems to be on cancer treatment and diagnostics.

The utilization of magnetic nanoparticles for the delivery of drugs to the specific site is getting so much attention due to its ease of use and higher success possibilities. The drug molecule is encapsulated or attached to magnetic nanoparticle having surface coating of biocompatible agents (gold, polymers, etc.). These magnetic nanoparticle-drug molecule conjugates after administration are guided to the specific site through external magnetic field. Once the conjugates reach the specific site, the drug molecule is released via change in temperature, pH, osmolality, or enzyme activity (McNamara and Tofail 2017). Nanomedicine researchers also focused on the treatment of neurodegenerative diseases like Parkinson's and Alzheimer's. The work is leading toward minimization of side effects involved with presently available therapies and improvement in drug delivery with maximum success rate. The nanomaterials are also of great choice because of their excellent capability to cross the blood-brain barrier and deeply reach the brain cells. The design, optimization, and biologically simulated development of intracranial nano-enabled scaffold device (NESD) for the precise delivery of dopamine to the brain are part of the research for Parkinson's disease therapy. Nanoparticles are also found to be effective for diagnosis and curing of Alzheimer's disease. Nanomaterials have attraction for circulating amyloid-β (Aβ) systems, and therefore it may induce "sink effect" and effective progress in the patient condition (Nikalje 2015). Another application of nanotechnology is for the delivery of drugs against specific viral or bacterial infections. The encapsulation of drug molecule in nanocarrier and its release at the target site (on or inside the microbial cell which leads to the damage of cellular DNA and inhibition of cellular processes, ultimately resulting in cell lysis) facilitate maximum output in terms of bacterial and viral infection treatments of diseases with high mortality like TB, HIV, hepatitis, etc. Liposomes, nanolipids, dendrimers, nanocrystals, nanoalbumins, silver, and gold are some of the examples of nanomaterials under clinical trial for use as nanomedicines for the treatment of cancers, tumors, and fungal and bacterial infections, for HIV therapy, and for use as immunosuppressants (Hall et al. 2007). There are already many nano-based medicines that are under the developmental phase and in clinical trials, and some of them are in the market too with exceptional results. The ongoing research in the arena of drug delivery under the umbrella of nanomedicine brings hope and healthy life in the near future.

22.5 Nanoparticles for Gene Therapy

Genetic diseases are another challenge in medicine due to the lack of treatment and tiresome and painful therapies of lifelong illness. The presently available and extensively studied gene therapy for the treatment of genetic diseases using recombinant

DNA technologies or genetic engineering did not at all found much scope and has not been proven effective till date. The gene therapy using nanotechnology has more chances of promising output. The reasons for higher success rate with nanoparticle-mediated gene therapy are the maximum reach inside the cell due to the nanoparticles' small size and also its ability to carry smaller fragments of the required gene to the target site for correction in gene sequence responsible for genetic diseases. The extensive research for use of nanoparticles in gene therapy is continued, and some of the gene therapy trials have already been completed. The mostly utilized nanomaterials for gene therapy are organic nanoparticles like liposomes. The application of liposomes (SGT 53) is reported for the successful renewal of function of human suppressor gene p^{53} by transporting a plasmid containing wild-type p^{53} sequence (Senzer et al. 2013). Another report reported liposomal siRNA preparation (Atu027) which employs a proprietary AtuRNAi for targeting and suppressing or removing PKN3, the gene widely known for instigating malignant cell growth (Leenders et al. 2004; Anselmo and Mitragotri 2016). Beside this, metallic nanoparticles like gold, platinum, and silver also have applications in gene therapy as vehicles for the delivery of gene segment, nucleotides, and protein fragment.

22.6 Organic and Polymeric Nanomaterials for Medicinal Use

In nanomedicine organic and polymeric nanomaterials are found to have an important role due to their phenomenal biocompatibility, acceptability, non-toxicity, and least side effects. Polymeric nanoparticles with great potential and application are completely changing the approach and execution of medicine. Initially non-biodegradable polymers, e.g., polyacrylates, polyacrylamide, polystyrene, and poly(methyl methacrylate) (PMMA), were used for drug delivery, wound healing, and antimicrobial treatment (Shastri 2003; Bettencourt and Almeida 2012). The side effects associated with non degradable polymers such as inflammation and chronic toxicity, safe and compatible biodegradable polymers get more attention and value. Biodegradable polymers like poly(lactide-co-glycolide) (PLGA), poly(lactide) (PLA), and poly(amino acid) copolymers in addition to natural polymers such as alginate, chitosan, albumin, and gelatin are safe and biocompatible (Elsabahy and Wooley 2012; Zhang et al. 2013; Banik et al. 2016). Polymeric organic nano-polymer carrier facilitate controlled release of drug molecule at target site, stable, encapsulation prevents degradation of molecules and having option for modification of surface and ligands (Ahmad et al. 2014). Organic nanoparticles such as chitosan, silk fibroin, and PLGA are used for bone regeneration, tissue proliferation, toothpaste, osteoporosis treatment, wound healing, osteomyelitis treatment, orthodontics, microbial infection, and cartilage tissue engineering in oral and dental treatment (Virlan et al. 2016). As discussed in the above sections, organic nanoparticles like liposomes, PLGA, etc. have high potential application in nanomedicine as drug carriers for drug delivery to the specific target site and also in diagnostics and imaging.

22.7 Nanoparticle-Based Medicines Approved for Clinical Trials and Applications

There are several nanomaterial-based medicines that have been developed and are under various phases of clinical trial, whereas some of the medicines got approval from FDA and other regulatory bodies. Clinical translation of nanomedicine is the challenging part after successful analysis (both in vitro and in vivo) and research. Toxicity assessment, self-degradation, specificity, and surface modifications are some of the significant points of concern. In Table 22.3, some of the nanomedicines with their status and application are depicted.

The nano-drugs mentioned in Table 22.3 are only a few from the long existing list of under-trial and marketed drugs. There are possibilities to have better health in the presence of smart and effective nanotechnology-based medicines.

22.8 Commercialization and Challenges in Nanomedicine

The above sections and Table 22.3 clearly indicate the importance of nanotechnology in medicine and clinical translation of nanomedicines. The commercialization of first-generation nano-based drugs is already completed and contributing for better human health. The second generation of nanomedicines is definitely going to improve quality of life after commercialization. The increasing technological advancements and day-to-day inventions in nanotechnology make it more interesting and hopeful for present and future medicines. The 3D structure of material at nanoscale with high surface-to-volume ratio is undoubtedly providing opportunities for the development of highly specific and smartly controlled drugs. Presently, the major focus of pharmaceutical companies is on scientific research in nanomedicine and for the development of most reliable products.

Doxil is known as the first liposome nanoparticle-based drug that reached the market in 1995, and after it, many more drugs got commercialized (Svenson 2012). It was predicted that the nanotechnology-based medicine market will grow by more than 300 billion dollars in the next 12 years for the USA alone (Flynn and Wei 2005). The combination of nanotechnology with biotechnology brings many opportunities not only in medicine or pharmaceutical sectors but also in revolutionizing allied areas of medical diagnostics, genetic engineering, and novel drug discovery. The major challenge with nanomedicine is robust synthesis of competent 3D nanostructure, as size, surface modification, and biocompatibility are the major requirements for nanomaterials in nanomedicine. Additionally, the cost of nanomedicine is another facet to be taken into consideration. Nanomaterials showed high potential as medicine by triggering protein production or silencing specific gene or targeting specific sites, but somehow, they are found difficult to cross the cell membrane, and the development of efficient drug delivery system is a critical point of consideration (Ragelle et al. 2017). The efficient and facile removal of nanoparticles from the body and protection of vital organs from the accumulation of nanomaterials used in drug delivery are other major areas of concern in nanomedicine. The development

Table 22.3 Nanomedicines: their status and applications

Drug molecule	Nanocarrier	Company/product	Status	Application	References
Doxorubicin	Polymeric micelle	Supratek Pharma	Phase 3	Advanced adenocarcinoma	
Amphotericin B	Liposome	Gilead Sciences/AmBisome	US FDA approved	Fungal infections	
FeO	FeO	Advanced Magnetics/ Combidex	Phase 3	Tumor imaging	
Onco TCS	Liposome	Inex	Phase 2	Lymphoma	
Myocet	Liposome-encapsulated doxorubicin	Elan Pharmaceuticals	Approved in Canada and Europe	Breast cancer	Havel et al. (2016)
Paclitaxel	Polymer drug conjugate	CT-2103/Xyotax	Phase 1	Primary peritoneal carcinoma, recurrent ovarian carcinoma	
Adenosine deaminase (ADA)	Polymer drug conjugate	Adagen	Approved	Immunodeficiency	
PEGylated liposomal doxorubicin	PEGylated liposomal	Doxil/Janssen Pharmaceutica	Approved	Kaposi's sarcoma, ovarian and metastatic breast cancers	
Paclitaxel	PEG-poly(D,L-lactide)	Genexol-PM/Samyang Biopharmaceuticals	Approved	Breast cancer	
Oxaliplatin	Polymeric micelle (NC-4016/NanoCarrier)	(NC-4016)	Phase 1	Solid tumors, lymphoma	
NK-911	mPEG-poly (aspartic acid)-doxorubicin conjugate	Nippon Kayaku Co. Ltd.	Phase 1	Solid tumors	
Taxoprexin	Polymer drug conjugate	Taxoprexin/Luitpold Pharmaceuticals	Phase 2	Metastatic melanoma	

Source: Available literatures and websites of FDA and pharmaceutical companies

of smart drugs under nanomedicine is underway, and it will certainly change the prospect of medicinal science and bring better opportunity in diagnostics and therapeutics. Another important aspect in nanomedicine is the proper understanding of protein absorption on nanomaterial and change in their behavior due to interaction with biomolecules. The understanding of basic phenomena and behavioral changes of nanoparticles while interacting with biological components may boost our knowledge and substantially boost existing technology.

22.9 Conclusions

Nanotechnology is revolutionizing each and every branch of science and technology. Nanomedicine is one integral part of nanotechnology that brings high expectations and hope for drug discovery and innovation, drug development, and drug delivery. The application of nanomaterials in diagnostics and imaging opens up clearer and wider doors for the diagnosis and treatment of diseases. Nanomedicine on the other hand provides opportunities for the effective delivery of drug molecule to the target site specifically with minimal side effects and higher biocompatibility for maximum acceptability of drug by the body system. Gold, silver, copper, zinc oxide, iron oxide, magnetic, and supramagnetic nanoparticles come up as efficient vehicles for the delivery of drug to more precise diagnosis and imaging. The development of smart diagnostic tools such as lab-on-chip and/or nano-biosensors can facilitate fast and precise diagnostics and monitoring of antigens, viral proteins, allergens, and metabolic disorders like diabetes, saving time in diagnosis and providing efficient and timely treatment. Similarly, development of smart drugs will improve absorption of drug molecules at the target site. There are still many challenges in front of nanomedicine scientists, and extensive study of responsible attributes for therapeutic performance of nanomaterials will contribute in bringing some fruitful output. The understanding of physicochemical properties of nanomaterials is essential for achieving the goal of clinical translation and commercialization.

Acknowledgments The authors are grateful to the Vice Chancellor, King George's Medical University (KGMU), Lucknow, for the encouragement and support for this work. SK Saxena is also supported by CCRH, Government of India, and US NIH grants: R37DA025576 and R01MH085259. The authors have no other relevant affiliations or financial involvement with any organization or entity with a financial interest in or financial conflict with the subject matter or materials discussed in the manuscript apart from those disclosed.

References

Abdel-Raouf N, Al-Enazi NM, Ibraheem IBM (2017) Green biosynthesis of gold nanoparticles using Galaxaura elongata and characterization of their antibacterial activity. Arab J Chem 10:S3029–S3039

Ahmad T, Wani IA, Manzoor N, Ahmed J, Asiri AM (2013) Biosynthesis, structural characterization and antimicrobial activity of gold and silver nanoparticles. Colloids Surf B: Biointerfaces 107:227–234

Ahmad Z, Shah A, Siddiq M, Kraatz H-B (2014) Polymeric micelles as drug delivery vehicles. RSC Adv 4(33):17028–17038

Anselmo AC, Mitragotri S (2016) Nanoparticles in the clinic. Bioeng Trans Med 1(1):10–29

Bakhtiar R (2013) Surface plasmon resonance spectroscopy: a versatile technique in a biochemist's toolbox. J Chem Educ 90(2):203–209

Banik BL, Fattahi P, Brown JL (2016) Polymeric nanoparticles: the future of nanomedicine. Wiley Interdiscip Rev Nanomed Nanobiotechnol 8(2):271–299

Bertrand N, Wu J, Xu X, Kamaly N, Farokhzad OC (2014) Cancer nanotechnology: the impact of passive and active targeting in the era of modern cancer biology. Adv Drug Deliv Rev 66:2–25

Bettencourt A, Almeida AJ (2012) Poly(methyl methacrylate) particulate carriers in drug delivery. J Microencapsul 29(4):353–367

Bogdanović U, Lazić V, Vodnik V, Budimir M, Marković Z, Dimitrijević S (2014) Copper nanoparticles with high antimicrobial activity. Mater Lett 128:75–78

Boisseau P, Loubaton B (2011) Nanomedicine, nanotechnology in medicine. C R Phys 12(7):620–636

Buszewski B, Railean-Plugaru V, Pomastowski P, Rafińska K, Szultka-Mlynska M, Golinska P, Wypij M, Laskowski D, Dahm H (2018) Antimicrobial activity of biosilver nanoparticles produced by a novel Streptacidiphilus durhamensis strain. J Microbiol Immunol Infect 51(1):45–54

Cozad MJ, Bachman SL, Grant SA (2011) Assessment of decellularized porcine diaphragm conjugated with gold nanomaterials as a tissue scaffold for wound healing. J Biomed Mater Res Part A 99A(3):426–434

de Aragão AP, de Oliveira TM, Quelemes PV, Perfeito MLG, Araújo MC, Santiago JdAS, Cardoso VS, Quaresma P, de Souza de Almeida Leite JR, da Silva DA (2016) Green synthesis of silver nanoparticles using the seaweed Gracilaria birdiae and their antibacterial activity. Arab J Chem https://doi.org/10.1016/j.arabjc.2016.04.014

Dobrucka R, Długaszewska J (2016) Biosynthesis and antibacterial activity of ZnO nanoparticles using Trifolium pratense flower extract. Saudi J Biol Sci 23(4):517–523

Elsabahy M, Wooley KL (2012) Design of polymeric nanoparticles for biomedical delivery applications. Chem Soc Rev 41(7):2545–2561

Emerich DF (2005) Nanomedicine – prospective therapeutic and diagnostic applications. Expert Opin Biol Ther 5(1):1–5

Flynn T, Wei C (2005) The pathway to commercialization for nanomedicine. Nanomedicine 1(1):47–51

Godymchuk A, Frolov G, Gusev A, Zakharova O, Yunda E, Kuznetsov D, Kolesnikov E (2015) Antibacterial properties of copper nanoparticle dispersions: influence of synthesis conditions and physicochemical characteristics. IOP Conf Ser Mater Sci Eng 98(1):012033

Gopinath V, Priyadarshini S, Loke MF, Arunkumar J, Marsili E, Mubarak Ali D, Velusamy P, Vadivelu J (2017) Biogenic synthesis, characterization of antibacterial silver nanoparticles and its cell cytotoxicity. Arab J Chem 10(8):1107–1117

Goreham RV, Mierczynska A, Smith LE, Sedev R, Vasilev K (2013) Small surface nanotopography encourages fibroblast and osteoblast cell adhesion. RSC Adv 3(26):10309–10317

Gunalan S, Sivaraj R, Rajendran V (2012) Green synthesized ZnO nanoparticles against bacterial and fungal pathogens. Prog Nat Sci Mater Int 22(6):693–700

Hall JB, Dobrovolskaia MA, Patri AK, McNeil SE (2007) Characterization of nanoparticles for therapeutics. Nanomedicine 2(6):789–803

Havel H, Finch G, Strode P, Wolfgang M, Zale S, Bobe I, Youssoufian H, Peterson M, Liu M (2016) Nanomedicines: from bench to bedside and beyond. AAPS J 18(6):1373–1378

Heo DN, Ko W-K, Bae MS, Lee JB, Lee D-W, Byun W, Lee CH, Kim E-C, Jung B-Y, Kwon IK (2014) Enhanced bone regeneration with a gold nanoparticle–hydrogel complex. J Mater Chem B2(11):1584–1593

Holubnycha V, Pogorielov M, Korniienko V, Kalinkevych O, Ivashchenko O, Peplinska B, Jarek M (2017) Antibacterial activity of the new copper nanoparticles and Cu NPs/chitosan solution. 2017 IEEE 7th International Conference Nanomaterials: Application & Properties (NAP)

Jyoti K, Baunthiyal M, Singh A (2016) Characterization of silver nanoparticles synthesized using Urtica dioica Linn. leaves and their synergistic effects with antibiotics. J Radiat Res Appl Sci 9(3):217–227

Ko W-K, Heo DN, Moon H-J, Lee SJ, Bae MS, Lee JB, Sun I-C, Jeon HB, Park HK, Kwon IK (2015) The effect of gold nanoparticle size on osteogenic differentiation of adipose-derived stem cells. J Colloid Interface Sci 438:68–76

Leenders F, Möpert K, Schmiedeknecht A, Santel A, Czauderna F, Aleku M, Penschuck S, Dames S, Sternberger M, Röhl T, Wellmann A, Arnold W, Giese K, Kaufmann J, Klippel A (2004) PKN3 is required for malignant prostate cell growth downstream of activated PI 3-kinase. EMBO J 23(16):3303–3313

Markman JL, Rekechenetskiy A, Holler E, Ljubimova JY (2013) Nanomedicine therapeutic approaches to overcome cancer drug resistance. Adv Drug Deliv Rev 65(13):1866–1879

McNamara K, Tofail SAM (2017) Nanoparticles in biomedical applications. Adv Phys X2(1):54–88

Meeto DD (2013) Nanomedicine: the revolution of the big future with tiny medicine bionanotechnology. Wiley, pp 163–178

Nikalje A (2015) Nanotechnology and its applications in medicine. Med Chem 5:081–089

Patra S, Mukherjee S, Barui AK, Ganguly A, Sreedhar B, Patra CR (2015) Green synthesis, characterization of gold and silver nanoparticles and their potential application for cancer therapeutics. Mater Sci Eng C53:298–309

Ragelle H, Danhier F, Préat V, Langer R, Anderson DG (2017) Nanoparticle-based drug delivery systems: a commercial and regulatory outlook as the field matures. Expert Opin Drug Deliv 14(7):851–864

Raja S, Ramesh V, Thivaharan V (2017) Green biosynthesis of silver nanoparticles using Calliandra haematocephala leaf extract, their antibacterial activity and hydrogen peroxide sensing capability. Arab J Chem 10(2):253–261

Salem W, Leitner DR, Zingl FG, Schratter G, Prassl R, Goessler W, Reidl J, Schild S (2015) Antibacterial activity of silver and zinc nanoparticles against Vibrio cholerae and enterotoxic Escherichia coli. Int J Med Microbiol 305(1):85–95

Sanjoy Kumar P (2015) The Ayurvedic Bhasma: the ancient science of nanomedicine. Recent Pat Nanomed (Discontinued) 5(1):12–18

Senzer N, Nemunaitis J, Nemunaitis D, Bedell C, Edelman G, Barve M, Nunan R, Pirollo KF, Rait A, Chang EH (2013) Phase I study of a systemically delivered p53 nanoparticle in advanced solid tumors. Mol Ther 21(5):1096–1103

Shamaila S, Zafar N, Riaz S, Sharif R, Nazir J, Naseem S (2016) Gold nanoparticles: an efficient antimicrobial agent against enteric bacterial human pathogen. Nano 6(4):71

Shastri VP (2003) Non-degradable biocompatible polymers in medicine: past, present and future. Curr Pharm Biotechnol 4(5):331–337

Solano-Umaña VV-B, Roberto J (2015) Gold and silver nanotechology on medicine. J Chem Biochem 3(1):21–33

Song Y, Lu F, Li H, Wang H, Zhang M, Liu Y, Kang Z (2018) Degradable carbon dots from cigarette smoking with broad-spectrum antimicrobial activities against drug-resistant bacteria. ACS Appl Bio Mat 1:1871–1879

Srivastava N, Mukhopadhyay M (2015a) Biosynthesis and characterization of gold nanoparticles using Zooglea ramigera and assessment of Its antibacterial property. J Clust Sci 26(3):675–692

Srivastava N, Mukhopadhyay M (2015b) Green synthesis and structural characterization of selenium nanoparticles and assessment of their antimicrobial property. Bioprocess Biosyst Eng 38(9):1723–1730

Sul O-J, Kim J-C, Kyung T-W, Kim H-J, Kim Y-Y, Kim S-H, Kim J-S, Choi H-S (2010) Gold nanoparticles inhibited the receptor activator of nuclear factor-κB ligand (RANKL)-induced osteoclast formation by acting as an antioxidant. Biosci Biotechnol Biochem 74(11):2209–2213

Svenson S (2012) Clinical translation of nanomedicines. Curr Opinion Solid State Mater Sci 16(6):287–294

Tang Y, Zeng X, Liang J (2010) Surface plasmon resonance: an introduction to a surface spectroscopy technique. J Chem Educ 87(7):742–746

Turkin O (2010) The application of nanomaterials in cancer diagnostics and treatment. University of Groningen, Groningen

Usman MS, El Zowalaty ME, Shameli K, Zainuddin N, Salama M, Ibrahim NA (2013) Synthesis, characterization, and antimicrobial properties of copper nanoparticles. Int J Nanomedicine 8:4467–4479

Vial S, Reis RL, Oliveira JM (2017) Recent advances using gold nanoparticles as a promising multimodal tool for tissue engineering and regenerative medicine. Curr Opinion Solid State Mater Sci 21(2):92–112

Vieira S, Vial S, Maia FR, Carvalho M, Reis RL, Granja PL, Oliveira JM (2015) Gellan gum-coated gold nanorods: an intracellular nanosystem for bone tissue engineering. RSC Adv 5(95):77996–78005

Virlan M, Miricescu D, Radulescu R, Sabliov C, Totan A, Calenic B, Greabu M (2016) Organic nanomaterials and their applications in the treatment of oral diseases. Molecules 21(2):207

Wang J (2016a) Cd-containing quantum dots for biomedical imaging. The world scientific encyclopedia of nanomedicine and bioengineering I. WORLD SCIENTIFIC 5:111–158

Wang Y-X (2016b) Superparamagnetic iron oxide nanoparticles for cellular imaging and targeted therapy: opportunities and challenges for clinical translation. The world scientific encyclopedia of nanomedicine and bioengineering I. World Scientific:29–51

Wang R, Billone PS, Mullett WM (2013) Nanomedicine in action: an overview of cancer nanomedicine on the market and in clinical trials. J Nanomater 2013:12

Zan X, Feng S, Balizan E, Lin Y, Wang Q (2013) Facile method for large scale alignment of one dimensional nanoparticles and control over myoblast orientation and differentiation. ACS Nano 7(10):8385–8396

Zhang Z, Tsai P-C, Ramezanli T, Michniak-Kohn BB (2013) Polymeric nanoparticles-based topical delivery systems for the treatment of dermatological diseases. Wiley Interdiscip Rev Nanomed Nanobiotechnol 5(3):205–218

MIX
Papier aus verantwortungsvollen Quellen
Paper from responsible sources
FSC® C105338
FSC
www.fsc.org

If you have any concerns about our products,
you can contact us on
ProductSafety@springernature.com

In case Publisher is established outside the EU,
the EU authorized representative is:
Springer Nature Customer Service Center GmbH
Europaplatz 3, 69115 Heidelberg, Germany

Printed by Libri Plureos GmbH
in Hamburg, Germany